ALS
Advances in Life Sciences

Somesthesis and the Neurobiology of the Somatosensory Cortex

Edited by
O. Franzén
R. Johansson
L. Terenius

Birkhäuser Verlag
Basel · Boston · Berlin

Editors

Prof. O. Franzén
Department of Human Anatomy
Biomedical Center (BMC)
Uppsala University
S-75123 Uppsala
Sweden

Prof. R. Johansson
Department of Physiology
Umeå University
S-90187 Umeå
Sweden

Prof. L. Terenius
Department of Clinical Neuroscience
Karolinska Institutet
S-17176 Stockholm
Sweden

A CIP catalogue record for this book is available from the Library of Congress, Washington D.C., USA

Deutsche Bibliothek Cataloging-in-Publication Data
Somesthesis and the neurobiology of the somatosensory cortex
/ ed. by O. Franzén ... - Basel ; Boston ; Berlin : Birkhäuser, 1996
(Advances in life sciences)

ISBN-13: 978-3-0348-9868-3 e-ISBN-13: 978-3-0348-9016-8
DOI: 10.1007/978-3-0348-9016-8

NE: Franzén, Ove [Hrsg.]

The publisher and editor can give no guarantee for the information on drug dosage and administration contained in this publication. The respective user must check its accuracy by consulting other sources of reference in each individual case.

The use of registered names, trademarks etc. in this publication, even if not identified as such, does not imply that they are exempt from the relevant protective laws and regulations or free for general use.

Softcover reprint of the hardcover 1st edition 1996

Printed on acid-free paper produced from chlorine-free pulp. TCF ∞
Camera-ready copy prepared by the authors
Cover illustration: Model of the cerebral hemispheres (last quarter of the 18th century) from the Museo di Storia Naturale dell'Università, sezione di Zoologia "La Specola", Florence: no. 598 and Representation of form in the cortex of the brain from the chapter by S.S. Hsiao, K.O. Johnson, A. Twombly and J. DiCarlo, this volume.

9 8 7 6 5 4 3 2 1

Acknowledgement

This symposium was generously supported by grants from
Medical Research Council, Swedish Natural Science Research Council,
Ministry of Social Affairs and Pharmacia AB.

Scientific Committee

Contents

Cortical representation of somatosensation

Sensory-motor interface

Neuronal population behavior: Imaging techniques

Preface

The field of neuroscience is experiencing a period of unprecedented growth. The range of approaches used to investigate the fundamental mechanisms underlying different aspects of somatosensory perception has expanded substantially since the last International Symposium on Information Processing in the Somatosensory System. Diversity of approach has become the hallmark of somatosensory research, and I believe that the impact of this diversity continues to be positive. Comparison of this symposium with previous symposia will show that researchers are employing an array of new methods and technologies and that they are offering new insights into old questions. The techniques range from studies of ionic channels, single unit recordings, functional brain imaging to psychophysics. A careful reading of the chapters reveals subtle and, on occasion, also obvious differences of opinion about experimental outcomes, approaches, and views. These differences are healthy signs of venture into new and uncharted territories. It is my hope that readers will appreciate both the diversity of the research addressing somesthesis, and the personal commitment, enthusiasm, and scientific rigor of the participants who made this symposium and volume meaningful for all involved. We all look forward to tracking the progress of somatosensory research at our next Symposium at the turn of the century.

Ove Franzén

November 1995

List of Contributors

David T. Blake
Krieger Mind/Brain Institute
Department of Neuroscience
The Johns Hopkins University
Baltimore, MD 21218, USA

Henning Boecker
Neurologische Klinik der Technischen Universität München
81675 München, Germany

Jörgen Boivie
Department of Neurology
University Hospital
Linköping, Sweden

Stanley J. Bolanowski, Jr.
Institute for Sensory Research and Department of Bioengineering and Neuroscience
Syracuse University
Syracuse, NY 13244, USA

A. S. Browning
Department of Anatomy and Cell Biology
University of Melbourne
Parkville, Victoria 3052, Australia

Harold Burton
Department of Anatomy and Neurobiology
Washington University
School of Medicine
St. Louis, MO 63110, USA

C. Elaine Chapman
Centre de Recherche en Sciences Neurologiques, École de Réadaptation
and
Département de Physiologie
Université de Montréal
Québec H3C 3J7, Canada

Andrew C. N. Chen
Human Physiology and Pain Research Laboratory
Rheumatic Diseases Centre
Hope Hospital
Manchester University School of Medicine
Salford M6 8 HD, UK

Arthur D. Craig
Division of Neurobiology
Barrow Neurological Institute
St. Joseph's Hospital and Medical Center
Phoenix, Arizona 85013, USA

Wayne E. Crill
Department of Physiology and Biophysics
University of Washington
School of Medicine
Seattle, WA 98195-7290, USA

K. A. Delemos
Department of Psychology
University of North Carolina
Chapel Hill, NC 27599, USA

Stuart W. G. Derbyshire
Human Physiology and
Pain Research Laboratory
Rheumatic Diseases Centre
Hope Hospital
Manchester University School of Medicine
Salford M6 8 HD, UK

James DiCarlo
The Krieger Mind/Brain Institute
Department of Neuroscience
The Johns Hopkins University
MD 21218, USA

Benoni B. Edin
Department of Physiology
Umeå University
901 87 UMEÅ, Sweden

Greg K. Essick
School of Dentistry
Dental Research Center
The University of North Carolina
Chapel Hill, NC 27599 – 7455, USA

Oleg V. Favorov
Department of Biomedical Engineering
University of North Carolina
Chapel Hill
N. C. 27599 – 7545, USA

Nina Forss
Low Temperature Laboratory
Helsinki University of Technology
SF – 021 50 Espoo
Finland

Jens Frahm
Biomedizinische NMR
Forschungs GmbH
Max-Planck Institut für
Biophysikalische Chemie
37018 Göttingen, Germany

Ove Franzén
Department of Human Anatomy
Biomedical Center
Uppsala University
Uppsala, Sweden

Hans-Joachim Freund
Medicinische Enrichtungen der Universität
Düsseldorf
Neurologische Klinik
40225 Düsseldorf, Germany

A. K. Goble
Department of Psychology
Green Hall
Princeton University
Princeton
NJ 08544 – 1010, USA

Antony W. Goodwin
Department of Anatomy and Cell Biology
University of Melbourne
Parkville, Victoria 3052, Australia

Mark Hollins
Department of Psychology
University of North Carolina
Chapel Hill, North Carolina 27599, USA

Steven S. Hsiao
Krieger Mind/Brain Institute
Department of Neuroscience
The Johns Hopkins University
Baltimore, MD 21218, USA

A. James
School of Dentistry
Dental Research Center
University of North Carolina
Chapel Hill
North Carolina, USA

Wan Jiang
Centre de Recherche en Sciences Neurologiques and
Département de Physiologie
Université de Montréal
Québec H3C 3J7, Canada

Roland S. Johansson
Department of Physiology
Umeå University
901 87 Umeå, Sweden

Kenneth O. Johnson
Krieger Mind/Brain Institute
Department of Neuroscience
The Johns Hopkins University
Baltimore, MD 21218, USA

Anthony K. P. Jones
Human Physiology and Pain Research Laboratory
Rheumatic Diseases, Centre Hope Hospital
Manchester University School of Medicine
Salford M6 8HD, UK

Edward G. Jones
Department of Anatomy and Neurobiology
University of California
Irvine, CA 92717, USA

K. Jowers
School of Dentistry
Dental Research Center
University of North Carolina,
Chapel Hill
North Carolina, USA

Jon H. Kaas
Department of Psychology
Vanderbilt University
Nashville
TN 37240, USA

Douglas G. Kelly
Department of Statistics
University of North Carolina
Chapel Hill,
N. C. 27599 – 7545, USA

Roberta L. Klatzky
Department of Psychology
Carnegie Mellon University
Pittsburgh, PA 15213, USA

Leah A. Krubitzer
Center for Neuroscience and Department of Psychology
University of California Davis
Davis, CA 95616, USA
and
Vision, Touch and Hearing Research Centre, Department of Physiology and Pharmacology
University of Queensland, Australia

Robert H. LaMotte
Department of Anesthesiology
Yale University School of Medicine
New Haven
Connecticut 06510, USA

Susan J. Lederman
Department of Psychology and Department of Computing and Information Science
Queen's University
Kingston, Ontario
Canada K7L3N6

W. Lin
Washington University
Department of Radiological Sciences
St. Louis, MO 63110, USA

Charles Lu
Department of Anesthesiology
Yale University School of Medicine
New Haven CT 06510, USA

A. K. MacLeod
Washington University
Department of Radiological Sciences
St. Louis, MO 63110, USA

Michael M. Merzenich
Keck Center for Integrative Neuroscience
Coleman Memorial Laboratory
University of California
San Francisco, California 94143, USA

R. A. Murray
School of Dentistry
Dental Research Center
University of North Carolina
Chapel Hill
North Carolina, USA

G. M. Murray
School of Physiology and Pharmacology
The University of New South Wales
Sydney, N.S.W. 2052
Australia

T. A. Nguyen
North Carolina School of Science and Mathematics
Durham, North Carolina, USA

Ulf Norrsell
Department of Physiology
Göteborg University
413 90 Göteborg, Sweden

R. Nudo
Department of Neurobiology and Anatomy
University of Texas Medical School
Box 20708
Houston, TX 77225, USA

Cheryl O'Neill
Department of Psychology
Queen's University
Kingston, Ontario
Canada K7L3N6

Håkan Olausson
Department of Physiology
Göteborg University
413 90 Göteborg, Sweden

Guiseppe Pellizzer
Department of Physiology
University of Minnesota Medical School
Minneapolis, MN 55455, USA
and
Brain Sciences Center
Veterans Affairs Medical Center
Minneapolis, MN 55417, USA

Tim P. Pons
Department of Neurosurgery
Bowman Gray School of Medicine
Wake Forest University
Winston-Salem
North Carolina, 27157, USA

E. M. Rath
School of Dentistry
Dental Research Center
University of North Carolina
Chapel Hill
North Carolina, USA

M. J. Rowe
School of Physiology and Pharmacology
The University of New South Wales
Sydney, N.S.W. 2052, Australia

Harris D. Schwark
Department of Biological Sciences
University of North Texas
Denton, Texas 76203, USA

Jafar Shenasa
Centre de Recherche en Sciences Neurologiques and Département de Physiologie
Université de Montréal
Québec H3C 3J7, Canada

J. W. Shores
School of Dentistry
Dental Research Center
University of North Carolina
Chapel Hill
North Carolina, USA

R. J. Sinclair
Department of Anatomy and Neurobiology
Washington University
St. Louis, MO 63110, USA

Mandayam A. Srinivasan
Research Laboratory of Electronics
Massachusetts Institute of Technology
Cambridge
Massachusetts 02139, USA

Mark Tommerdahl
Departments of Biomedical Engineering and Physiology
University of North Carolina
Chapel Hill, N.C. 27599 – 7545, USA

Guilio Tononi
The Neurosciences Institute
San Diego
California 92121, USA

A. B. Turman
School of Physiology and Pharmacology
The University of New South Wales
Sydney, N.S.W. 2052, Australia

Alexander Twombly
Krieger Mind/Brain Institute
Department of Neuroscience
The Johns Hopkins University
Baltimore, MD 21218, USA

Åke B. Vallbo
Department of Physiology
Göteborg University
413 90 Göteborg, Sweden

Xiaoqin Wang
Department of Biomedical Engineering
Krieger Mind/Brain Institute
The Johns Hopkins University
Baltimore, MD 21205, USA

Johan Wessberg
Department of Physiology
Göteborg University
413 90 Göteborg, Sweden

H. E. Wheat
Department of Anatomy and Cell Biology
University of Melbourne
Parkville, Victoria 3052, Australia

Barry Whitsel
Department of Physiology
University of North Carolina
Chapel Hill, N.C. 27599 – 7545, USA

Stephan R. Williams
Centre de Recherche en Sciences Neurologiques and Département de Physiologie
Université de Montréal
Québec H3C 3J7, Canada

C. Xerri
Lab de Psychophysiologie
Faculté de St Jerome
13397 Marseille Cedex 13, France

H. Q. Zhang
School of Physiology and Pharmacology
The University of New South Wales
Sydney, N.S.W. 2052, Australia

Iolanda C. Zompa
Centre de Recherche en Sciences Neurologiques and Département de Physiologie
Université de Montréal
Québec H3C 3J7, Canada

Somesthesis and the Neurobiology of the Somatosensory Cortex
O. Franzén, R. Johansson and L. Terenius (eds.)

Divergence of thalamocortical projections and limits on somatosensory cortical plasticity

E.G. Jones

Department of Anatomy and Neurobiology, University of California Irvine, Irvine, California 92717, USA

Summary. Paired intracortical injections of Fast Blue (FB) and Diamidino Yellow (DY) separated by 100-1500 μm were made in the first somatosensory area of monkeys by inserting crystals of dye into tracks of microelectrodes used to record receptive fields of neurons in area 3b. Each injection retrogradely labeled one or more elongated clusters of cells extending anteroposteriorly through the ventral posterior medial (VPM) or ventral posterior lateral (VPL) nucleus. Double labeled cells occurred only when distances between centers of injections were less than 600 μm. With separations greater than 600 μm, most clusters of retrogradely labeled cells contained mainly cells labeled by one dye. However, even with distances of 1-1.5 mm, labeled clusters contained small but significant numbers of cells labeled with the other dye.

The double labeling shows that the extent of intracortical arborization of single thalamocortical axons in area 3b is no greater than 600 μm. The extended distribution and overlap of single labeled cells shows that adjoining cells in the same thalamic representation can project to cortical targets as distant as 1.5 mm. The fine grain of the area 3b representation probably depends upon inputs from the majority population of each thalamic cell group. However, divergent and convergent projections from minority populations of the group form a pre-existing basis for rapid changes in the representational map when lemniscal activity is heightened or reduced.

Introduction

The maps of the contralateral half body surface in the somatosensory cortex of primates (Powell and Mountcastle, 1959b; Paul et al., 1972; Dreyer et al., 1975; Kaas et al., 1979; Merzenich et al., 1978; Nelson et al., 1980; Mountcastle, 1984) display a high degree of somatotopic order, as well as reflecting peripheral innervation density (Mountcastle and Powell, 1959) and submodality of afferent input (Powell and Mountcastle, 1959; Phillips et al., 1970; Sur et al., 1980, 1981,1984; Iwamura et al., 1983, 1988, 1993), but they are not static. Activity dependent changes in the extent of parts of the representation can be elicited by reductions or modifications of input from the periphery (Kaas et al., 1983; Merzenich et al., 1983a,b; Wall et al., 1986, 1992a,b; Cusick et al., 1990; Kaas, 1991) and under behavioral conditions (Clark et al., 1988; Jenkins et al., 1990a,b; Allard et al., 1991; Recanzone et al., 1992) These changes are commonly manifested by expansions or contractions of the representation of a small body part in relation to representations of adjacent parts. With total loss of sensory input from an upper limb maintained over a protracted period, representations of parts of the face can expand into territory normally occupied by the limb (Pons et al., 1992).

Activity-dependent alterations in the body representation in somatosensory cortex may depend upon functional expression of pre-existing but hitherto functionally suppressed thalamic and/or intracortical connections (Kaas et al., 1983; Jacobs and Donoghue, 1991; Lund et al., 1994) and/or upon sprouting of thalamocortical or corticocortical connections into new territories (Pons et al., 1992). Activity-dependent up- and down-regulation of gene expression for transmitter- and receptor-related molecules leading to alterations in the balance of excitation and inhibition in the cortex may also play a role (Jones, 1990, 1993). Virtually none of these potential mechanisms has been explored in any depth.

In the case of the thalamocortical projection, the extent of intracortical distribution of individual thalamocortical fibers and of groups of fibers arising from thalamic cells with the same receptive fields and submodality properties is still not fully known. In macaque monkeys, injected afferent fibers thought to be of thalamic origin have been described with single, focal domains of termination approximately 500 μm in extent in area 3b (Garraghty and Sur, 1990) but the extent of branching in the white matter and thus the potential divergence in the projection has not been determined.

Anterograde and retrograde tracing studies of projections from the ventral posterior nucleus of macaques have not been conducted at sufficiently high resolution to provide the detail necessary for an evaluation of the extent of cortical projection of groups of cells that represent a small part of the body in the ventral posterior nucleus (Jones et al., 1979; Nelson and Kaas, 1981; Darian-Smith et al., 1990; Darian-Smith and Darian-Smith, 1993). Injections of tiny amounts of tritiated amino acids at physiologically defined sites in the ventral posterior nucleus, however, resulted in labeling of axonal terminations in multiple foci separated by gaps in area 3b (Jones et al., 1982). This and the widespread distribution of retrogradely labeled cells after injections of tracer in SI (Darian-Smith and Darian-Smith, 1993) suggest that divergent thalamocortical projections not revealed in conventional multiunit mapping studies, could form a basis for the expansion of part of the cortical representation under conditions in which inputs are enhanced.

The present study was focused on the extent to which thalamic cells providing input to physiologically defined, representations in the somatic sensory cortex diverge and the extent of overlap in the projections of individual cells (Rausell and Jones, 1995).

Methods

Two normal *Macaca fuscata* and four normal *Macaca mulatta* monkeys were used (Table 1).

Table I Analysis of individual cases (Rausell and Jones, 1995)

Case number	Receptive field at injection site	Injection site	Inter-injection site	Thalamic nucleus	Double labeling	Anteroposterior extent of thalamic labeling (μm)
51L	DY: Upper peroneal region FB: Upper peroneal region	Area 3b	100	VPL	+++	675
53L	DY: Zygomatic region FB: Zygomatic region	Area 3b	250	VPM	++	850
51R	DY: Upper lip	Area 3b	280	VPM	++	1200
53R	DY: Medial aspect of the knee FB: Medial aspect of the knee	Area 3b	300	VPL	++	850
58R	DY: Great toe FB: Lateral plantar digital pad	Area 3b	600	VPL	0	850
52L	DY: Lateral inguinal region FB: Medial inguinal region	Area 3b	1100	VPL	0	850
43R	DY: Face FB: Face	Area 3b	1160	VPM	0	900
45R	DY: Face FB: Face	Area 3b	1310	VPM	0	1120
52R	DY: Upper lip FB: Upper lip	Area 3b	1500	VPM	0	850

The animals were anesthetized with ketamine, and maintained on Nembutal and the medial and/or lateral aspects of the postcentral gyrus were exposed. Tungsten microelectrodes were inserted into area 3b where it comes to the surface at the medial and lateral ends of the central sulcus (Powell and Mountcastle, 1959a; Roberts and Akert, 1963). Multiunit discharges were recorded in response to light tactile stimuli and receptive fields were mapped. Two loci, 1.5mm or less apart, in which responses were obtained from stimulation of overlapping or non-overlapping receptive fields were identified. Micropipettes containing 8% Fast Blue (FB) or 5% Diamidino Yellow (DY) were dipped into the dyes so that small crystals of the dye adhered to the tip. The crystals were then pushed into the two identified tracks left by the metal microelectrodes, FB at one site and DY at the other.

The animals survived for two weeks after which they were perfused with normal saline followed by 4% paraformaldehyde and 0.2% glutaraldehyde in 0.1 M phosphate buffer. The brain was removed and blocked; the blocks were infiltrated with 30% sucrose and frozen in dry ice.

The cerebral cortex was sectioned parasagittally on a sliding microtome at 50 μm and the thalamus frontally at 25 μm or alternating at 20 μm and 40 μm. Sections were mounted on clean slides, and after partial drying were coverslipped in 60% glycerol in 0.1M phosphate buffer.

Sections were examined under epifluorescence using a broad band ultraviolet excitation filter. The outlines of the sections, landmarks such as blood vessels, and the locations of dye deposits in the cerebral cortex and of retrogradely labeled cells in the thalamus were plotted using a Minnesota Datametrics MD2 plotting system. Sections were then remounted on gelatin coated slides, dried, and stained with thionin or stained prior to mounting for cytochrome oxidase (CO, Wong-Riley, 1979). The re-stained sections were scanned into Adobe Photoshop using a Nikon LS-3510 film scanner and a MacIntosh Quadra 950. Scans of the maps of labeled cell distribution were made using a Hewlett Packard ScanJet IICX flatbed scanner and the two scans were superimposed in the computer to generate maps in which all retrogradely labeled cells were plotted in relation to Nissl-stained neurons located in the same area or to densities of CO staining.

The maps of the distributions of retrogradely labeled cells in the thalamus were copied on acetate sheets and put in register by using the profiles of sectioned blood vessels as landmarks. FB, DY and double labeled neurons were color coded and the coordinates of each labeled neurons were transferred to a data sheet (QuatroPro) by means of a digitizing tablet, and 3-dimensional reconstructions of the whole labeled population were made (Acrospin, Deltagraph). The reconstructions were rotated to provide frontal, parasagittal and dorsal views of retrogradely labeled cells in the ventral posterior (VP) nucleus of the thalamus. The total number of retrogradely labeled neurons, the relative proportions of FB, DY and double labeled neurons were calculated and correlations were made between the relative proportions and the distance between the centers of the dye deposits.

Results

All injection sites were located in area 3b as determined cytoarchitectonically and were centered in layers III-IV with no extension into the white matter, relatively little spread over the cortical surface and no involvement of area 1 (Fig. 1).

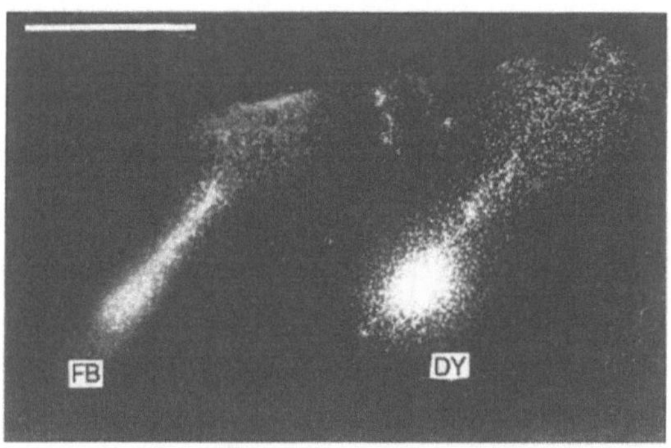

Figure 1. Fluorescence photomicrograph, showing locations and extent of deposits of Diamidino Yellow (DY) and Fast Blue (FB) in tracks left by tungsten microelectrodes used to record receptive field data on neurons located in the foot representation of area 3b. Bar represents 500 μm.

The distance between the FB and DY deposits of each pair varied from 100 μm to 1,500 μm (Table 1). The absolute and relative numbers of FB, DY and double labeled cells in the VP nucleus of each case are shown in table 2.

Table 2 Numbers and percentages of labeled neurons (Rausell and Jones, 1995)

Case number	inter-injection distance	total labeled neurons	FB neurons	DY neurons	Double labeled neurons
51L	100	343	46(13.4%)	51 (14.9%)	146 (71.7%)
53L	250	376	23 (6.1%)	266 (70.7%)	87 (23.1%)
51R	280	594	279 (46.9%)	184 (30.9%)	131 (22.1%)
53R	300	878	505 (57.5%)	150 (17.1%)	223 (25.4%)
58R	600	3935	2298 (58.4%)	1371 (34.8%)	266 (6.8%)
52L	1100	350	113 (32.3%)	237 (67.7%)	0
43R	1160	224	56 (25.0%)	168 (75.0%)	0
45R	1310	209	145 (69.4%)	64 (30.6%)	0
52R	1500	370	19 (5.1%)	351 (94.9%)	0

Retrogradely labeled neurons were located in the ventral posterior medial nucleus (VPM) or in the ventral posterior lateral nucleus depending on the locations of the injections in relation to the representation in area 3b. The number of neurons retrogradely labeled with each dye varied according to the size of the corresponding injection. Double labeled neurons were found only in cases in which the pairs of dye deposits were separated by 600 μm or less (Fig. 2).

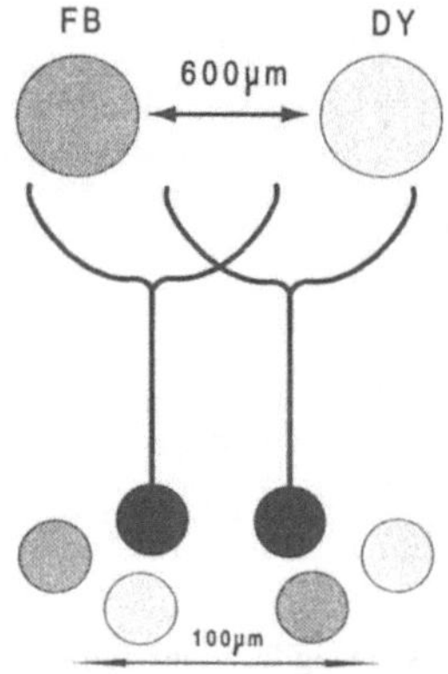

Figure 2. Schematic representation of the double labeling results. Injections of fast blue (FB) and diamidino yellow (DY) placed within 600 μm of one another retrogradely label two overlapping populations of thalamic cells in which significant numbers are double labeled. Injections separated by distances greater than 600 μm do not result in double labeling. This indicates that the extent of terminal ramification of the axons of individual thalamocortical cells is approximately 600 μm.

Comparison with the CO-stained sections showed that retrogradely labeled neurons appeared in both the CO-rich and CO-weak compartments of either VPM or VPL (Rausell and Jones, 1991a; Rausell et al., 1992) but the majority were concentrated in the CO-rich compartments. There, most of the retrogradely labeled neurons clearly belonged to the clusters of Nissl stained neurons normally found in that compartment. The smaller cells that form the matrix compartment of weaker staining in CO preparations contained only a few labeled cells. The percentage of neurons labeled in each cluster depended on the extent of the injection site but even in cases with the highest number of retrogradely labeled neurons, no more than 60-75% of the neurons of a cluster were labeled.

Even with the smallest injections, two or three clusters of cells were labeled. Invariably these were aligned anteroposteriorly in the ventral posterior nucleus and extended through most of its anteroposterior extent. In VPM, the cells occupied one or more of the CO-dense rods of that nucleus while in VPL, the anteroposteriorly aligned clusters of retrogradely labeled neurons occupied parts of the CO dense sheets or lamellae of that nucleus. The occurrence of double labeled neurons depended on the distance separating the two injection sites but there was a high degree of intermingling of neurons retrogradely labeled with different dyes in the same cluster well beyond inter injection distances that gave no double labeling. Double labeled neurons and a high degree of intermingling of neurons labeled with different dyes in the same cluster occurred when the distance between deposits was less than 600 μm (Table 1) (Fig. 3).

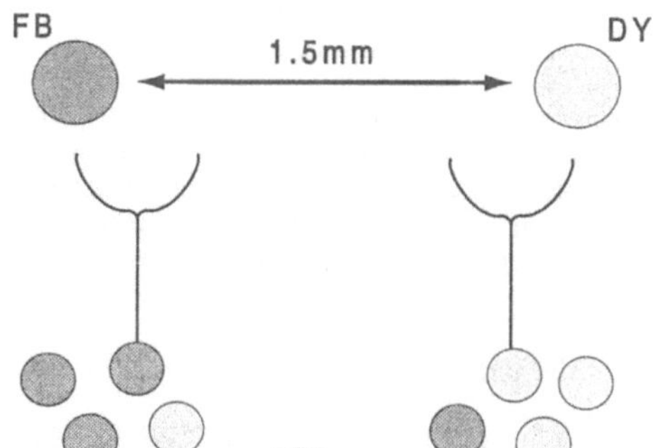

Figure 3. **Schematic representation of the results from injections placed more than 600 μm apart. Injections of fast blue (FB) and diamidino yellow (DY) placed up to 1.5 mm apart each retrogradely label a majority population of cells without double labeling, but there is substantial overlap. The two labeled populations only become completely separated with inter-injection distances greater than 1.5 mm.**

With inter-injection distances greater than 600µm, no double labeled neurons were found, and the degree of intermingling of cells labeled with different dyes in the same cluster was reduced. With inter-injection distances up to 1.5mm, neurons retrogradely labeled with FB or DY only tended to be concentrated in neighboring clusters but here was still a high degree of inter-mingling (Fig. 3). Beyond 1.5mm inter-injection distances, the number of neurons labeled with the seemingly inappropriate dye became zero and the clusters of DY and FB labeled cells were clearly separate.

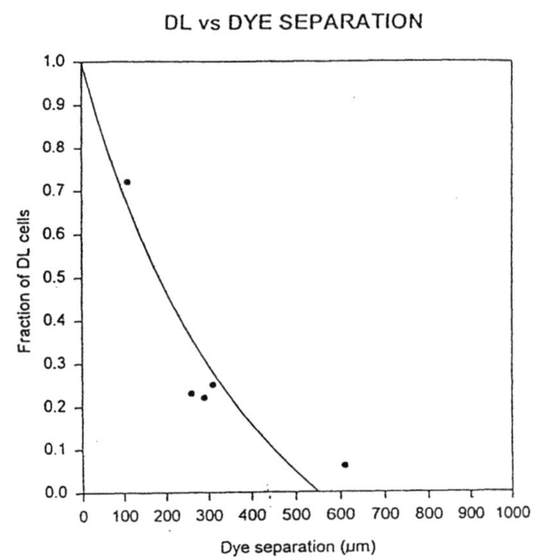

Figure 4. The mean of the percentage of double labeled cells (ordinate) in the five cases in which double labeled cells were found (dots), plotted against the distance between the centers of the injections in each case (abscissa). The derived line, represents the probability of finding double labeled cells at a particular inter-injection distance. For a probability of 0 (no double labeled cells) the distance between the centers of the injections must exceed 500 µm. (From Rausell and Jones, 1995).

In Fig. 4 the average of the percentage of double labeled neurons among the population retrogradely labeled with each dye in five cases is plotted against the distance between the cortical dye deposits. The intracortical extent of the terminal arborizations of the double labeled neurons can be predicted to be at least as large as the separation between the two dyes for the neurons to be double labeled. The probability (Q) of finding a double labeled neuron, out of the total population of labeled neurons, depends on the distance (D) between the injections and the extent of the arborization of a single thalamocortical axon

(A), if A is radially symmetric. The probability of either dye deposit occurring within a particular arborization is proportional to A+D while the probability of both dye deposits occurring within a particular arborization is proportional to A-D. Therefore, Q= (A-D) / (A+D). The continuous line in Fig. 4 is the best fit of this equation to the experimental data points provided that for D=0 (two dye deposits overlapping), Q=1 (all cells double labeled). The best fit is achieved if A=550 μm, which gives the average extent of a single thalamocortical axon termination in the cortex.

Discussion

The present study demonstrated not only the extent of terminal arborization of individual thalamocortical axons in area 3b but also the extent of convergence in the projection of populations of thalamic neurons lying within a particular part of the thalamic body representation. Ventral posterior neurons could only be double labeled by retrograde transport from paired injections in area 3b when the inter-injection distance was less than 600 μm. This and the accompanying regression analysis indicate that the extent of intracortical arborization of single thalamocortical axons in area 3b is no greater than 550-600 μm. However, even after the smallest injections, neurons were labeled throughout the full anteroposterior extent of VPM or VPL and paired dye injections, separated by up to 1.5 mm, still resulted in overlapping sets of labeled neurons. The degree of overlap was inversely proportional to inter-injection distance and was greatest when cortical sites containing neurons with different receptive fields received dye applications (Rausell and Jones, 1995). Intermingling of neurons labeled with different dyes was still consistently observed even with injections at different receptive field locations. Most neurons of a VP cluster and thus of a defined part of the thalamic representation project to a corresponding cortical domain in area 3b, defined in these experiments by the locus of one DY injection, but a small group of cells of the same cluster project to surrounding domains, defined by the locus of the second injection. These offshoot projections are to loci that can represent the same or different parts of the body surface in area 3b. This divergence could not be attributed to involvement of area 1 in any of the injections.

The projection of the representation of a small part of the body in VP to the corresponding somatotopic locus in area 3b is formed by axons of neurons located in one elongated set of VP clusters whose axons converge on the restricted cortical locus. This cortical locus, however, also receives axons of a few neurons located in adjacent, non-somatotopically related groups of VP cells. That is, neurons in a set of VP clusters primarily related to one cortical locus project to surrounding cortical loci in which neurons have receptive fields on the same or different parts of the body. These loci form the principal cortical targets of other, usually adjacent VP clusters, commonly in different parts of the VP body map.

The cortical projection of a single elongated group of cells is substantially greater than the 550 - 600 μm subtended by the axonal arborizations of its individual cells. The projection pattern is one of shifted overlap. A majority of the relay neurons in an elongated cluster of VP cells projects to restricted region of cortex but a significant minority projects to neighboring regions which receive most of their input from loci in VP neighboring the first but still containing many neurons projecting to the first region, and so on throughout the nucleus. Neurons unlabeled after cortical injections but lying within the clusters of labeled neurons probably represent both interneurons (Hunt et al., 1991; Benson et al., 1991) and relay neurons that project to cortex outside the dye-injected foci.

The majority of the cells projecting to a cortical locus are aggregated in a concentrated, elongated set of clusters in VPL or VPM. This accounts for the correspondence between physiological maps in the ventral posterior nucleus and somatic sensory cortex of macaques (Mountcastle and Henneman, 1952; Poggio and Mountcastle, 1963; Jones and Friedman, 1982; Jones et al., 1982; Jones et al., 1986a,b; Rausell and Jones, 1991a). Descriptions of lamellar arrangements of retrogradely labeled cells in VPL after relatively large injections of retrograde tracers in the cortex (Jones et al., 1979; Nelson and Kaas, 1981; Ma and Juliano, 1991) undoubtedly reflect the size of the injections and the organization of the clusters as parts of the lamellae.

Rods of cells with the same submodality properties and closely overlapping receptive fields, and extending anteroposteriorly through VPL or VPM were previously shown to project upon one or a small number of focal domains in monkey somatosensory cortex (Jones et al., 1982; Jones, 1985). Darian-Smith et al. (1990) and Darian-Smith and

Darian-Smith (1993) showed that larger and more widespread populations of thalamic cells in parts of VP apparently unrelated somatotopically to the injection site, could be retrogradely labeled after relatively small cortical injections. The present results, show that focal injections of DY or FB tend to label cells mostly concentrated in a single rod-like domain in VPL or VPM. However, outlying cells that are members of other anatomical and functional rods are also labeled and there is invariably a significant amount of mixing of cells retrogradely labeled from injections of dye into two non-somatotopically related cortical loci, more than 600 μm distant from one another.

Physiologically, the projections from a primary thalamic focus to a seemingly inappropriate cortical domain may be weaker and not detected by methods customarily used to map representations in somatosensory cortex for these less dense thalamocortical projections may escape detection with multiunit methods. In cats, single thalamic cells with receptive fields on a toe could be antidromically activated by intracortical microstimulation applied to cortical sites in which an adjacent toe was represented (Snow et al., 1988). Hence, it may be necessary to use single unit methods to reveal what may be weaker synaptic connections. Two or more foci of enhanced metabolic activity can be demonstrated in the somatosensory cortex of monkeys following repetitive stimulation of a finger (Juliano et al., 1981; Juliano and Whitsel, 1987) and this may reflect better the divergence seen in the present results. These foci probably reflect the presence of a primary focus of thalamocortical projection, similar to that demonstrated by the major concentration of retrogradely labeled cells following a single injection in the present study, and the divergent projections from cells that co-mingle with the major concentration.

Single fibers and groups of fibers arising from dorsal column or trigeminal nuclear cells with the same submodality and receptive fields properties terminate on a relatively small population of ventroposterior cells mainly oriented in anteroposterior rods (Jones et al., 1982; Jones, 1983) Most of these cells will, thus, project to the same locus in the cortex, as demonstrated in the present study and the organized projection of place and submodality from the periphery to the cortical map will be preserved. The termination of lemniscal afferents on elongated clusters of thalamic cells similar to those demonstrated by retrograde labeling in the present study appears to account for the physiological observation that single units and multiunit clusters recorded over long anteroposterior distances in VPL and VPM tend to have the same receptive fields and submodality properties (Jones and Friedman, 1982; Jones et al., 1982, 1986a; Rausell and Jones, 1991a). This reflects the predominant input to each elongated group of cells and it provides the basis, via the projection of the majority population to a defined cortical locus, for the transfer of the thalamic somatotopic map to that in the cortex. However, a substantial minority of cells in the thalamic group projects to regions of cortex up to 1.5 mm distant from that receiving input form the majority population. If we assume that all these cells are synaptically engaged by the same lemniscal afferents as the majority population, then a route is open for the place and submodalilty information carried by the lemniscal afferents to reach an apparently non-somatotopically related locus in the cortex. At this second locus, the majority input should be from cells that form the majority population of another cell grouping potentially in a somatotopically unrelated part of the thalamic representation. In cases of peripheral deafferentation in which the majority input to the second cortical locus is reduced, the divergent projection could be revealed. Because the divergent projections arise from fewer thalamic cells than those which form the principal input to a cortical locus, under normal conditions the input from the primary population probably predominates. The existence of the divergent projections could, however, explain the immediacy with which cortical representations mapped by multiunit recording can expand (Merzenich et al., 1983b; Cusick et al., 1990; Wall et al., 1992a,b). It could be the primary mechanism of short term plasticity, other mechanisms such as axonal sprouting only being engaged later.

When activity is enhanced in a set of lemniscal fibers as the result of increased stimulation of part of the body surface (Jenkins et al., 1990a,b; Recanzone et al., 1992) the divergent projections may come to dominate the principal projection to a cortical locus, thus giving the appearance of part of the representation spreading at the expense of another. A "cortical distance limit" of 1-1.5 mm was proposed as the limit on spread of a representation after forelimb digit amputation (Kaas et al., 1983). This may be accounted for by the lack of overlap in thalamo-cortical projections demonstrated by injections more than 1.5mm apart in the present study. The distance limit can be exceeded and reach 3-3.5 mm after combinations of peripheral nerve section (Garraghty et al., 1994), or 12 mm after upper limb deafferentation (Pons et al., 1992). This could reflect both magnification of an affect due to divergence similar to that demonstrated here in the projections of lower centers to the thalamus, as well as sprouting in brainstem, thalamus and/or cortex.

Acknowledgments: Supported by Grant numbers NS21377 and NS22317 from the National Institutes of Health, United States Public Health Service. I thank Drs. Estrella Rausell, Ariel Agmon, Marco Molinari and Angel Viñuela for their contributions.

References

Allard, T. Clark, S.A. Jenkins, W.M. Merzenich, M.M. (1991) Reorganization of somatosensory area 3b representations in adult owl monkeys after digital syndactyly. Journal of Neurophysiology. 66:1048-1058.
Benson, D.L. Isackson, P.J. Hendry, S.H.C. Jones, E.G. (1991) Differential gene expression for glutamic acid decarboxylase and type II calcium-calmodulin-dependent protein kinase in basal ganglia, thalamus and hypothalamus of the monkey. Journal of Neuroscience 11:1540-1564.
Clark, S.A. Allard, T. Jenkins, W.M. Merzenich, M.M. (1988) Receptive fields in the body-surface map in adult cortex defined by temporarily correlated inputs. Nature 332:444-445.
Cusick, C.G. Wall, J.T. Whiting, J.H. Wiley, R.G. (1990) Temporal progression of cortical reorganization following nerve injury. Brain Research 537:355-358.
Darian-Smith, C. Darian-Smith, I. (1993) Thalamic projections to areas 3a, 3b and 4 in the sensorimotor cortex of the mature and infant macaque monkey. Journal of Comparative Neurology 335:173-199.
Darian-Smith, C. Darian-Smith, I. Cheema, S.S. (1990) thalamic projections to sensorimotor cortex in the macaque monkey: use of multiple retrograde fluorescent tracers. Journal of Comparative Neurology 299:17-46.
Dreyer, D.A. Loe, P.R. Metz, C.B. Whitsel, B.L. (1975) Representation of head and face in postcentral gyrus of the macaque. Journal of Neurophysiology 38:714-734.
Garraghty, P.E. Hanes, D.P. Florence, S. Kaas, J.H. (1994) Pattern of peripheral deafferentation predicts reorganizational limits in adult primate somatosensory cortex. Somatosensory and Motor Research. 11:109-117.
Garraghty, P.E. Sur, M. (1990) Morphology of single intracellularly stained axons terminating in area 3b of macaque monkeys. Journal of Comparative Neurology 294:583-593.
Hunt, C.A., Pang, D.Z., Jones, E.G. (1991) Distribution and density of GABA cells in intralaminar and adjacent nuclei of monkey thalamus. Neuroscience 43:185-196.
Iwamura, Y., Tanaka, M., Sakamoto, M., Hikosaka, O. (1983) Functional subdivisions representing different finger regions in area 3 of the first somatosensory cortex of the conscious monkey. Experimental Brain Research. 51:315-326.
Iwamura, Y., Tanaka, M., Sakamoto, M., Hikosaka, O. (1993) Rostocaudal gradients in the neuronal receptive field complexity in the finger region of the alert monkey's postcentral gyrus. Experimental Brain Research. 92:360-368.
Jacobs, K.M., Donoghue, J.P. (1991) Reshaping the cortical map by unmasking latent intracortical connections. Science 251:94-945.
Jenkins, W.M., Merzenich, M.M., Ochs, M.T., Allard, T., Guic-Robles, E. (1990) Functional reorganization of primary somatosensory cortex in adult owl monkeys after behaviorally controlled tactile stimulation. Journal of Neurophysiology 63:82-104.
Jones, E.G. (1983) Distribution patterns of individual medial lemniscal axons in thalamic ventrobasal complex of monkeys. Journal of Comparative Neurology 215:1-16.
Jones, E.G. (1985) The Thalamus. New York: Plenum.
Jones, E.G. (1990) The role of afferent activity in the maintenance of primate neocortical function. Journal of Experimental Biology 153:155-176.
Jones, E.G. (1993) GABAergic neurons and their role in cortical plasticity in primates. Cerebral Cortex. 3:361-372.

Jones, E.G., Friedman, D.P. (1982) Projection pattern of functional components of thalamic ventrobasal complex on monkey somatosensory cortex, Journal of Neurophysiology 48:521-544.

Jones, E.G., Friedman, D.P., Hendry, S.H.C. (1982) Thalamic basis of place and modality-specific columns in monkey somatosensory cortex: a correlative anatomical and physiological study. Journal of Neurophysiology, 48:545-568.

Jones, E.G. Hendry, S.H.C., Brandon, C. (1986a) Cytochrome oxidase staining reveals functional organization of monkey somatosenory thalamus. Experimental Brain Research 62:438-442.

Jones, E.G., Schwark, H.D., Callahan, P.J. (1986b) Extent of the ipsilateral representation in the ventral posterior medial nucleus of the monkey thalamus. Experimental Brain Research 63:310-320.

Jones, E.G., Wise, S.P., Coulter, J.D. (1979) Differential thalamic relationships of sensory-motor and parietal cortical fields in monkeys. Journal of Comparative Neurology 183:833-882.

Juliano, S.L., Hand, P.J., Whitsel, B.L. (1981) Patterns of increased metabolic activity in the somatosensory cortex of monkeys (Macaca fascicularis) subjected to controlled cutaneous stimulation a 2-deoxyglucose study. Journal of Neurophysiology 46:1260-1284.

Juliano, S.L., Whitsel, B.L. (1987) A combined 2-deoxyglucose and neurophysiological study of primate somatosensory cortex. Journal of Comparative Neurology 263:514-525.

Kaas, J.H., Merzenich, M.M., Killackey, H.P. (1983) Multiple representations of the body within the primary somatosensory cortex of primates. Science 204:521-523.

Kaas, J.H., Pons, T.P., Wall, J.T., Garraghty, P.E., Cusick, C.G. (1987) Consistent features of the representation of the hand in area 3b of macaque monkeys. Somatosensory and Motor Research 4:309-331.

Lund, J.P., Sun, G.-D., Lamarre, Y. (1994) Cortical reorganization and deafferentation in adult macaques. Science 265:546-548.

Ma, W., Juliano, S. (1991) The relationship between thalamocortical connections and stimulus-evoked metabolic activity in the ventroposterior nucleus of the monkey. Somatosensory and Motor Research 8:77-86.

Merzenich, M.M., Kaas, J.H., Sur, M., Lin, C.S. (1978) Double representation of the body surface within cytoarchitectonic areas 3b and 1 in SI in the owl monkey (Aotus trivirgatus). Journal of Comparative Neurology 181:41-74.

Merzenich, M.M., Kaas, J.H., Wall, J., Nelson, R.J., Sur, M., Felleman, D. (1983) Progression of change following median nerve section in the cortical representation of the hand in areas 3b and 1 in adult owl and squirrel monkeys. Neuroscience 10:639-296.

Mountcastle, V.B., (1984) Central nervous mechanisms in mechanoreceptive sensibility. In: Handbook of Physiology section I: The Nervous System volume III Sensory Processes, part 1, JM Brookhart, VB Mountcastle, V.B. and I Darian-Smith (eds) Bethesda, MD, American Physiological Society, pp 789-878.

Mountcastle, V.B., Henneman, E. (1952) The representation of tactile sensibility in the thalamus of the monkey. Journal of Comparative Neurology 97:409-439.

Mountcastle, V.B., Powell, T.P.S. (1959) Central nervous mechanisms subserving position sense and kinesthesis. Bulletin of Johns Hopkins Hospital 105:173-200.

Nelson, R.J., Kaas, J.H. (1981) Connections of the ventroposterior nucleus of the thalamus with the body surface representations in cortical areas 3b and 1 of the cynomolgus macaque (Macaca fascicularis). Journal of Comparative Neurology 199L29-64.

Nelson, R.J., Sur, M., Felleman, D.J., Kaas, J.H. (1980) Representations of the body surface in post-central parietal cortex of Macaca fascicularis. Journal of Comparative Neurology 192:611-643.

Paul, R.L., Merzenich, M., Goodman, H. (1972) Representation of slowly and rapidly adapting cutaneous mechanoreceptors of the hand in Journal of Neurophysiology. 26:775-806.

Pons, T.P., Garraghty, P.E., Ommaya, A.K., Kaas, J.H., Taub, E., Mishkin, M. (1991) Massive cortical reorganization after sensory deafferentation in adult macaques. Science 252:1857-1860.

Powell, T.P.S., Mountcastle, V.B. (1959) The cytoarchitecture of the postcentral gyrus of the monkey Macaca mulatta. Bulletin of Johns Hopkins Hospital 105:133-162.

Rausell, E., Bae, C.S., Viñuela, A., Huntley, G.W., Jones, E.G. (1992) Calbindin and parvalbumin cells in monkey VPL thalamic nucleus: distribution, laminar cortical projections and relations to spinothalamic terminations. Journal of Neuroscience 12:4088-4111.

Rausell, E., Jones, E.G. (1991) Histochemical and immunoctyochemical compartments of the thalamic VPM nucleus in monkeys and their relationship to the representational map. Journal of Neuroscience 11:210-225.

Rausell, E., Jones, E.G. (1991) Chemically distinct compartments of the thalamic VPM nucleus in monkeys relay principal and spinal trigeminal Brodmann's areas 3 and 1 of Macaca mulatta. Brain Research, 36:229-249.

Phillips, C.G., Powell, T.P.S., Wiesendanger, M. (1970) Projection from low-threshold muscle afferents of hand and forearm to area 3a of baboon's cortex. Journal of Physiology, London 210:59-60.

Poggio, G.F., Mountcastle, V.B. (1963) The functional properties of ventrobasal thalamic neurons studied in unanesthetized monkeys. pathways to different layers of the somatosensory cortex. Journal of Neuroscience 11:226-237.

Rausell, E., Jones, E.G. (1995) Extent of intracortical aroborization of thaamocortical axons as a determinant of representational plasticity in monkey somatic sensory cortex. Journal of Neuroscience. (in press).

Recanzone, G.H., Merzenich, M.M., Jenkins, W.M., Grajski, K.A., Dinse, H.R. (1992) Topographic reorganization of the hand representation in cortical area 3b of owl monkeys trained in a frequency-discrimination task. Journal of Neurophysiology 67:1031-1056.

Roberts, T.S., Akert, K. (1963) Insular and opercular cortex and its thalamic projection in Macaca mulatta. Schweizische Archives Neurologische Neurochirurugie und Psychiatrie. 92:1-43.

Rose, J.E., Mountcastle, V.B. (1959) Touch and kinesthesis. In: Handbook of Physiology. J Field, HW Magoun, VE Hall (eds) American Physiological Society. Washington DC. pp387-429.

Snow, P.J., Nudo, R.J., Rivers, W., Jenkins, W.M., Merzenich, M.M. (1988) Somatotopically inappropriate projections from thalamocortical neurons to the SI cortex of the cat demonstrated by the use of intracortical microstimulation. Somatosensory and Motor Research 5:349-372.

Sur, M., Merzenich, M.M., Kaas, J.H. (1980) Magnification, receptive-field area, and "hypercolumn" size in areas 3b and 1 of somatosensory cortex in owl monkeys. Journal of Neurophysiology. 44:295-311.

Sur, M., Wall, J.T., Kaas, J.H. (1981) Modular segregation of functional cell classes within the postcentral somatosensory cortex of monkeys. Science 212:1059-1061.

Sur, M., Wall, J.T., Kaas, J.H. (1984) Modular distribution of neurons with slowly adapting and rapidly adapting responses in area 3b of somatosensory cortex in monkeys. Journal of Neurophysiology 51:724-744.

Wall, J.T., Huerta, M.F., Kaas, J.H. (1992) Changes in the cortical map of the hand following postnatal median nerve injury in monkeys modification of somatotopic aggregates. Journal of Neuroscience 12:3445-3455.

Wall, J.T., Kaas, J.H., Sur, M., Nelson, R.J., Felleman, D.J., Merzenich, M.M. (1986) Functional reorganization in somatosensory cortical areas 3b and 1 of adult monkeys after median nerve repair: possible relationships to sensory recover in humans. Journal of Neuroscience 6:218-233.

Wall, J.T., Nepomuceno, V., Rasey, S.K. (1993) Nerve innervation of the hand and associated nerve dominance aggregates in the somatosensory cortex of a primate (squirrel monkey). Journal of Comparative Neurology 337:191-207.

Wong-Riley, M. (1979) Changes in the visual system of monocularly sutured or enucleuated cats demonstrable with cytochrome oxidase histochemistry. Brain Research. 171:11-28.

Somesthesis and the Neurobiology of the Somatosensory Cortex
O. Franzén, R. Johansson and L. Terenius (eds.)

Inhibitory circuitry in relation to the functional organization of somatosensory cortex

Harris D. Schwark

University of North Texas, Dept. of Biological Sciences, P.O. Box 5218, Denton, TX 76203, USA

Summary. Inhibition plays an important role in control of receptive field size in the somatosensory cortex (SI). In the present studies, the distributions of GABA-immunoreactive neurons and $GABA_A$ receptors were determined in relation to the reported functional properties of inhibition. In each of the cytoarchitectonic areas of SI, densities of GABA neurons and receptors were highest in layer II. Approximately 26% of the neurons in SI were GABAergic. The distributions of these GABA elements did not parallel the functional effects of inhibition in controlling receptive field size. Instead, preliminary experiments suggest that the extent of intrinsic, horizontal connections is more closely correlated with the extent of receptive field expansion when inhibition is blocked. In other experiments, the role of afferent activity in regulating receptive field size was examined. Such regulation begins as early as the dorsal column nuclei, and involves activity of capsaicin-sensitive primary afferents. These afferents may act in the DCN through substance P receptors, which were localized to the cell nest regions of the nuclei. Thus, receptive field size appears to be regulated at multiple levels of the somatosensory system by local inhibitory mechanisms.

Introduction

Inhibition is a hallmark of the response of many cortical neurons to afferent stimulation. This inhibitory response is usually disynaptic (Ferster and Lindström, 1983; Yamamoto et al., 1990), suggesting that it arises from intrinsic cortical inputs. Inhibitory neurons of the cortex fall into a variety of morphological classes (Houser et al., 1984), and typically account for 20-25% of the total population of neurons (Gabbot & Somogyi, 1986; Hendry et al., 1987).

The precise function of cortical inhibition is not yet clear. Inhibition in the visual cortex has been implicated in lending specificity to neurons' responses to visual stimuli (Sillito, 1984), perhaps by controlling excitation at the level of thalamocortical inputs (Berman et al., 1992). In somatosensory cortex (but apparently not in visual and auditory cortex), cortical inhibition controls the size of receptive fields (Dykes et al., 1984; Alloway et al., 1989). Since discrimination of multiple stimuli is an important function of sensory systems, changes in receptive field size will significantly impact the performance of the system.

Factors controlling receptive field size have been studied by iontophoresis of bicuculline to block local inhibition (e.g. Dykes et al., 1984; Alloway et al., 1989), resulting in receptive field expansion in most SI neurons. A second approach to studying mechanisms that control receptive field size is to block inputs from peripheral receptive fields by injecting local anesthetic. When this is done, new (expanded) receptive fields appear within minutes (e.g., Nakahama et al., 1966; Calford and Tweedale, 1991b). The mechanisms underlying this change are not known, but it has been suggested that tonic inhibition may play a role (Calford and Tweedale, 1991a). Because of the importance of inhibition in somatosensory function, we undertook a series of experiments designed to reveal the organization of the inhibitory circuitry in cat SI, and to relate that organization to reported functional effects of inhibition (Li and Schwark, 1994; Schwark et al., 1994). In other experiments, we have begun to study the mechanisms involved in the control of receptive field size by peripheral inputs (Pettit and Schwark, 1993).

The inhibitory circuitry of somatosensory cortex

GABAergic neurons and receptors

The distribution of inhibitory neurons was studied in series of 16 μm parasagittal sections through the forearm/forepaw representation of cat SI (Li and Schwark, 1994). Neurons were stained by ABC-peroxidase immunocytochemical methods with antibodies to GABA. Densities and distribution patterns of GABA-immunoreactive (GABA+) neurons were determined following staining of free-floating, frozen sections. The areal density of GABA+ neurons was highest in layer II, and declined steadily to layer VI. This pattern was consistent across the four cytoarchitectonic areas of SI (Fig.

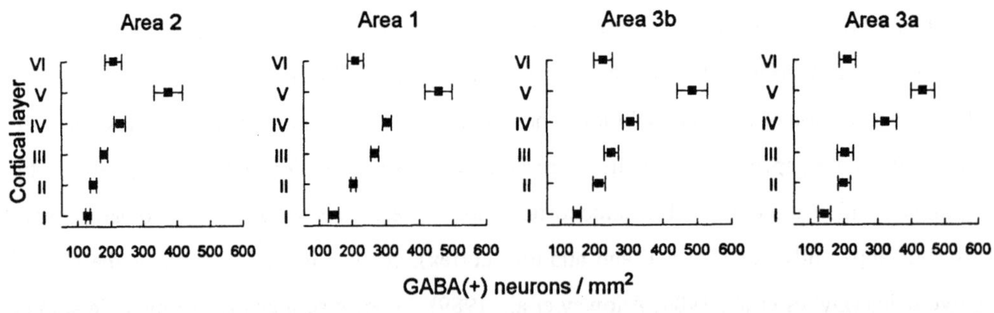

Figure 1. Mean areal density of GABA+ neurons in each of the four cytoarchitectonic areas of cat SI. Areal density is the number of GABA+ neurons per unit area of the section (corrected for shrinkage). Error bars represent standard errors of the means derived from five hemispheres. Modified from Li and Schwark (1994) and used with permission of John Wiley and Sons.

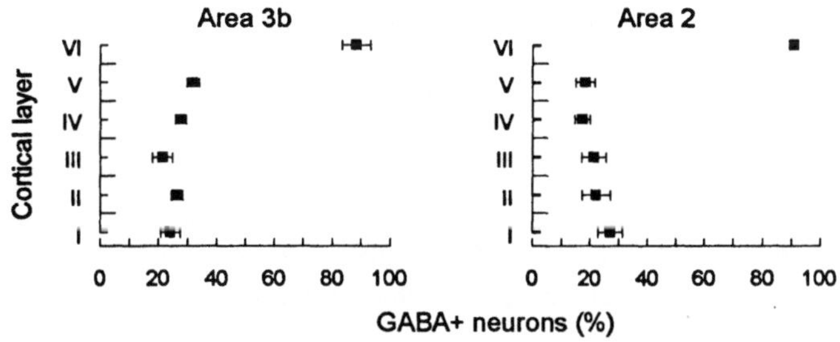

Figure 2. Proportion of GABA+ neurons as a function of cortical layer in areas 3b and 2. Error bars represent standard errors of the means derived from six hemispheres. Modified from Li and Schwark (1994) and used with permission of John Wiley and Sons.

1). In general, the laminar distribution of GABA+ neurons in cat SI was similar to that reported for cat visual cortex (Gabbott and Somogyi, 1986), but lacked the distinct peak in layer IV which has been reported for the monkey (Hendry et al., 1987).

Because differences in areal densities can arise from differences in total neuron numbers, the proportions of GABA+ neurons in areas 3b and 2 were determined following post-embedding staining of 1 μm plastic sections. The numbers and sizes of GABA+ and GABA- somata were determined, and a correction for differences in soma size applied (DeHoff and Rhines, 1961). The proportion of GABA+ neurons was then obtained by dividing the number of GABA+ neurons by the total number of neurons. In layer I, over 90% of the neurons were GABA+, while in the remaining layers GABA+ neurons accounted for approximately 25% of all neurons (Fig. 2). The overall proportion of GABA(+) neurons in SI (26.4%) was somewhat higher than in visual cortex (20.6%; Gabbott and Somogyi, 1986), a difference which also exists in the monkey (Hendry et al. 1987). Adjusting for the total number of neurons in these areas (Beaulieu and Colonnier, 1989), this difference in the cat appears to be due primarily to differences in the numbers of non-GABAergic neurons.

In a second set of experiments, we characterized the inhibitory circuitry in SI by examining the distribution of $GABA_A$ receptors (the ones blocked by bicuculline). Sections through SI were incubated in [^{3}H]muscimol and then exposed, along with radioactive standards, to tritium-sensitive film (Schwark et al., 1994). The distribution of $GABA_A$ receptors closely resembled the distribution of GABA+ neurons. Layer II contained the highest densities of $GABA_A$ receptors, and layer VI, the lowest (Fig. 3). This pattern was consistent across the four cytoarchitectonic areas of SI. Primary auditory and visual cortex had similar patterns of $GABA_A$ binding (see also Mower et al., 1986),

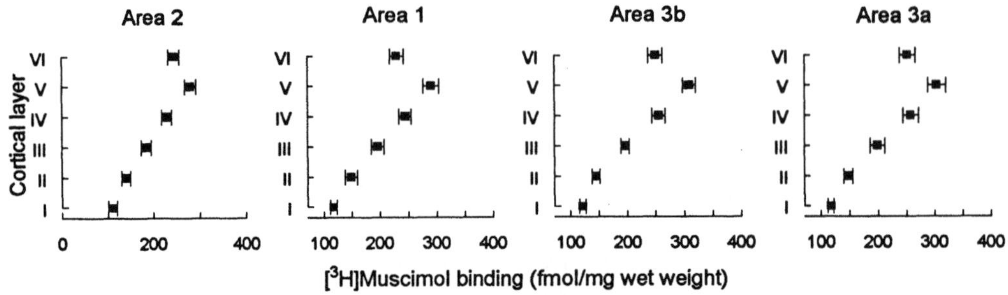

Figure 3. Average levels of [^{3}H]muscimol binding by cortical layer in the four cytoarchitectonic areas of cat SI. Each point represents the average of 32-42 density profiles. Modified from Schwark et al. (1994) and used with permission of John Wiley and Sons.

except binding levels in layers IV-VI were twice as dense in visual cortex as in SI. By using the data from these two experiments, we were able to compare the distributions of GABA neurons and receptors in SI. The pattern of [^{3}H]muscimol binding in SI correlated very well with the areal density of GABA+ neurons (Fig. 4), perhaps reflecting the limited extent of the axonal arborizations of most cortical GABAergic neurons.

Inhibitory circuitry in relation to the effects of bicuculline iontophoresis

The extent of receptive field enlargement following bicuculline iontophoresis varies across cortical layers and with neuronal response type. Neurons in the middle layers of SI undergo the smallest RF enlargement (Dykes et al., 1984), as do slowly-adapting neurons (Alloway et al., 1989). Also, neurons in rapidly adapting regions of SI (defined by multiunit recordings in layer IV) appear to undergo larger receptive field expansions than those in slowly adapting regions (Dykes et al., 1984). We compared our quantitative measures of the inhibitory circuitry in SI to these functional

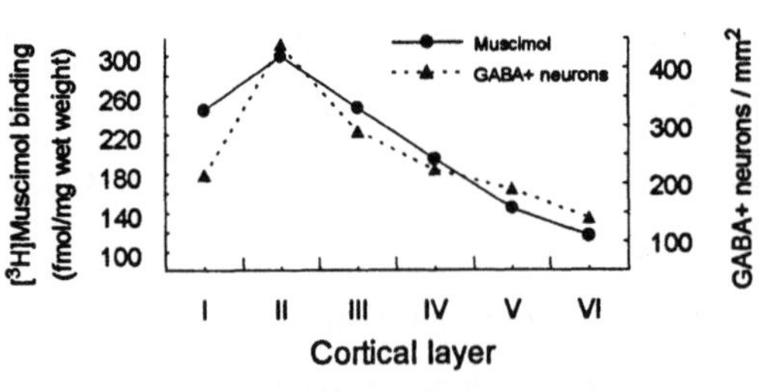

Figure 4. Comparison of the laminar distributions of [^{3}H]muscimol binding and GABA+ neuron densities in area 3b. Modified from Schwark et al. (1994) and used with permission of John Wiley and Sons.

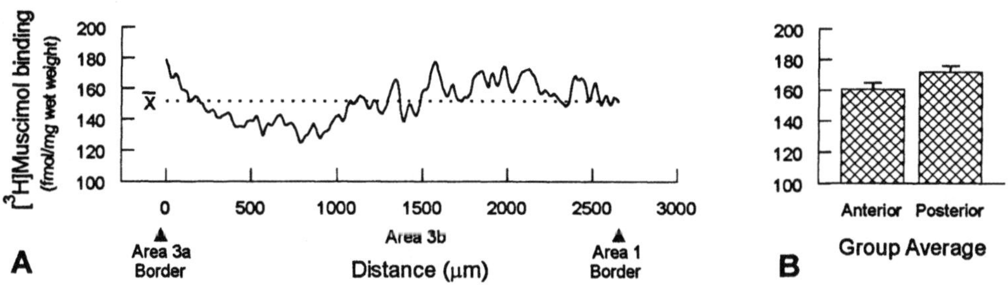

Figure 5. [^{3}H]Muscimol binding in layer IV of area 3b. **A.** An averaged density profile from six sections, smoothed using a 5-bin moving boxcar average. **B.** Overall binding density in layer IV, derived from six hemispheres, was significantly lower in anterior than in posterior regions. Modified from Schwark et al. (1994) and used with permission of John Wiley and Sons.

differences.

There was no obvious relation between the laminar distributions of GABA neurons or receptors (illustrated in Figs 1 - 3) and the laminar differences in bicuculline effects. To look for features of the inhibitory circuitry that might correspond to submodality regions of SI (Sretavan and Dykes, 1983), we measured GABA neuron and $GABA_A$ receptor densities in series of six closely-spaced sections. Areal density patterns in area 3b that are consistent within each series might correspond to submodality regions. However, analyses revealed no such consistent areal patterns in the distributions of GABA neurons or receptors. Further analysis of the distribution of receptors in layer IV of area 3b revealed significantly higher numbers of binding sites in posterior area 3b compared to anterior 3b (6.3%, $p<0.05$) (Fig. 5). This difference may correspond to the submodality regions of SI, such that rapidly adapting regions contain higher receptor densities than slowly adapting regions.

If the inhibitory circuitry of SI does not correspond well with the degree to which receptive fields expand when inhibition is blocked, what other features of cortical organization might? During bicuculline iontophoresis, the expanded portions of the receptive fields have longer latencies of activation than do the original receptive fields, suggesting that the expansions arise from connections intrinsic to SI (Alloway et al., 1989). To test this idea, we have analyzed the distribution of retrogradely labeled neurons following small injections of tracers into various layers of SI. Our preliminary results reveal laminar differences in the extent of these connections that resemble the laminar pattern reported for bicuculline's effects (Dykes et al., 1984). The intrinsic, horizontal connections are shortest in layer IV, where receptive field expansions are smallest, and longest in the superficial and deep layers (Fig. 6). Thus, receptive field expansion which results from bicuculline

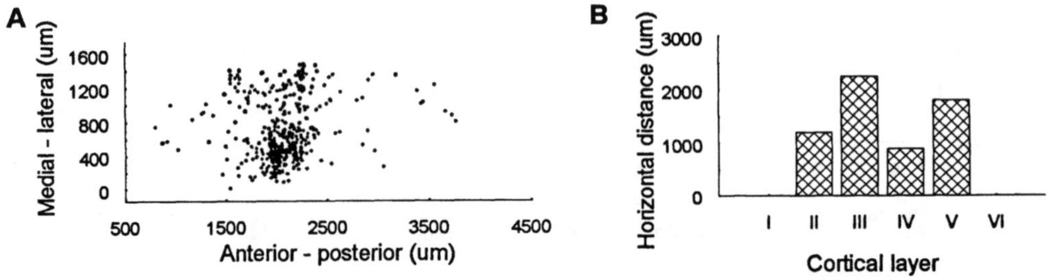

Figure 6. Distribution of retrogradely labeled neurons following small injections of tracers into area 3b. **A.** An example of the labeling pattern following an injection into layer V, as viewed from the cortical surface. **B.** The furthest extent of local labeling within laminae reflects the enlargement of receptive fields produced by bicuculline iontophoresis.

iontophoresis may arise from unmasking of normally ineffective inputs to the SI neurons. These inputs appear to be of cortical origin, from intrinsic, horizontal connections. Because the source of cortical inhibition is located near the target neurons, it might be predicted that afferent activity from within the receptive field would also influence receptive field size. Evidence for this is described below.

Control of receptive field size by peripheral inputs

Receptive field sizes of somatosensory neurons can be affected by changes in excitatory inputs from the peripheral receptive fields. Nakahama et al. (1966) reported that receptive fields of neurons in cat VPL expanded following injection of the local anesthetic procaine into excitatory receptive fields. Similar changes occur in the receptive fields of SI neurons (Calford and Tweedale, 1991b). Moreover, blockade of a subset of peripheral afferents by capsaicin injection results in similar receptive field changes (Calford and Tweedale, 1991a), leading the authors to suggest that C fibers might maintain tonic inhibition which restricts receptive field size.

To begin to identify the mechanisms involved in afferent control of receptive field size, we conducted experiments similar to those of Calford and Tweedale (1991a; 1991b) while recording from single neurons in the cat dorsal column nuclei (DCN) (Pettit and Schwark, 1993). Injection of lidocaine into the peripheral receptive field resulted in the appearance of newly responsive areas for every neuron tested (Fig. 7A). In many of the neurons, the expanded portions of the receptive fields had response properties that were different from those of the original receptive field, suggesting that

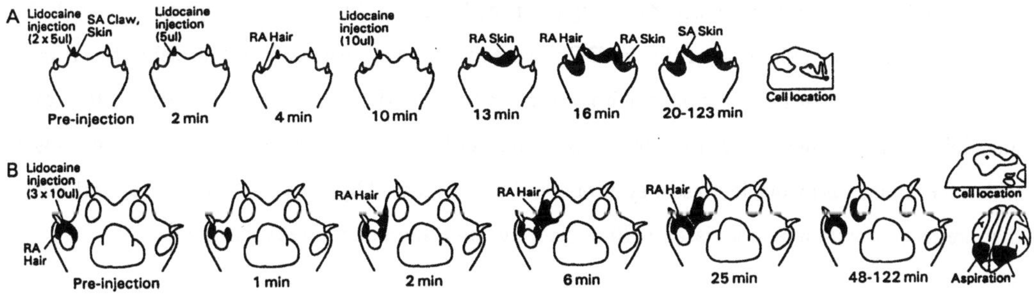

Figure 7. Reorganization of receptive fields of DCN neurons following disruption of peripheral inputs by lidocaine injection. **A.** Lidocaine injection into a peripheral receptive field resulted in the appearance of new receptive fields within 4 min. The newly responsive fields persisted throughout the recording period (123 min). **B.** Similar changes in receptive fields were observed in animals from which somatosensory and motor cortical areas had been removed prior to the experiment. Excerpted with permission from Pettit and Schwark (1993). Copyright 1993 American Association for the Advancement of Science.

the new fields arose from an unmasking of normally ineffective inputs. The mechanisms that underlie these changes apparently can originate at low levels of the somatosensory system, because similar results were obtained in animals in which cortical inputs to the DCN had been removed before injection (Fig. 7B). In other experiments, we have found that capsaicin injection into peripheral receptive fields resulted in the same types of receptive field changes in DCN neurons as lidocaine injection. Thus, it appears that activity in a subset of peripheral afferents (most likely C or Aδ fibers) is sufficient to mask some of the inputs to DCN neurons and thereby restrict receptive field size. To look for signs that tonic inhibition might play a role in this process, we measured the spontaneous firing rates of these neurons before and after injection of lidocaine or capsaicin. There were no consistent changes in spontaneous firing rates following these injections. It may be that activity in C or Aδ fibers changes the excitability of DCN neurons through other mechanisms, or that inhibitory processes involved do not affect spontaneous firing rates. Regardless of the mechanisms involved, the fact that such receptive field changes occur in the DCN means that similar changes at higher levels (e.g., thalamus and SI) must be interpreted with caution. It seems likely that receptive field changes in the DCN will be reflected at higher levels of the somatosensory system.

A role for small caliber afferents in restricting receptive field size of DCN neurons is somewhat surprising, because these fibers appear to make up only a small proportion of the afferents to the DCN. At least some of these afferents contain substance P (Conti et al., 1990), and although these fibers terminate primarily outside the cell nest regions of the DCN, iontophoresis of substance P causes long-lasting excitation in 50% of DCN neurons (Krnjević and Morris, 1974). Moreover, we have recently found dense substance P binding in the cell nests of the cat, monkey, and human DCN. Binding levels in the cell nests of the cat DCN were as high as those of the rat dorsal horn. It may be that the small caliber fibers act on inhibitory neurons through these substance P receptors to create tonic inhibition and thereby restrict the receptive fields of DCN neurons. In this manner, responses within the cutaneous medial lemniscal system could be modulated by activity of fine caliber afferents that are more commonly associated with nociceptive and thermoceptive responses.

Conclusions

An important function of inhibition in the somatosensory system is to modulate receptive field size. This modulation is accomplished through local inhibitory actions on excitatory connections, and appears to take place at multiple levels of the somatosensory system. For example, blockade of local inhibition by bicuculline iontophoresis results in expansion of receptive fields of neurons in SI as well as in the DCN (unpublished data from this laboratory). In the DCN, local inhibition appears to be maintained by excitatory inputs from within the neuron's receptive field, since loss of these inputs results in receptive field expansion. Activity in capsaicin-sensitive peripheral afferents (mostly C and Aδ fibers) is necessary to maintain restricted receptive fields. Intrinsic activity at higher levels in the somatosensory system may give rise to additional modulation of receptive fields.

As a result of this organization, it appears that inhibitory circuitry is well positioned to detect changes in environmental demands, and to regulate the operation of the somatosensory system in response to these changes. It remains to be determined if the inhibitory organization that regulates receptive field size in the DCN resembles the organization at higher levels of the somatosensory system.

Acknowledgements
This research was supported by NIH grant NS 25729 and NSF grant IBN-9221956. The contributions of Drs. J.L. Fuchs, J. Li, and M.J. Pettit to this research are gratefully acknowledged.

References

Alloway, K.D., Rosenthal, P. and Burton, H. (1989) Quantitative measurements of receptive field changes during antagonism of GABAergic transmission in primary somatosensory cortex of cats. *Experimental Brain Research* 78:514-532.
Beaulieu, C. and Colonnier, M. (1989) Number of neurons in individual laminae of areas 3b, 4γ, and 6aα of the cat cerebral cortex: A comparison with major visual areas. *Journal of Comparative Neurology* 279:228-234.
Berman, N.J., Douglas, R.J. and Martin, K.A.C. (1992) GABA-mediated inhibition in the neural networks of visual cortex. *Progress in Brain Research* 90:443-476.

Calford, M.B. and Tweedale, R. (1991a) C-fibres provide a source of masking inhibition to primary somatosensory cortex. *Proceedings of the Royal Society of London [Biology]* 243:269-275.

Calford, M.B. and Tweedale, R. (1991b) Acute changes in cutaneous receptive fields in primary somatosensory cortex after digit denervation in adult flying fox. *Journal of Neurophysiology* 65:178-187.

Conti, F., DeBiasi, S., Giuffrida, R. and Rustioni, A. (1990) Substance P-containing projections in the dorsal columns of rats and cats. *Neuroscience* 34:607-621.

Dehoff, R.T. and Rhines, F.E. (1961) Determination of the number of particles per unit volume of measurements made on random plane sections: The general cylinder and the ellipsoid. *Transactions AIME* 221:975-982.

Dykes, R.W., Landry, P., Metherate, R. and Hicks, T.P. (1984) Functional role of GABA in cat primary somatosensory cortex: shaping receptive fields of cortical neurons. *Journal of Neurophysiology* 52:1066-1093.

Ferster, D. and Lindstrom, S. (1983) An intracellular analysis of geniculo-cortical connectivity in area 17 of the cat. *Journal of Physiology (London)* 342:181-215.

Gabbott, P.L.A. and Somogyi, P. (1986) Quantitative distribution of GABA-immunoreactive neurons in the visual cortex (area 17) of the cat. *Experimental Brain Research* 61:323-331.

Hendry, S.H.C., Schwark, H.D., Jones, E.G. and Yan, J. (1987) Numbers and proportions of GABA immunoreactive neurons in different areas of monkey cerebral cortex. *Journal of Neuroscience* 7:1503-1519.

Houser, C.R., Vaughn, J.E., Hendry, S.H.C., Jones, E.G. and Peters, A. (1984) GABA neurons in the cerebral cortex. *In:* E.G. Jones and A. Peters (eds): *Cerebral Cortex. Vol. 2. Functional Properties of Cortical Cells*, Plenum Press, New York, pp 63-89.

Krnjević, K. and Morris, M.E. (1974) An excitatory action of substance P on cuneate neurons. *Canadian Journal of Physiology and Pharmacology* 52:736-744.

Li, J. and Schwark, H.D. (1994) Distribution and proportions of GABA-immunoreactive neurons in cat primary somatosensory cortex. *Journal of Comparative Neurology* 343:353-361.

Mower, G.D., White, M.F. and Rustad, R. (1986) [^{3}H]Muscimol binding of GABA receptors in the visual cortex of normal and monocularly deprived cats. *Brain Research* 380:253-260.

Nakahama, H., Nishioka, S. and Otsuka, T. (1966) Excitation and inhibition in ventrobasal thalamic neurons before and after cutaneous input deprivation. *In:* T. Tokizane and J.P. Schade (eds): *Progress in Brain Research. Vol 21A. Correlative Neurosciences*, Elsevier, Amsterdam, pp 180-196.

Pettit, M.J. and Schwark, H.D. (1993) Receptive field reorganization in dorsal column nuclei during temporary denervation. *Science* 262:2054-2056.

Schwark, H.D., Li, J. and Fuchs, J.L. (1994) Regional distribution of $GABA_A$ receptor binding sites in cat somatosensory and motor cortex. *Journal of Comparative Neurology* 343:362-369.

Sillito, A.M. (1984) Functional considerations of the operation of GABAergic inhibitory processes in the visual cortex. *In:* E.G. Jones and A. Peters (eds): *Cerebral Cortex. Vol 2. Functional Properties of Cortical Cells*, Plenum, New York, pp 91-117.

Sretavan, D. and Dykes, R.W. (1983) The organization of two cutaneous submodalities in the forearm region of area 3b of cat somatosensory cortex. *Journal of Comparative Neurology* 213:381-398.

Yamamoto, T., Samejima, A. and Oka, H. (1990) The mode of synaptic activation of pyramidal neurons in the cat primary somatosensory cortex: An intracellular HRP study. *Experimental Brain Research* 80:12-22.

Somesthesis and the Neurobiology of the Somatosensory Cortex
O. Franzén, R. Johansson and L. Terenius (eds.)

Pain, temperature, and the sense of the body

A. D. (Bud) Craig

Division of Neurobiology, Barrow Neurological Institute, 350 West Thomas Road, Phoenix, AZ 85013, USA

Summary. This essay discusses new evidence supporting the concept that the distinct sensibilities for pain and temperature are based in part on specific 'labeled lines' that interact centrally as components of an ascending projection system subserving homeostasis and the generalized sense of the condition of the body itself (common sensation, or "Gemeingefühl").

The somatic sensations of touch, temperature and pain have historically been regarded and studied as separate modalities (or sensory channels), based on: 1) the distinctiveness of the sensory impressions and the stimulus modalities, 2) the early demonstrations of separate touch, warmth, cold, and pain spots in the skin, and 3) the more recent identification of primary receptors and second order dorsal horn neurons with specific response properties (Weber, 1846; Sherrington, 1900; Boring, 1942; Darian-Smith, 1984; Hellon, 1983; Mountcastle, 1984; Perl, 1984a). Yet, there are also indications of physiological overlap between somatic modalities under normal (as well as pathological) conditions. Some of these, for example the elicitation of pain by extreme mechanical or thermal stimuli and the effect of temperature on perceived pressure intensity, have been explained by the properties of peripheral receptors. However, other findings, such as the inhibition of pain by cold or vibration (for refs., see Craig and Hunsley, 1991), the lowering of tactile acuity by painful heat (Apkarian et al., 1994), and the presence of central neurons responsive to two or even all of these modalities (e.g., so-called thermal-plus-mechanical [T+M], wide dynamic range [WDR] and multireceptive neurons), indicate that these modalities must be not only separate but also integrated centrally. The recognition of this dual nature obviates the long-standing controversy of specificity versus convergence in somatosensation, which has particularly distracted pain and temperature research. This recognition compels investigation of the interactions *between* the specific channels, and it affords the opportunity for new insights into the fundamental principles of the central processing of somatic afferent activity. In this contribution I focus on specificity and integration in the channels for pain and temperature sensibilities.

Considerable functional anatomical evidence supports the psychophysical identification of pain and temperature as specific sensations. Primary afferent fibers include Aδ and C nociceptors, which respond selectively and linearly to noxious mechanical pressure or to noxious heat, and Aδ (cold) thermoreceptors, which respond selectively to innocuous cold. Both types of primary afferents terminate in the superficial dorsal horn (laminae I-II). Within lamina I, the main output layer of the superficial dorsal horn, there are nociceptive-specific (NS) and thermoreceptive-(cold)-specific (COLD) neurons that appear to receive input selectively from these classes of primary receptors (Christensen and Perl, 1970). There are also C thermoreceptors that respond selectively to innocuous warming, but very few warm-responsive central neurons have yet been recorded (Kumazawa and Perl, 1975; Dostrovsky and Hellon, 1978; Craig and Hunsley, 1991), supporting the notion that inhibition of the COLD channel may underly the sensation of warmth (Poulos and Benjamin, 1968). Although the nociceptive- and thermoreceptive-specific neurons are found intermingled in lamina I, they are nonetheless differentiated in several ways. The NS and COLD lamina I cells that project to the thalamus in the cat have opposite responses to morphine (Craig and Hunsley, 1991) and to descending inhibition (Dostrovsky et al., 1983), and they have significantly different axonal conduction velocities (Craig and Kniffki, 1985) and distinct (albeit overlapping) patterns of axonal projections to different nuclei in the thalamus (Dostrovsky and Craig, 1993). These properties indicate anatomical differences, and recent intracellular labeling data have shown that the somata of these cells are morphologically distinct: NS lamina I cells are primarily fusiform neurons, and COLD lamina I cells are pyramidal neurons (Han and Craig, 1994). Thus, NS and COLD lamina I neurons provide specific ascending sensory channels, or "labeled lines", for pain and temperature sensation. The presence of NS- and COLD-like neurons in the main lamina I spino-thalamo-cortical relay nucleus of the primate thalamus confirms this conclusion (see below; Craig et al., 1994).

There is a third major class of lamina I spinothalamic (STT) neurons whose response properties overlap with those of NS and COLD cells in cat and monkey; they respond to noxious heat, noxious pinch, and noxious cold (hence called HPC cells; Craig and Kniffki, 1985; Craig and Serrano, 1994; Dostrovsky and Craig, 1995). (There are also WDR lamina I STT cells in monkey [Price et al., 1978; Ferrington et al., 1987] that are pertinent to a discussion of the integration of pain with touch

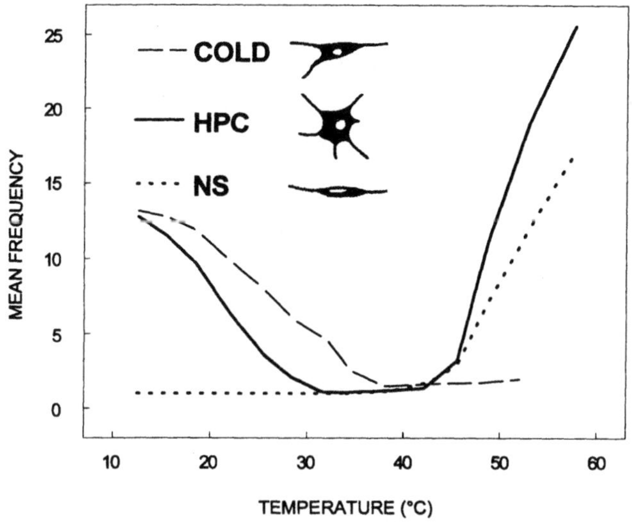

Figure 1. Stimulus-response functions of the three types of lamina I STT cells.

[see Perl, 1984b].) The HPC cells are similar to NS cells with respect to their heat threshold, their suppression by morphine and their thalamic axonal projection pattern. But, they have a U-shaped, bimodal temperature response function (Fig. 1). Their sensitivity to cold monotonically increases at temperatures below ca. 24°C (i.e., room temperature), which overlaps with the range of COLD cell responses (linear from 36° to 15°C). The HPC neurons appear to receive input from C polymodal and cold-sensitive C nociceptors (Georgopolous, 1976; LaMotte and Thalhammer, 1982; Sugiura et al., 1986; see also Strassman et al., 1993, 1994). Some HPC cells are also metaboreceptive, and some are sensitive to the itch-inducing agent histamine (Craig and Kniffki, 1985; R.H. LaMotte and A.D. Craig, unpublished). Intracellular labeling has shown that HPC cells morphologically are multipolar lamina I neurons (Han and Craig, 1994). Thus, there are at least three distinct ascending channels within the lamina I STT projection system. Nonetheless, the overlap in the characteristics and the projections of HPC cells with those of NS and COLD lamina I STT neurons clearly suggests that their activity may be integrated in some fashion at supraspinal and thalamo-cortical levels.

Integration between channels in sensory systems can be revealed by illusions, caused by stimuli that activate such channels in an unusual, differential manner. A striking psychophysical demonstration of an interaction between temperature and pain sensations in normal humans is

Thunberg's (1896) thermal grill illusion. Thunberg showed that the simultaneous presentation of spatially interlaced warm and cool stimuli evokes a sensation of burning heat, which he compared to the burning quality of a painfully cold stimulus. That is, a stimulus composed of innocuous temperatures in an unusual spatial arrangement evokes a sensation of pain.

This phenomenon resembles the effect of a pressure block of A fiber conduction in a peripheral nerve. As Aδ fibers are progressively blocked, cold sensibility is eliminated and the threshold at which a cold stimulus evokes a painful, burning sensation rises from the normal 15°C to as high as 24°C (clearly, due to C fibers; notably, a burning sensation is then also evoked by pinch: Fruhstorfer, 1974; LaMotte and Thalhammer, 1982; Wahren et al., 1989; Yarnitsky and Ochoa, 1990). Similarly, injecting cold water (below 24°C) in a vein causes a sensation of burning pain (Fruhstorfer and Lindblom, 1983; Klement and Arndt, 1992), presumably by activating C fibers but not cutaneous Aδ cold-specific fibers. The burning pain caused by lukewarm water applied to feet numb from prolonged cold exposure could have the same basis. These effects are consistent with the hypothesis that differential activation of cold-sensitive C nociceptors (i.e., the lamina I HPC channel) and Aδ innocuous cold thermoreceptors (the lamina I COLD channel) underlies the thermal grill illusion.

Recent experiments have shown that integration of a differential shift in HPC and COLD activity can indeed explain the thermal grill illusion (Craig and Bushnell, 1994). The cold temperature used in the grill stimulus (20°C) activates both COLD cells and HPC cells, but the addition of the intermixed warm bars (40°C) reduces only the COLD cell activity. This reduction effectively exposes (or unmasks) the activity of HPC lamina I STT cells. The stimuli used in the grill do not activate any other types of ascending STT neurons. The proportionate reduction in cold sensation in this paradigm confirms that activity in the COLD channel is associated with cold sensation, and the unmasking of the activity in the HPC channel (driven by cold-sensitive C nociceptors) corresponds with the appearance of the sensation of painful, burning heat. These observations substantiate the sensory roles of these distinct channels for temperature and pain sensibilities. Moreover, the unmasking reveals the integration of pain and temperature at thalamo-cortical levels. The grill illusion can be explained as a disinhibitory phenomenon, a simple form of subtractive integration in which COLD channel activity normally inhibits HPC channel activity. The quantitative difference in activity between HPC and COLD cells responding to the grill successfully predicts the noxious cold temperature (10°C) that evokes a perceived level of pain similar to the grill. Disinhibition of the HPC channel can also explain the progressive effects of an

A fiber block on cold and cold pain sensibilities. Thus, there is both specificity and integration in the representation of pain and temperature at thalamo-cortical levels.

The presence of both specific pain and temperature activity within the lamina I projection system and the integration between these channels at thalamo-cortical levels reflects the fundamental association of these modalities, which is both anatomical and functional. Pain and temperature sensibilities are closely coupled structurally at all levels of the central nervous system; lesions of the brain rarely affect one and not the other (see Perl, 1984). Their classic association with the anterolateral (spinothalamic) system contrasts with the association of touch with the lemniscal (dorsal column) system. Pain and temperature sensibilities are related functionally in an essential way; both modalities inform the central nervous system about the condition of all of the somatic tissues, including skin, muscles, joints, teeth, cornea, and viscera (for evidence that the viscera are thermosensitive, see Hellon, 1984). That is, they are enteroceptive, in contrast to touch, which is an exteroceptive or "projicient" sense (Sherrington, 1900). Although noxious and thermal stimuli applied to the skin can to some degree be projected, or referred, to an external object, it is the change in status of the skin itself that is sensed and that drives adjustments of the internal milieau. Whereas touch guides highly adaptive motor activity, both noxious and thermal stimuli characteristically generate strong autonomic and motoric reflex responses, and in addition, a strong, differentiating affect that motivates appropriate behavioral responses. These responses on each level serve to ensure the maintenance of the integrity of the body and the organism. Similarly, each of the variety of sensations that can arise from the body itself, i.e., temperature, pain, itch, tickle, vasomotor flush, shudder, deep (muscle, joint), visceral, and urogenital sensations, shares these common characteristics. In addition, each typically localizes relatively poorly and displays spatial summation, spatial radiation, and temporal persistence. Most importantly, all of the "feelings" from the body are related to the physiological status of the tissues and organs of the body, that is, to the maintenance and well-being of the organism. These are in fact the characteristics of the "common sensation" ("Gemeingefühl") discussed at length by Weber (1846) and Sherrington (1900) as the sense of the physical body, or the "material me" ("affizierte Leiblichkeit"). The concept of common sensation was abandoned some years ago in favor of different somatic submodalities (see Boring, 1942), but the global perspective of a sense of the condition of the body provides a parsimonious explanation of the functional association of pain and temperature and their integration. Further, it can be modernized today by its correspondence with the functional anatomy of the lamina I projection system.

A mammalian organism requires ongoing information on the status of all components of its physical body to drive autonomic, motoric, motivational and sensory mechanisms for maintenance and survival. The connectivity of lamina I indicates that its fundamental role is to process and distribute functionally selective afferent information relevant to the condition of the body, i.e., the physiological status of the tissues and organs of the entire organism (for details and references for the following, see Craig, 1991, 1993, 1995, 1996, and Craig et al., 1994). Lamina I receives input from small-diameter Aδ and C fibers that innervate all tissues of the body (with free nerve endings; see Heppelmann et al., 1995). These fibers respond to changes in the status of the tissues and organs of the body, including not only mechanical or thermal stress or damage (noxious stimuli) and changes in temperature, but also changes in metabolic conditions, such as pH, pCO_2, pO_2, glucose concentration, osmolarity, interstitial inflammatory agents, and muscle exercise (e.g., Moskowitz, 1991; MacIver and Tanelian, 1992; Steen et al. 1995; Pickar et al. 1994). Such fibers terminate as a group in laminae I-II and more sparsely in laminae V and X; however, thermoreceptive Aδ fibers, cold-sensitive C nociceptors, and C afferents from muscle, joint, respiratory and visceral tissues all apparently terminate nearly exclusively in lamina I. Notably, in common with the thoracolumbar sympathetic nuclei, lamina I also receives descending terminations from the paraventricular n. of the hypothalamus and from the pre-sympathetic brainstem nuclei.

In the spinal cord, propriospinal lamina I projections to the thoracolumbar sympathetic nuclei provide a basis for spinal somato-sympathetic reflex arcs, the first level of response to changes in the status of the body. Interestingly, the physiological characteristics of COLD and HPC cells parallel those of sympathetic cutaneous and muscle vasoconstrictor efferents, respectively; it is probably no coincidence that a comfortable ambient temperature is at or above the threshold of HPC cells. Other propriospinal or segmental lamina I connections that might be involved in spinal motoric reflexes have yet to be identified; lamina V neurons may be of importance for this function (see Schouenborg and Sjölund, 1983; Perl, 1984b). In the brainstem, lamina I projections to cardio-respiratory and pre-sympathetic regions (A2, C1, A5, A7) provide a basis for more generalized somato-autonomic reflex arcs, the second level of homeostatic response. Lamina I brainstem terminations in general visceral integrative sites (the A1 cell group, the parabrachial nucleus) provide relays for nociceptive, thermoreceptive and metaboreceptive activity to the hypothalamus and the central amygdala. Lamina I also projects topographically to the periaqueductal gray, a midbrain limbic integration and effector site for homeostasis, motivational state, and coordinated behavioral responses. Thus, the spinal and brainstem projections of lamina I provide substrates appropriate for afferent input associated with the condition of the body to drive homeostasis.

Lamina I spinothalamic projections provide substrates for the sensory and motivational aspects of the sense of the condition of the body. Ascending lamina I STT fibers form the lateral spinothalamic tract, classically associated with pain and temperature on the basis of lesions in humans. In monkeys and humans, there is a specific, topographically organized lamina I thalamo-cortical relay nucleus (the posterior ventral medial nucleus, VMpo) that contains NS- and COLD-like neurons and that projects topographically to the dorsal anterior insula. This lateral thalamo-cortical somatosensory pathway could be associated with the perceived quality, intensity, and location, and the memory of enteroceptive sensations (including pain and temperature) from the body. (VMpo also has a corollary topographic projection to area 3a, providing a path for integration with somatomotor activity.) Stimulation in the region of VMpo in human thalamus evokes discrete sensations of pain or cold (Hassler and Riechert, 1959; Dostrovsky et al., 1992; Lenz et al., 1993). Noxious or cold stimuli strongly activate the dorsal anterior insula in human PET studies (Coghill et al., 1994; Bushnell et al., 1995), and lesions of the posterolateral thalamus or the anterior insular cortex can produce analgesia and thermanesthesia in humans (Head and Holmes, 1911; Biemond, 1956; Berthier et al., 1988). Significantly, VMpo lies adjacent to the basal VM nucleus, which is the thalamo-cortical relay for the gustatory and general visceral solitario-parabrachial afferent system (see Allen et al., 1991), and which projects to the most anterior part of the insula, in parallel with VMpo. Together, these insular areas receive sensory input from the body that mirrors the organization of the efferent sympathetic and parasympathetic systems. (Interestingly, parasympathetic afferents may inhibit rather than produce pain.) Thus, the lamina I - VMpo - anterior insula pathway provides the basis for specific representations of pain and temperature sensibilities that appear to be embedded within a general cortical representation of the condition of the body.

Lamina I STT neurons also terminate topographically in the ventral caudal part of the medial dorsal nucleus (MDvc) in primates, which projects to the anterior cingulate cortex (Ray and Price, 1993). Both acute and chronic pain activate the anterior cingulate in human PET studies (Coghill et al., 1994; Derbyshire et al., 1994), and nociceptive neurons have been recorded there in rabbits (Sikes and Vogt, 1992). Because this limbic cortical area is associated with attention, motivation, avoidance conditioning, and affect, this lamina I thalamo-cortical pathway could be involved in the behaviorally motivating affect (urgency) generated by sensations of the body such as pain. The recent demonstration that the anterior cingulate cortex is also activated by itch is consistent with this conclusion (Hsieh et al., 1994).

Thus, the functional and anatomical characteristics of lamina I projection neurons indicate that they distribute nociceptive, thermoreceptive and metaboreceptive activity from all tissues of the body to spinal, brainstem, and telencephalic areas concerned with the autonomic, motivational, sensory and behavioral responses that maintain the integrity of the mammalian organism. Lamina I and its projections form the basis of a sensory system that represents the condition of the body. Notably, the most anterior parts of this system, i.e., the anterior insula, anterior cingulate, central amygdala, ventromedial and anterior hypothalamus, and the lateral and ventrolateral columns of the periaqueductal gray, are all interconnected with each other, and all of these limbic areas are involved in homeostasis and motivation/emotion. These lamina I projection targets together form a network for integration of the afferent input from the body. Homeostatic responses are in general autonomic and unconscious, nonetheless this entire forebrain and midbrain network, not just the anterior insular and anterior cingulate cortices, must be involved in the sense of the condition of the body. In this way, what humans perceive as "feelings" from the body may be both a reflection of integrated homeostatic processing and direct afferent activity from the same ascending projection system affecting homeostasis, as hypothesized earlier (Weber, 1846; Sherrington, 1900).

Because there is both specificity and integration within this system, lesions that disrupt this network could have profound and paradoxical sensory (and homeostatic) effects. For example, disruption of the integration of HPC and COLD activity by a loss of specific COLD input, resulting in disinhibition of HPC activity like that produced by the thermal grill illusion, can be predicted to result in thermosensory (cold) dysfunction accompanied by cold allodynia and a burning dysesthetic pain at cool room temperatures. These characteristics describe the thalamic (central post-stroke) pain syndrome, which can be caused by lesions along the lateral lamina I spino-thalamo-cortical pathway (posterolateral thalamus, internal capsule, anterior insula, and lower sites: Boivie et al., 1989). Not surprisingly, the loss of cold sensibility is generally accompanied by a paradoxical analgesia to pin prick in the painful area as well. Clinical findings implicating mediodorsal thalamus and anterior cingulate cortex in the generation of such pain (Head and Holmes, 1911; Richardson, 1967; Jeanmonod et al., 1993; Derbyshire et al., 1994) suggest the inference that the medial lamina I - MDvc - anterior cingulate pathway conveys the HPC activity that is normally integrated with COLD activity. Strong support for this hypothesis is provided by recent PET findings demonstrating that the thermal grill activates the anterior cingulate and the anterior insular cortices, whereas its component innocuous thermal stimuli activate only the insular cortex (Bushnell et al., 1995). Furthermore, histamine-induced itch, which drives HPC lamina I neurons but does not affect COLD cell activity, also activates anterior cingulate cortex (Hsieh et al., 1994). The localization of integration within this portion of the network representing the sense of the condition of the body offers a substantive basis for observations that lesions in posteromedial thalamus or anterior cingulate can be efficacious for the alleviation of central and chronic pain (Hassenbusch et al., 1990; Jeanmonod et al., 1993).

Finally, the comparative anatomy of this system provides an intriguing epilogue to these considerations. Although lamina I projections to spinal and brainstem regions are probably similar in rat, cat, monkey and human, there is a marked phylogenetic shift in the organization of lamina I projections to the thalamus. In the rat and cat, there is dense input to the nucleus submedius, which is developmentally related to the primate MDvc, but which projects to orbitofrontal cortex rather than to the anterior cingulate. In rat and cat, a thin region along the ventral aspect of VMb may be the primordium of the primate VMpo, but in rat this region receives parabrachial and not lamina I input (which terminates instead in the ventrobasal complex proper and in the hypothalamus), and in cat the lamina I input to this region is diffuse and only crudely topographic. The dense, topographic and specific lamina I - VMpo - anterior insula pathway is present only in primates, and it is even more pronounced in the human than in the monkey. This suggests that the direct pathway subserving the sense of the condition of the body is most highly developed in human beings. This insight gains fundamental significance from the neurological concept that insular cortex is integral for the mental generation of the image of the self that underlies the basic emotional states required for the motivation to make rational decisions affecting survival and quality of life (Damasio, 1993). It provides direct support for the notion that the evolution of the recognition of self (i.e., awareness) is based in part on the emergent sense of the condition of the self (Sherrington, 1900; Damasio, 1993).

In conclusion, the present considerations indicate that lamina I neurons are part of a larger limbic network involved in homeostatic and behavioral maintenance of the integrity of the self. This network in humans engenders the sense of the condition of the body, or the "feelings" from the body that provide the underpinning for basic emotional and motivational states. The lamina I projection system provides afferent input to this network related to the ongoing condition of the body. Embedded within this integrative system are the processes responsible for the specific sensations of pain and temperature.

Acknowledgements

I am very grateful to several esteemed colleagues who commented privately on this essay.

References

Allen, G. V., Saper, C. B., Hurley, K. M. and Cechetto, D. F. (1991) Organization of visceral and limbic connections in the insular cortex of the rat. *Journal of Comparative Neurology* 311: 1-16.

Apkarian, A. V., Stea, R. A. and Bolanowski, S. J. (1994) Heat-induced pain diminishes vibrotactile perception: A touch gate. *Somatosensory and Motor Research* 11: 259-267.

Berthier, M. L., Starkstein, S. E. and Leiguarda, R. C. (1988) Asymbolia for pain:a sensory-limbic disconnection syndrome. *Annals of Neurology* 24: 41-49.

Biemond, A. (1956) The conduction of pain above the level of the thalamus opticus. *Arch. Neurol. Psychiat.* 75: 231-244.

Boivie, J., Leijon, G. and Johansson, I. (1989) Central post-stroke pain--A study of the mechanisms through analyses of the sensory abnormalities. *Pain* 37: 173-185.

Boring, E.G. (1942) Sensation and Perception in the History of Experimental Psychology. Appleton-Century-Crofts, New York.

Bushnell, M. C., Craig, A. D., Reimann, E. M., Yun, L.-S. and Evans, A. (1995) Cerebral activation in the human brain by pain, temperature and an illusion of pain. *Society for Neuroscience Abstracts* 21: 1637.

Christensen, B. N. and Perl, E. R. (1970) Spinal neurons specifically excited by noxious or thermal stimuli: marginal zone of the dorsal horn. *Journal of Neurophysiology* 33: 293-307.

Coghill, R. C., Talbot, J. D., Evans, A. C., et al. (1994) Distributed processing of pain and vibration by the human brain. *Journal of Neuroscience* 14: 4095-4108.

Craig, A. D. (1991) Spinal distribution of ascending lamina I axons anterogradely labeled with *Phaseolus vulgaris* leucoagglutinin (PHA-L) in the cat. *Journal of Comparative Neurology* 313: 377-393.

Craig, A. D. (1993) Propriospinal input to thoracolumbar sympathetic nuclei from cervical and lumbar lamina I neurons in the cat and the monkey. *Journal of Comparative Neurology* 331: 517-530.

Craig, A. D. (1995) Distribution of brainstem projections from spinal lamina I neurons in the cat and the monkey. *Journal of Comparative Neurology* (in press).

Craig, A. D. (1996) Supraspinal projections of lamina I neurons. In: J.-M. Besson (ed.): *Forebrain Processing of Pain*, John Libbey, Paris, (in press).

Craig, A. D. and Bushnell, M. C. (1994) The thermal grill illusion: unmasking the burn of cold pain. *Science* 265: 252-255.

Craig, A. D., Bushnell, M. C., Zhang, E.-T. and Blomqvist, A. (1994) A thalamic nucleus specific for pain and temperature sensation. *Nature* 372: 770-773.

Craig, A. D. and Hunsley, S. J. (1991) Morphine enhances the activity of thermoreceptive cold-specific lamina I spinothalamic neurons in the cat. *Brain Research* 558: 93-97.

Craig, A. D. and Serrano, L. P. (1994) Effects of systemic morphine on lamina I spinothalamic tract neurons in the cat. *Brain Research* 636: 233-244.

Craig, A. D. and Kniffki, K.-D. (1985) Spinothalamic lumbosacral lamina I cells responsive to skin and muscle stimulation in the cat. *Journal of Physiology (Lond)* 365: 197-221.

Damasio, A.R. (1993) *Descartes' Error: Emotion, Reason, and the Human Brain.* Putnam, New York.

Darian-Smith, I. (1984) Thermal Sensibility. In: I. Darian-Smith (ed.): *Handbook of Physiology, Section 1, The Nervous System, Volume III, Sensory Processes*, American Physiological Society, Bethesda, pp 879-913.

Derbyshire, S. W. G., Jones, A. K. P., Devani, P., Friston, K. J., Feinmann, C., Harris, M., and Frackowiak, R. (1994) Cerebral responses to pain in patients with atypical facial pain measured by positron emission tomography. *Journal of Neurology, Neurosurgery, and Psychiatry* 57: 1166-1172.

Dostrovsky, J. O. and Craig, A. D. (1993) Antidromic identification of nociceptive lamina I cell terminations in the cat thalamus. *Society for Neuroscience Abstracts* 19: 1572.

Dostrovsky, J. O. and Hellon, R. F. (1978) The representation of facial temperature in the caudal trigeminal nucleus of the cat. *Journal of Physiology (Lond)* 277: 29-47.

Dostrovsky, J. O., Shah, Y. and Gray, B. G. (1983) Descending inhibitory influences from periaqueductal gray, nucleus raphe magnus, and adjacent reticular formation: II. Effects on medullary dorsal horn nociceptive and nonnociceptive neurons. *Journal of Neurophysiology* 49: 948-960.

Dostrovsky, J. O., Wells, F. E. B. and Tasker, R. R. (1992) Pain sensations evoked by stimulation in human thalamus. In: R. Inoka, Y. Shigenaga and M. Tohyama (eds.): *Processing and Inhibition of Nociceptive Information, International Congress Series 989*, Excerpta Medica, Amsterdam, pp 115-120.

Ferrington, D. G., Sorkin, L. S. and Willis, W. D. (1987) Responses of spinothalamic tract cells in the superficial dorsal horn of the primate lumbar spinal cord. *Journal of Physiology (Lond)* 388: 681-703.

Fruhstorfer, H. and Lindblom, V. (1983) Vascular participation in deep cold pain. *Pain* 17: 235-242.

Georgopoulos, A. P. (1976) Functional properties of primary affernt units probably related to pain mechanisms in primate glabrous skin. *Journal of Neurophysiology* 39: 71-83.

Han, Z.-S. and Craig, A. D. (1994) Morphological characteristics of physiologically identified lamina I cells in cats. *Society for Neuroscience Abstracts* 20: 547.

Hassenbusch, S. J., Pillay, P. K. and Barnett, G. H. (1990) Radiofrequency cingulotomy for intractable cancer pain using stereotaxis guided by magnetic resonance imaging. *Neurosurgery* 27: 220-223.

Hassler, R. and Riechert, T. (1959) Klinische und anatomische Befunde bei stereotaktischen Schmerzoperationen im thalamus. *Archives in Psychiatry* 200: 93-122.

Head, H. and Holmes, G. (1911) Sensory disturbances from cerebral lesions. *Brain* 34: 102-254.

Hellon, R. (1983) Thermoreceptors. In: J.T. Shepard and F.M. Abboud (eds.): *Handbook of Physiology, Section 2, The Cardiovascular System*, American Physiological Soc., Bethesda, pp 659-673.

Heppelmann, B., Messlinger, K., Neiss, W. F. and Schmidt, R. F. (1995) Fine sensory innervation of the knee joint capsule by group III and group IV nerve fibers in the cat. *Journal of Comparative Neurology* 351: 415-428.

Hsieh, J.-C., Hägermark, S., Ståhle-Bäckdahl, M., Ericson, K., Ericksson, L., Stone-Elander, S., and Ingvar, M. (1994) Urge to scratch represented in the human cerebral cortex during itch. *Journal of Neurophysiology* 72: 3004-3008.

Jeanmonod, D., Magnin, M. and Morel, A. (1993) Thalamus and neurogenic pain: physiological, anatomical and clinical data. *NeuroReport* 4: 475-478.

Klement, W. and Arndt, J. O. (1992) The role of nociceptors of cutaneous veins in the mediation of cold pain in man. *Journal of Physiology (Lond.)* 449: 73-83.

Kumazawa, T., Perl, E. R., Burgess, P. R. and Whitehorn, D. (1975) Ascending projections from marginal zone (lamina I) neurons of the spinal dorsal horn. *Journal of Comparative Neurology* 162: 1-12.

LaMotte, R. H. and Thalhammer, J. G. (1982) Response properties of high-threshold cutaneous cold receptors in the primate. *Brain Research* 244: 279-287.

Lenz, F. A., Seike, M., Richardson, R. T., Lin, Y. C., Baker, F. H., Khoja, I., Jaeger, C. J., and Gracely, R. H. (1993) Thermal and pain sensations evoked by microstimulation in the area of human ventrocaudal nucleus. *Journal of Neurophysiology* 70: 200-212.

MacIver, M. B. and Tanelian, D. L. (1992) Activation of C fibers by metabolic perturbations associated with tourniquet ischemia. *Anesthesiology* 76: 617-623.

Moskowitz, M. A. (1991) The visceral organ brain: implications for the pathphysiology of vascular head pain. *Neurology* 41: 182-186.

Mountcastle, V. B. (1984) Central nervous mechanisms in mechanoreceptive sensibility. In: I. Darian-Smith (ed.): *Handbook of Physiology, Section 1, The Nervous System, Volume III, Sensory Processes*, American Physiological Society, Bethesda, pp 789-878.

Perl, E. R. (1984a) Pain and nociception. In: I. Darian-Smith (ed.): *Handbook of Physiology, Section 1, The Nervous System, Volume III, Sensory Processes*, American Physiological Society, Bethesda, pp 915-975.

Perl, E. R. (1984b) Why are selectively responsive and multireceptive neurons both present in somatosensory pathways? In: C. v. Euler, O. Franzen, U. Lindblom, and D. Ottoson (eds.): *Somatosensory Mechanisms*, MacMillan Press, London, pp 141-161.

Pickar, J. G., Hill, J. M. and Kaufman, M. P. (1994) Dynamic exercise stimulates group III muscle afferents. *Journal of Neurophysiology* 71: 753-760.

Poulos, D. A. and Benjamin, R. M. (1968) Response of thalamic neurons to thermal stimulation of the tongue. *Journal of Neurophysiology* 31: 28-43.

Price, D. D., Hayes, R. L., Ruda, M. and Dubner, R. (1978) Spatial and temporal transformations of input to spinothalamic tract neurons and their relation to somatic sensations. *Journal of Neurophysiology* 41: 933-947.

Richardson, D. E. (1967) Thalamotomy for intractable pain. *Confinia neurologica* 29: 139-145.

Schouenborg, J. and Sjölund, B. H. (1983) Activity evoked by A- and C-afferent fibers in rat dorsal horn neurons and its relation to a flexion reflex. *Journal of Neurophysiology* 50: 1108-1121.

Sherrington, C. S. (1900) Cutaneous sensations. In: E.A. Schäfer (ed.): *Text-book of Physiology*, Pentland, Edinburgh, pp 920-1001.

Sikes, R. W. and Vogt, B. A. (1992) Nociceptive neurons in area 24 of rabbit cingulate cortex. *Journal of Neurophysiology* 68: 1720-1732.

Steen, K. H., Steen, A. E. and Reeh, P. W. (1995) A dominant role of acid pH in inflammatory excitation and sensitization of nociceptors in rat skin, *in vitro*. *Journal of Neuroscience* 15: 3982-3989.

Strassman, A. M., Mineta, Y. and Vos, B. P. (1994) Distribution of fos-like immunoreactivity in the medullary and upper cervical dorsal horn produced by stimulation of dural blood vessels in the rat. *Journal of Neuroscience* 14: 3725-3735.

Strassman, A. M. and Vos, B. P. (1993) Somatotopic and laminar organization of fos-like immunoreactivity in the medullary and upper cervical dorsal horn induced by noxious facial stimulation in the rat. *Journal of Comparative Neurology* 331: 495-516.

Sugiura, Y., Lee, C. L. and Perl, E. R. (1986) Central projections of identified, unmyelinated (C) afferent fibers innervating mammalian skin. *Science* 234: 358-361.

Thunberg, T. (1896) Förnimmelserne vid till samma ställe lokaliserad, samtidigt pågående köld- och värmeretning. *Uppsala Läkareförenings Förhandlingar*, Första bandet, P. Hedenius and J.A. Hammar (eds.), Akademiska boktryckeriet, pp. 489-495.

Wahren, L. K., Torebjörk, E. and Jörum, E. (1989) Central suppression of cold-induced C fibre pain by myelinated fibre input. *Pain* 38: 313-319.

Weber, E. H. (1846) Der Tastsinn und das Gemeingefühl. In: R. Wagner (ed.): *Handwörterbuch des Physiologie mit Rücksicht auf physiologische Pathologie, Bd 3, Abt 2*, Biewig und Sohn, Braunschweig, pp 481-588.

Yarnitsky, D. and Ochoa, J. L. (1990) Release of cold-induced burning pain by block of cold-specific afferent input. *Brain* 113: 893-902.

Somesthesis and the Neurobiology of the Somatosensory Cortex
O. Franzén, R. Johansson and L. Terenius (eds.)

The functional role of a noninactivating sodium current in neocortical neurons

W. E. Crill

Department of Physiology and Biophysics Box 357290, University of Washington, Seattle WA 98195-7290, USA

Summary. Neocortical neurons have a tetrodotoxin (TTX) sensitive noninactivating current. Whole cell voltage clamp shows this current is activated about 10 mV negative to spike threshold. On-cell patches show transient sodium channels that fail to activate in less than 1% of the depolarizations which is compatible with the magnitude of the persistent cuurrent. This rare modal change in inactivation provides the cell with a small persistent current mechanism that amplifies depolarizing synaptic currents. Its presence in dendrites effectively shortens the electrical length of the dendrites.

Introduction

Neurons are the basic integrative unit of the nervous system. They are responsible for transmitting signals over long distances and serve as nodes for the summation of excitatory and inhibitory synaptic signals. For a number of years we have been trying to understand the rules by which synaptic currents are transduced into a spike trains for transmission to the next synaptic site. For these experiments we use layer V pyramidal cells from rat.

Layer V pyramidal cells, like other neurons, have many types of voltage-gated ion channels (Hille, 1992). The transient sodium channels give the neuron all-or-nothing spike characteristics. These neurons also have at least four different types of calcium channels and five types of potassium channels. Two other types of channels are activated in the subthreshold range of membrane potentials—the hyperpolarizing inward rectifier, usually called the h-conductance, and a noninactivating sodium conductance (I_{NaP}) mechanism activated by depolarization. I_{NaP} is the topic of this presentation.

Results and discussion

It has been known since the work of by David Prince and Barry Connors (Hotson, Prince and Schwartzkroin, 1979; Connors, Gutnick and Prince, 1982) and also Rudolfo Llinas (Llinás and Sugamori 1980; Llinás, 1988) that mammalian neurons have a tetrodotoxin (TTX) sensitive noninactivating sodium electrogenic response. The persistent sodium current, I_{NaP} , is activated about 10 mV negative to resting potential and has rapid activation kinetics that are nearly as fast as the transient sodium conductance (Stafstrom, Schwindt, Chubb and Crill, 1985). The voltage dependence has great functional significance because even though the noninactivating sodium current is small, compared to the transient sodium current, it dominates cell properties because there are no other voltage dependent currents in the subthreshold range of membrane potentials (Brown, Schwindt and Crill, 1994). Furthermore, this is the voltage range where the critical integration of synaptic potentials occurs that either does or does not lead to neuron firing.

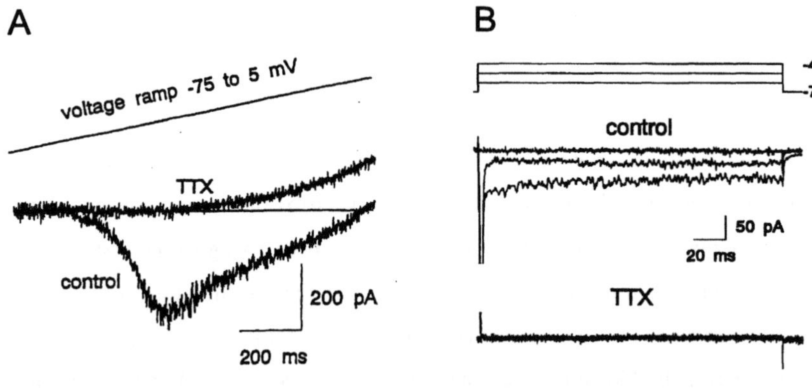

Figure 1. Whole cell voltage clamp of rat layer V pyramidal neuron. A. Steady-state current (bottom traces) before and after TTX, evoked by a slowly depolarizing ramp command voltage (top traces). B. Depolarizing voltage-clamp steps (top traces) evoke large transient initial inward current (clipped) and persistent inward current (middle traces). Both transient and persistent current are blocked by TTX (bottom traces).

Figure 1A shows the noninactivating sodium current evoked in an acutely isolated pyramidal neuron in the whole cell voltage clamp mode. The voltage command is a slowly rising ramp that only shows gradually or noninactivating currents. The steady-state current is blocked by TTX.

Figure 1B demonstrates I_{NaP} evoked by voltage clamp steps. The peak transient current has been clipped to show the smaller persistent TTX sensitive current.

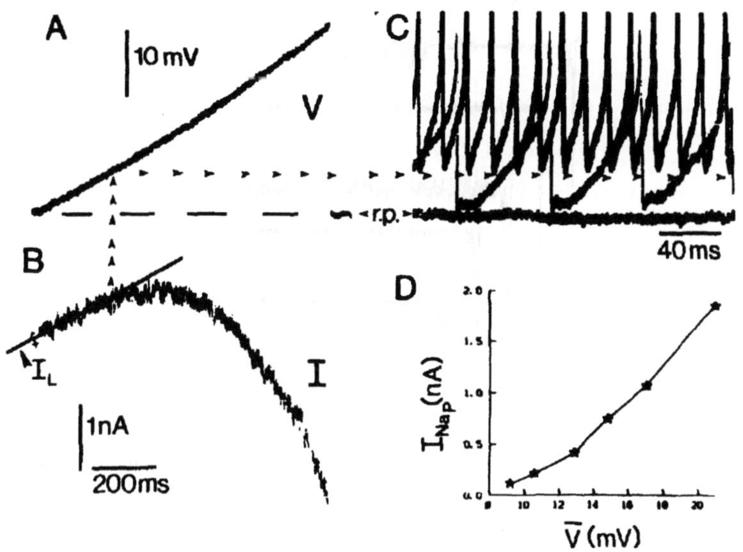

Figure 2. Relationship of interspike membrane trajectory and activation of I_{NaP}. All traces are from the same neocortical neuron. A. Ramp voltage clamp command. Dashed line is resting potential. B. Membrane current measured during ramp command in A. C. Superimposed traces of resting neuron and two rates of repetitive firing in response to injected constant current. Top of spikes are clipped. The arrows in A-C identify membrane potential where I_{NaP} is first detected. D. Plot of magnitude of I_{NaP} a different measured interspike potentials. From Stafstrom, Schwindt and Crill, 1984.

I_{NaP} is activated in the voltage range traversed by the interspike interval (Fig 2) and will add to any tonic synaptic depolarization (Stafstrom, Schwindt and Crill, 1984). A and B show the voltage clamp command and current respectively. C is superimposed repetitive firing at two rates and the arrows show identical voltages. I_{NaP} is activated at the voltages occurring during the interspike interval. The contribution of I_{NaP} is difficult to demonstrate experimentally because the pharmacology of I_{NaP} is virtually identical to I_{Na}. I_{NaP} is a depolarizing current and would add to either synaptic or experimentally applied current.

I_{NaP} is activated with fast kinetics and even a single EPSP can activate it. That is a single EPSP is more effective (more inward current flows) at the potentials just negative to action potential threshold than at resting potential. I_{NaP} is also the primary depolarizing current underlying the voltage dependent oscillations seen in cortical neurons studied by Alonzo and Llinás (1989) and Amitai (1994).

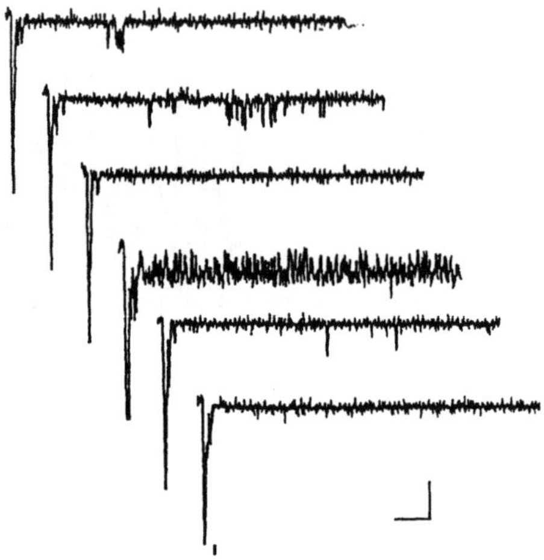

Figure 3. Selected traces of on-cell channel activity evoked by a step voltage command to -55 mV. Traces 1 and 2 show late bursts and trace 4 shows persistent openings. Transient current is clipped. Time calibration is 40 msec and vertical bar is 2 pA.

I_{NaP} could represent one subtype of sodium channel different from those responsible for the action potential. It was also possible it only reflected the overlap of the Hodgkin and Huxley activation and inactivation curves of the transient sodium conductance; however the presence of I_{NaP} over a very large range of voltages strongly indicates this was not the mechanism (Stafstrom, Schwindt, Chubb and Crill, 1985; French, Sah, Buckett and Gage, 1990).

To identify the properties of I_{NaP} channels we used on cell patch studies with the hope of recording noninactivating channels (Alzheimer, Schwindt and Crill, 1993). To our surprise we only found one functional type of behavior as shown in Figure 3. In about 1% of the depolarizations one channel would fail to inactivate. The simplest interpretation is that channels which normally have classical activation and inactivation properties also show another rare mode of inactivation where they continue to open during a depolarizing stimulus. This noninactivating mode of sodium channel behavior has been found in skeletal muscle and heart (Patlak and Ortiz, 1986; Nilius, 1988; Taylor, 1993; Böhle and Benndorf, 1995).

But why should persistent sodium channel behavior have a more negative voltage activation than the transient current if a single channel type is responsible for both persistent and transient behavior? We gained some insight from earlier biophysical studies on sodium channel behavior by Aldrich, Corey and Stevens (1983) and Gonoi and Hille (1987). Aldrich, Corey and Stevens

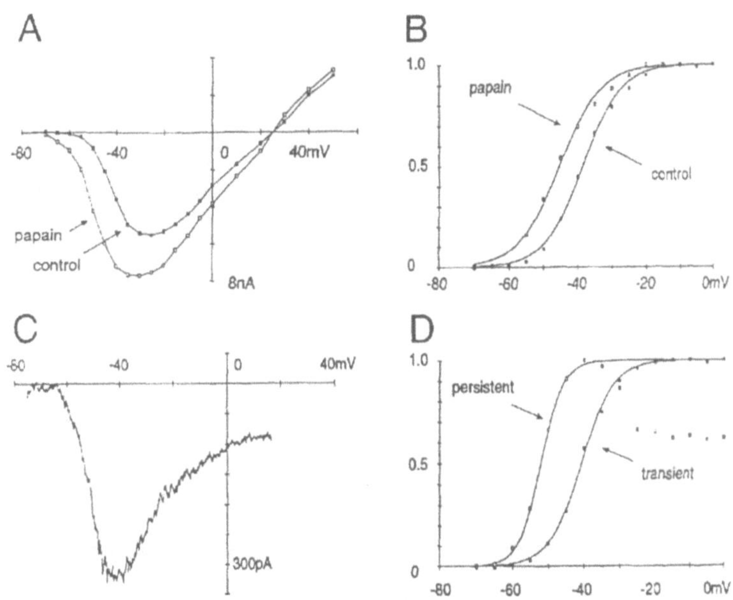

Figure 4. Current-voltage and conductance voltage relationships before and after papain (A and B) and for peristent and transient sodium current (C and D). A. Initial I-V relation for transient sodium current immediatly after gaining access to neuron (■) and after 10 minutes using patch pipette filled with papain (□). B. Activation curves derived from neuron data shown in A. C. I-V relationship for persistent current in another cell evoked by slowly rising voltage command ramp. D. Activation curve for persistent current from cell shown in C (□) compared to average activation curve for transient sodium current from 33 other neurons (■). From Brown, Schwindt and Crill, 1994.

(1983) provided strong evidence in cultured neuroblastoma cells that, in contrast to the Hodgkin and Huxley model, sodium activation and inactivation were linked and inactivation was largely voltage independent and much faster than activation. These properties would functionally shift the whole cell activation curve in the positive direction because when only small numbers of sodium channels were slowly activated at activation threshold, the rapid inactivation would prevent any appreciable inward transient current. An appreciable inward current flow only at more depolarized potentials where many channels are rapidly activated. However, if inactivation was blocked, a detectable slowly activating sodium current could be identified at more negative potentials, revealing the true activation curve. Gonoi and Hille (1987) showed this effect by blocking inactivation in neuroblastoma cells. We postulated that if the Aldrich, Corey and Stevens model (1983) held for central mammalian neurons, the sodium current generated by the modal properties of inactivation would reveal the persistent current at potentials negative to spike threshold. To test this hypothesis we predicted that when inactivation of the transient channels was blocked by papain as shown in Figure 4, the negative shift in activation curve would be identical to the activation curve for I_{NaP} measured without pharmacological alteration of inactivation (Brown, Schwindt and Crill, 1994). The half activation of I_{NaP} was on the average 6.9 mV negative to the activation of the transient current. When inactivation was blocked by papain the activation of the transient channels shifted 6.4 mV negative, virtually identical to the activation curve for I_{NaP} . These experiments show that the negative shift in activation does not require a subtype of sodium channel with more negative activation properties.

If modal changes in inactivation are a general property of sodium channels and sodium channels are in the dendrites which have been shown by several laboratories (Huguenard, Hamill and Prince, 1989; Stuart and Sakmann, 1994), even with a density that does not support action potentials, the persistent current will still have a major influence on synaptic effectiveness. A synaptic depolarization will first activate the persistent current and will decrease the electrical length of the dendrites making synapses closer to the soma. Jack, Noble and Tsien (1975) first showed that depolarization of dendrites is electrically much closer to the soma if there is a low threshold inward current in the dendrites. That is, it is not necessary to activate dendritic spikes to bring the synaptic sites electrically closer to the soma. Moreover, if the noninactivating modal change is modifiable this would provide even more plasticity to the postsynaptic properties responsible for synaptic transduction. Schwindt and Crill (1995) have recently demonstrated that tonic current flowing to the soma from glutamate iontophoresis far out on the dendrites is markedly amplified by a TTX-sensitive component.

To summarize, the subthreshold depolarizing current caused by a modal change in sodium inactivation has a significant influence on cell behavior — largely because it is active in a voltage range where it can dominate the cell's response properties. Its influence becomes even greater since persistent sodium channels are also located in the dendritic tree where inward rectification of the dendrites will control synaptic effectiveness.

Acknowledgments
This work was supported by NIH grant NS16792 and the W. M. Keck Foundation.

References

Aldrich, R. W., Corey, D.P. and Stevens, C.F. (1983) A reinterpretation of mammalian sodium channel gating based on single channel recording. *Nature* 306:436-441.

Alonso, A. and Llinás, R.R. (1989) Subthreshold Na^+ dependent theta-like rhythmicity in stellate cells of entorhinal cortex layer II. *Nature* 342: 1175-1177.

Alzheimer C., Schwindt, P.C. and Crill, W.E. (1993) Modal gating of Na^+ channels as a mechanism of persistent Na^+ current in pyramidal neurons from rat and cat sensorimotor cortex. *Journal of Neuroscience* 13:660-673.

Amitai, Y. (1994) Membrane potential oscillations underlying firing patterns in neocortical neurons. *Neuroscience* 63:151-161.

Böhle, T. and Benndorf, K. (1995) Multimodal action of single Na^+ channels in myocardial mouse cells. *Biophysical Journal* 68:121-130.

Brown, A. M., Schwindt, P. C. and Crill, W. E. (1994) Different voltage dependence of transient and persistent Na^+ currents is compatible with modal-gating hypothesis for sodium channels. *Journal of Neurophysiology* 71: 2562-2565.

Connors, B.W., Gutnick, M.J. and Prince, D.A. (1982) Electrophysiological properties of neocortical neurons *in vitro*. *Journal of Neurophysiology* 48:1302-1320.

French, C.R., Sah, P., Buckett, K.J. and Gage, P.W. (1990) A voltage-dependent persistent sodium current in mammalian hippocampal neurons. *Journal of General Physiology* 95:1139-1157.

Gonoi, T.and Hille, B. (1987) Gating of sodium channels: inactivation modifiers discriminate among models. *Journal of General Physiology* 89: 253-274.

Hille, B. (1992) *Ionic Channels of Excitable Membranes*. Sinauer, Sunderland, MA.

Hotson, J.R., Prince, D.A. and Schwartzkroin, P.A. (1979) Anomalous inward rectification in hippocampal neurons. *Journal of Neurophysiology* 42:889-895.

Huguenard, J.R., Hamill, O.P. and Prince, DA. (1989) Sodium channels in dendrites of rat cortical pyramidal neurons. *Proceedings of the National Academy of Sciences USA* 86:2473-2477.

Jack, J.J.B., Noble, D. and Tsien, R.W. (1975) *Electric current flow in excitable cells*. University Press, London, Oxford.

Llinás, R. and Sugimori, M. (1980) Electrophysiological properties of *in vitro* Purkinje cell somata in mammalian cerebellar slices. *Journal of Physiology (London)* 305:171-195.

Llinás, R.R. (1988) The intrinsic electrophysiological properties of mammalian neurons: insights into central nervous system function. *Science* 242:1654-1664.

Nilius, B. (1988) Modal gating behavior of cardiac sodium channels in cell-free membrane patches. *Biophysics Journal* 53: 857-862.

Patlak, J.B. and Ortiz, M. (1986) Two modes of gating during late Na^+ channel currents in frog sartorius muscle. *Journal of General Physiology* 87:305-326.

Schwindt, P.C. and Crill, W. E. (1995) Amplification of synaptic current by persistent sodium conductance in apical dendrite of neocortical neurons. *Journal of Neurophysiology, in press.*

Stafstrom, C.E., Schwindt, P.C., Chubb, M.C. and Crill, W.E. (1985) Properties of persistent sodium conductance and calcium conductance of layer V neurons from cat sensorimotor cortex *in vitro*. *Journal of Neurophysiology* 53:153-170.

Stafstrom, C. E., Schwindt, P. C. and Crill, W. E. (1984) Repetitive firing in layer V neurons from cat neocortex *in vitro*. *Journal of Neurophysiology* 52: 264-277.

Stuart, G.J. and Sakmann, B. (1994) Active propagation of sodium action potentials into neocortical pyramidal cell dendrites. *Nature* 367: 69-72.

Taylor, C.P. (1993) Na^+ currents that fail to inactivate. *Trends in Neuroscience* 116:455-460.

Somesthesis and the Neurobiology of the Somatosensory Cortex
O. Franzén, R. Johansson and L. Terenius (eds.)

Information processing channels in the sense of touch

S.J. Bolanowski

Institute for Sensory Research and Department of Bioengineering and Neuroscience, Syracuse University, Syracuse, NY 13244, USA

Summary. The concept of channels and their properties as defined in terms of mechanisms is discussed with respect to the multichannel model for taction. Analysis of neurophysiological results and the establishment of psychophysical-physiological linking hypothesis show how the functional basis of the multichannel model was established. A discussion of how the effects of multimodal interactions, attention and learning affect the information carried by the channels is also included.

Mechanisms and properties of channels

Based on psychophysical, physiological and anatomical experiments we have proposed that four channels mediate tactile sensations resulting from stimulation of glabrous skin (Bolanowski et al., 1988). A **channel** is defined here as a low-level analyzer responsible for detecting and encoding particular information about a physical stimulus. That is, a channel can be thought of as a functional/structural pathway in which specific information about the external world is passed to some location in the brain where the perception of a particular sensory event occurs. The combination of all relevant channels contributes to the entity commonly referred to as a **system**. One constraint on the channels making up a system is that they must, in combination, account for the entire range of capabilities of that system (e.g., the threshold-frequency characteristics of the channels in combination must fit the overall threshold capabilities of the observer as in, for example, the frequency response of the auditory system). As such, the channels that we have uncovered are perhaps similar to those found in other sensory systems: the critical bands of audition, the spatial frequency channels of vision and even possibly the conduits of the four primary tastes (sweet, sour, salty, bitter).

In discussing the channel concept, it should be noted that a channel's functional foundation is based on **mechanisms** inherent to that channel, these mechanisms being defined by the specific manner in which externally-obtained information is manipulated. The types of mechanisms include: filtering, non-linear compressions, analog-to-digital conversions, comparisons, phase delays, negative feedback, and automatic gain control, to name only a few. In its simplist form, a channel may be no more than a linear filter followed by analog-to-digital, non-linear conversion.

Furthermore, the concept of a channel does not in itself define the **properties** of a channel, these instead being determined by the channel's underlying mechanisms. For example, channels typically possess the property called "threshold-independence" which requires that detections by one channel receive no contribution from other channels, and that detection by a given channel must occur when threshold is exceeded. Another property is "frequency selectivity" which, as will be shown subsequently, can be produced in part by simple filtering at the receptor level. The property called "labeling" implies that an observer can always distinguish among the responses of different channels, whereas the property of "univariance" implies that a channel's output varies along only a single perceptual dimension such that a change in stimulus quality can be completely compensated by an appropriate change in stimulus intensity. Lastly, the properties of temporal summation and spatial summation mean that increases in either the duration or size of the stimulus, respectively, can result in the improvement of sensitivity.

The multichannel model for taction

Most of the psychophysical studies upon which the multi-channel tactile model (Bolanowski et al., 1988) is based were conducted on the thenar eminence using vibratory displacements applied perpendicularly to the skin surface via a single stimulus probe (Verrillo, 1962). The threshold-frequency characteristics of the four channels were determined by manipulating stimulus parameters such as the frequency and amplitude of the vibration (Verrillo, 1966a, 1968; Gescheider, 1976), contactor size (Verrillo, 1963), stimulus duration (Verrillo, 1965; see also Checkosky and Bolanowski, 1992), skin-surface temperature (Bolanowski and Verrillo, 1982; Verrillo and Bolanowski, 1986), and the application of various masking techniques (Verrillo and Gescheider, 1977; Gescheider et al., 1979; Capraro et al., 1979; Gescheider et al., 1982; Gescheider et al., 1983; Hamer et al., 1983; Gescheider et al., 1985). Figure 1 shows the psychophysical portion of the model as it pertains to stimulus-intensity/frequency space (i.e. frequency characteristics). Shown in the Figure are the threshold-frequency characteristics of the four channels, P (Pacinian), NPI (non-Pacinian I), NPII and NPIII, as presently understood.

There are four major tenets to this portion of the model: 1) stimuli at threshold levels are signaled by the channel that is the most sensitive (i.e., threshold or channel independence), each channel constrained in terms of frequency selectivity; 2) the individual channels possess the properties of labeling and possibly univariance, with the P channel additionally displaying temporal and spatial summation; 3) suprathreshold sensations are the result of a combination of the neural activity transmitted by all activated channels; and 4) the overall shape of the threshold - frequency

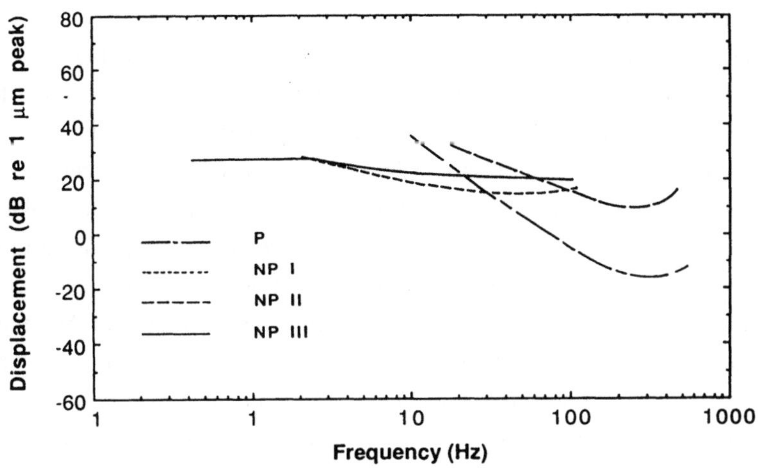

Figure 1. The four-channel model of vibrotaction showing the threshold frequency characteristics of the various channels as defined by the legend in the Figure.

characteristic that is obtained during experimentation is a function of the stimulus parameters. One fundamental point regarding the model is that not all of the four channels need to be activated for a suprathreshold sensory experience. Indeed, for certain suprathreshold conditions only one or two channels will be activated (Bolanowski et al., 1992a,b). In that instance a specific tactile sensation will result, albeit of a quality different from that observed if all four channels are driven. A second equally important point, is that the model intrinsically does not require the existence of four channels as we have found on the hairy skin of the forearm (Bolanowski et al., 1994). The model simply hypothesizes that the sense of touch utilizes information being transmitted by separate, independent channels. Of course integration of the activity arriving over the different channels must ultimately be combined to provide a unified percept (see Fig. 3), especially at suprathreshold levels. Where this occurs, however, has not yet been determined.

In the formulation of the complete model (Bolanowski et al., 1988), the psychophysical results were correlated with relevant physiological and anatomical data available in the literature (e.g. Lindblom, 1965; Lindblom and Lund, 1966; Talbot et al., 1968; Jänig, 1971; Munger, 1971; Mountcastle et al., 1972; Johansson and Vallbo, 1979; Freeman and Johnson, 1982; Torebjörk and Ochoa, 1980; Ochoa and Torebjörk, 1983; Vallbo and Johansson, 1984; Bolanowski and

Zwislocki, 1984). In particular, we have relied heavily on the microneurographic results obtained by Johansson et al., (1982). Using vibratory stimuli presented in a manner similar to that used by us for defining the psychophysical multichannel model and percutaneous recordings in humans, these investigators measured average equi-amplitude response areas of the four fiber types innervating the glabrous skin. We replotted their results obtained from each class of fibers (RA, PC, SAI and SAII) in terms of rate-intensity characterisitics. Using extrapolations and interpolations, we then estimated equi-response frequency characterisitics relating the amplitude to the frequency of the stimulus for a fixed criterion response. In the past, we (Bolanowski and Verrillo, 1982; Bolanowski and Zwislocki, 1984) have stressed the importance that response criterion plays in establishing psychophysical/physiological linking hypotheses, especially as they relate to determining the shapes of frequency-response functions. Because of this, equi-response frequency characterisitics were derived for the four fiber types using a number of different criteria as shown in Fig. 2. The results shown in Fig. 2 have not been published previously. It should be evident for each fiber class shown in the Figure, that the shapes of the frequency response functions are strongly dependent upon response criterion. For example, in Fig. 2D the functions are U shaped for criteria of 1 or 2 spikes/stimulus bursts, but linear with negative slopes on the log-log coordinate system when firing rates (spikes/sec) are considered. In formulating the psychophysical/physiological link for the multichannel model, it was necessary to assume a critierion response for each fiber class. One factor contributing to the selection of response criteria was the assumption that the underlying neural and psychophysical frequency-response characterisictics should have similar shapes in their frequency responses, given that the psychophysical channels could be modeled most simply as a filter matching the receptors' frequency selectivity.

The complete multichannel model proposed that each psychophysical channel has a specific substrate consisting of specific end organs innervated by select groups of peripheral nerve fibers, the isolated activation of which can produce unique and "unitary" sensations. The psychophysically defined channels, their end-organ/nerve-fiber substrates, the proposed neural code, their respective unitary suprathreshold sensations, and the nominal frequency range over which they operate at threshold levels are: 1) the P channel, mediated by Pacinian corpuscles (PC) and PC fibers using a code of 4 spikes/stimulus burst, yields the sensation of "vibration" in the frequency range of 40->500 Hz ; 2) NPI, mediated by Meissner corpuscles and rapidly adapting fibers (RA) using a code of 1 spike/stimulus burst, produces the sensation of "flutter" in the frequency range of 2-40 Hz; 3) NPII, mediated by Ruffini end organs and slowly adapting, type-two fibers (SAII) using a code of 5 spikes/sec, produces a "buzz-like" sensation in the frequency

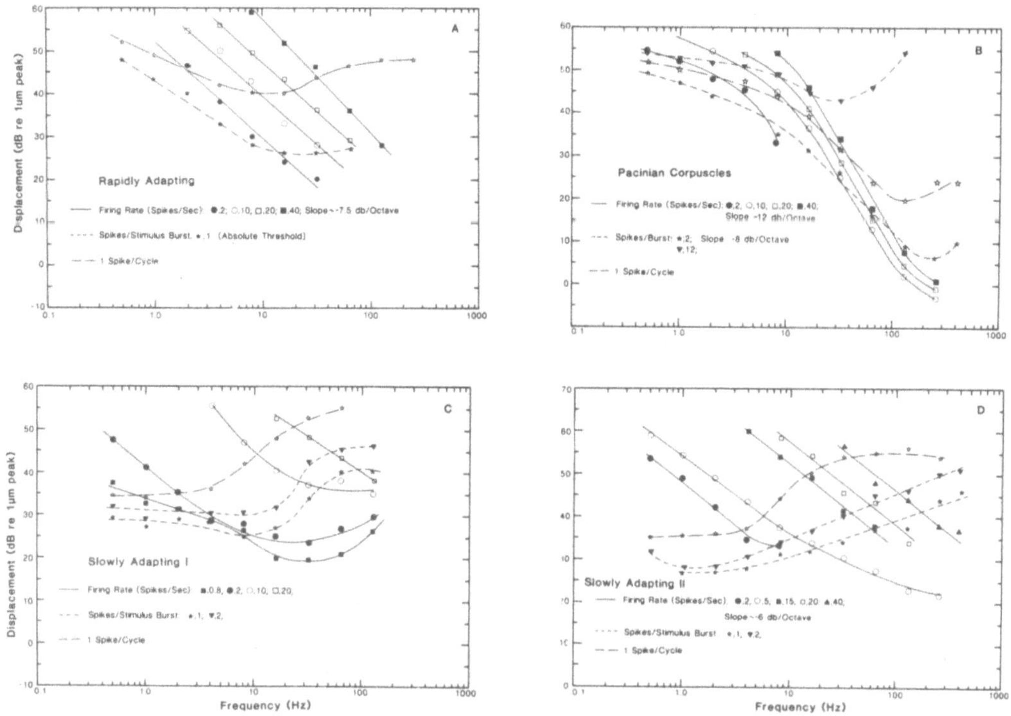

Figure 2. Average neurophysiological functions relating stimulus amplitude and frequency for various response criteria as measured on the four peripheral fiber classes found in glabrous skin. The data upon which the functions were derived were obtained by Johansson et al., (1982). A) Rapidly adapting; B) Pacinian corpuscles; C) Slowly adapting, type I; and D) Slowly adapting, type II.

range of 100- > 500Hz; and 4) NPIII, mediated by Merkel-cell neurite complexes and slowly adapting, type-one fibers (SA I) using a code of 0.8 spikes/sec produces the sensation of "pressure" in the frequency range of 0.4- 2 Hz. Since that time it has been determined that the assumed code for the P channel was incorrect; the P channel requires activity over more than one PC fiber (Checkosky and Bolanowski, 1992). We are continuing to perform experiments designed to further refine and modify the model, and to test other aspects of taction using the model as the theoretical basis. In particular we have been conducting experiments to ascertain how higher-order analyses affect transmission of signals passed through the low-level channels.

Low-level versus high-level analysis

It should be clear from the foregoing discussion that progress in understanding the sense of touch has benefitted by employing tasks directed at understanding low-level processing. In particular, threshold tasks have allowed us to determine the low-level processing schemes (channels, see Fig. 3), that mediate tactile sensations both physiologically and psychophysically, without having to deal with higher-level cognitive processing (i.e., object recognition) that occurs at suprathreshold

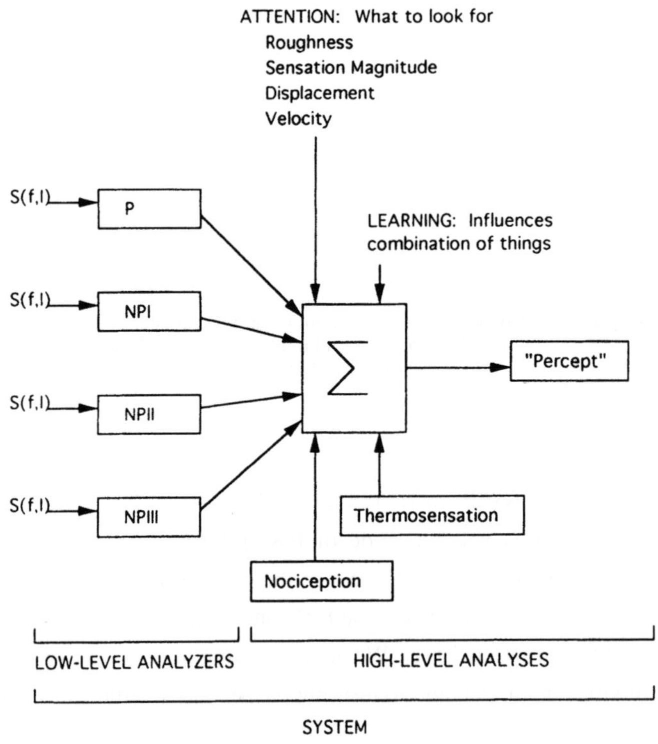

Figure 3. Diagram defining the low-level analyzers (channels) and high-level analyses that combine to form the overall tactile system. S(f,I) signifies inputs to the system in terms of stimulus displacements as a function of frequency and intensity. The unified "percept" results from integration of information arriving over the low-level analyzers which is interpreted and modified by higher-level analyses and inputs from additional peripheral and central nervous-system locations.

levels. It should be clear, however, that the information arriving over the four low-level analyzers must somehow be combined centrally to produce a unified percept as in, for example, roughness (e.g., Connor and Johnson, 1992). Furthermore, there is earlier evidence suggesting, as shown in Fig. 3, that factors such as attention, cross-modal interactions (touch-pain; touch-temperature) and learning can also play a role in tactile perception. For example, Burgess and colleagues (e.g., Burgess et al., 1983) have shown that subjects can attend to different aspects of the stimulus (depth of indentation, velocity), and Craig and Evans (1995) have made some progress in uncovering attentional effects. Furthermore it has been found that touch can influence the perception of pain (gate theory of pain, Melzack and Wall, 1965). We (Apkarian et al., 1994) have

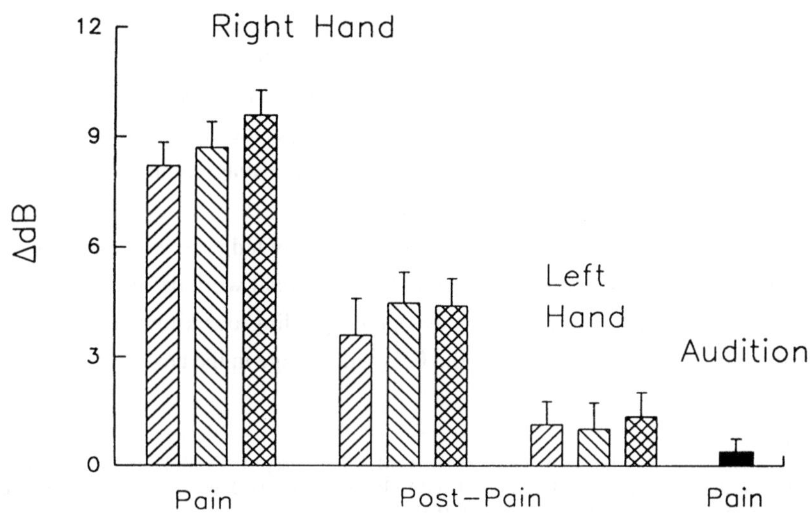

Figure 4. Average change in vibratory thresholds between baseline (no pain) and during thermal stimulation resulting in sustained pain for the right hand (first three bars on the left of the Figure); between baseline and recovery from pain for the right hand (second set of three bars from the left); and between left-hand baseline and left-hand postpain thresholds (third set of three bars from the left). Each condition was tested at three vibration frequencies that independently activated three of the four channels (1 Hz, NPIII, left-slanted bars; 10 Hz, NPI, right-slanted bars; and 100 Hz, P, cross-hatched bars). The filled bar on the right shows average auditory threshold changes during right-hand pain. Vertical bars represent standard errors. Postpain thresholds for the left (contralateral) hand were not significantly changed. Auditory thresholds measured at 1 kHz were also not significantly changed during painful thermal stimulation of the right hand. Reprinted with permission from Apkarian et al., (1994).

recently found the opposite effect, that painfully hot stimuli can dramatically reduce tactile sensitivity, a so-called "touch gate". In these experiments, subjects were presented to the thenar eminence vibratory stimuli selected to individually activate the various psychophysical channels. Vibrotactile thresholds were determined before, during and after pain co-localized with the vibratory stimuli. As shown in Fig. 4, regardless of the channel activated, the painfully hot stimuli reduced tactile sensitivity by, on average, 7.3 dB, without affecting the contralateral tactile thresholds nor the auditory thresholds. Thus, the effect is probably not attentional, but does demonstrate the high-level processing aspects present in the sense of touch. Lastly, learning has been found to affect cortical representations of the stimulus (Lee and Whitsel, 1992; Recanzone et al., 1992a,b,c; Mogilner et al., 1993). Because of this we have begun experiments to determine if learning can affect tactile sensations. In particular, we have found that subjects trained on an intensity-discrimination task can improve performance by a factor of approximately 50% (Bolanowski et al., 1995) after 23 days of training. Although no change in threshold sensitivity occurred, intensity discrimination for the homologous, contralateral, untrained site were also affected, but in a complicated way that may be channel and intensity specific. Training in the intensity-discrimination task also may have produced changes in spatial discrimination. This vibratory-based spatial effect, and the bilateral and stimulus-specific effects, may implicate cortical neural-plasticity and suggest that the perception of sensation magnitude comes about through complex (possibly learned), spatio-temporal mechanisms.

In closing, it is true that a substantial body of knowledge exists about the low-level processing schemes utilized by the tactile system, but additional work defining the neural codes, etc. must still be performed. The field is wide open regarding our knowledge of higher-level processing and it will be exciting to see how the field will change in the next few years.

Acknowledgements
The author thanks Drs. R.T.Verrillo, G.A. Gescheider, C.M. Checkosky, J. C. Makous for input regarding the manuscript and the technical support of Mss. K. Hall and L. Pawson. The work was partially supported by the U.S. National Institutes of Health, grants DC00380 and DC00098.

References

Apkarian, A.V., Stea, R.A. and Bolanowski, S.J. (1994) Heat-induced pain diminishes vibrotactile perception: A touch gate. *Somatosensory and Motor Research* 11: 259-267.

Bolanowski, S.J., Gescheider, G.A. and Verrillo, R.T. (1992a) Individual magnitude estimates correlate with neural intensity characteristics. *Society for Neuroscience Abstracts* 18: 1544.

Bolanowski, S.J., Gescheider, G.A. and Verrillo, R.T. (1992b) Individual magnitude estimation functions and their relation to the Pacinian (P) and non-Pacinian (NP) channels. *Journal of the Acoustical Society of America* 92: 2377.

Bolanowski, S. J., Jr., Gescheider, G.A., Verrillo, R.T. and Checkosky, C.M. (1988) Four channels mediate the mechanical aspects of touch. *Journal of the Acoustical Society of America* 84: 1680-1694.

Bolanowski, S. J., Jr., Gescheider, G.A. and Verrillo, R.T. (1994) Expansions of the four-channel model for touch: Hairy skin. *Somatosensory and Motor Research* 11: 279-290.

Bolanowski, S.J., Hall, K.L., Makous, J.C. and Merzenich, M.M. (1995) Intensity discrimination of vibratory stimuli can be improved by training. *Society for Neuroscience Abstracts* 21: 1443.

Bolanowski, S. J., Jr. and Verrillo, R.T. (1982) Temperature and criterion effects in a somatosensory subsystem: A neurophysiological and psychophysical study. *Journal of Neurophysiology* 48: 836-855.

Bolanowski, S. J., Jr. and Zwislocki, J.J. (1984) Intensity and frequency characteristics of Pacinian corpuscles. I. Action potentials. *Journal of Neurophysiology* 451: 793-811.

Burgess, P.R., Mei, J., Tuckett, R.P., Horch, K.W., Ballinger, C.M. and Poulos, D.A. (1983) The neural signal for skin indentation depth. I. Changing indentations. *Journal of Neuroscience* 3: 1572-1585.

Capraro, A.J., Verrillo, R.T. and Zwislocki, J.J. (1979) Psychophysical evidence for a triplex system of mechanoreception. *Sensory Processes* 3: 334-352.

Checkosky, C.M. and Bolanowski, S.J. (1992) Effects of stimulus duration on the response properties of Pacinian corpuscles: Implications for the neural code. *Journal of the Acoustical Society of America* 91: 3372-2280.

Connor, C.E. and Johnson, K.O. (1992) Neural coding of tactile texture: Comparison of spatial and temporal mechanisms for roughness perception. *Journal of Neuroscience* 12: 3414-3426.

Craig, J.C. and Evans, P.M. (1995) Tactile selective attention and temporal masking. *Perception and Psychophysics* 57:511-518.

Freeman, A.W. and Johnson, K.O. (1982) A model accounting for the effects of vibratory amplitude on responses of cutaneous mechanoreceptors in macaque monkey. *Journal of Physiology (Lond.)* 323: 43-64.

Gescheider, G. A. (1976) Evidence in support of the duplex theory of mechanoreception. *Sensory Processes* 1: 68-76.

Gescheider, G. A., Frisina, R.D. and Verrillo, R.T. (1979) Selective adaptation of vibrotactile thresholds. *Sensory Processes* 3: 37-48.

Gescheider, G. A., O'Malley, M.J. and Verrillo, R.T. (1983) Vibrotactile forward masking: Evidence for channnel independance. *Journal of the Acoustical Society of America* 74: 474-485.

Gescheider, G. A., Sklar, B.F., Van Doren, C.L. and Verrillo, R.T. (1985) Vibrotactile forward masking: Psychophysical evidence for a triplex theory of cutaneous mechanoreception. *Journal of the Acoustical Society of America* 74: 534-543.

Gescheider, G.A., Verrillo, R.T., and Van Doren, C.L. (1982) Prediction of vibrotactile masking functions. *Journal of the Acoustical Society of America* 72: 1421-1426.

Hamer, R.D., Verrillo, R.T. and Zwislocki, J.J. (1983) Vibrotactile masking of Pacinian and non-Pacinian channels. *Journal of the Acoustical Society of America* 73: 1293-1303.

Jänig, W. (1971) Morphology of rapidly and slowly adapting mechanoreceptors in the hairless skin of the cat's hind foot. *Brain Research* 28: 217-231.

Johansson, R. S., Landström, U. and Lundström, R. (1982) Responses of mechanoreceptive afferent units in the glabrous skin of the human hand to sinusoidal skin displacements. *Brain Research* 244: 17-25.

Johansson, R. S. and Vallbo, A.B. (1979) Detection of tactile stimuli. Thresholds of afferent units related to psychophysical thresholds in the human hand. *Journal of Physiology (Lond.)* 297: 405-42.

Lee, C.-J. and Whitsel, B. (1992) Mechanisms underlying somatosensory cortical dynamics: I. In vivo studies. *Cerebral Cortex* 2: 81-106.

Lindblom, U. (1965) Properties of touch receptors in distal glaborous skin of the monkey. *Journal of Neurophysiology* 428: 966-985.

Lindblom, U. and Lund, L. (1966) The discharge from vibration sensitive receptors in the monkey foot. *Experimental Neurology* 15: 401-417.

Melzak, R. and Wall, P. (1965) Pain mechanisms: A new theory. *Science* 150: 971-978.

Mogilner, A., Grossman, J.A.I., Ribary, U., Joliot, M., Volmann, J., Rapaport, D., Beasley, R.W. and Llinás, R.R. (1993) Somatosensory cortical plasticity in adult humans revealed by magnetoencephalography. *Proceedings of the National Academy of Science* 90: 3593-3597.

Mountcastle, V. B., LaMotte, R.H. and Carli, G. (1972) Detection thresholds for stimuli in humans and monkeys: Comparison with threshold events in mechanoreceptive afferent nerve fibers innervating the monkey hand. *Journal of Neurophysiology* 435: 122-136.

Munger, B. C. (1971) Patterns of organization of peripheral sensory receptors. In: W. R. Loewenstein (ed.) *Handbook of Sensory Physiology*. Vol. 1, Springer, Berlin, pp 523-556.

Ochoa, J. L. and Torebjork, H.E. (1983) Sensations evoked by intraneural microstimulation of single mechanoreceptor units innervating the human hand. *Journal of Physiology (Lond.)* 342: 633-653.

Recanzone, G.H., Jenkins, W.M., Hradek, G.H. and Merzenich, M.M. (1992a) Progressive improvement in discriminative abilities in adult owl monkeys performing a tactile discrimination task. *Journal of Neurophysiology* 467: 1015-1030.

Recanzone, G.H., Merzenich, M.M., Jenkins, W.M. (1992b) Frequency discrimination training engaging a restricted skin surface results in an emergence of a cutaneous response zone in cortical area 3a. *Journal of Neurophysiology* 467: 1057-1070.

Recanzone, G.H., Merzenich, M. M. and Schreiner, C.S. J. (1992c) Changes in the distributed temporal response properties of SI cortical neurns reflect improvements in performance on a temporally based tactile discrimination task. *Journal of Neurophysiology* 467, 1071-1091.

Talbot, W.H., Darian-Smith, I. Kornhuber, H.H. and Mountcastle, V.B. (1968) The sense of flutter-vibration: Comparison of the human capacity with response patterns of mechanoreceptive afferents from the monkey hand. *Journal of Neurophysiology* 431: 301-334.

Torebjörk, H. E. and Ochoa, J.L. (1980) Specific sensations evoked by activity in single identified sensory units in man. *Acta Physiologica Scandinavica* 110: 445-447.

Vallbo, A. B. and Johansson, R.S. (1984) Skin mechanoreceptors in the human hand related to touch sensation. *Human Neurobiology* 3: 3-14.

Verrillo, R. T. (1962) Investigation of some parameters of the cutaneous threshold for vibration. *Journal of the Acoustical Society of America* 34: 1768-1773.

Verrillo, R. T. (1963) Effect of contactor area on the vibrotactile threshold. *Journal of the Acoustical Society of America* 35: 1962-1966.

Verrillo, R. T. (1965) Temporal summation in vibrotactile sensitivity. *Journal of the Acoustical Society of America* 37, 843-846.

Verrillo, R. T. (1966) Effects of spatial parameters on the vibrotactile threshold. *Journal of Experimental Psychology* 71: 570-574.

Verrillo, R. T. (1968) A duplex mechanism of mechanoreception. In: D.R. Kenshalo, (ed.) *The Skin Senses*, Thomas, Springfield, IL, pp 135-159.

Verrillo, R. T. and Bolanowski, S.J., Jr. (1986) The effects of skin temperature on the psychophysical responses to vibration on glabrous and hairy skin. *Journal of the Acoustical Society of America* 80: 528-532.

Verrillo, R. T. and Gescheider, G.A. (1977) Effects of prior stimulation on vibrotactile thresholds. *Sensory Processes* 1:292-300.

Somesthesis and the Neurobiology of the Somatosensory Cortex
O. Franzén, R. Johansson and L. Terenius (eds.)

A novel approach for studying direction discrimination

G.K. Essick[1,2,3], E.M. Rath[1,2,3], D.G. Kelly[4], A. James[2] and R.A. Murray[1,2]

[1]*School of Dentistry,* [2]*Dental Research Center,* [3]*Curriculum in Neurobiology,* [4]*Departments of Statistics and Mathematics, University of North Carolina, Chapel Hill, North Carolina, USA*

Summary. The use of frictionless moving stimuli to study direction discrimination limits the peripheral neural mechanisms by which information can be encoded. The utility of such stimuli, provided by a novel dense-array tactile stimulator, to address basic psychophysical questions regarding cutaneous motion processing is demonstrated. Data collected to date suggest that direction of simulated motion is discriminated when the product of the length of skin traversed and the duration of stimulation attains or exceeds a criterion -- reflecting a perfect space-time tradeoff or reciprocity at threshold.

Introduction

Over the past twenty years, we have been studying the percepts evoked by a natural stimulus object moving across the skin (e.g., see Essick et al., 1991). Of particular interest have been the percept of direction and its dependency on the parameters of stimulus motion. The impetus for these studies arose from early observations of the behavior of individual somatosensory cortical neurons in monkey (Whitsel et al., 1972, 1979). Specifically, many neurons exhibited notably different responses to different directions of motion across their receptive fields (RFs). Moreover, the directional selectivity varied as a function of the velocity of movement, the length of skin traversed, and RF location. The striking degree of directional selectivity was not observed, however, in the responses of individual afferents of primates. These and similar observations led us and others to hypothesize that the onsets of activity evoked in linear sequences of mechanoreceptors underlie the directional selectivity of individual cortical neurons (Whitsel et al., 1972; Costanzo and Gardner, 1980; Essick and Whitsel, 1985a,b; Gardner et al., 1989). Moreover, it was suggested that populations of these neurons represent the substrate for the percept of direction of motion across the skin. Indeed, many of the psychophysical findings we have reported are in agreement with these suggestions.

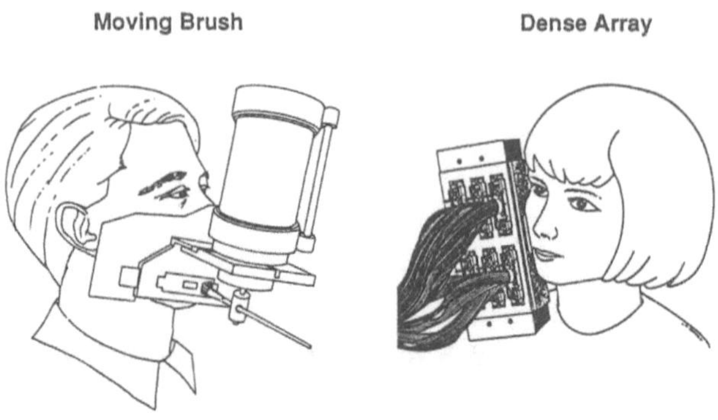

Figure 1. The stimulators positioned on the right side of the face.

Most recently, we have investigated the capacity of individual mechanoreceptive afferents in the human median, radial and inferior alveolar nerves to encode information about the direction of motion across their receptive fields (Edin et al., 1995; Essick and Edin, 1995). Similar to psychophysical experiments (see Fig. 1, left panel), each stimulus consisted of a single constant-velocity brush movement over a thin Teflon plate secured to the skin. The RF was exposed through an aperture in the plate, whose dimensions defined the length and width of skin to be stimulated. With few exceptions, the mechanoreceptors exhibited distinct differences in response to stimuli moving in different directions. These differences were often most pronounced in the spatial patterns of discharge and were shown to reflect asymmetrical sensitivity to skin stretch. Consistent directional differences in mean discharge rate were also observed for over 70% of the mechanoreceptors. Similar findings were reported by Greenspan (1992) for mechanoreceptors supplying the feline hairy hindlimb.

These observations suggested that information about direction of motion may be available to central neural mechanisms not only from the times at which spatially discrete sets of receptors are sequentially activated (i.e., by a spatiotemporal code), but by other mechanisms which, in part, depend on the mechanoreceptors' sensitivity to lateral skin strain. For example, (i) directionally specific patterns of discharge evoked in individual mechanoreceptors, (ii) directionally specific intensities at which [the population of] excited receptors discharge, and (iii) the spatial distribution of receptors activated at any instant in time all represent hypothetical encoding mechanisms for direction (see Edin et al., 1995). Central processing of information provided by these latter mechanisms may involve integration of inputs over time (Olausson and Norrsell, 1993); however, use of the term 'spatiotemporal' will be restricted to that noted above.

To better appreciate the role of spatiotemporally distributed tactile information in subjects' psychophysical performance, we developed a novel tactile stimulator to eliminate, or at least minimize, the contributions of alternate encoding mechanisms for stimulus direction. The dense-array tactile stimulator (Fig. 1, right panel) was developed by Zedek Corporation (Durham, North Carolina) and consists of a transducer module, driving electronics, and host PC computer. The transducer module consists of 288 (24 rows by 12 columns) 0.01-inch-diameter probes. The probe center-to-center is nominally 0.12 cm across rows and columns such that the total array is

2.7 cm x 1.3 cm. The probes extend through a display plate, and the distance between the probes at rest and the skin can be easily adjusted. The plate is covered with a plastic film (Saran Wrap; Dow Chemical Corp.) in order to protect the array and maintain clinical asepsis (Cholewiak and Collins, 1990). Thin, double-sided tape is used to secure the covered array to the skin.

The transducer module connects to the driving electronics by six 50-wire ribbon cables. The driving electronics are housed in a cabinet 10.5" high, 19" long and 12" deep. The back plane accommodates twelve driver cards (1 card for each column of probes) and a power supply. The free edge of each driver card supports 24 LEDs which signify the status (active/nonactive) of a particular probe. The driving electronics interface to an AT-compatible computer by way of a custom-made interface card.

Each probe is attached to a piezoelectric bender/bimorph (see Bliss et al., 1970). The bimorph momentarily deflects in proportion to an imposed voltage, which in turn moves the probe into the skin. A selection of the 288 probes for activation (i.e., a frame) is specified by 36 bytes of data (1 bit per probe). Frames are downloaded from the computer to the interface at experimenter-specified intervals ranging from 1.67 to 5 msec. The digital signals operate electronic latches on the driver cards which power the individual bender/bimorphs. This operating strategy ensures that the probes are effectively driven by a square wave pulse (frame

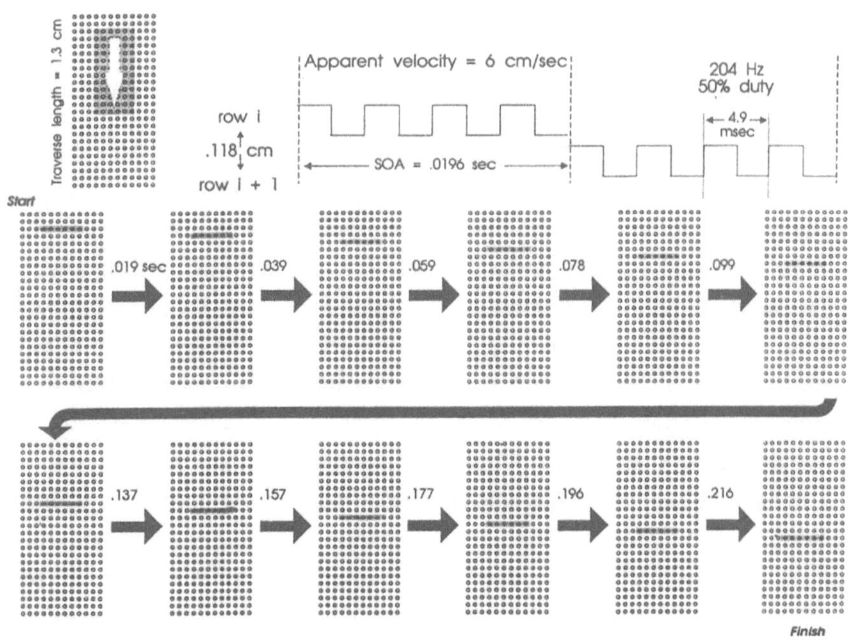

Figure 2. Mechanics of simulating motion over a 1.3-cm-long chord of skin. Each frame illustrates the active group of probes (shaded) at a particular time. Note in this example that the edge-like stimuli are composed of probes from only the six central columns of the array.

refresh rate or driving frequency = 100 to 300 Hz, 50% duty cycle). The power supply is digitally controlled so that the amplitude of the pulses can be varied from 0 to 40 volts with 8-bits of resolution. The option allows the perceived intensity of the patterns to be adjusted. The displacement of the probes in a similar device, but one with different driving electronics, the OPTACON, has been estimated to attain about 65μ when loaded by the digital skin (Bliss et al., 1970).

Frames can be designed to generate any moving tactile image within the limits of the 24 x 12 array of probes. For the experiments described below, edge-like patterns consisting of rows of probes were generated. For such stimulus patterns, the duration that each row was displayed (i.e., the 'exposure' or EXPO) as well as the time between the activation of successive rows (i.e., the 'stimulus onset asynchrony' or SOA) was restricted to multiples of the frame refresh period. For SOA=EXPO, the $(i+1)^{th}$ row rose and the ith row fell in a highly regular, step-locked rhythm (see Fig. 2). The velocity of motion was defined as the distance between adjacent rows divided by the SOA. To illustrate, a frame refresh rate of 204 Hz (period = 4.9 msec) allowed attainment of five octaves of velocity between 0.75 and 24 cm/sec: A SOA=EXPO of 32, 16, 8, 4, 2, and 1 cycles simulated velocities of 0.75, 1.5, 3, 6, 12 and 24 cm/sec, respectively. The moving edge could be widened in the direction of motion to N rows by making the EXPO equal to N times the SOA.

Application 1. Comparison of subjects' capacity to discriminate real and simulated motion

Systematic studies by Gould et al. (1979) and Norrsell and Olausson (1992) have shown that the threshold distance for discriminating direction of motion over the forearm can vary greatly as a result of variation in the forces applied to the skin. Thresholds from ca 1 to 3 cm are expected for a stylus or airjet that exerts only low vertical forces (viz., 1--2 gm-wgt). Whereas the threshold for frictionless airjet stimuli remains constant, the threshold for a stylus or probe that laterally stretches the skin falls precipitously in response to an increase in the applied force. Indeed, the lowest thresholds (often ≤ 0.1 cm) have been reported for stimuli that are glued to and only stretch the skin. These observations led to the conclusion that mechanisms which rely on the sensitivity of the mechanoreceptors to lateral skin stretch exhibit "superior resolution" in comparison to those which rely on spatiotemporally distributed information (Olausson and Norrsell, 1993).

To evaluate this hypothesis, we recruited six neurologically normal young adults (one male, age 38 yrs; five females, ages 18 to 22 yrs) to participate in six sensory testing sessions. In each of three sessions, selected at random, the stimuli were provided by the brushing stimulator and the dense-array tactile stimulator (Fig. 1). During all testing, the same cutaneous region, 0.70 cm long and 0.94 cm wide overlying the right mental foramen, was traversed by stimuli moving at each of 0.5, 2, 6, 12 and 24 cm/sec. The brush delivered about 5.5 gm-wgt of force at an angle 46 degrees to the skin surface (Edin et al., 1995). The driving voltages for the dense-array stimuli were adjusted for each subject so that stimulation by all velocities felt equally intense (see Appendix).

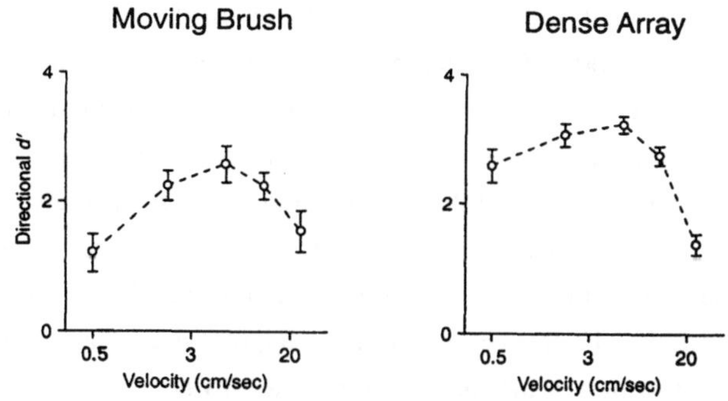

Figure 3. The directional d'-versus-velocity relationship observed with the moving brush and dense-array stimuli.

A temporal single-interval, forced-choice paradigm was employed for data collection. Forty-eight trials for each of the five test velocities were delivered in random order. Stimulus motion was either toward or away from the subject's chin. Feedback as to correctness of response was provided. At the completion of each session, an index of directional sensitivity, d', was calculated for each velocity of stimulation (see Essick et al., 1988). Values of d' equal to 0.0 and 4.0 indicate chance and perfect performance, respectively.

Figure 3 illustrates mean $d' \pm 1$ standard error plotted as a function of stimulus velocity for natural stimulus motion (left panel) and simulated motion (right panel). Directional sensitivity varied as a function of velocity ($p<0.001$) and differed for the two modes of stimulation ($p<0.051$). In contrast to the hypothesis, mean d' was greater (rather than less) for the simulated motion (2.63 units) than for the real motion (1.97 units). Moreover, a significant interaction between velocity and mode of stimulation was detected ($p<0.001$).

A gamma-curve-fitting procedure was applied to the data obtained from each session to quantitatively characterize the d'-versus-velocity relationship (Essick et al., 1988; Essick et al., 1991). Four functions of the parameters of the gamma function completely described the d'-versus-velocity relationship. Of foremost note, subjects' peak or maximum directional sensitivity was slightly greater for the simulated motion (3.46 units) than for real motion (2.72 units; $p<0.003$). This suggests that if the affected skin area is the same, central mechanisms dependent on the lateral strain sensitivity of receptors may not exhibit greater sensitivity than central mechanisms dependent on the spatiotemporal order of their activation. Given that the dense array tactile stimulator may not activate slowly adapting mechanoreceptors (Gardner and Palmer, 1989), more mechanoreceptors were likely recruited by the natural stimulus, yet its direction of motion was not better discriminated.

A second observation of interest was that the degree to which stimulus velocity influenced directional sensitivity was the same for both real and simulated motion. Specifically, neither the global velocity-tuning (the width of the inverted U-shaped curve relating d' to velocity) nor the local velocity-tuning (the "peakedness" of the curve) differed for the two modes of stimulation ($p>0.57$ and 0.064, respectively). In addition, data from sessions with both stimulators conformed equally well to the gamma-function model ($p>0.11$; mean $R^2=0.87$). Given these similarities, we concluded that the curved nature of the d'-versus-velocity relationship cannot be attributed to velocity-dependent variations in the perceived intensity of the stimuli.

A third observation of interest was that the velocity at which the subjects exhibited maximum directional sensitivity was less for the simulated motion (3.90 cm/sec) than for real motion (7.17 cm/sec; $p<0.05$). We hypothesized that this was due to the width of the stimulus in the direction of motion (the brush was always wider in the direction of motion than a single row of activated probes). A related finding reported in the visual literature has been attributed to variation in stimulus width (Burr and Ross, 1982). Specifically, contrast sensitivity for discriminating the direction of moving bar stimuli was found to be velocity-tuned. The greater the width of the bar in the direction of motion, the higher was the velocity at which maximum contrast sensitivity occurred. In the experiments reported in this paper, the brush was always wider in the direction of motion than a single row of activated probes.

To evaluate the hypothesis, four new subjects (three males and 1 female) participated in a study during which systematic changes in the width of a simulated edge were investigated. During all testing the same cutaneous region, 0.6 cm long and 0.94 cm wide, was traversed by stimuli moving at each of 0.45, 1.1, 2.6, 6.1 and 14.7 cm/sec. In each of two sessions, the width of the edge attained a width of 1, 2, 4 and 6 rows. For example, the 6-row-wide stimulus was realized as an edge that successively changed in number of rows according to the following sequence as it translated across the skin: 1, 2, 3, 4, 5, 6, 5, 4, 3, 2, 1 rows. The stimuli were delivered at driving voltages about four times their thresholds (Murray and Essick, unpublished). The experimental sessions were otherwise identical to those described above.

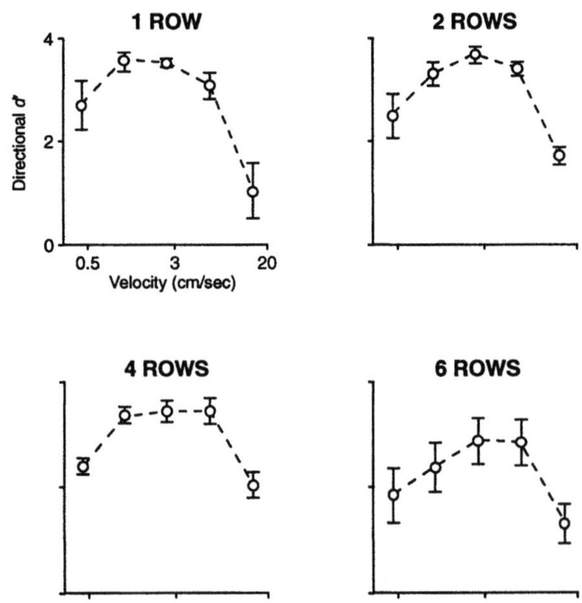

Figure 4. The directional d'-versus-velocity relationship observed with dense-array stimuli whose width differed in the direction of motion.

Figure 4 illustrates mean $d' \pm 1$ standard error plotted as a function of velocity for stimuli 1, 2, 4 and 6 rows wide. Directional sensitivity varied as a function of velocity ($p<0.001$) and differed for stimuli of different widths ($p<0.05$). On the average, directional sensitivity was less for the widest (6-row) stimuli than for stimuli 1, 2 or 4 rows wide. Consistent with our hypothesis, a significant interaction between velocity and stimulus width was detected ($p<0.05$). For stimuli that were consistently 1-row wide, directional sensitivity was greatest at test velocities 1.1 and 2.6 cm/sec. In contrast, for the widest stimulus, directional sensitivity was greatest at test velocities 2.6 and 6.1 cm/sec. These observations suggest that the subtle differences in the d'-versus-velocity relationships reported above for real and simulated motion may be attributed, in part, to differences in the width of the stimuli employed.

Application 2. Rapid estimation of the threshold length for direction discrimination

Most recently, we have evaluated a psychophysical procedure to "track" the threshold traverse length for direction discrimination (viz., sequential testing based on maximum likelihood parameter estimation). With this approach, reliable and accurate estimates can be obtained with only a fraction of the number of trials required by conventional approaches (see Dreyer et al., 1978; Whitsel et al., 1979; Essick et al., 1990).

The basic paradigm was developed by Harvey (1986). After each trial, the algorithm calculates the likelihood that the subject's prior responses resulted from each of 100 psychometric functions, whose points of maximum slope (i.e., the threshold) are distributed within the range 0.1 -- 2.75 cm. Each Weibull function is assumed to have a maximum slope corresponding to Harvey's $\beta = 4$ and to describe the relationship between the frequency at which the subject correctly discriminates direction of motion and the logarithm of the traverse length. The number of rows whose total span approximates the threshold predicted by the most likely function (adjusted by an additive constant $\varepsilon = 0.0652$; see Harvey, 1986) defines the length of skin for the next trial. The threshold estimates the traverse length predicted to result in correct discrimination of direction 81.6% of the time. Due to the constant ε, the adjusted length over which motion is simulated approximates 92% correct performance after a sufficient number of trials. The first stimulus traverses a distance about half the length of the array (1.53 cm).

All testing to date has been conducted with stimuli delivered to the middle of the left side of the face. Direction of motion was either up or down. The position of the stimuli within the array was varied from trial to trial to minimize subjects' use of positional cues to determine direction of motion. Sixty stimulus trials were delivered at each of 5 octaves of velocity between 0.75 and 24 cm/sec. The driving voltage was adjusted so that stimuli at different velocities and traversing different lengths of skin felt equally intense (see Appendix). Stimulus velocity and direction were selected in random order for delivery. After each trial, the subject reported the direction. Feedback as to correctness was provided.

At the completion of each session, the length of skin traversed at each velocity was plotted as a function of trial number, as illustrated in Fig. 5A. Second, the proportion of trials on which the subject discriminated direction correctly was calculated and plotted as a function of the length (Fig. 5B). Third, the best fitting Weibull function, i.e., the maximum likelihood function with no assumption regarding the maximum slope, was determined (solid curve, Fig. 5B). Harvey's algorithm failed to find a unique curve for ca 7% of the datasets (see Chen et al., 1995). For these datasets, the maximum likelihood psychometric function constrained by the assumptions adopted for data collection (dashed curve, Fig. 5B) was used to obtain an estimate of the threshold length. As illustrated by Fig. 5B, the threshold predicted by the constrained curve (1.72 cm) was similar to the threshold predicted by the best fitting curve (1.83 cm) in cases for which the latter could be determined. Moreover, the threshold predicted by the best fitting curve was remarkably similar to that obtained if only the first half of the data was analyzed (compare Fig. 5C with 5B).

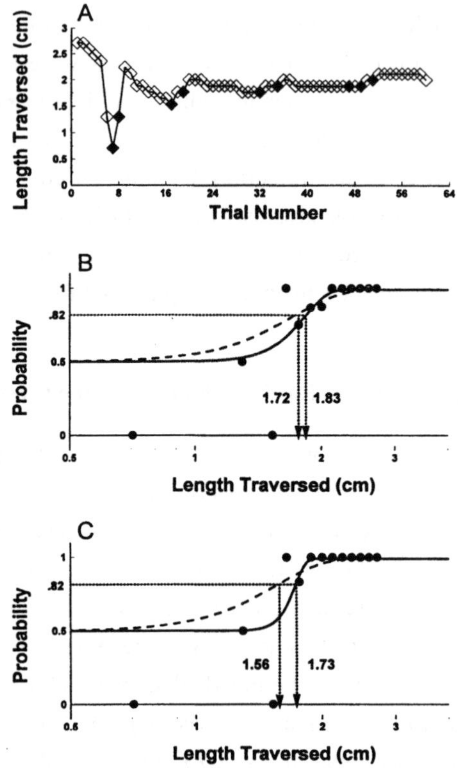

Figure 5. Example of post-session analyses for data obtained at one velocity from one session from one subject. [A] Length of skin traversed plotted as a function of trial number. Unfilled (filled) symbols indicate that the subject's response as to stimulus direction was correct (incorrect). [B] Best fitting psychometric function (solid curve), and the maximum likelihood psychometric function constrained by assumptions made during data collection (dashed curve). The predicted thresholds are indicated by arrows. [C] A second set of psychometric functions, but based on only the first 30 of the 60 trials of data.

To date, twelve neurologically normal subjects (five males ages 20.7 -- 23.7 yrs; seven females ages 19.2 -- 24.4 yrs; median age of subjects = 20.9 yrs) have each participated in four testing sessions. Eight of the subjects were tested once per day for four days. In Fig. 6, the means and range of estimates of the threshold are plotted as a function of stimulus velocity. Note that both axes are scaled logarithmically, and that the upper and lower dashed reference lines indicate slopes of 1.0 and 0.5, respectively. Analyses revealed the following. First, subjects'

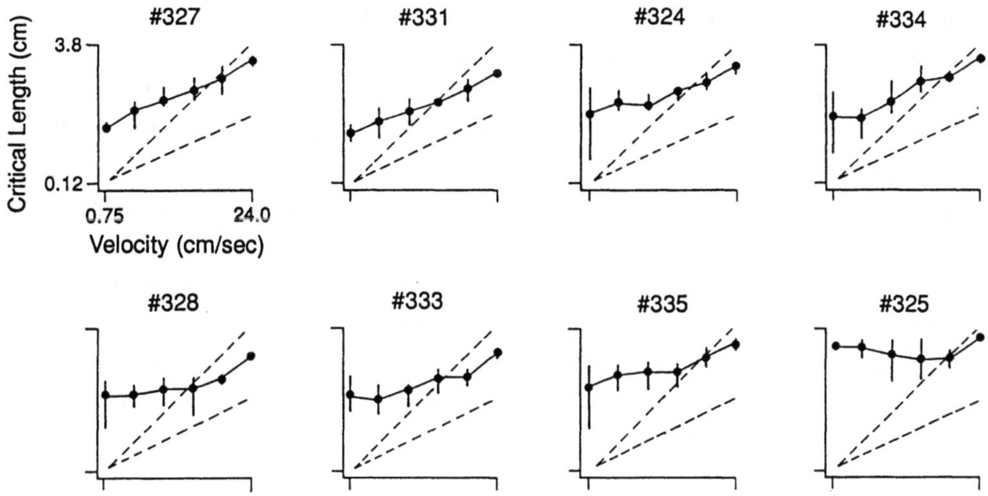

Figure 6. The threshold or 'critical' length for direction discrimination plotted as a function of stimulus velocity for each of eight subjects. Means and ranges from four testing sessions are shown for six velocities between 0.75 and 24 cm/sec. Moving stimuli were simulated on the middle of the left side of the face. Note that both axes are scaled logarithmically.

thresholds decreased by an average of 0.11 cm per session, indicating a modest learning effect ($p<0.004$). The effect was observed for all six velocities of simulated motion ($p>0.95$ for session-by-velocity interaction). Second, estimates based on the first 30 trials were 0.03 cm less than estimates based on all 60 trials of data, independently of stimulus velocity ($p>0.98$ and $p>0.65$ for an effect of velocity and a number of trials-by-velocity interaction, respectively). The within-subject variance (corrected for the learning effect; see Lederman and Taylor, 1972), however, was twice as great for the smaller sample estimates (0.061 versus 0.031 cm), independently of stimulus velocity ($p>0.65$ and $p>0.57$ for effects of velocity and number of trials-by-velocity interaction, respectively). Third, among-subject variability in the threshold lengths could be reduced by calculating the means of estimates from more than one session. The reduction afforded by use of the means of two estimates versus one was notably greater than the reduction afforded by use of the means of three versus two.

The accuracy of estimates based on only 30 trials is more difficult to evaluate. To assess the *maximum* accuracy and precision at which the algorithm tracked the threshold, we simulated 30 trials 300 times for each value of threshold between 0.1 and 2.7 cm in 1-mm increments. A maximum slope of the psychometric function corresponding to Harvey's $\beta=4$ was specified for the simulated subject and for the tracking algorithm. The difference between the mean of the distribution of 300 values provided by the algorithm and the simulated-subject's actual threshold defined the constant error of the method; the estimation reliability was characterized by the standard deviation of the distribution. The constant error always fell within the interval bounded by ± 0.1 cm and the standard deviation within the range 0.14 cm to 0.28 cm. These findings imply that reliable and accurate estimates can be obtained with as few as 30 stimulus trials. Subsequent simulations, however, revealed that both the constant error and standard deviation may be twice these values if the length of skin traversed on the first trial is very short (0.12 cm) or very long (2.7 cm), or if estimates are obtained from only 15 trials. Thresholds also tend to be systematically underestimated by as much as 0.26 cm if the constant ε is omitted from the algorithm during data collection.

The data from the four subjects tested more than once per day were indistinguishable from those in Fig. 6. The question of interest was whether subjects provide lower or higher thresholds on the second of two back-to-back sessions. On the average, the thresholds were 0.10 cm greater on the second session, but the means for individual subjects varied from 0.22 cm less to 0.34 cm greater (with no difference for thresholds at different velocities; $p>0.83$).

Discussion

Use of the dense-array tactile stimulator to selectively study spatiotemporal processing depends on the applicability of a number of assumptions regarding the responses evoked in individual and populations of mechanoreceptive afferents. Although we have not investigated the neural response, the assumptions appear reasonable based on the frictionless nature of the stimuli and the extensive observations reported by Gardner et al. (1989; Gardner and Palmer, 1989) using the OPTACON to study fast adapting type I units (FAIs) supplying the monkey hand. Most fast adapting mechanoreceptors supplying the human facial skin are type I mechanoreceptors (Johansson et al., 1988; Edin et al., 1995).

First, unlike natural stimuli, different directions of simulated motion do not evoke either complex or consistent asymmetric *patterns* of discharge in most mechanoreceptors. Indeed, FAI mechanoreceptors in the monkey median and ulnar nerves were shown to exhibit uniform sensitivity throughout their receptive fields to OPTACON stimuli. One nerve impulse per pulse of the probes was evoked in most afferents for frequencies from 25 to 100 Hz, independently of the location of the probes in relation to the RF center. Our stimuli differ most notably from those employed by Gardner et al. in that different velocities are attained at the same, but a substantially higher, temporal frequency.

Second, unlike natural stimuli, different directions of simulated movement do not evoke notably different intensities of discharge in most mechanoreceptors. Only one-quarter of the FAI mechanoreceptors studied by Gardner et al. exhibited two spikes to probe indentation and retraction at the RF center. Moreover, this was typically observed in only one direction of motion. Since the temporal frequency of our stimuli is two times greater than the highest

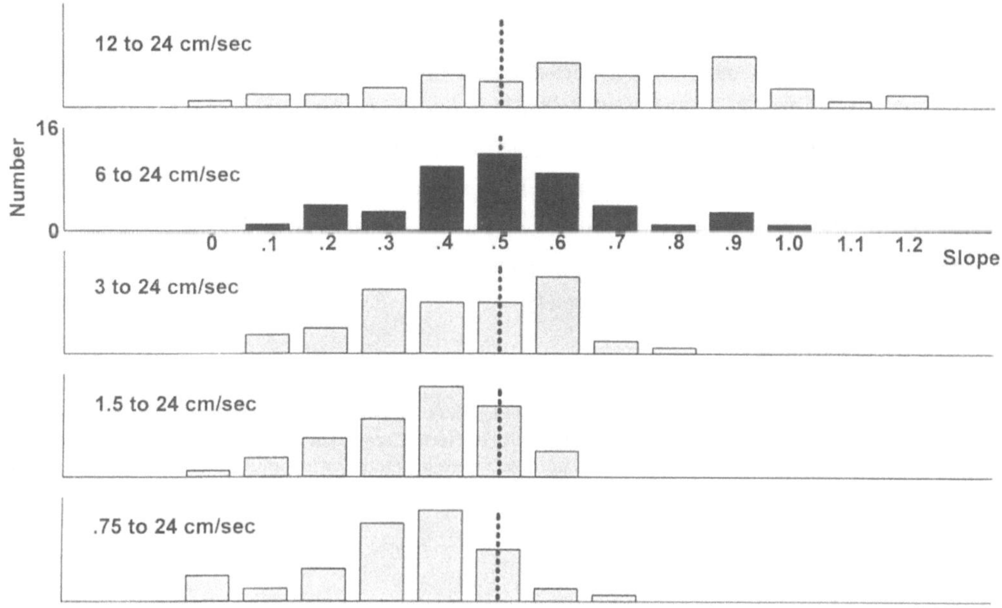

Figure 7. Distributions of slope of the (log-log) relationship between the critical length and velocity. Each histogram of 48 values is based on data obtained from the same 12 subjects, each tested four times. The panels are ordered according to the range of velocities over which the slopes were calculated. For example, the top panel shows the distribution of slopes calculated from data at only 12 and 24 cm/sec.

employed by Gardner, less than one nerve impulse per cycle of the probes is predicted (Johansson et al., 1982; Lundström, 1986). Although we believe that the same mean firing rate is evoked by stimuli moving in all directions at all velocities, experimentation is required to resolve this important issue.

Third, the spatial distribution of receptors activated by opposing directions of simulated motion is expected to be about the same. Gardner et al. demonstrated that RF boundaries shift one row (ca 0.12 cm) in the direction of motion, suggesting the presence of weak lateral facilitation between adjacent receptor terminals and/or postspike facilitation. In contrast, SAI and SAII mechanoreceptors can be stimulated by natural stimuli 7--8 cm from their receptive fields as a result of skin stretch (Edin and Abbs, 1991). As such, natural moving stimuli, unlike those provided by the OPTACON or dense array, produce spatially complex and extensive patterns of mechanoreceptor activation which may provide information as to the direction of motion.

That we are studying central neural processes that rely on spatiotemporally distributed inputs is perhaps demonstrated most convincingly by the data. Examination of the plots shown in Fig. 6, as well as those from the four subjects testing during back-to-back sessions, revealed that the (log-log) relationship between the threshold length and stimulus velocity paralleled the reference line with slope=0.5 for all subjects, at least over the range 6--24 cm/sec (see Fig. 7). This suggests that for velocities greater than about 6 cm/sec, direction is discriminated when the product of the length of skin traversed ("space") and the duration of stimulation ("time") attains or exceeds a criterion -- reflecting a perfect space-time tradeoff or reciprocity at threshold (Henderson, 1973). We hypothesize that if this criterion is not attained, information about direction (or spatiotemporal order) is unavailable to central neural processes. This can be tested with biologically based, neural network models of somatosensory cortical function (Whitsel et al., 1991). Specifically, the psychophysical data predict that the same degree of directionality will be demonstrated by these models in response to different sets of simulated inputs whose spatial and temporal attributes covary inversely.

Acknowledgments
This research was supported by the National Institutes of Health Grant No. DE07509 and No. DE00165, the Frank Porter Graham Child Development Center at the University of North Carolina at Chapel Hill, and the University of North Carolina Dental Alumni Association. Stimulator R&D were generously supported by the School of Dentistry. The authors would like to thank Ms. Lisa Buford for expert assistance in preparing the illustrations.

References

Bliss, J.C., Katcher, M.H., Rogers C.H. and Shepard, R.P. (1970) Optical-to-tactile image conversion for the blind. *IEEE Transactions on Man-Machine Systems* MMS-11: 58-65.

Burr, D.C. and Ross, J. (1982) Contrast sensitivity at high velocities. *Vision Research* 22: 479-484.

Chen, C.C., Essick, G.K., Kelly, D.G., Young, M.G., Nestor J.M. and Masse, B. (1995) Gender-, side- and site-dependent variations in perioral spatial resolution. *Archives of Oral Biology* 40: 539-548.

Cholewiak, R.W. (1979) Spatial factors in the perceived intensity of vibrotactile patterns. *Sensory Processes* 3: 141-156.

Cholewiak, R.W. and Collins, A.A. (1990) The effects of a plastic-film covering on vibrotactile pattern perception with the Optacon. *Behavior Research Methods, Instruments, & Computers* 22: 21-26.

Costanzo, R.M. and Gardner, E.P. (1980) A quantitative analysis of responses of direction-sensitive neurons in somatosensory cortex of awake monkeys. *Journal of Neurophysiology* 43: 1319-1341.

Dreyer, D.A., Hollins M. and Whitsel, B.L. (1978) Factors influencing cutaneous directional sensitivity. *Sensory Processes* 2: 71-79.

Edin, B.B. and Abbs, J.H. (1991) Finger movement responses of cutaneous mechanoreceptors in the dorsal skin of the human hand. *Journal of Neurophysiology* 65: 657-670.

Edin, B.B., Essick, G.K., Trulsson M. and Olsson, K.Å. (1995) Receptor encoding of moving tactile stimuli in humans. I. Temporal pattern of discharge of individual low-threshold mechanoreceptors. *Journal of Neuroscience* 15(1): 830-847.

Essick, G.K., Afferica, T., Aldershof, B., Nestor, J., Kelly D. and Whitsel, B.L. (1988) Human perioral directional sensitivity. *Experimental Neurology* 100: 506-523.

Essick, G.K., Bredehoeft, K.R., McLaughlin D.F. and Szaniszlo, J.A. (1991) Directional sensitivity along the upper limb in humans. *Somatosensory and Motor Research* 8: 13-22.

Essick, G.K., Dolan, P.J., Turvey, T.A., Kelly D.G. and Whitsel, B.L. (1990) Effects of trauma to the mandibular nerve on human perioral directional sensitivity. *Archives of Oral Biology* 35: 785-794.

Essick, G.K. and Edin, B.B. (1995) Receptor encoding of moving tactile stimuli in humans. II. The mean response of individual low-threshold mechanoreceptors to motion across the receptive field. *Journal of Neuroscience* 15(1): 848-864.

Essick, G.K. and Whitsel, B.L. (1985a) Assessment of the capacity of human subjects and S-I neurons to distinguish opposing directions of motion across the skin. *Brain Research Reviews* 10: 187-212.

Essick, G.K. and Whitsel, B.L. (1985b) Factors influencing cutaneous directional sensitivity. A correlative psychophysical and neurophysiological investigation. *Brain Research Reviews* 10: 213-230.

Gardner, E.P., Hämäläinen, H.A., Palmer C.I. and Warren, S. (1989) Touching the outside world: Representation of motion and direction within primary somatosensory cortex. *In:* J. Lund (ed.): *Sensory Processing In the Mammalian Brain: Neural Substrates and Experimental Strategies*, Oxford University press, New York, pp 49-66.

Gardner, E.P. and Palmer, C.I. (1989) Simulation of motion on the skin. I. Receptive fields and temporal frequency coding by cutaneous mechanoreceptors of OPTACON pulses delivered to the hand. *Journal of Neurophysiology* 62: 1410-1436.

Gould, W.R., Vierck C.J. and Luck, M.M. (1979) Cues supporting recognition of the orientation or direction of movement of tactile stimuli. *In:* D.R. Kenshalo (ed.): *Sensory Function of the Skin of Humans,* Plenum Press, New York, pp 63-73.

Greenspan, J.D. (1992) Influence of velocity and direction of surface-parallel cutaneous stimuli on responses of mechanoreceptors in feline hairy skin. *Journal of Neurophysiology* 68: 876-889.

Harvey, L.O. Jr. (1986) Efficient estimation of sensory thresholds. *Behavior Research Methods, Instruments, & Computers* 18(6): 623-632.

Henderson, D.C. (1973) Visual discrimination of motion: Stimulus relationships at threshold and the question of luminance-time reciprocity. *Perception & Psychophysics* 1: 121-130.

Johansson, R.S., Landström U. and Lundström, R. (1982) Responses of mechanoreceptive afferent units in the glabrous skin of the human hand to sinusoidal skin displacements. *Brain Research* 244: 17-25.

Johansson, R.S., Trulsson, M., Olsson K.Å. and Westberg, K.G. (1988) Mechanoreceptor activity from the human face and oral mucosa. *Experimental Brain Research* 72: 204-208.

Lederman, S.J. and Taylor, M.M. (1972) Fingertip force, surface geometry, and the perception of roughness by active touch. *Perception & Psychophysics* 12: 401-408.

Levitt, H. (1970) Transformed up-down methods in psychoacoustics. *Journal of the Acoustical Society of America* 49: 467-477.

Lundström, R.J.I. (1986) Responses of mechanoreceptive afferent units in the glabrous skin on the human hand to vibration. *Scandinavian Journal of Work, Environment and Health* 12: 413-416.

Norrsell, U. and Olausson, H. (1992) Human, tactile, directional sensibility and its peripheral origins. *Acta Physiologica Scandinavica* 144: 155-161.

Olausson, H. and Norrsell, U. (1993) Observations on human tactile directional sensibility. *Journal of Physiology (London)* 464: 545-559.

Sherrick, C.E. (1975) The art of tactile communication. *American Psychologist* 4: 353-360.

Verrillo, R.T. and Gescheider, G.A. (1975) Enhancement and summation in the perception of two successive vibrotactile stimuli. *Perception & Psychophysics* 18: 128-136.

Whitsel, B.L., Dreyer, D.A., Hollins M. and Young, M.G. (1979) The coding of direction of tactile stimulus movement: correlative psychophysical and electrophysiological data. *In:* D.R. Kenshalo (ed.) *Sensory Functions of the Skin of Humans*, Plenum Press, New York, pp 79-107.

Whitsel, B.L., Favorov, O.V., Kelly, D.G. and Tommerdahl, M. (1991) Mechanisms of dynamic peri- and intra-columnar interactions in somatosensory cortex: Stimulus-specific contrast enhancement by NMDA receptor activation. *In:* O. Franzén and J. Westman (eds.): *Information Processing in the Somatosensory System*, Macmillan Press, London, pp 353-369.

Whitsel, B.L, Roppolo J.R. and Werner, G. (1972) Cortical information processing of stimulus motion on primate skin. *Journal of Neurophysiology* 35: 691-717.

Appendix: Protocol to match perceived intensities of different patterns generated on the dense array

Vibrotactile patterns that vary in their spatial (e.g., the number of activated probes; Cholewiak, 1979) and temporal (e.g., the SOA; Verrillo and Gescheider, 1975) characteristics vary in their subjective intensity. In order to minimize variations in subjective intensity, we established a protocol to adjust all test stimuli in a session to the same perceived intensity. Briefly summarized, the experimenter specifies a standard stimulus -- such as a briefly activated, small

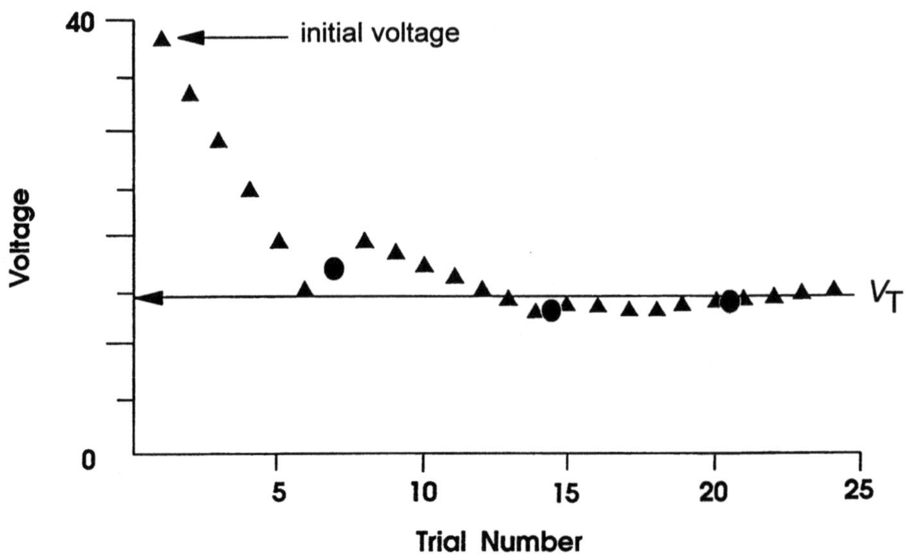

Figure 8. Graph illustrating the loudness-matching procedure. The test stimulus voltage is plotted as a function of trial number. Note that the voltage is systematically decreased (increased) by the algorithm until the test stimulus feels weaker (stronger) than the standard stimulus. The voltage of the test stimulus at which it and the standard stimulus pattern are perceived equally intense, V_T, is calculated as the average of the mid-points (denoted by solid circles) of the 2nd, 4th and 6th 'runs' of trials as suggested by Levitt (1970).

rectangle of probes in the center of the transducer display. The method of limits is used to determine the threshold voltage V_t for detecting the standard stimulus. The loudness of the standard stimulus is then raised an experimenter-specified number of decibels sensation level, b, by increasing V_t to

$$V_s = 10^{[b/20 + \log(V_t)]}$$

(see Sherrick, 1975). A temporal two-interval forced-choice tracking procedure is employed to match the loudness of each test stimulus to that of the standard stimulus. The order of delivery of the test and standard stimuli is randomized. In addition, trials associated with different test stimuli are interleaved. At the completion of each trial the subject reports the stimulus interval that was perceived most intense, and the voltage of the test stimulus is modified accordingly (see Fig. 8).

Somesthesis and the Neurobiology of the Somatosensory Cortex
O. Franzén, R. Johansson and L. Terenius (eds.)

Tactile directional sensibility; theoretical and functional aspects

Ulf Norrsell and Håkan Olausson

Göteborg University, Department of Physiology, Medicinaregatan 11, S-41390 Göteborg, Sweden

Summary. The ability to discriminate the direction of movement across the skin utilizes several information categories. Spatial details, i.e., information regarding receptive field position and size of successively engaged primary afferents can be appraised relative to temporal details regarding the order of engagement. This procedure can be improved by utilization of information from other afferents signalling relative changes of lateral skin tension which are caused by friction and will appear around the moving object in the skin. The integrative nature of the directional sensibility can perhaps explain its particular susceptibility to deterioration compared to other tactile functional components.

Normally, it is easy for people to tell the direction of a movement across the hairy skin of the body. The relevant direction is recognized after motion over a short distance, whether the stimulus is an examiner's fingertip, or a hand-held dowel, or some sort of electro-mechanical appliance. However, functions may deteriorate, and one reason for testing tactile directional sensibility is that it is fairly selectively vulnerable to damage of the somatosensory pathway all the way between receptors and cerebral cortex (Foerster, 1936; Sperry et al., 1969; Norrsell, 1973; Wall and Noordenbos, 1977; Bender et al., 1982; Hankey and Edis, 1989; Essick, 1992). The midline 'split' of the tactile directional sensibility which is produced by cerebral commissurotomy (Sperry et al., 1969; Norrsell, 1973) illustrates the phenomenon.

Cerebral commissurotomies are made for treatment of certain types of pharmacologically unmanageable epilepsy (Bogen et al., 1965). In the present cases they consisted of surgical transections of the corpus callosum and the anterior and hippocampal commissures. Apart from anti-epileptic effects this surgery may produce so called 'split-brain' symptoms, i.e., unilateral agnosia, aphasia or dyspraxia depending on the method of examination. A major reason for the symptoms is that many types of behaviour depend on a nervous afferent input, which is available from only one half of the peripheral field to each hemisphere, and that the field halves fail to unite in the absence of the cerebral commissures. Consequently, search for 'split-brain' symptoms provides one way to document the mode of administration of sensory messages by the nervous system. In the present cases it had been observed that the surgery had caused a split down the midline of the normal ability to tell the direction of motion of a dowel, which was dragged across the skin. This finding contrasted to the commissurotomized patients' ability to recognize other simple tactile stimuli fairly accurately across the midline (cf. Norrsell, 1973).

The two diagrams in Fig. 1 show such a patient's ability to identify direction of motion on either cheek. The results from stimulation of the right cheek have been linked with solid lines, and those from the left cheek with interrupted lines. The identification was performed in two different ways. The results which are shown with triangles in the diagrams were obtained when the patient performed the identification non-verbally by mimicking the movement with an ipsilateral finger. For that type of response there were afferent and efferent pathways available for each hemisphere. Alternatively, the identification was made verbally, and these results are shown with circles in the diagrams. For that type of response, there was no efferent pathway from the right hemisphere due to the lateralization of speech. The testing procedure was conventional for the type of problem. Non-verbal test responses were used for establishing a normal, crossed, cerebral projection of the relevant tactile information. Usage of verbal test responses exploited the lateralization of expressive language in order to measure the left cerebral hemisphere's competence when limited to input of afferent information from the same side of the body.

The distances from start to finish of the different moving stimuli are shown on the abscissae of the diagrams in Fig. 1. The ordinates show per cent correct responses. The patient had a choice of three directions; upwards, downwards and sideways, and the two diagrams show the results after two tests involving different types of mobile stimulation. For the results shown in diagram A the moving stimulus had made continuous contact with the skin. The triangles in the upper left corner of diagram A show that there was little difference between the cheeks when non-verbal responses had been delivered. For verbal responses, however, there was a marked difference between the two cheeks. The solid line at the top of the diagram indicates that the patient had little trouble to tell the direction of movements on the right cheek for distances as short as 5 mm. The interrupted line in the lower part of the diagram shows that she had notable problems with the left cheek throughout, but that her

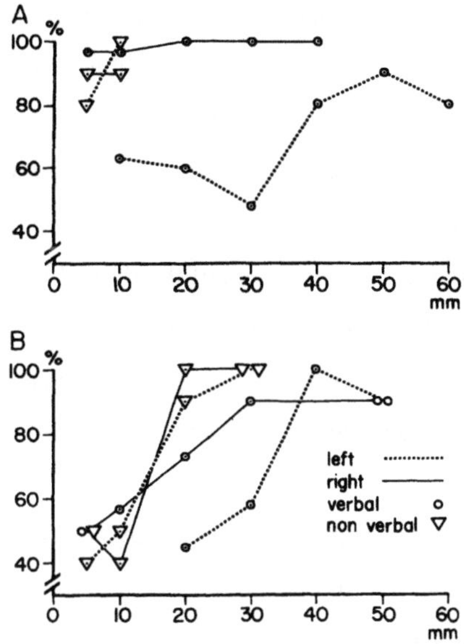

Figure 1. Tactile directional sensibility on the right (solid lines) and left (interrupted lines) cheeks of a 35 years old woman 5 years after transection of the corpus callosum and the anterior and hippocampal commissures for treatment of intractable epileptic seizures. A. The subject's ability, in per cent correct responses, to indicate the direction of a stimulus moving continuously over the distance marked on the abscissa. B. The ability to indicate the directional relation between two discrete touch stimuli made sequentially at points separated by the distance marked on the abscissa. Each point represents 10 to 40 trials. The circles indicate tests involving a verbal readout, and the triangles tests involving a non-verbal readout. The figure was published previously (Norrsell, 1973).

competence improved for distances exceeding 40 mm. The observation indicated that normal directional sensibility required the somatosensory input from the contralateral body half.

The results shown in diagram B were obtained with a moving stimulus that consisted instead of two sequential, touch stimulations. Comparison of the two diagrams motivates two observations. First, overall greater distances between start and finish were necessary for the patient to be able to ascertain the correct, directional associations with the stimulation procedure of diagram B. Presumably, the continuous motion of the stimuli used for diagram A had imparted more or perhaps special information. Second, this change of performance depending on the stimulation procedure did not apply to the patient's accuracy when reporting verbally the spatial relations of stimuli on the left cheek. Comparison between diagrams A and B shows that the patient's results for the left cheek, with a verbal readout, was similar for either type of stimulation (cf. interrupted lines connecting circles). Therefore, the verbal reports of moving stimuli on the left cheek appeared to depend on the same type of afferent information for both situations; afferent information that did not include any possibly special cues evoked by continuous movement across the skin. The patient actually failed to observe any difference between the two procedures when applied to the left cheek (cf. Norrsell, 1973).

Special characteristics of the directional sensibility have been noted also in connexion with lesions of the spinal cord's dorsal funiculi (Foerster, 1936; Wall and Noordenbos, 1977). Wall and

Noordenbos (1977) described three patients with lesions of varying sizes of the dorsal spinal cord. The patient with the smallest lesion had been operated for a "small hourglass hemangioblastoma in the L1 segment," one and a half years before the examination, and "had a restricted lesion of one dorsal column." It could not be excluded, however, that the lesion "encroached to a certain extent upon the medial part of the dorsolateral area *(of the ipsilateral lateral funiculus)*, known to contain ascending pathways." The sensory changes were "limited to the anterior aspect of the thigh." Inside this area the ability to localize stimuli, for two different methods of examination, were completely normal, whereas: "Movement on the skin could be detected but the speed and direction of movement were poorly identified," and "figure identification *(i.e., numbers drawn on the skin)* was all but impossible."

A test for the ability to recognize the single direction of a straight movement may perhaps be considered to be a reduced version of the well-known tactile, neurological tests for graphesthesia, or figure identification (Bender et al., 1982). It carries the advantage, however, of using a more easily defined stimulus, which in addition is less cognitively demanding for the patient.

The cited examples indicated categorical differences between the behavioural act of recognizing the directions of motions of continuously moving stimuli, and that involving the when and where of static touch stimulations. Against that background we decided, some time ago, to investigate the psychophysics of this simple, neurological test which consists of asking a patient to describe the direction of motion of a finger or instrument which is drawn across the skin. It had already been suggested, in one instance over a hundred years ago, that skin receptors which registered frictionally induced changes of lateral tension were involved in directional sensibility (Aubert and Kammler, 1858; Hall and Donaldson, 1885; Gould et al., 1979). Preliminary experiments had confirmed the idea, which anyway ought to be easy to accept, since normal people can easily tell in which direction their skin is being pulled (Norrsell and Olausson, 1992). In addition, slowly adapting receptors which can reproducibly measure small changes of lateral skin tension have been demonstrated in man (Edin and Abbs, 1991; Edin, 1992).

As a next step, in order to evaluate the importance of the friction dependent, cutaneous stretch receptor system, we decided to determine if the directional sensibility was influenced by the basic skin tension of the stimulated area (Olausson and Norrsell, 1993). Subjects were examined with regard to their ability to discriminate whether a metal rod which was resting on the radial surface of their left arm was moving upwards or downwards. The metal rod could slide easily between plastic bushes mounted in a holder, and therefore was considered to deliver a constant load on the skin. The holder was itself attached to a modified X–Y recorder, which generated different movements on the command of a computer. The load that the rod exerted on the skin could be altered by changing weights which were attached to the rod inside the holder. The forearm's basic skin tension was altered by asking the subjects to straighten or bend the elbow joint. The examinations were performed under two conditions, either with the forearm's skin relatively loose due to a 90° angle at the elbow, or with the forearm's skin relatively stretched due to a 180° angle at the elbow.

The experiment was made with seven persons of both genders with ages between 21 and 29 years. The median age was 23 years. The results are summarized in Fig. 2. The two panels of the figure show the average results of all subjects for the two different testing conditions. The results to the left, in panel A, were obtained for the bent elbow condition, when the forearm's skin was relatively more supple and tractable. The results to the right, in panel B, were obtained for the

A

load (g) \ distance (mm)	2	4	8	16	32
8	89	93	99	99	99
4	82	88	88	91	97
2	66	73	78	87	93
1	54	62	69	76	84
0,5	57	55	64	75	79

B

load (g) \ distance (mm)	2	4	8	16	32
8	88	89	96	95	97
4	68	72	75	72	90
2	56	57	62	73	86
1	51	63	62	67	81
0,5	52	46	59	62	82

Figure 2. Per cent average correct responses about the direction of movement of a tactile stimulus for 7 subjects in a two-choice task. The load of the stimulus (ordinate), and the span of the movement (abscissa) were changed step-wise, and 25 different load/distance combinations were used. The figure in each box shows the result for each combination. The shaded boxes in the lower left hand corners of the diagrams contain values which did not deviate significantly from the 50 per cent chance level ($p>0.05$, Student's *t* test). The shaded boxes in the upper right hand corners contain values which did not deviate significantly from the 100 per cent maximal result ($p>0.05$, Student's *t* test). Diagram A. Values for the bent elbow condition when the skin was relatively supple. Diagram B. Values for the straight elbow condition when the skin was braced. The figure was published previously (Olausson and Norrsell, 1993).

straight elbow condition, with the skin braced. The subjects had been instructed to discriminate up from down, and the testing was made according to the two-alternative forced-choice procedure of Sekuler, Nash and Armstrong (1973). That is, for each stimulation the metal rod moved up and down, or *vice versa*, and the subjects had been instructed to report their best estimate of the direction of the opening movement. Stimuli covering 5 different distances were used: 2 mm, 4 mm, 8 mm, 16 mm and 32 mm. In addition, five different stimulus loads were used: 0.5 g, 1 g, 2 g, 4 g and 8 g. In such a way 25 different load–distance combinations were obtained, and each subject endured 16 trials for each combination.

The chance response level was 50 per cent since the task was a two-choice problem. Notice that the boxes in the lower left corner of panel A of Fig. 2 are shaded in grey. These boxes show the results for the shortest distances, and the lowest loads, as listed along the outside borders of the panel. The figures inside the boxes are close to 50, and the shading indicates that these values did not deviate significantly from chance according to Student's *t* test. The boxes in the opposite corner of the same panel are also shaded. These boxes show the results for the longest distances, and the highest loads. The figures inside the boxes are close to the maximal value of 100 per cent, and in this case the shading indicates that the values did not deviate significantly from the maximum. The chance results, and the maximal results are fairly balanced in panel A, and permit a suggestion that the stimulus variables had covered relevant ranges for the bent elbow condition.

This was no longer true, on the other hand, after the skin had been braced by straightening the elbow. The results for that condition are shown in panel B. The shadings of that diagram have the same implication as before, and show that the threshold had increased for both variables. It can also be noticed that maximal results were only obtained for the highest load under that condition. The distensibility of the skin of the stimulated area was measured separately, and found to decrease for all subjects when their elbows were straight (Olausson and Norrsell, 1993). The total numbers of correct responses also decreased for all subjects under the straight elbow condition. Thus, it seemed reasonable to assume that the directional sensibility depended on the distensibility of the skin, and therefore on afferent messages coming from receptors which register changes of lateral tension of the skin.

However, changes of lateral skin tension is not the only way in which the direction of motion can be signalled. People are able to locate static touch stimuli spatially, and the brain is able to determine direction of motion from spatial information expressed as a function of time. The spatial information

A

load (g)					
8	58	58	71	76	82
4	60	56	74	81	92
2	58	63	66	77	83
1	61	63	67	74	89
0,5	48	52	62	62	74
	2	4	8	16	32

distance (mm)

B

load (g)					
8	50	54	68	75	82
4	52	61	69	75	81
2	49	46	65	84	88
1	53	61	63	84	84
0,5	48	60	61	67	77
	2	4	8	16	32

distance (mm)

Figure 3. Average percentage of correct responses about the direction of movement of a frictionless tactile stimulus for 7 subjects in a two-choice task. The load of the stimulus (ordinate), and the span of the movement (abscissa) were changed step-wise, and 25 different load/distance combinations were used. The number in each box shows the result for each combination. The shaded boxes contain values which did not deviate significantly from the 50 per cent chance level ($p>0.05$, Student's *t* test). Diagram A. Values for the bent elbow condition when the skin was relatively supple. Diagram B. Values for the straight elbow condition when the skin was braced. The figure was published previously (Norrsell and Olausson, 1994).

may be obtained from a stimulus making a purely vertical indentation of the skin. Consequently, the altered mechanical conditions following our bracing of the skin may have diminished information about such vertical forces (Pubols, 1982; Pubols and Pubols, 1983). The utilization of spatial data under static conditions may depend on the quantity of information; for instance Johnson and Phillips (1981) observed that increased skin indentations improved the performances of people who were identifying the orientation of gratings with their fingertips..

For that reason we repeated our experiment with a small, but critical modification (Norrsell and Olausson, 1994). We replaced the metal rod and its holder in the X–Y recorder with a nozzle emitting a thin, perpendicular air-stream of body temperature, which caused approximately the same amount of spatial skin indentation. The skin-load exerted by the air-stream could be changed, and set reproducibly with different input air-pressures for the nozzle. Thus, the experiment illustrated in Fig. 3 was similar to that previously illustrated in Fig. 2 with one exception. The friction between the moving stimulus and the skin had been eliminated. The same protocol and the same number and type of subjects were used. The results, however, were very different. A comparison between panels A and B of Fig. 3 reveals that more or less the same results were now obtained for the two different states of skin compliance. That is, when the subjects' elbows were bent, and the skin more tractable like in A, or when the elbows were straight, and the skin braced like in B. The lack of difference confirmed that frictionally induced information about changes of lateral skin tension may contribute to the normal directional sensibility of the hairy skin. Furthermore, the shaded boxes on the left side of the panels, like in Fig. 2, indicate results that did not deviate significantly from chance. It is notable that in the absence of friction two mm had become too short a distance for discrimination of directions at any load. The lack of any shaded boxes indicating maximal results at the right-hand side of the panels in Fig. 3 is also notable. Thus, the frictionally induced information had not only contributed, but seemed to be required for the optimal tactile directional sensibility of the forearm.

Nevertheless, the data in Fig. 3 also reiterates the assertion that the brain has got capacity to determine the direction of motion by means of spatial information expressed as a function of time. We were interested to determine the spatial resolution of that function, since it may set the lower limit for the directional sensibility when the conditions for frictionally induced information are unfavourable. Therefore, in a second experiment using the same frictionless stimulus, we determined the shortest distance for which six subjects could discriminate up from down. The air-stream load was set at 1 g (Norrsell and Olausson, 1994). We found that on average they were able to perform the discrimination for distances not longer than 8 mm. It cannot be excluded that the true value may be even lower. The obtained value, on the other hand, was sufficiently low to indicate that moving stimuli can reveal a spatial acuity for the skin which is finer than that established for tests involving static stimuli, i.e., traditional tests for tactile function like two-point discrimination, or absolute or relative point localisation. That observation was already made by Loomis and Collins (1978), and is supported by the findings of Essick and colleagues (1992). Therefore, it seems legitimate to suggest that the sensory apparatus of the hairy skin may have evolved for appraising mobile rather than static stimuli.

The ability to report directions of motion on the skin may appear to be a crude function because it is conceptually simple. The underlying information processing, on the other hand, may be far from simple. From what we now know it ought to involve comparisons between simultaneous, differential activations and disengagements of multiple cutaneous stretch receptors (Olausson et al., 1995). In addition, the deductions obtained from these comparisons are attuned to estimates of shifts of spatial data relative to time. Such eclectic demands could explain why the tactile directional sensibility is particularly sensitive to disturbances of the somatosensory neuronal apparatus.

Nevertheless, the directional sensibility does not seem likely to have developed for the convenience of neurologists. Neither can ability to communicate verbally the direction of a movement across the skin have high ethological value. It appears to be very important, on the other hand, to be able to determine with high precision whether the movement of an object over the skin conforms with movements which can be predicted from personal motor activity. This is true, for example, for the precision grip of fingers (Johansson et al., 1992 a, 1992 b, 1992 c). With regard to hairy skin, this function will enable an individual to realize, whether a contact is self-inflicted or not. When people are moving around, they have clothes scraping against the skin, or they may brush against various objects. They are able to ignore such predictable contacts, however, because they know beforehand how they are going to move. The direction and speed of a contact's motion, on the other hand, may diverge from that predicted by autogenous movements. Such would be the case when it signified the collapse of a support, or revealed the intentions of another being, friend or foe.

Acknowledgements
Supported by the Swedish Medical Research Council (project no. 2857).

References

Aubert, H. and Kammler, A. (1858) Untersuchungen über den Druck- und Raumsinn der Haut. *Moleschott's Untersuchungen zur Naturlehre*, 5: 145-179.

Bender, M.B., Stacy, C. and Cohen, J. (1982) Agraphesthesia. A disorder of directional cutaneous kinesthesia or a disorientation in cutaneous space. *Journal of the Neurological Sciences*, 53: 531-555.

Bogen, J.E., Fisher, E.D. and Vogel, P.J. (1965) Cerebral commissurotomy: a second case report. *Journal of the American Medical Association*, 194: 1328-1329.

Edin, B.B. (1992) Quantitative analysis of static strain sensitivity in human mechanoreceptors from hairy skin. *Journal of Neurophysiology*, 67: 1105-1113.

Edin, B.B. and Abbs, J.H. (1991) Finger movement responses of cutaneous mechanoreceptors in the dorsal skin of the human hand. *Journal of Neurophysiology*, 65: 657-670.

Essick, G.K. (1992) Comprehensive clinical evaluation of perioral sensory function. *Oral and Maxillofacial Surgery Clinics of North America*, 4: 503-526.

Essick, G.K., McGuire, M., Joseph, A. and Franzén, O. (1992) Characterization of the percepts evoked by discontinuous motion over the perioral skin. *Somatosensory and Motor Research*, 9: 175-184.

Foerster, O. (1936) Die Bedeutung des Hinterstranges für die taktile Sensibilität. *In*: Bumke, O. and Foerster, O (eds): *Handbuch der Neurologie.* Verlag von Julius Springer, Berlin, pp 359-363.

Gould, W.R., Vierck Jr, C.J. & Luck, M.M. (1979) Cues supporting recognition of the orientation or direction of movement of tactile stimuli. *In*: Kenshalo, D.R (ed): *Sensory function of the skin of humans. Proceedings of the second international symposium on the skin senses, held at Florida State University, Tallahassee, Florida, June 5-7, 1978.* Plenum Press, New York, pp 63-73.

Hall, G.S. and Donaldson, H.H. (1885) Motor sensations on the skin. *Mind*, 10: 557-572.

Hankey, G.J. and Edis, R.H. (1989) The utility of testing tactile perception of direction of scratch as a sensitive clinical sign of posterior column dysfunction in spinal cord disorders. *Journal of Neurology, Neurosurgery and Psychiatry*, 52: 395-398.

Johansson, R.S., Häger, C. and Bäckström, L. (1992) Somatosensory control of precision grip during unpredictable pulling loads: III. Impairments during digital anaesthesia. *Experimental Brain Research*, 89: 204-213.

Johansson, R.S., Häger, C. and Riso, R. (1992) Somatosensory control of precision grip during unpredictable pulling loads: II. Changes in load force rate. *Experimental Brain Research*, 89: 192-203.

Johansson, R.S., Riso, R., Häger, C. and Bäckström, L. (1992) Somatosensory control of precision grip during unpredictable pulling loads: I. Changes in load force amplitude. *Experimental Brain Research*, 89: 181-191.

Johnson, K.O. and Phillips, J.R. (1981) Tactile spatial resolution. I. Two-point discrimination, gap detection, grating resolution, and letter recognition. *Journal of Neurophysiology*, 46: 1177-1191.

Loomis, J.M. and Collins, C.C. (1978) Sensitivity to shifts of a point stimulus: an instance of tactile hyperacuity. *Perception and Psychophysics*, 24: 487-492.

Norrsell, U. (1973) Defects in tactile directional sensitivity after forebrain commissurotomy in man. *Nature*, 245: 41-42.

Norrsell, U. and Olausson, H. (1992) Human, tactile, directional sensibility and its peripheral origins. *Acta Physiologica Scandinavica*, 144: 155-161.

Norrsell, U. and Olausson, H. (1994) Spatial cues serving the tactile directional sensibility of the human forearm. *Journal of Physiology*, 478: 533-540.

Olausson, H., Kakuda, N. and Vallbo, Å. (1995) Afferent messages serving tactile directional sensibility in man. *IBRO IV World Congress of Neuroscience.* Kyoto, Japan.

Olausson, H. and Norrsell, U. (1993) Observations on human tactile directional sensibility. *Journal of Physiology*, 464: 545-559.

Pubols Jr, B.H. (1982) Factors affecting cutaneous mechanoreceptor response. I. Constant-force versus constant-displacement stimulation. *Journal of Neurophysiology*, 47: 515-529.

Pubols Jr, B.H. and Pubols, L.M. (1983) Tactile receptor discharge and mechanical properties of glabrous skin. *Federation Proceedings*, 42: 2528-2535.

Sekuler, R., Nash, D. and Armstrong, R. (1973) Sensitive, objective procedure for evaluating response to light touch. *Neurology*, 23: 1282-1291.

Sperry, R.W., Gazzaniga, M.S. & Bogen, J.E. (1969) Interhemispheric relationships: The neocortical commissures; syndromes of hemisphere disconnection. *In*: Vinken, P.J. and Bruyn, G.W (eds): *Handbook of Clinical Neurology*. North-Holland, Amsterdam, pp 273-290.

Wall, P.D. and Noordenbos, W. (1977) Sensory functions which remain in man after complete transection of dorsal columns. *Brain*, 100: 641-653.

Somesthesis and the Neurobiology of the Somatosensory Cortex
O. Franzén, R. Johansson and L. Terenius (eds.)

Experimental assessment of the temporal hypothesis of velocity scaling

G.K. Essick[1,2,3], O. Franzén[4], T.A. Nguyen[5], K. Jowers[1], J.W. Shores[2], A. James[2] and J. Boivie[6]

[1]*School of Dentistry,* [2]*Dental Research Center and* [3]*Curriculum in Neurobiology, University of North Carolina, Chapel Hill, North Carolina, USA;* [4]*Department of Human Anatomy, Biomedical Center, Uppsala University, Uppsala, Sweden;* [5]*North Carolina School of Science and Mathematics, Durham, North Carolina, USA; and* [6]*Department of Neurology, University Hospital, Linköping, Sweden*

Summary. Previously published psychophysical observations suggest that human subjects use temporal cues, i.e., information about stimulus duration, to scale the velocity of motion across the skin. The results of three new experiments provide additional support for this temporal hypothesis of velocity scaling. In an additional experiment, the robustness of the neural mechanism(s) underlying this perceptual capacity was explicitly tested in three patients who had sustained supratentorial cerebrovascular lesions.

Introduction

Ten years ago at our somatosensory symposium, we described the capacity of human subjects to scale the velocity of tactile stimuli (Franzén et al., 1984). Free magnitude estimation was employed to obtain the response of seven subjects to stimuli moving distal-to-proximal over the dorsal forearm and volar phalange of the index finger. Stimuli were provided by a small brush connected to a servo-motor which allowed precise control of its velocity. The subjects' responses to nine octaves of velocity from 0.5 to 256 cm/sec were obtained to stimuli traversing each of 1, 4, 6 and 8 cm of skin on the forearm and of 0.5, 3 and 6 cm of skin on the finger. We found that subjective velocity increased as a power function of physical velocity with exponent of 0.6. Differences in neither innervation density (as existed between the forearm and finger) nor the length of skin traversed notably influenced the response function. These findings suggested that increases in the density and/or total number of mechanoreceptors activated did not alter subjects' capacity to scale velocity, a finding that stands in marked contrast to other somatosensory capacities. For example, Whitsel and colleagues (Dreyer et al., 1978; Whitsel et al., 1979) showed that subjects' ability to discriminate direction of stimulus motion improves substantially in response to these very same changes in stimulus conditions.

Since the 1984 symposium, we published the results of two additional studies which sought to better understand the apparent insensitivity of velocity percepts to spatial changes in the population of receptors activated by moving tactile stimuli. The hypothesis evaluated in these experiments was that subjective velocity R_V is computed from subjective distance R_{TL} and subjective duration R_D in the same way that physical velocity is calculated from physical distance and time. In the first study, the same seven subjects whose velocity percepts had been studied earlier, rated the perceived distance over which the stimuli moved (Whitsel et al., 1986) and their duration on the skin (Essick et al., 1988b). The stimuli were delivered only over one length of skin (4 cm) to the dorsal forearm. In the second study (Essick et al., 1991) five different subjects scaled (i) the velocity, (ii) the total length of skin traversed, and (iii) the total duration of brushing stimuli which moved in either of two different manners (continuously or discontinuously) over the face. Stimuli were applied in the lateral-to-medial direction at each of 10 velocities from 0.5 to 64 cm/sec to a site overlying the mental foramen. The data from the two groups of subjects led to the same result: R_{TL}/R_D remained constant with change in physical velocity, suggesting that subjects *do not* compute velocity from a ratio of percepts of distance and time. Rather, a direct proportionality between $1/R_D$ and R_V (slope = 1.0) was observed for continuous motion over the forearm (Essick et al., 1988b) and for continuous and discontinuous motion over the face (Essick et al., 1991). These observations imply that subjects use temporal cues to judge the velocity of moving tactile stimuli -- a notion that we refer to as the temporal hypothesis of velocity scaling.

This paper reports four new experiments conducted on 20 subjects (ages ranging from 16.8 to 69 years) that advance our understanding of the capacity to scale stimulus velocity and its underlying neural mechanisms. In the fourth experiment, we tested the hypothesis that CNS mechanisms underlying the percepts of velocity and direction are differentially sensitive to central insults by examining the capacities of three patients who had sustained supratentorial cerebrovascular lesions, to discriminate the direction and to scale the velocity of stimulus motion on the skin.

Experiment 1

Experiment 1 sought to evaluate the magnitude of subjective velocity evoked by moving tactile stimuli delivered to body sites that differ in innervation. The rationale was to extend our previous findings from the forearm and finger (Franzén et al., 1984). Eleven neurologically healthy, young adult subjects (six males ages 21.8 -- 34.7 yrs; five females ages 22.2 -- 29.5 yrs) with no prior experience of psychophysical scaling participated. During a single testing session, each subject

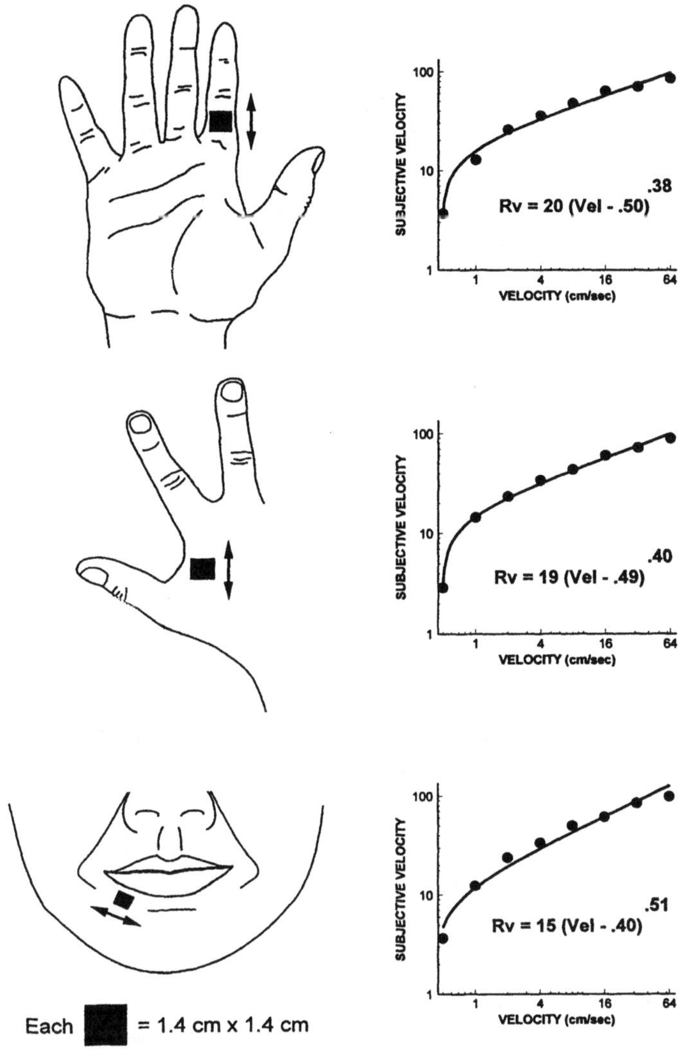

Figure 1. Psychometric relationship (right panel) between subjective velocity and physical velocity for each test site studied in Experiment 1 (left panel). Averaged data from 11 subjects are shown. Note that both axes of the graphs are scaled logarithmically.

scaled the velocity of moving brush stimuli applied to the different skin sites shown in Fig. 1: (i) a glabrous site on the proximal phalanx of the right index finger, (ii) a hairy site on the interdigital webbing between the thumb and right index finger, and (iii) a facial site below the right vermilion border. These sites corresponded to the cutaneous regions for which the peripheral neural response to the very same stimuli in human subjects had been studied (Edin et al., 1995; Essick and Edin, 1995).

For Experiment 1, a computer-controlled servomotor moved a brush in each of two opposing directions at each site at each of eight velocities (7 octaves) between 0.5 and 64 cm/sec. The brush delivered 151--257 mN of normal force to the skin (and 84--161 mN of tangential force; Edin et al., 1995). A Teflon aperture plate, through which a 1.4 cm x 1.4 cm chord of skin was exposed, was lightly taped to the region. The order in which the three sites were studied was randomized, as were stimulus direction and velocity. The subject was instructed to assign a number to each stimulus in accord with absolute magnitude estimation (Gescheider, 1988):

> You have an inherent notion about the size or 'largeness' of numbers. You have an inherent notion about the rate, velocity, or speed at which something moves. You should assign a number to each brush stroke in such a manner that your impression of how large the number is matches your impression of how fast the brush moved across the skin. You may use any positive number deemed appropriate. Do not expect the brush or its velocity to feel the same at the different skin sites. Most importantly, do not let what you feel or how you respond to any stimulus at any test site affect how you respond to any subsequent stimulus at any site.

Two numerical estimates, the first of which was excluded as practice, were obtained from each subject at each site for each velocity and direction of motion. Linear regression was applied to the data from each subject to analyze the relationship between R_V and physical velocity. Since neither the slope nor intercept of the linear (log-log) relationship varied significantly with direction of motion, the data were geometrically averaged. Over the velocity range 2 to 32 cm/sec, power functions fit the data from individual subjects quite well (median $r^2 = 0.90$, 25th percentile = 0.78 for the 33 functions).

For each test site, the data from all subjects were averaged (see Kalikow, 1967; Engen, 1971) and modeled with the function, $R_V = c \times (V - a)^n$ where V is velocity, to accommodate the downward curvilinear trend observed at low velocities (Ekman, 1961; Franzén et al., 1984; Essick et al., 1991). As shown in Fig. 1, excellent fits were obtained to the averaged data (r^2 = 0.99, 0.99 and 0.97 for curves shown from top to bottom). Consistent with our previous work (Franzén et al., 1984), neither the exponent n of the psychophysical relationship nor the numerical range of estimates provided by the subjects varied among the three test sites (p>0.40 and 0.61, respectively). Moreover, no effect of gender was noted (all p-values>0.13). The exponent n varied among subjects from 0.26 to 0.70.

Experiment 2

Experiment 2 sought to evaluate the magnitude of subjective velocity evoked by moving tactile stimuli delivered over different lengths of skin. The rationale for this experiment was as follows. If subjects rely on temporal cues to scale stimulus velocity (Essick et al., 1988b, 1991), then the same physical velocity should feel progressively faster as the length of skin over which the stimulus is delivered is shortened. Since the goal of our earlier work was to evaluate the impact of the length of skin traversed on the exponent *n* of the psychophysical relationship, these experiments were not designed for comparison of the responses assigned to stimuli delivered over different lengths of skin (see Franzén et al., 1984). That is, the instructions permitted the subject to use the same range of numbers even though his(her) impression of the velocities may have varied with the length of skin traversed.

Four neurologically healthy, young adult subjects (two males ages 19.8 and 21.7 yrs; two females ages 16.8 and 17.0 yrs; two subjects were psychophysically inexperienced and two subjects were highly trained) participated in three experimental sessions. During each session, each subject scaled the velocity of moving stimuli delivered to the left mid-cheek area of the face and the left dorsal forearm. The computer-interfaced motor moved a small brush down the face or down the arm at each of six velocities (5 octaves) between 0.75 and 24 cm/sec. The brush used delivered 107--128 mN of normal force to the skin (and 56--88 mN of tangential force; Edin et al., 1995). Subjects' responses were obtained with stimuli traversing 0.6, 1.2, 2.4 and 4.8 cm of skin. The width of skin contacted by the stimuli was always 0.6 cm. Four replications of each velocity were delivered during each of the eight combinations of site x length-of-skin. The subjects' responses to the first two replications were regarded as practice and were not recorded. The eight combinations were tested in random order for each subject during each experimental session. The six velocities also were delivered in random order. Only data from the second and third days of testing were analyzed (the first session constituted practice).

The brushing-stimulus trials were alternated with visual-stimulus trials in accord with the modified method of magnitude matching (Duncan et al., 1988). An electric lamp was mounted inside a small metallic box. The box was positioned 75 cm in front of the subject at eye level such that its circular window subserved 3 degrees of visual arc. Prior to experimentation, one of the experimenters (A.J.) identified six voltages which resulted in roughly equal increments of brightness. The subjects were carefully instructed to rate the velocity of the moving tactile stimuli and brightness of the visual stimuli on a common numerical scale (see Duncan et al., 1988; Lee and Essick, 1993).

The degree to which each subject assigned the same numbers to the same moving brush and visual stimuli upon replication was assessed by the intraclass correlation coefficient, *ICC* (Koch, 1985). The mean *ICC* for the velocity percepts was 0.80. The estimates of *ICC* for the velocity percepts did not vary for stimuli delivered to the different sites ($p>0.09$) or over different lengths of skin ($p>0.38$), nor did they vary from one experimental session to another ($p>0.83$). The mean *ICC* for the brightness percepts was 0.94. The estimates of *ICC* for the brightness percepts did not vary from one experimental session to another ($p>0.58$). Thus, the subjects exhibited a high degree of consistency in assigning numbers to the same tactile or visual stimulus upon replication.

A power function described the relationship between subjective and physical velocity extremely well. Since the estimate of the parameter *a* (see Experiment 1) was not statistically significant for 72% of the datasets, these data were fit to the reduced model, $R_V = c \times V^n$. For the 32 relationships characterized (4 subjects x 2 sites tested x 4 lengths of skin traversed), the median r^2 was 0.97 (range = 0.81 -- 0.99).

Similar to Franzén et al. (1984), the exponent *n* of the psychophysical relationship was found to be independent of the test site ($p>0.45$) and of the length of skin traversed ($p>0.74$). Although geometric mean *n* equaled 0.46, the estimates from individual subjects varied from 0.26 to 0.98. The exponents for the brightness-versus-voltage relationship varied relatively little (from 1.76 to 3.10).

Large systematic differences in the estimates of subjective velocity were observed for stimuli delivered over different length of skin: The highest estimates were obtained from use of the shortest aperture and the lowest estimates from use of the longest aperture. The left two panels of Fig. 2 illustrate the relationships between subjective velocity and physical velocity for

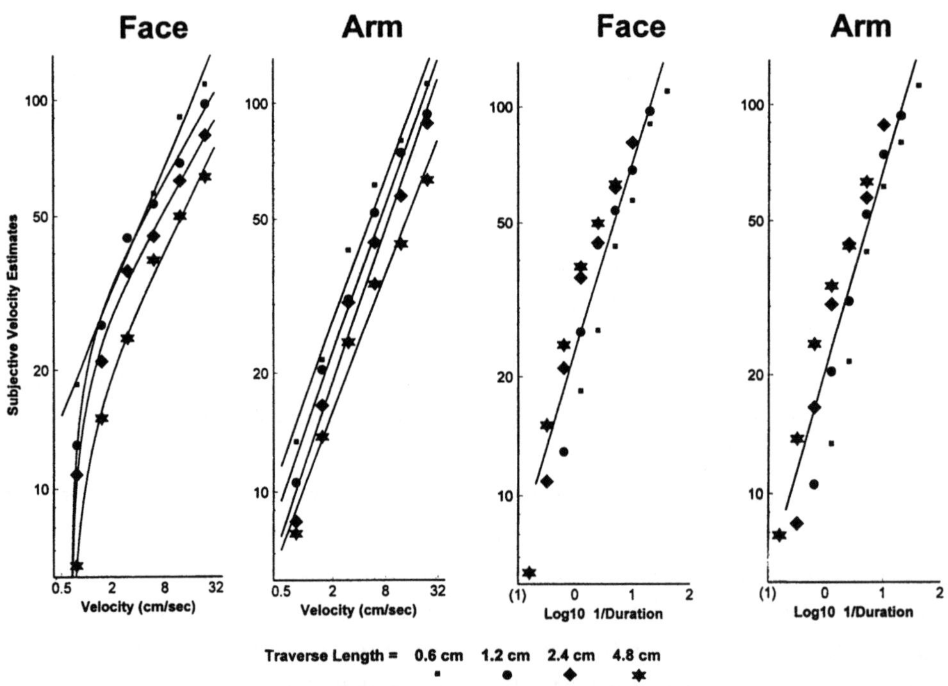

Figure 2. [Left two panels] Psychophysical relationship between subjective velocity and physical velocity for each test site, for each length of skin traversed in Experiment 2. Data averaged over subjects. [Right two panels] Data from the left two panels replotted as a function of the (log) reciprocal of stimulus duration.

the face and arm, respectively. Repeated measures analysis-of-variance confirmed that estimates of velocity varied significantly in magnitude among the four lengths of skin ($p<0.05$) and between sites ($p<0.04$). On both sites, the same physical velocity delivered over the shortest length of skin was assigned a number 80.3% higher, on average, than when it was delivered over the longest length. The same physical velocity over the same length of skin was assigned a number that was 9.5% higher on average for the face than for the arm. The extent to which the magnitude matching contributed to the identification of these effects is unclear (see Appendix).

The right two panels of Fig. 2 illustrate the averaged data from all four subjects plotted as a function of (log) reciprocal of stimulus duration. If the subjects' responses had been based solely on temporal cues, all data points would have converged to a single function (e.g., the best fitting line, indicated). Clearly, this was not the case. Instead, for each site, data from the longest length of skin lie above the best fitting line and data from the shortest length of skin lie below the best fitting line. That is, the estimates from use of the longest length were relatively greater, and those from use of the shortest length relatively less than those predicted by the temporal hypothesis.

Experiment 3

Experiment 3 investigated the temporal hypothesis using a dense-array tactile stimulator (see Essick et al., this volume for detailed description of the device). Use of the dense-array tactile stimulator was anticipated to provide two advantages. First, trials during which stimuli are delivered over different lengths of skin could be readily interleaved (with the brushing stimulator different lengths of skin can only be stimulated by changing the aperture plate in the physical apparatus). Second, the total duration of stimulation could be made the same for all velocities and lengths of skin traversed by widening the edge in the direction of motion.

Five neurologically healthy, young adult subjects (three males ages 19.8 to 42.6 yrs; two females ages 16.8 and 31.7 yrs; two subjects also participated in Experiment 2) participated in nine experimental sessions. During each session, each subject scaled the velocity of moving stimuli delivered to the left mid-cheek area. The velocities, directions, and the traverse lengths of the moving stimuli were identical to those employed in Experiment 2 with the exception that the longest length (4.8 cm) used in the brushing experiments could not be presented using the dense array. Similar to Experiment 2, the width of skin stimulated perpendicular to the direction of motion was always 0.6 cm. Three replications of each velocity were delivered during each session. The response to the first replication was regarded as practice and was not recorded.

In three sessions, the stimulus set closely resembled that used in Experiment 2. As such, in these experiments the stimuli varied in duration as determined by their velocity and the length of skin traversed. In another three sessions, the edge was widened to make all 18 stimuli (6 velocities x 3 lengths of skin) of equal total duration, 3.3 sec. And in a final three sessions, stimuli from both sets (variable-duration and equal-duration) were randomly interleaved. These three types of sessions allowed the following questions to be addressed. First, is the temporal hypothesis applicable to simulated motion across the skin? Second, to what extent are the percepts of velocity altered when moving stimuli at different velocities over different lengths of skin are made equal in total duration? And third, does the experimental context (viz., only variable-duration stimuli, only equal-duration stimuli, interleaved stimuli of both sets) have an effect on the psychophysical relations? Each of the three types of sessions (i.e., variable-duration stimuli, equal-duration stimuli, interleaved stimuli of both sets) was employed for each subject in random order, three times. The data for the first session of each type were regarded as practice and discarded. As in Experiment 1, the subject was instructed to assign a number to each stimulus in accord with absolute magnitude estimation (Gescheider, 1988).

The subjects exhibited a high degree of consistency in assigning the same number to the same stimulus upon replication during each session (mean $ICC = 0.84$). Consistency did not differ among the different types of session ($p>0.88$). Once again, a power function was found to describe the relationship between subjective velocity and physical velocity extremely well. Since the estimate of the parameter a (see Experiment 1) was not statistically significant for 65% of the datasets, these data were fit to the reduced model, $R_V = c \times V^n$. For the 60 relationships characterized (5 subjects x 4 conditions x 3 lengths of skin traversed), the median r^2 was 0.96 (10th percentile = 0.85).

Similar to Experiment 2, the exponent n of the psychophysical relationship was found to be independent of the length of skin traversed ($p>0.66$), and did not vary for the variable-duration and equal-duration stimuli ($p>0.87$). Moreover, the exponent n was the same for variable-duration and for equal-duration stimuli regardless of whether the two stimulus sets were presented separately or in the same testing session ($p>0.22$). Although n averaged 0.34, the estimates varied among subjects from 0.20 to 0.63.

Similar to Experiment 2, large systematic differences in the magnitude of subjective velocity were evoked by stimuli delivered over different lengths of skin: Higher numbers were assigned to the same physical apparent velocity when they were delivered to a shorter length of skin. The averaged data from all five subjects are illustrated in Fig. 3. Since the data did not differ significantly for stimulus sets presented in different or in the same testing session (p-values>0.2), panels A and C were generated from all the data obtained with variable-duration stimuli and equal-duration stimuli, respectively. Repeated measures analysis-of-variance revealed that estimates of velocity varied in magnitude among the three lengths of skin traversed ($p<0.001$). The same physical velocity delivered over the shortest length of skin was assigned a number 29.9% higher, on average, than when delivered over the longest length (viz., 2.4 cm) of skin. This estimate approximates that observed with the natural brushing stimuli used in Experiment 2

(37.7%). The differences among the three lengths of skin appeared slightly less for the equal-duration stimuli than for the variable-duration stimuli ($p<0.11$; compare panel C with panel A).

Subjects reported that they obtained their impression of velocity from the leading edge of the equal-duration stimuli, i.e., as the stimulus pattern first developed on the skin. As such, velocity was most difficult to discern at the shortest length of skin at the fast velocities. The data presented in panel C are consistent with these reports: The deviation of the points from the psychophysical function (see dotted line) attests to subjects' increased difficulty in scaling these equal-duration stimuli when the leading edge traversed the skin quickly. The fact that the stimuli had been made equal in total duration, however, did not eliminate the effect of the length of skin traversed on R_V.

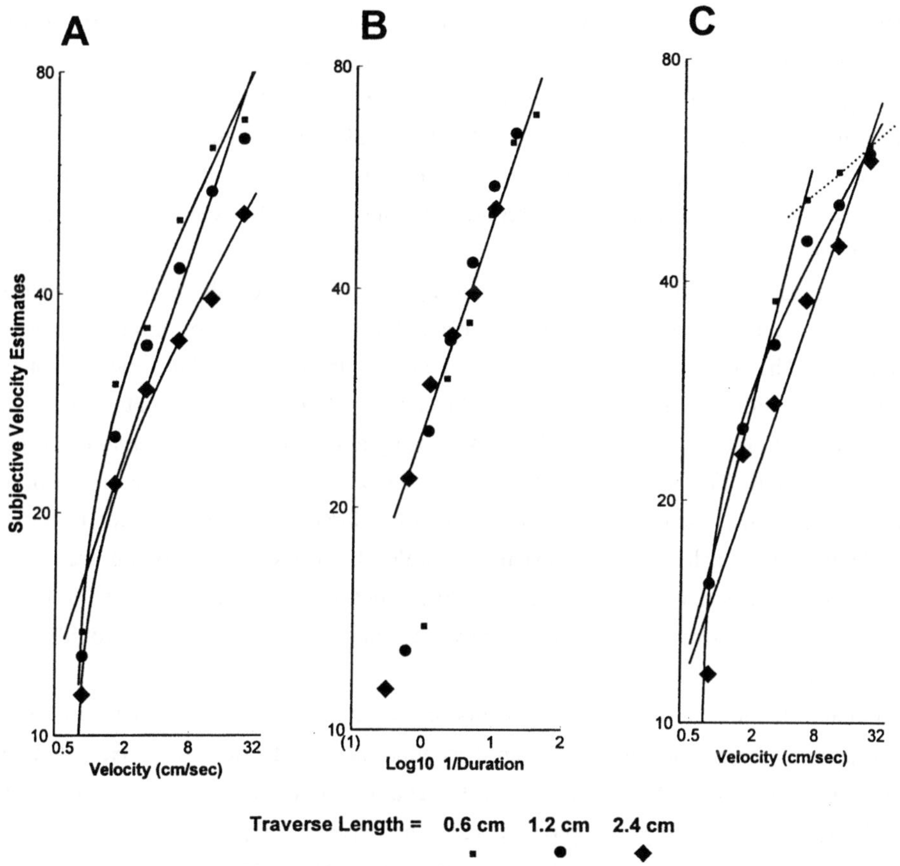

Figure 3. [A, C] Psychophysical relationship between subjective velocity and physical velocity for each length of skin traversed in Experiment 3. Data obtained from variable-duration stimuli (panel A) and equal-duration stimuli (panel C) averaged over subjects. [B] Averaged data from panel A replotted as a function of the (log) reciprocal of stimulus duration.

Panel B of Fig. 3 illustrates the data from panel A replotted as a function of (log) reciprocal of stimulus duration. In contrast to the data obtained with natural brushing stimuli (see Fig. 2, right two panels), the data obtained with simulated motion converged to a single function. Thus, the data are fully consistent with the notion that the subjects based their percepts of velocity solely on temporal cues.

Experiment 4

It is well established that somatosensory function may be impaired by injuries to the central nervous system (e.g., see Bender et al., 1982). Little attention, however, has been devoted to the capacity of patients with such injuries to scale the velocity of tactile stimuli moving across the skin (Wall and Noordenbos, 1977). To this end, we evaluated the somatosensory function of three patients (two males ages 69 and 50 yrs; one female age 39 yrs) at University Hospital, Linköping, Sweden. Each had experienced cerebral infarction involving the parietal cortex and/or subjacent white matter. The cerebral lesion of patient #21 was limited to the right hemisphere; and of patient #51, to the left hemisphere. Although patient #39 had recently experienced a hemorrhagic stroke on the right side, he reported having previously experienced a similar insult on the left side.

Each patient's capacity to distinguish opposing directions of brush stimuli delivered to the long axis of the mid-dorsal hand was evaluated during either two or three sensory-testing appointments. The brush applied 22--68 mN of normal force to the skin (and 19--54 mN of tangential force; Edin et al., 1995). Forty-eight or 96 trials of data were obtained for stimuli moving over each of two or three lengths of skin, using a temporal single-interval paradigm (see Essick et al., 1988a). The stimuli always moved at 10 cm/sec, either toward the fingers or toward the wrist. The hand ipsilateral to the recent cerebral infarction was tested prior to testing the contralateral hand. Each patient's capacity to scale the velocity of moving tactile stimuli presented to the same test sites was evaluated. Seven octaves of velocity from 0.5 to 64 cm/sec were delivered over 2-cm chords of skin. Ample learning/practice trials were provided prior to data collection.

The data obtained from the three patients are summarized in Fig. 4. The left side of the figure illustrates the relationships between directional sensitivity in d' units based on sensory decision theory (Essick, 1991; Essick et al., 1992) and the length of skin traversed. The right side of the figure illustrates the psychophysical relations between subjective velocity and physical velocity. Data from testing skin sites ipsi- and contralateral to the infarction are shown. Consistent with previous work, directional sensitivity increased with the length of skin traversed (Essick et al., 1990), but this capacity differed appreciably on the two sides of the body. For example, patients #21 and #51 could not discriminate direction of motion for stimuli delivered over any length tested contralateral to their infarctions; patient #39 was impaired to a greater extent on the hand contralateral to his old infarct than on the hand contralateral to the recent infarct.

In contrast to direction discrimination, each patient scaled the velocity of the moving tactile stimuli with no difficulty ipsi- and contralateral to the cortical injury. Moreover, the psychophysical functions generated from the data obtained from the two sides were almost identical. The r^2 values ranged from 0.93 to 0.99. The growth in subjective velocity with increase

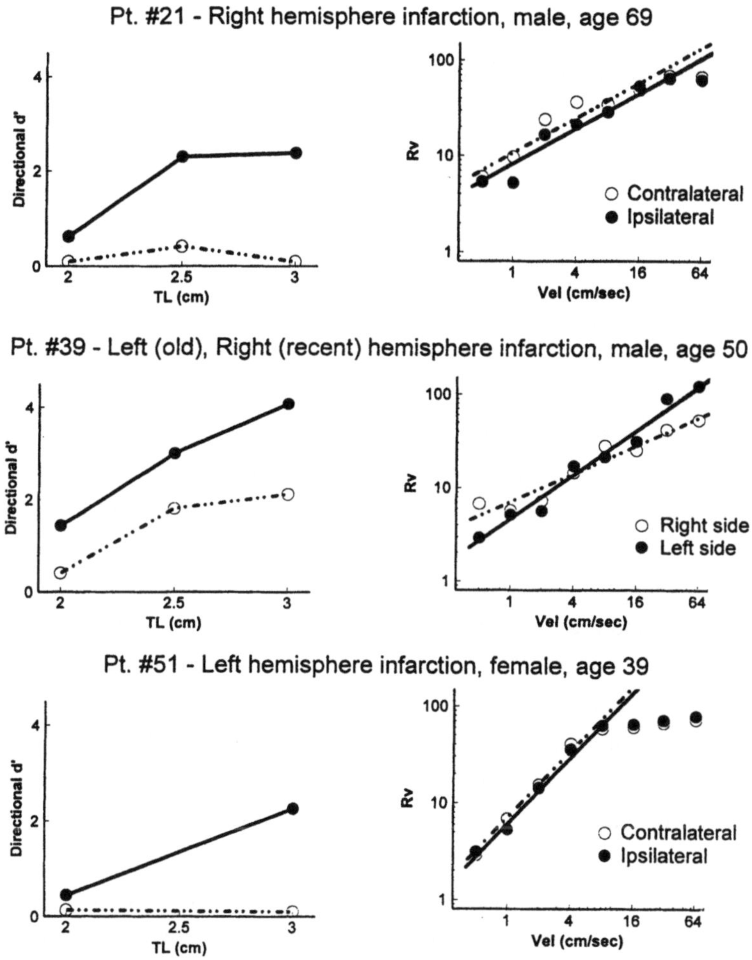

Figure 4. Directional sensitivity (left panels) and subjective velocity (right panels) for three patients who had experienced cerebral infarctions involving parietal cortex and/or subjacent white matter. Each panel presents data collected from the dorsal hand ipsi- and contralateral to the infarction.

in physical velocity was comparable for patients #21 and #39 (mean $n = 0.64$), and similar to that we have observed from neurologically normal subjects. Patient #51 did not understand the range of numbers available on the response device, and her responses saturated for velocities above 8 cm/sec. Identical functions were generated, however, over the range 0.5 to 8 cm/sec ($n = 1.1$). Based on these tendencies, it is evident that cerebral cortical injuries that greatly impair a patient's capacity to discriminate the direction of moving tactile stimuli appear to not affect his/her capacity to scale stimulus velocity.

Discussion

The data obtained in the new experiments described in this paper support the following conclusions:

1. *The exponent n of the psychophysical relationship for stimulus velocity is invariant with respect to the number and density of mechanoreceptors activated.* Consistent with the findings of Franzén et al. (1984), subjects' response to change in velocity, n, was the same from stimuli moving over chords of skin that differed 8-fold (from 0.6 to 4.8 cm; Experiment 2) and 4-fold (from 0.6 to 2.4 cm; Experiment 3) in length. Similarly, n was the same for stimuli moving over skin fields differing in innervation density: The exponent did not vary among sites on the ventral index finger, dorsum of the hand, and chin area of the face (Experiment 1) or between the dorsal forearm and mid-cheek area of the face (Experiment 2). Although precise estimates are not available, the large differences in the two-point discrimination threshold observed at these sites have been widely interpreted to indicate that the sites differ in the density of their peripheral innervation (Weinstein, 1968; see also Johnson and Hsiao, 1992).

These findings imply that neither the number nor density of mechanoreceptors activated has appreciable impact on subjects' capacity to scale stimulus velocity. The findings should not, however, be interpreted to indicate that these spatial factors have no impact on subjects' capacity to discriminate stimuli differing in velocity. Relevant to this possibility is the recent demonstration that discriminative capacity cannot be inferred from the results of studies that employ scaling techniques (suggesting that the outcomes of scaling and discrimination tasks reflect the operation of two basically different sensory-perceptual processes). To illustrate, the size of the just-noticeably-different intensity for both auditory and vibrotactile stimuli has been shown independent of the growth of sensation magnitude for stimuli of equal subjective intensity (see Johnson et al., 1993; Gescheider et al., 1994). Similarly, patients who have experienced injuries to the median nerve often exhibit steeper subjective intensity functions when tested with highly controlled tactile pulses (e.g., Franzén and Lindblom, 1976), although the value of the

Weber fraction for discriminating amplitudes of suprathreshold pulses on the intact and injured (sutured) hands is the same (Franzén, unpublished).

Interestingly, the temporal hypothesis predicts that two stimuli which are moving at different velocities but are perceived to be of equal speed, require the same proportional change in velocity to be discriminated. This is because if those stimuli are perceived to be of the same velocity, they must have the same duration. The same proportional change in velocity is thus required to attain the same change in stimulus duration. As such, the Weber fraction for velocity discrimination for stimuli applied over different lengths of skin is predicted to be the same for stimuli perceived to have the same velocity.

2. *In contrast to the exponent n, the magnitude of subjects' estimates varies in response to variations in the length of skin traversed.* The same physical velocity delivered over a 0.6-cm chord of skin was assigned a number 29.9%, 37.7% and 80.3% higher, on average, than when delivered over a chord of skin 4 times (Experiment 3), 4 times (Experiment 2) and 8 times (Experiment 2) longer, respectively. Similarly, reanalysis of the data presented in Franzén et al. (1984) with emphasis on the magnitude of the estimates revealed a similar effect (viz., percepts of velocity on the forearm were perceived to be 32% slower when delivered over a 8-cm-long chord of skin than when delivered over a 1-cm-long chord). We argue that these effects can be attributed largely to variation in the total duration of stimulation (see points 3 and 4 below). However, the data obtained in Experiment 2 demonstrated that spatial factors exerted an influence on subjects' estimates of velocity: The same physical velocity over the same length of skin (and thus the same duration) was assigned a number 9.5% higher on the face than on the arm. Although the basis for this subtle effect is unclear, it may reflect differences in the magnitude of the mechanoreceptor afferent firing rates evoked by the same moving stimulus when applied to facial versus forearm skin (Essick and Edin, 1995).

3. *The psychophysical impression of the velocity of simulated movement (apparent motion) over different lengths of skin is consistent with the temporal hypothesis.* The data obtained in Experiment 3 fully support the temporal hypothesis of Essick and colleagues (1988b, 1991) because the subjects' estimates of velocity were predictable solely on the basis of the duration of stimulation. The stimuli were provided by the dense-array tactile stimulator and vibrated adjacent skin sites at 204 Hz in a sequential, linear order. As described earlier in the symposium (cf. Essick et al., this volume), the device was driven in a mode to attain different apparent velocities by varying the number of pulses applied to the closely spaced skin sites.

4. *The psychophysical impression of the velocity of natural brushing stimuli delivered over different lengths of skin is largely consistent with the temporal hypothesis.* Experiment 2 showed that subjects' impression of brush velocity is influenced by the duration of stimulation. However, the estimates obtained at the longest traverse length were relatively greater, and those with the

shortest traverse length were relatively less than those predicted by the hypothesis. Thus, a source of information partially compensated for the error inherent in judging velocity solely by stimulus duration. We attribute this compensation to information contained within the discharge of the population of afferents stimulated (Essick and Edin, 1995).

5. *The mechanisms underlying the temporal hypothesis are robust in spite of the appreciable among-subject variation in the exponent n of the relationship between perceived and actual stimulus velocity.* In previous studies we reported that the exponent n of the subjective velocity-versus-physical velocity relationship approximated 0.6 (Franzén et al., 1984) and 0.33 (Essick et al., 1991). At the time of the latter study, we attributed this difference to the different skin sites tested and lengths of skin stimulated. The findings of the four experiments described earlier in this paper make it evident that these two stimulus parameters have no impact on the exponent (see point 1 above), even though appreciable among-subject variation in n was noted in each experiment.

In spite of among-subject variation, the data from our studies consistently support the temporal hypothesis of velocity scaling. A possible interpretation is that the different subjects relied on different sources of information which do not signal duration with the same fidelity. For example, Essick and Edin (1995) have shown that individual mechanoreceptors in the human median, radial and inferior alveolar nerves reliably encode information about the velocity and duration of stimulus motion across their receptive fields. Since the same brushing stimuli and range of velocities used in our psychophysical experiments were used in the neurophysiological studies, it seems evident that the subjects (like the afferents studied by Essick and Edin) could have extracted temporal information directly from the duration of activity in the population of mechanoreceptors stimulated. Information about duration could also have been obtained by a summating process which integrated the times between the onsets of activation of adjacent mechanoreceptors along the path of motion. Alternatively, information about total duration of stimulation might have been obtained from other percepts evoked by the stimulation. For example, the perceived distance over which a stimulus is applied contracts systematically as velocity is increased (Whitsel et al., 1986; Essick et al., 1988b, 1991). Thus, the distance between the perceived positions of stimulus onset and offset might have been used as a cue for stimulus duration. The point we wish to emphasize here is that between-subject differences in kinds of information used to infer stimulus duration may account, in part, for the among-subject variation we observed in the exponent n.

6. *Cerebral cortical injuries that greatly impair human capacity to discriminate the direction of moving tactile stimuli may not affect the capacity to scale their velocity.* The findings of Experiment 4 are consistent with those of Wall and Noordenbos (1977) who reported that patients could detect and grossly judge velocity of movement after lesions to the dorsal columns,

but could not discriminate direction of motion. More generally, the results indicate that injuries to central neural structures involved in the analysis of certain types of spatiotemporal information (e.g., that regarding direction of motion) may not impair sensory capacities based on other information (e.g., that regarding duration) provided by the same moving tactile stimulus.

Given that velocity information over the range we have studied psychophysically is directly available at both the peripheral and central levels of the somatosensory nervous system (Whitsel et al., 1979; Franzén et al., 1984; Essick et al., 1985a,b; Essick and Edin, 1995), one may ask: Why is it that the human perceptual system largely relies on temporal cues (see also Essick et al., 1988b; 1991)? One plausible explanation is that the clear-cut velocity sensitivity of both peripheral and central somatosensory neurons may not be the only neural substrate used to construct a unique velocity percept, but merely a reflection of the operation of mechanisms by which the percepts of direction, texture, and force are mediated. That this might be the case is illustrated by the dependency of the relationship between the discharge rate of individual neurons and stimulus velocity on both the direction of movement across and the forces applied to the receptive field. The possibility that temporal cues provide the best solution to the problem of velocity scaling can be addressed objectively and directly using biologically based, neural network models of somatosensory nervous system function (Whitsel and Franzén, 1989; Whitsel et al., 1991).

Acknowledgments
This research was supported by the National Institutes of Health Grant No. DE07509, the Swedish Medical Research Council Grant No. 9058, the University of North Carolina School of Dentistry Grover Hunter Student Research Fund, and the NIH National Center for Research Resources Grant No. 2SO3RR03267-13. The authors thank Drs. Barry Whitsel and Oleg Favorov for providing useful comments regarding the interpretation of the data presented in this paper, and thank Mr. Mike Young for preparation of figures.

References

Bender, M.B., Stacy C. and Cohen, J. (1982) Agraphesthesia: A disorder of directional cutaneous kinesthesia or a disorientation in cutaneous space. *Journal of the Neurological Sciences* 53: 531-555.

Dreyer, D.A., Hollins M. and Whitsel, B.L. (1978) Factors influencing cutaneous directional sensitivity. *Sensory Processes* 2: 71-79.

Duncan, G.H., Feine, J.S., Bushnell M.C. and Boyer, M. (1988) Use of magnitude matching for measuring group differences in pain perception. *In:* R. Dubner, G.F. Gebhart and M.R. Bond (eds.): *Proceedings of the Vth World Congress on Pain,* Elsevier, Amsterdam, pp 383-390.

Edin, B.B., Essick, G.K., Trulsson M. and Olsson, K.Å. (1995) Receptor encoding of moving tactile stimuli in humans. I. Temporal pattern of discharge of individual low-threshold mechanoreceptors. *Journal of Neuroscience* 15(1): 830-847.

Ekman, G. (1961) A simple method for fitting psychophysical power functions. *Journal of Psychology* 51: 343-350.

Engen, T. (1971) Psychophysics II. Scaling methods. *In:* J.W. Kling and L.A. Riggs (eds.): *Woodworth and Schlosberg's Experimental Psychology,* Holt, Rinehart and Winston, Inc., New York, pp 47-86.

Essick, G.K. (1991) Human capacity to process directional information provided by tactile stimuli which move across the skin: Characterization and potential neural mechanisms. *In:* O. Franzén and J. Westman (eds.): *Information Processing in the Somatosensory System*, Macmillan Press, London, pp 329-339.

Essick, G.K., Afferica, T., Aldershof, B., Nestor, J., Kelly D. and Whitsel, B. (1988a) Human perioral directional sensitivity. *Experimental Neurology* 100: 506-523.

Essick, G.K., Dolan, P.J., Turvey, T.A., Kelly D.G. and Whitsel, B.L. (1990) Effects of trauma to the mandibular nerve on human perioral directional sensitivity. *Archives of Oral Biology* 35: 785-794.

Essick, G.K. and Edin, B.B. (1995) Receptor encoding of moving tactile stimuli in humans. II. The mean response of individual low-threshold mechanoreceptors to motion across the receptive field. *Journal of Neuroscience* 15(1): 848-864.

Essick, G.K., Franzén, O., McMillian A. and Whitsel, B. (1991) Utilization of temporal and spatial cues to judge the velocity and traverse length of a moving tactile stimulus. *In:* O. Franzén and J. Westman (eds.): *Information Processing in the Somatosensory System*, Macmillan Press, London, pp 341-352.

Essick, G.K., Franzén O. and Whitsel, B.L. (1988b) Discrimination and scaling of velocity of stimulus motion across the skin. *Somatosensory and Motor Research* 6(1): 21-40.

Essick, G.K., McGuire M., Joseph, A. and Franzén, O. (1992) Characterization of the percepts evoked by discontinuous motion over the perioral skin. *Somatosensory and Motor Research* 9: 175-184.

Essick, G.K. and Whitsel, B.L. (1985a) Assessment of the capacity of human subjects and S-I neurons to distinguish opposing directions of motion across the skin. *Brain Research Reviews* 10: 187-212.

Essick, G.K. and Whitsel, B.L. (1985b) Factors influencing cutaneous directional sensitivity. A correlative psychophysical and neurophysiological investigation. *Brain Research Reviews* 10: 213-230.

Franzén, O. and Lindblom, U. (1976) Tactile intensity functions in patients with sutured peripheral nerve. *In:* Y. Zotterman (ed.): *Sensory Functions of the Skin of Primates with Special Reference to Man*, Pergamon Press, Oxford, pp 113-128.

Franzén, O., Thompson, F., Whitsel B. and Young, M. (1984) Peripheral coding mechanisms of touch velocity. *In:* C. von Euler, O. Franzén, U. Lindblom and D. Ottoson (eds.): *Somatosensory Mechanisms*, Plenum Press, New York, pp 213-226.

Gescheider, G.A. (1988) Psychophysical scaling. *Annual Review of Psychology* 39: 169-200.

Gescheider, G.A., Bolanowski, S.J., Zwislocki, J.J., Hall, K.L. and Mascia, C. (1994) The effects of masking on the growth of vibrotactile sensation magnitude and on the amplitude difference limen: A test of the equal sensation magnitude-equal difference limen hypothesis. *Journal of the Acoustical Society of America* 96(3): 1479-1488.

Johnson, K.O. and Hsiao, S.S. (1992) Tactual form and texture perception. *Annual Review of Neuroscience* 15: 227-250.

Johnson, J.H., Turner, C.W., Zwislocki, J.J. and Margolis, R.H. (1993) Just noticeable differences for intensity and their relation to loudness. *Journal of the Acoustical Society of America* 93(2): 983-991.

Kalikow, D.N. (1967) Psychofit. Unpublished computer program (Fortran) for the analysis of magnitude estimates. Brown University, RI, USA.

Koch, G.G. (1985) Intraclass correlation coefficient. *In:* S.L. Kotz and N.L. Johnson (eds.): *Encyclopedia of Statistical Sciences*, John Wiley & Sons, New York, pp 212-217.

Lee, J. and Essick, G.K. (1993) Spatial and gender-dependent variations in perioral pinprick sensitivity. *Journal of Dental Research* 72(8): 1206-1214.

Wall, P.D. and Noordenbos, W. (1977) Sensory functions which remain in man after complete destruction of dorsal columns. *Brain* 100: 641-653.

Weinstein, S. (1968) Intensive and extensive aspects of tactile sensitivity as a function of body part, sex, and laterality. *In:* D.R. Kenshalo (ed.): *The Skin Senses*, Charles C. Thomas, Springfield, Ill., pp 193-218.

Whitsel, B.L., Dreyer, D.A., Hollins M. and Young, M. (1979) The coding of direction of tactile stimulus movement: correlative psychophysical and electrophysiological data. *In:* D.R. Kenshalo (ed.): *Sensory Functions of the Skin of Humans*, Plenum Press, New York, pp 79-107.

Whitsel, B.L., Favorov, O.V., Kelly, D.G. and Tommerdahl, M. (1991) Mechanisms of dynamic peri- and intra-columnar interactions in somatosensory cortex: Stimulus-specific contrast enhancement by NMDA receptor activation. *In:* O. Franzén and J. Westman (eds.): *Information Processing in the Somatosensory System*, Macmillan Press, London, pp 353-369.

Whitsel, B.L. and Franzén, O. (1989) Dynamics of information processing of the somatosensory cortex. *In:* C. von Euler, I. Lundberg and G. Lennerstrand (eds.): *Brain and Reading*, Macmillan Press, London, pp 129-137.

Whitsel, B.L., Franzén, O., Dreyer, D.A., Hollins, M., Young, M., Essick G.K. and Wong, C. (1986) Dependence of subjective traverse length on velocity of moving tactile stimuli. *Somatosensory Research* 3(3): 185-196.

Appendix: Utility of magnitude-matching in Experiment 2

The modified method of magnitude matching has been used to increase the likelihood of detecting a small between-group difference in subjects' estimates of a perceptual attribute in the presence of high within-group (viz., among-subject) variabilities. In Experiment 2, we attempted to utilize magnitude matching in a novel manner, i.e., to reduce or account for within-subject variability. Since (i) the subjects were instructed to use a common numerical scale for percepts of velocity and brightness and (ii) the very same set of visual stimuli were used with all eight combinations of site x length-of-skin, we anticipated that the velocity magnitude estimates obtained with different combinations could be compared in a meaningful manner.

With the modified magnitude-matching procedure, it is customary to divide each subject's numerical response to the stimuli under investigation by the geometric mean of his(her) responses to all the visual stimuli. This approach was not warranted for the dataset we collected for the following reasons.

First, for three of the four subjects less than 2% of the variance in the estimates of brightness could be attributed to assignment of different numbers during testing of different combinations of site and length. Accordingly, dividing the velocity estimates by the mean brightness estimate would have no effect on the results. For the other subject, 36.2% of the variance in the estimates of brightness could be attributed to assignment of different numbers during testing of different combinations of site x length-of-skin. However, the mean estimates of brightness and velocity were highly correlated (Pearson's $r = 0.93$, $p = .0001$), suggesting that the subject used the same range of numbers, rather than the same numerical scale, for the two types of percepts.

Second, the fourth subject did not assign greater numbers to velocity percepts over shorter lengths of skin. Analysis of her velocity estimates, divided by the mean brightness estimates, failed to reveal an effect that was not observed from analysis of the raw data. Therefore, we cannot determine whether inclusion of the visual stimuli influenced the results regarding the effects of site and length of skin traversed on percepts of velocity.

Somesthesis and the Neurobiology of the Somatosensory Cortex
O. Franzén, R. Johansson and L. Terenius (eds.)

Vibrotactile adaptation of the RA system: A psychophysical analysis

M. Hollins, K. A. Delemos and A. K. Goble

Department of Psychology, University of North Carolina, Chapel Hill, NC 27599, U.S.A.

Summary. Its capacity for precise frequency discrimination, and the fact that its detection threshold is nearly independent of frequency, suggest that the RA system is importantly involved in the perceptual representation of vibratory waveform. This possibility is supported by experiments comparing the threshold-raising abilities of vibrotactile adapting stimuli that differ along dimensions of amplitude, frequency, and complexity. These studies show that information about the peak amplitude of a vibratory stimulus, which is essential for a non-distorted representation of waveform, is preserved by the RA system.

Introduction

One of the most remarkable aspects of vibrotaction is its susceptibility to adaptation. If a vibrating probe is held against the skin for several minutes, the vibratory sensation declines to a fraction of its original intensity, and detection threshold rises substantially (Cohen and Lindley, 1938; Wedell and Cummings, 1938; Hahn, 1966, 1968; Gescheider and Wright, 1968, 1969; Berglund and Berglund, 1970; Verrillo and Gescheider, 1977; Gescheider et al., 1979; Lundström and Johansson, 1986; Goble and Hollins, 1993). The sizes of these effects depend in predictable ways on the amplitudes and frequencies of the adapting and test stimuli. The magnitude and orderliness of vibrotactile adaptation, as well as its extended time course, which permits investigators to study the cumulative effects of temporal patterns of stimulation, make it a powerful experimental tool for studying somatosensory information processing.

It is now clear that the physiological changes underlying vibrotactile adaptation are distributed across levels of the somatosensory system, for first-order mechanoreceptive afferents show orderly changes in responsivity following extended exposure to suprathreshold vibration (Lundström and Johansson, 1986; Leung, 1994; Leung et al., 1994), and brainstem structures show further changes that are not attributable to peripheral factors (O'Mara et al., 1988). Dynamic cortical population events revealed by optical imaging (Tommerdahl and Whitsel, this volume) and other techniques (Lee and Whitsel, 1992; Lee et al., 1992) may also contribute to vibrotactile adaptation. A salient property of these cortical events is that their spatial distribution is elaborated over time, becoming more internally complex, crisply defined, and stimulus-specific with extended exposure to a stimulus (Tommerdahl and Whitsel, this volume).

In the context of these physiological changes, it is interesting to ask how the distinctiveness of *perceptual* responses is affected by vibrotactile adaptation. To answer this question, amplitude difference threshold was measured on the thenar under various conditions of adaptation (Goble and Hollins, 1993), using two-interval forced-choice tracking (Zwislocki et al., 1958).

The resulting variations in difference threshold demonstrated that adaptation, while it tends to make vibratory sensations weaker, can also make them more distinctive. Whether it does so or not depends on the relation between the amplitudes of the adapting and test stimuli, the clearest improvement occurring when the adapting stimulus is comparable in amplitude to the test stimuli. Delemos and Hollins (in press) have shown that this enhancement occurs even when the test and adapting stimuli differ in frequency, provided that they comparably excite the same channel.

Interestingly, this same adapted state improved subjects' ability to make frequency, as well as amplitude, discriminations (Goble and Hollins, 1994); this was particularly true on the fingertip, where frequency perception is extremely acute. Because frequency discrimination has been convincingly shown not only to correlate closely with cortical events (Mountcastle et al., 1969) but also to depend on an intact cortex (LaMotte and Mountcastle, 1979), it appears likely that at least some aspects of vibrotactile adaptation involve cortical processing.

Many of the questions that can be addressed with adaptation paradigms derive their context from landmark studies of vibrotaction by Verrillo, Mountcastle and others, which demonstrated the importance of distinguishing among the roles played by different classes of mechanoreceptive afferents and their central connections. Verrillo (1963, 1968) showed that there were at least two independent vibrotactile channels with different frequency characteristics: One, clearly identifiable as the Pacinian system (Verrillo, 1966), had the greater sensitivity at high frequencies, while a second, termed the non-Pacinian system, was the more sensitive of the two at low frequencies (typically, 10-50 Hz). Talbot et al. (1968) showed that the main non-Pacinian vibrotactile channel was the one now designated either the RA system (e.g. Johnson and Hsiao, 1992) or NP I (Bolanowski et al., 1988). But differences between the RA and Pacinian (PC) channels go far beyond frequency tuning: Verrillo (1963, 1968) found that these channels differ in their susceptibility to temporal and spatial summation; Gescheider et al. (1978) demonstrated that their sensitivities are differently affected by the presence of a surround; and Talbot et al. (1968) reported that RA and PC signals give rise to qualitatively different sensations ("flutter" and "vibration", respectively).

The picture that is beginning to emerge is of channels that differ not only in their formal properties, but in their functional roles as well. For example, a paramount role for the slowly-adapting type I (SA I) channel in constructing sensations of roughness based on spatial information is now well established (Connor et al., 1990; Connor and Johnson, 1992), and it is widely agreed that the RA system contributes to perception of slippage across a surface (Johansson and Westling, 1987; Srinivasan et al., 1990; Johnson and Hsiao, 1992). Johnson and Hsiao (1992) further propose that a major role of the PC system is to produce a faithful temporal representation of stimulus waveforms, and use it to mediate vibratory discrimination.

As regards this last hypothesis, however, there are grounds for believing that it is the RA channel, rather than the Pacinian, which is the premier system mediating the perceptual representation of vibratory amplitude, frequency, and waveform. This conclusion is suggested by a number of arguments, no one of which is compelling by itself, but which are persuasive in combination. Our main aim in this chapter is to develop the case for this alternative view by reviewing several lines of evidence, the last of which are derived from experiments in which adaptation has been used as an analytical tool.

Ability of the RA System to Represent Vibratory Waveform

The first reason for believing that the RA system keeps close track of the repetitive temporal events that constitute vibration, is that frequency discrimination is apparently based on temporal coding at low frequencies, where the RA system plays the major role in vibrotaction, but not at high frequencies, where the Pacinian system predominates. Somatosensory cortex is necessary for good frequency discrimination in primates (LaMotte and Mountcastle, 1979), and entrainment of activity in cortical neurons that receive an RA input has been proved sufficiently precise at low frequencies to mediate this perceptual ability (Mountcastle et al., 1969). Above 100 Hz, however, neither cortical neurons receiving RA input nor [with occasional exceptions in SII (Burton and Sinclair, 1991)] those receiving PC input are well entrained (Mountcastle et al., 1969). The idea that the Pacinian system may, instead, be able to base frequency discrimination on spatial cues (Mountcastle et al., 1969) gains support from the fact that frequency discrimination in the Pacinian range is apparently more acute when vibrations are permitted to spread laterally (Mountcastle et al., 1969, 1990; Franzén and Nordmark, 1975) than when a stationary surround prevents them from doing so (Goff, 1967). In any case, there seems little doubt that the susceptibility to entrainment of cortical neurons receiving RA input is consistent with the hypothesis that they are critically involved in the representation and analysis of the temporal structure of vibratory stimuli.

A second reason for suspecting that the RA system contributes importantly to representing the waveform of vibratory stimuli is that its psychophysical threshold-versus-frequency function is better suited than that of the Pacinian system for such a role. The threshold functions of both systems are U-shaped, but the Pacinian function is steep, dropping along much of its low-frequency limb with a slope of -12 dB/octave (Verrillo, 1966); the RA threshold function, on the other hand, is shallow, varying little more than 1 dB over the range of 20 and 100 Hz (Bolanowski et al., 1988). Indifference to frequency within a critical range is an important qualification for a system that is to represent complex waveforms, since if this property is lacking the waveforms will be distorted by differential attenuation of different frequency components. This constraint applies in principle to any linear system, but is more severe in vibrotaction, with its broadly tuned channels, than in hearing or (in a spatial domain) vision, where different frequencies are processed semi-independently by narrowly tuned mechanisms. The contrasting shapes of the RA and PC threshold functions therefore suggest that, other things being equal, the former system is able to represent complex vibratory waveforms more veridically than the latter.

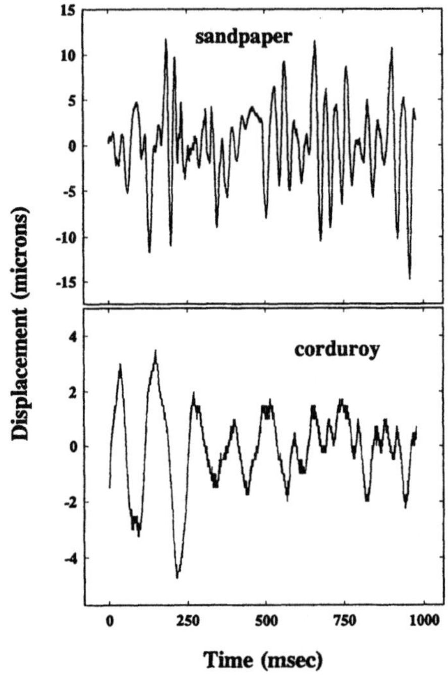

Figure 1. Recordings of vibration set up in the earlobe by drawing sandpaper (upper record) or corduroy (lower record) across its posteromedial surface at a rate of about 2.5 cm/sec. Recordings were obtained from the linear variable differential transformer (LVDT) associated with the vibrotactile contactor, which in this experiment was applied to the anterolateral surface of the lobe, and used as a probe for the pickup, rather than the delivery, of vibration.

It is worth emphasizing that the stimulus waveforms generated in everyday life when a textured object brushes across the skin can be very complex, as recordings made in collaboration with Elizabeth Roy in our lab demonstrate (Figure 1). The anatomical region used in these recording experiments was the flexible lower portion of the earlobe, which was drawn forward and laterally into a horizontal orientation. Stimulus surfaces were attached to Plexiglas blocks, and drawn by hand across the lower surface of the supinated earlobe at a rate of about 2.5 cm/sec. Recordings were made with the vibrotactile stimulator used in the other studies reported here, with the end of the rod held in firm contact (about 50 g) with the upper surface of the earlobe: The contactor now picked up vibrations from, rather than delivering them to, the skin. Amplitude was measured using a signal from a linear variable differential transformer, which was passed through an analog-to-digital converter to a computer that sampled at just over 2 kHz.

The two representative recordings shown in Figure 1 are from the same individual, differing only in the stimuli used to produce them: 220-grit sandpaper (upper panel) and corduroy (lower panel). The records show a clear difference between the vibrations induced by the two stimuli: That from the corduroy has a prominent low-frequency periodicity, which presumably reflects the spatial frequency of the wales, while the vibration produced by the sandpaper is less regular and appears to by dominated by higher frequencies. Fast Fourier transforms (not shown) confirm these impressions: That for the corduroy has a sharp peak at about 7 Hz, while that for the sandpaper shows a broader peak in the 20-30 Hz range. (Only negligible vibration was recorded above 100 Hz, perhaps because any such components were attenuated on their passage through the earlobe.) What the recordings demonstrate is that vibrations caused by lateral movement of a surface contain substantial information about texture. To fully capture this information, however, a mechanoreceptive channel would have to preserve information about vibratory waveform.

Amplitude Determines Adapting Effectiveness of Sinusoidal Vibrations

The third line of evidence that the RA channel is capable of preserving information about temporal waveforms has emerged only recently, and much of the remainder of this chapter will be devoted to describing it. Briefly, it is that the information-processing algorithms of the RA system, analyzed with adaptation paradigms, use as their primary index of magnitude an encoded representation of the *peak amplitude* of the stimulus waveform, whether simple or complex. The importance of their doing so is that amplitude information must be preserved in any linear representation of vibratory waveform; other intensitive codes, such as power, are unsuitable for such representations because they are not linearly related to amplitude, and are indifferent to certain changes in stimulus waveform (those produced by changes in the phase relationships among the components of a complex wave) that can affect amplitude.

This conclusion that the RA system both maintains and uses information about peak amplitude is based on the results of several studies designed to identify what factors determine the amount of adaptation produced by a particular stimulus. Earlier work (Verrillo and Gescheider, 1977; Gescheider et al., 1979) had shown that vibrotactile adaptation is, at least in the main, channel-specific. That is, a test stimulus is typically detected by one channel, and adaptation raises threshold only when the adapting stimulus is strong enough to activate that same channel. For example, Verrillo and Gescheider (1977) used a 10-Hz adapting stimulus to selectively stimulate non-Pacinian receptors; they found that, at low amplitudes, this adapting stimulus raised the threshold of low-frequency, but not high-frequency, test stimuli. Only at very high amplitudes did it raise the threshold for high frequency test stimuli, presumably because only then did it stimulate the Pacinian system, which was the channel detecting the high-frequency tests.

A question left unanswered by these earlier studies was how different frequencies would compare in their ability to adapt a given channel. Given that the system consisting of RA fibers and their central connections plays the major role in detecting low-frequency (10-60 Hz) vibrations (Talbot et al., 1968; Bolanowski et al., 1988), how do different frequencies compare in their ability to adapt this channel? If a test stimulus that is always detected by this channel is used, the question can be answered by measuring the degree to which different adapting stimuli raise the threshold of this test stimulus. Specifically, adapting stimuli of various frequencies can be adjusted in amplitude so that each is just strong enough to raise by (say) 10 dB the threshold of detection of the test stimulus, and the resulting set of amplitude values can then be compared with other known properties of the RA system.

We carried out this experiment (Hollins et al., 1990), by delivering to the index fingerpad a 10-Hz test stimulus -- a frequency at which the RA system is the most sensitive one. The adapting amplitudes needed to raise test threshold (measured in this case with Békésy tracking) by 10 dB were then determined, for adapting frequencies ranging from 10 to 90 Hz. The outcome of this study was that the requisite amplitude dropped slightly from 10 to 30 Hz, then remained essentially constant from 30 to 90 Hz. A plot of these amplitudes is comparable in shape to the threshold-versus-frequency function of the RA system [NP I of Bolanowski et al. (1988)], suggesting that two adapting stimuli that are themselves equal multiples of RA threshold are equally effective in *raising* RA threshold. This sounds unsurprising, until we consider the activity levels evoked in primary afferents by the adapting stimuli.

There is reason to believe that only a very few spikes are needed in an RA fiber -- perhaps one under optimal conditions -- in order for the subject to detect a stimulus event (Johansson and Vallbo, 1979). But when amplitude is gradually raised above this level, it soon reaches a point at which the number of impulses elicited per cycle is nearly independent of frequency, from 10 to more than 100 Hz (Johansson et al., 1982). This means that, in response to vibrations a few dB above threshold, RA afferents have a much higher firing rate (in impulses/second) for high frequency stimuli than for low frequency ones. Yet the psychophysical adapting effects of these stimuli are virtually equivalent. The conclusion to be drawn from this seems to be that adapting effectiveness within the RA channel is determined not by by total amount of neural activity in afferents of this class, but rather, by the information that this activity conveys to the brain about the amplitude of the adapting stimulus. We repeated our measurements of the adapting effectiveness of different frequencies on the face, and found a closely comparable result for a low-frequency vibrotactile channel there (Hollins et al., 1991), so apparently the finding has some generality.

It is interesting that, in contrast to these psychophysical findings, single-fiber recordings from RA afferents show that their absolute threshold is raised more by exposure to a high-frequency stimulus than to a low-frequency one of the same amplitude (Leung, 1994). Whether this disparity between the psychophysical and physiological findings results from differences in stimulus conditions in the two studies [e.g., the use of a stationary ring by Hollins et al. (1990); the gluing of the skin to the stimulator by Leung (1994)], or from the fact that the psychophysical measurements depend on central as well as peripheral processing, remains to be determined.

Amplitude Determines Adapting Effectiveness of Complex Vibrations

Recently, we began using another approach to address the question of how adapting effectiveness is determined -- namely, combining sinusoidal vibrations into more complex waveforms. One purpose of these experiments is to determine how the effectiveness of a complex wave is related to the effectiveness of its frequency components. This is an issue addressed by Marks (1979), who presented evidence that the Pacinian system works like an auditory critical band: He discovered that the vibrotactile "loudness" of a complex waveform is determined by the sum of the powers of its components, when both components activate the Pacinian channel. Recently, Makous et al. (1995) have extended this idea by showing that, within the Pacinian channel, power additivity also governs the masking effectiveness of complex stimuli. Their work makes the important point that, when total effective power is calculated, the power at each frequency should be expressed in multiples of (that is, in dB with respect to) threshold power at that frequency.

These results are compelling as regards the Pacinian channel, but leave open the question of how the RA channel computes effectiveness of a complex stimulus: Does it do so on the basis of total power (in accordance with the critical band model) or amplitude (as the data on detectability and adapting effectiveness of sinusoidal stimuli might suggest)? The issue is a complex one, for amplitude can be conceptualized in two ways: peak amplitude and root-mean-squared (rms) amplitude. When sine waves of different frequencies are combined, peak amplitude and rms amplitude do not increase by the same factor. Consider, for example, the case of two sine waves of equal amplitude, having frequencies of 90 Hz and 110 Hz. When these are combined, the result is a complex wave that resembles a carrier wave of 100 Hz that grows larger and smaller 20 times a second, giving rise to the perceptual phenomenon called beats. The maximum amplitude of this waveform, achieved whenever the components are in phase, is 6 db greater than the peak amplitude of either component alone. The combined rms amplitude, however, is only 3 dB greater than the rms amplitude of either component, for this measure takes into account out-of-phase as well as in-phase portions of the wave. The complex wave's power, which varies as the square of rms amplitude, is likewise only 3 dB greater than the power of each component.

Consider, then, an experiment in which a beat stimulus is formed by a combination of 90 and 110 Hz sinusoids, each having a peak amplitude of 14 dB re: the threshold of the mechanism under study. The beat waveform will have a peak amplitude of 20 dB, but a power of only 17 dB. Suppose now that the adapting effectiveness of this complex stimulus is compared with that of a 100 Hz sinusoid presented, in separate runs, at a series of amplitudes. At what amplitude will the 100-Hz adapting stimulus be equal to the beat stimulus in its ability to raise the threshold of detection at 20 Hz?

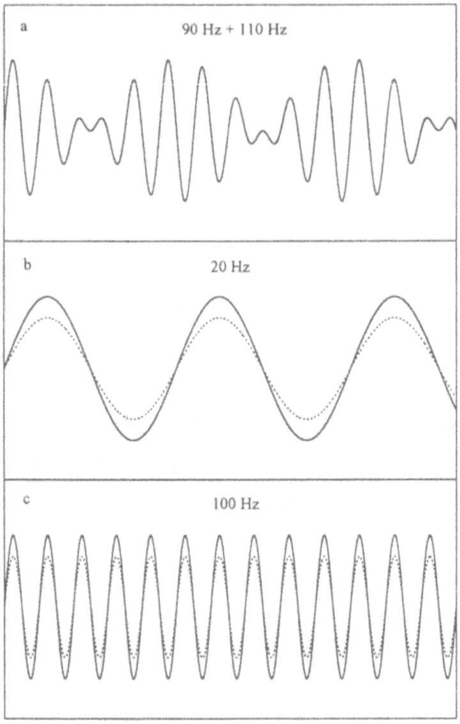

Figure 2. Diagram of (a) a complex stimulus waveform, and (b,c) sinusoidal stimuli that were compared with it in adapting effectiveness, i.e., ability to raise the detection threshold of a 20 Hz sinusoidal test stimulus. The complex stimulus, made by combining frequencies of 90 and 110 Hz, produced beats as well as a higher-pitched tactile sensation. Its adapting effectiveness was found to be equivalent to that of a sinusoidal vibration of the same peak amplitude (solid curves in b and c), rather than the same power (dotted lines).

To answer this question, a series of experiments were carried out (Delemos and Hollins, in preparation), in which detection threshold at 20 Hz was measured with 2IFC tracking, while the subject was adapted to either (a) the 90/110 Hz beat stimulus; (b) 100 Hz; or (c) 20 Hz. Each of these adapting stimuli was presented (in separate runs) at a series of amplitudes, and in each case, threshold of the test stimulus was a monotonically increasing function of adapting amplitude. Since the three functions were essentially parallel, the relationship between the relative effectiveness of the sinusoidal and beat stimuli was a simple one, illustrated in Figure 2: The adapting effectivess of a given beat stimulus (upper panel) was matched by a sinusoidal stimulus of the same peak amplitude (solid curves in the middle and lower plots), rather than by one of lesser amplitude having the same total power (dashed curves). In other words, when the magnitude of the adapting stimuli is expressed in terms of their power, the beat stimulus is consistently more effective than we would expect it to be, from a "critical band" perspective. When, however, the stimuli are described in terms of their peak amplitude, adapting effectiveness is independent of waveform.

Apparently, then, the adaptable CNS mechanisms that receive information from the RA channel encode not power, but peak amplitude. Consistent with this is the fact, discussed earlier, that RA threshold amplitude is virtually constant across the frequency spectrum: This channel is indifferent to the increasing power of higher frequencies. In this respect it differs qualitatively from the Pacinian system, which functions as a critical band.

A recent discovery by Leung (1994) suggests that differences in the rules governing adaptation within the RA and PC channels begin to appear at the earliest levels of information processing within these two systems -- before the site of spike generation in the primary afferents. Leung found, by varying adapting stimulus amplitude within a range over which firing rate of the afferent did not change (because it was entrained), that the amount of adaptation produced in RA afferents depends not on firing rate but on stimulus amplitude, and therefore presumably on the degree of activation of transduction-related processes. Adaptation in PC afferents, in contrast, is governed more by firing rate than by amplitude.

Converging lines of evidence thus indicate that, for the RA system, amplitude is the dimension that determines adapting effectiveness. The fact that it is keenly attuned to amplitude, however, does not compromise this channel's ability to register other parameters of vibratory stimulation. In addition to its ability to mediate precise frequency discrimination, the RA channel provides subjects with information about stimulus waveform -- enabling them not only to detect beats, but to feel a rasping quality when presented with a complex wave that is composed of a fundamental and a third or fifth harmonic (unpublished observations). The picture that emerges, then, is of a system that conveys to the brain detailed and faithful information about the temporal structure of a vibratory stimulus -- its amplitude, frequency, and complexity. Analysis of stimulus dynamics, rather than mere detection, appears to be a primary function of this channel.

Acknowledgement
This work was supported by NIDR Program Project grant DE07509.

References

Bolanowski, S. J., Jr., Gescheider, G. A., Verrillo, R. T. and Checkosky, C. M. (1988) Four channels mediate the mechanical aspects of touch. *Journal of the Acoustical Society of America* 84: 1680-1694.

Berglund, U. and Berglund, B. (1970) Adaptation and recovery in vibrotactile perception. *Perceptual and Motor Skills* 30: 843-853.

Burton, H. and Sinclair, R. J. (1991) Second somatosensory cortical area in macaque monkeys: 2. Neuronal responses to punctate vibrotactile stimulation of glabrous skin on the hand. *Brain Research* 538: 127-135.

Cohen, L. H. and Lindley, S. B. (1938) Studies in vibratory sensibility. *American Journal of Psychology* 51: 44-63.

Connor, C. E., Hsiao, S. S., Phillips, J. R. and Johnson, K. O. (1990) Tactile roughness: neural codes that account for psychophysical magnitude estimates. *Journal of Neuroscience* 10: 3823-3836.

Connor, C. E. and Johnson, K. O. (1992) Neural coding of tactile texture: Comparison of spatial and temporal mechanisms for roughness perception. *Journal of Neuroscience* 12: 3414-3426.

Delemos, K. A. and Hollins, M. Adaptation-induced enhancement of vibrotactile amplitude discrimination: The role of adapting frequency. Journal of the Acoustical Society of America; *in press*.

Delemos, K. A. and Hollins, M. The relative effectiveness of simple and complex (beating) vibrotactile adapting stimuli: The role of amplitude; *in preparation*.

Franzén, O. and Nordmark, J. (1975) Vibrotactile frequency discrimination. *Perception and Psychophysics* 17: 480-484.

Gescheider, G. A., Frisina, R. D. and Verrillo, R. T. (1979) Selective adaptation of vibrotactile thresholds. *Sensory Processes* 3: 37-48.

Gescheider, G. A., Capraro, A. J., Frisina, R. D., Hamer, R. D. and Verrillo, R. T. (1978) The effects of a surround on vibrotactile thresholds. *Sensory Processes* 2: 99-115.

Gescheider, G. A. and Wright, J. H. (1968) Effects of sensory adaptation on the form of the psychophysical magnitude function for cutaneous vibration. *Journal of Experimental Psychology* 77: 308-313.

Gescheider, G. A. and Wright, J. H. (1969) Effects of vibrotactile adaptation on the perception of stimuli of varied intensity. *Journal of Experimental Psychology* 81: 449-453.

Goble, A. K. and Hollins, M. (1993) Vibrotactile adaptation enhances amplitude discrimination. *Journal of the Acoustical Society of America* 93: 418-424.

Goble, A. K. and Hollins, M. (1994) Vibrotactile adaptation enhances frequency discrimination. *Journal of the Acoustical Society of America* 96: 771-780.

Goff, G. D. (1967) Differential discrimination of frequency of cutaneous mechanical vibration. *Journal of Experimental Psychology* 74: 294-299.

Hahn, J. F. (1966) Vibrotactile adaptation and recovery measured by two methods. *Journal of Experimental Psychology* 71: 655-658.

Hahn, J. F. (1968) Low-frequency vibrotactile adaptation. *Journal of Experimental Psychology* 78: 655-659.

Hollins, M., Delemos, K. A. and Goble, A. K. (1991) Vibrotactile adaptation on the face. *Perception and Psychophysics* 49: 21-30.

Hollins, M., Goble, A. K., Whitsel, B. L. and Tommerdahl, M. (1990) Time course and action spectrum of vibrotactile adaptation. *Somatosensory and Motor Research* 7: 205-221.

Johansson, R. S., Landström, U. and Lundström, R. (1982) Responses of mechanoreceptive afferent units in the glabrous skin of the human hand to sinusoidal skin displacements. *Brain Research* 244: 17-25.

Johansson, R. S. and Vallbo, A. B. (1979) Detection of tactile stimuli. Thresholds of afferent units related to psychophysical thresholds in the human hand. *Journal of Physiology* 297: 405-422.

Johansson, R. S. and Westling, G. (1987) Signals in tactile afferents from the fingers eliciting adaptive motor responses during precision grip. *Experimental Brain Research* 66: 141-154.

Johnson, K. O. and Hsiao, S. S. (1992) Neural mechanisms of tactual form and texture perception. *Annual Review of Neuroscience* 15: 227-250.

LaMotte, R. H. (1977) Psychophysical and neurophysiological studies of tactile sensibility. In N. Hollies and R. Goldman (Eds.), Clothing comfort: Interaction of thermal, ventilation construction and assessment factors (pp. 83-105). Ann Arbor: Ann Arbor Science.

LaMotte, R. H. and Mountcastle, V. B. (1979) Disorders in somesthesis following lesions of parietal lobe. *Journal of Neurophysiology* 42: 400-419.

Lee, C.-J. and Whitsel, B. L. (1992) Mechanisms underlying somatosensory cortical dynamics: I. *In vivo* studies. *Cerebral Cortex* 2: 81-106.

Lee, C.-J., Whitsel, B. L. and Tommerdahl, M. (1992) Mechanisms underlying somatosensory cortical dynamics: II. *In vitro* studies. *Cerebral Cortex* 2: 107-133.

Leung, Y. Y. (1994) Adaptation of mechanoreceptive afferents to continuous sinusoidal vibration. Master of Science in Engineering, Johns Hopkins University.

Leung, Y. Y., Hsiao, S. S. and Johnson, K. O. (1994) Adaptation of mechanoreceptive afferents to continuous sinusoidal vibration. *Society for Neuroscience Abstracts* 20: 1381.

Lundström, R. and Johansson, R. S. (1986) Acute impairment of the sensitivity of skin mechanoreceptive units casued by vibration exposure of the hand. *Ergonomics* 29: 687-698.

Makous, J. C., Friedman, R. M. and Vierck, C. J., Jr. (1995) A critical band filter in touch. *Journal of Neuroscience* 15: 2808-2818.

Marks, L. E. (1979) Summation of vibrotactile intensity: an analog to auditory critical bands? *Sensory Processes* 3: 188-203.

Mountcastle, V. B., Steinmetz, M. A. and Romo, R. (1990) Frequency discrimination in the sense of flutter: Psychophysical measurements correlated with postcentral events in behaving monkeys. *Journal of Neuroscience* 10: 3032-3044.

Mountcastle, V. B., Talbot, W. H., Sakata, H. and Hyvärinen, J. (1969) Cortical neuronal mechanisms in flutter-vibration studied in unanesthetized monkeys. Neuronal periodicity and frequency discrimination. *Journal of Neurophysiology* 32: 452-484.

O'Mara, S., Rowe, M. J. and Tarvin, R. P. C. (1988) Neural mechanisms in vibrotactile adaptation. *Journal of Neurophysiology* 59: 607-622.

Srinivasan, M. A., Whitehouse, J. M. and LaMotte, R. H. (1990) Tactile detection of slip: surface microgeometry and peripheral neural codes. *Journal of Neurophysiology* 63: 1323-1332.

Talbot, W. H., Darian-Smith, I., Kornhuber, H. H. and Mountcastle, V. B. (1968) The sense of flutter-vibration: Comparison of the human capacity with response patterns of mechanoreceptive afferents from the monkey hand. *Journal of Neurophysiology* 31: 301-334.

Verrillo, R. T. (1963) Effect of contactor area on the vibrotactile threshold. *Journal of the Acoustical Society of America* 35: 1962-1966.

Verrillo, R. T. (1966) Vibrotactile sensitivity and the frequency response of the Pacinian corpuscle. *Psychonomic Science* 4: 135-136.

Verrillo, R. T. (1968) A duplex mechanism of mechanoreception. In D. R. Kenshalo (Eds.), The skin senses (pp. 139-156). Springfield, IL: Charles C. Thomas.

Verrillo, R. T. and Gescheider, G. A. (1977) Effect of prior stimulation on vibrotactile thresholds. *Sensory processes* 1: 292-300.

Wedell, C. H. and Cummings, S. B., Jr. (1938) Fatigue of the vibratory sense. *Journal of Experimental Psychology* 22: 429-438.

Zwislocki, J., Maire, F., Feldman, A. S. and Rubin, H. (1958) On the effect of practice and motivation on the threshold of audibility. *Journal of the Acoustical Society of America* 30: 254-262.

O. Franzén, R. Johansson and L. Terenius (eds.)

Tactile neural codes for the shapes and orientations of objects

Robert H. LaMotte, Charles Lu and *Mandayam A. Srinivasan

*Department of Anesthesiology, Yale University School of Medicine, 333 Cedar St., New Haven CT 06510 and *Research Laboratory of Electronics, Massachusetts Institute of Technology, Cambridge, Massachusetts, USA*

Summary. A series of small, three dimensional objects of differing shape each mounted on a flat plate were constructed. The objects were a cylinder and sphere, each with a 5 mm radius, and two toroids each with a radius of 5 mm along one axis and differing radii of 1 or 3 mm along the orthogonal axis. Using a servocontrolled translation device with 3 degrees of freedom, each object was stroked back and forth across the skin under a maintained compressional force along a series of laterally shifted parallel linear trajectories oriented 90° with respect to the long axis of the finger. The toroid was oriented at 0,30,60 or 90°. Three dimensional "spatial event plots" (SEPs) -discharge rate vs location of object on the receptive field- were obtained for slowly- and rapidly-adapting mechanoreceptive type I fibers (SAs and RAs) with receptive fields centrally located on the fingerpad.

Interpreting the SEPs as responses of a spatially distributed population of fibers, the data supported the following conclusions and hypotheses: (1) The contour of the base of the SEPs of RAs and SAs representing the outline of the region of mechanoreceptor activity in the skin encoded the size, shape and orientation of the two dimensional outline of the object in contact with the skin; (2) The third dimension of shape was represented best in the shape of the distribution of spatial discharge rates (impulses/sec/mm) primarily in the SA fiber population. The shapes of the spatial discharge rates of RAs were more irregular and variable than those of SAs and exhibited poor or no representation of object shape. It was hypothesized that the distribution of slopes of the spatial discharge rate profile evoked by an object in the SA population encoded the distribution of curvatures on the surface of the object in contact with the skin. The constant curvature along the minor axis of a toroidally shaped object was found to be represented by a constant slope along the rising and/or the falling phase of a corresponding cross-section through the spatial discharge rate profile.

Introduction

The momentary "view" of the shape of an object as it sweeps across the fingerpad is more often described by the perceiver in terms of the object's features rather than its mode of contact with the skin. The spatial distribution of curvatures that define a shape (Gauss, 1827; Srinivasan and LaMotte, 1991) seem to the perceiver to remain constant in the presence of considerable variations in the object's orientation on the skin or its contact force and trajectory. The sensory information required for this perceptual constancy must be contained in the responses of a population of mechanoreceptive peripheral nerve fibers. So must the information representing the mode of tactile contact. In the following, we describe candidate neural mechanisms by which an object's shape and its orientation on the skin can be separately represented in the responses of these mechanoreceptive primary afferent fibers. The tactile stimuli are toroidal objects that are constructed to differ in radius along one axis and which are stroked across the passive fingerpad at differing orientations with respect to the stroke trajectory.

Methods

Stimulus objects and method of stimulation. The toroids had a radius of 5 mm along one axis and, for the orthogonal axis, a radius of 1 mm for one toroid and 3 mm for the other. A sphere and a cylindrical bar each had a radius of 5 mm. Thus, for all these specimens, curvature was the same along one axis but different for the other. The first step in producing the toroids and sphere was to use a radius cutting tool to cut a toroid from a rotating cylinder. Then, with the axis of the toroid horizontal, the top of the toroid was cut off along a section parallel to the axis, 0.5mm from the peak. The resulting bump, which had the shape of the top of an egg, was mounted on a flat plate resulting in a toroid "in relief". The outline of the base of this bump was an ellipse. A longitudinal section through the cylinder, 0.5 mm from the surface was mounted to a plate in similar fashion. A given plate containing one of these shapes was mounted on a three-axis load cell attached to a torque motor. The motor was mounted on a three-axis translation table that was used to control the motion of the stimulus object along a linear trajectory in the horizontal plane while the torque motor controlled the compressional force of the object against the skin.

Neural responses to each object stroked across the fingerpad of the anesthetized monkey (macaca fascicularis) were recorded from rapidly adapting type I and slowly adapting type I mechanoreceptive fibers (RAs and SAs respectively) in the median or ulnar nerve. All receptive fields were centrally located on the distal pad of the 2nd, 3rd or 4th fingers. The orientation of the cylindrical bar was held constant at 90° (parallel to the long axis of the finger) whereas the orientations of the two toroids were 90, 60, 30 or 0°. The object was stroked along a direction perpendicular to the axis of the finger from one side of the receptive field to the other and back along a series of linear parallel trajectories separated by 0.1 or 0.2 mm. The object was first placed proximal to the most sensitive spot in the receptive field and stroked back and forth and then shifted along the orthogonal axis, repeatedly, until reaching the most distal aspect of the receptive field. The overall compressional force was maintained at 10 gwt.

Results

Orientation

The responses of a typical SA and RA to each orientation of the thinnest toroid (1 x 5 mm radius) indicate that orientation is encoded in the two-dimensional orientation of the spatial response profiles of both fiber populations (Fig 1A and 1C). Each dot in each spatial event plot (SEP) is the location of the center of the toroid as the object, with indicated orientation, was stroked at 10 mm/s in a single direction in a series of parallel strokes oriented at 0°. We interpret each SEP as the spatially distributed activity of a population of fibers with identical functional properties viewed at the instant in time when the object was centrally located on the finger pad (LaMotte et al., 1994).

The cluster of activity in the center of each SEP changed its orientation in relation to the orientation of the toroid. To measure the orientation of this two-dimensional cluster, we first eliminated the background discharge evoked by the flat plate surrounding the toroid. Principal components analysis (Morrison, 1967) was then used to obtain the orientation of the principal component through the remaining central cluster of data points (Fig. 1B and 1D). In this analysis, the squared distance between each point and the principal component along the orthogonal line is minimized.

The orientation of the toroid was encoded equally well in the responses of RAs and SAs. These neural representations of orientation were closely related to physical orientation. The SEPs of some RAs, while retaining a poor representation of shape, nevertheless exhibited an accurate orientation due to their "off" response to the unloading of the skin as the object moved off the receptive field. The variability in principal components for each fiber type was lowest at 90 and greatest at 0°.

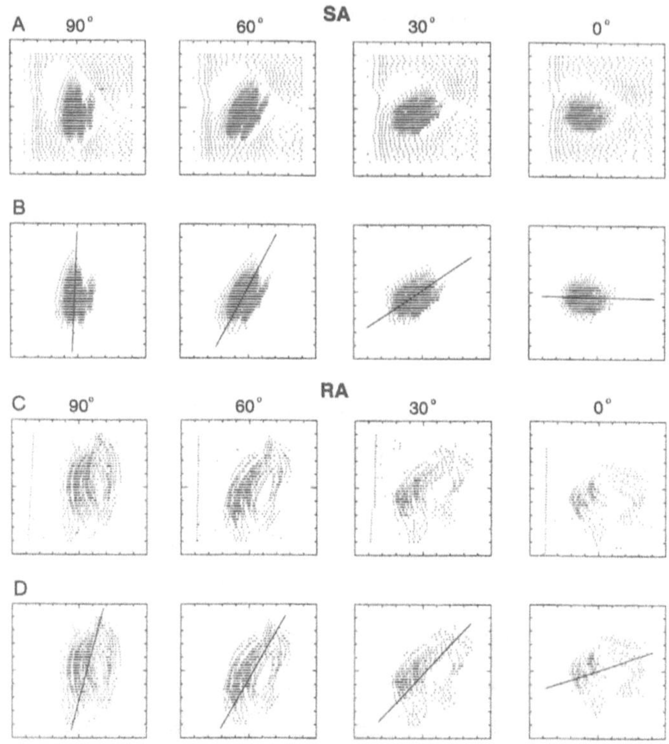

Figure 1. Peripheral neural representation of the orientation of an object stroked across the fingerpad. Spatial event plots of the responses of an SA and RA to different orientations of the thinnest toroid stroked in the same direction across their receptive fields on the fingerpad. The toroid (1 x 5 mm radii) was oriented 90, 60, 30 and 0° to the direction of stroking which was perpendicular to the long axis of the finger. Each dot is the location of the center of the toroid when an action potential occurred. After each stroke from left to right the object was shifted by 0.1 mm proximal to distal (upward in the figure). The contact force was 10 g-wt. and the stroke velocity 10 mm/s. Tic mark spacing on each axis is 1 mm. A and C: Raw data obtained for an SA and RA. B and D: The same data after the background discharges evoked by the pressure of the flat plate and/or skin stretch was eliminated. The solid line in each panel is the major vector, determined by a principle components analysis, that best describes the spatial orientation of the cluster of action potentials evoked by the toroid.

Shape

The discharge rate of each fiber increased as a function of the curvature of the minor and major axes of the stimulus objects stroked in the same manner over the skin. The cylinder, with zero curvature on its major axis, evoked the lowest rate, followed in order of increasing rate by the sphere and the toroids with 3 x 5 and 1 x 5 mm radii respectively. However, discharge rate did not provide a reliable code for the shape of an object since it could change readily with alterations in contact force, the orientation of the object and the direction and velocity of stroking.

Accordingly, it seemed likely that the shape of an object is encoded in the shape of the spatial response profile of a population of fibers (e.g. LaMotte and Srinivasan, 1987a; Goodwin et al., 1995). One representation of this profile is the two-dimensional SEPs, shown for an SA (Fig. 2A) and an RA (Fig. 2C). These were obtained in response to the two toroids, the sphere and the cylinder each stroked in the same direction along the 0° trajectory. The toroids were oriented at 90°, the contact force was 10 g-wt and the stroke velocity 10 mm/s. The outline of the base of each object mounted on the flat plate is superimposed on the impulse raster. The shape of this outline is the shape of a horizontal cross-section through the object 0.5 mm from the top.

The shape and size of the two dimensional outline of the object in contact with the skin are well represented by the borders of the fiber activity. This was true for RAs as well as SAs. The shape of the object was better defined by the leading side of a fiber's response (left borders for these SEPs) than by the trailing side. The overall curvature of the left border of the dot raster encoded the elliptical shape at the base of the object in contact with the skin. For the objects oriented at 90° it is greatest for the sphere (but was still greater for the toroids oriented at 0°) and least for the cylinder whose border has no curvature as shown in the figure. The vertical length of the region of fiber activity in each SEP and its width at the midportion encode the length and width of the contact area along the major and minor axes of the object. For the stimuli used in these experiments, the contact area, in turn, is directly related to the curvature of the object along these axes. Thus, the width is greatest for the sphere and cylinder with 5 mm radii for minor axes and least for the 1 x 5 mm toroid. The vertical length is greatest for the cylinder (limited to 8 mm by the number of trajectories) and roughly the same size for the toroid and the sphere each of which had a radius of 5 mm for the major axis.

The third dimension of the shape of each object is likely to be represented by the pressure distribution along the surface of the skin at a given instance in time (Srinivasan and LaMotte, 1991). To study the shape of the spatial response profile generated by this distribution, the spatial discharge rate, defined as the number of impulses/0.2 mm bin) was plotted on a vertical axis perpendicular to the SEP in the horizontal plane. An average of the middle five cross-sections (parallel to the direction of stroking) through this three-dimensional SEP is shown for the same SA in Fig. 2B. The peak spatial discharge rate was lowest for the cylinder and increased with increasing curvature of the minor axis from the sphere to the thinner toroid. When the same curves were normalized with respect to their respective peak discharge rates, the four spatial response profiles were still discriminable on the basis of their base widths and the rates of decline from peak to base on the trailing side of the profile (and for some SAs, the rates of ascent from base to peak on the leading side). As the curvature along the minor axis of an object increased, the base widths decreased. For most fibers, these response profiles were more or less triangular and a straight line appeared, as a first approximation, to be the most reasonable fit to the ascending or the declining phase of each profile. Accordingly, as a gross estimate of the rate of ascent and decline of each profile, a straight line was fitted to the data from base to peak, and another line from peak to base. The slopes of the lines for the declining phases were found to increase with the curvature of the minor axis for the four objects for the majority of the SAs tested. A similar relationship between slope and curvature was obtained in a minority of SAs for the rising phase as well. Similar results, however, were not obtained for either phase for RAs whose spatial response profiles were much more irregular and variable in comparison with those of SAs (Fig. 2D). The shape of the RA response profiles deteriorated further when the stroke velocity was lowered to 5 mm/s (not shown) which, for some RAs, evoked only a few impulses per stroke. In contrast, the shapes of the spatial response profiles of SAs were more robust in encoding object shape in the presence of changes in stroke velocity.

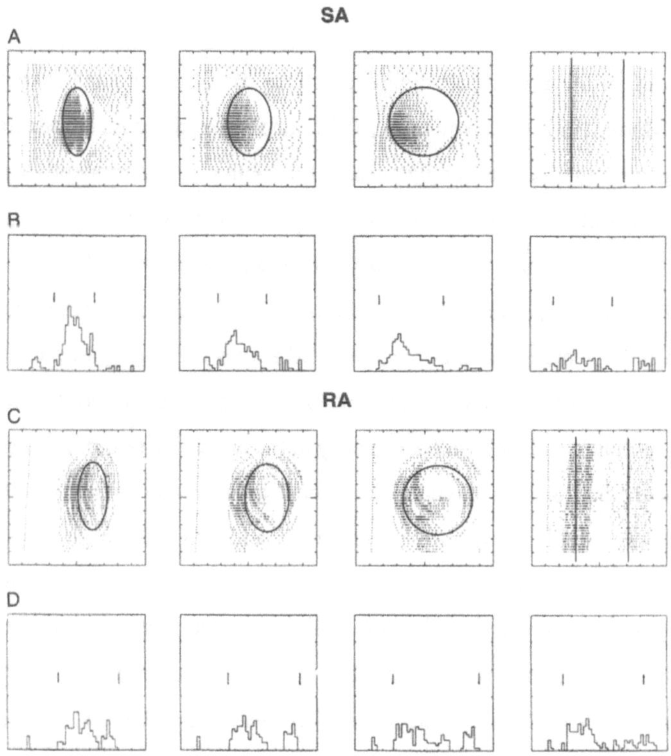

Figure 2. Peripheral neural representations of the shapes of objects stroked across the fingerpad. Spatial event plots of the responses of an SA and RA to objects of the same orientation but different shape stroked in the same direction across their receptive fields on the fingerpad. A and C: Same format as A and C in Fig. 1. The shape of the base of each object on the flat plate is superimposed. This is the outline of a horizontal cross-section through the shape 0.5 mm from the top. B and D: The spatial discharge rate (impulses/0.2 mm bin) on the vertical axis was averaged from the data obtained from five consecutive strokes located in the middle of the shape or cluster of action potentials evoked by the object (data in each panel obtained from panel directly above). The two vertical tic marks directly above each histogram delineate the beginning and end of the cluster of impulses evoked by the object itself as opposed to background discharge produced by the plate.

Discussion

Human subjects can readily discriminate differences in the curvature of small shapes pressed against (Srinivasan and LaMotte, 1987; Goodwin et al., 1991) and stroked across (LaMotte and Srinivasan, 1987a) the fingerpad on the basis of tactile cues alone. Similarly, proprioceptive information is not required to tactually identify, with little or no error, up to six categories of orientation of a thin (1 x 5 mm radii) toroid pressed against the stationary fingerpad (LaMotte et al., 1992).

The results obtained in the present experiments provide a basis for hypothesizing how the three-dimensional shape and orientation of an object stroked across the fingerpad are separately encoded in the evoked responses of SA and RA mechanoreceptive cutaneous afferent fibers. The features of shape and object orientation cannot be encoded solely in the pattern or rate of discharges in individual afferent fibers. These rates could be altered to the same degree by appropriate variations in curvature or orientation as they could by changes in shape. In addition, discharge rates are also greatly influenced by changes in the contact force of an object (Goodwin et al., 1995) as well its velocity and direction of stroking (LaMotte and Srinivasan, 1987a,b). Instead, the properties of shape and orientation must be coded in the responses of populations of mechanoreceptive afferent fibers.

The central portion of each two-dimensional SEP was interpreted as representing the spatial distribution of activity evoked by the object in a population of SAs and RAs at an instance in time. The area, two dimensional shape and orientation of this region of afferent activity encode, respectively, the size, shape and orientation of that region of the object in contact with and parallel to the surface of the skin. The parameters of size, two dimensional contact region and orientation of an object appeared to be represented equally well in the spatial response profiles of SA and RA populations. The curvature of the leading edge of the spatial response profiles increased and the width of the response decreased with increasing curvature of the object in the horizontal plane parallel to the skin. The neural representation of object orientation was better when the major axis was perpendicular rather than parallel to the direction of stroking. This is likely to be due to the greater magnitude and rate of change in the curvature of the skin produced by the object, and greater fiber discharge rates, for strokes along the axis of greater object curvature.

The third dimension of shape, responsible for a characteristic spatial distribution of pressures on the skin, was represented best in the shape of the distribution of spatial discharge rates (impulses/sec/mm) primarily in the SA fiber population. The shapes of the spatial distribution of discharge rates of RAs were more irregular and variable than those of SAs and exhibited poor or no representations of object shape. It is hypothesized that the distribution of slopes of the spatial discharge rate profile evoked by an object in the SA population encodes the distribution of curvatures on the surface of the object in contact with the skin. The curvature along the minor or major axis of toroidally shaped objects (including cylinders and spheres) is a constant. A constant curvature, then, is represented by a constant slope along the rising and/or falling phase of a corresponding cross-section through the spatial discharge rate profile. In the present experiments this profile was tent-shaped and was generated primarily by the leading edge of the object stroked across the skin. Thus, the representation of three-dimensional shape in the responses of primary afferent fibers is not isomorphically related to the physical shape of the object stroked across the skin. A greater degree of isomorphism in neural representation is obtained when pressing but not stroking a three-dimensional object (Goodwin et al., 1995) against the skin or when stroking a two-dimensional forms represented by a pattern of elements in a horizontal plane (Johnson and Lamb, 1981).

The progression of increasing curvatures from the sphere to the thinner of the two toroids produced correspondingly steeper, constant slopes on the falling phases of the spatial response profiles. Differences in object curvature were represented more consistently by changes in the slope of the declining portion of this profile more often than in the rising part. This is possibly due to a greater response variability from transient mechanical events generated by the sudden impact of the toroidal shapes (especially those with small radii) on a plate that has already reached the desired velocity by the time the object in relief reaches the most sensitive spot in a fiber's receptive field. More symmetrical "tent-shaped" profiles of spatial discharge rate were obtained from SAs in response to an aperiodic wavy surface stroked across the skin (unpublished observations and LaMotte et al., 1994). The wavy surface consisted of a sequence of alternating convex and concave cylinders (with radii of 3.1 to 16.9 mm) and constructed to have a continuously changing surface slope. These curvatures were generally larger and more smoothly changing than those used in the present experiments. Similarly, symmetrical tent shaped spatial discharge rate profiles were obtained from SAs in response to spheres of differing curvature pressed against the fingerpad (Goodwin et al., 1995).

References

Gauss, C.F. (1827) Disquisitiones generales circa superficies curvas. *Commentationes Societatis Regiae ScientiarumGottingenesis Recentiores.* 6:99-146.

Goodwin, A.W., Browning, A.S. and Wheat, H.E. (1995) Representation of curved surfaces in responses of mechanoreceptive afferent fibers innervating the monkey's fingerpad. *Journal of Neuroscience* 15:798-810.

Goodwin, A.W., K.T. John, and A.H. Marceglia (1991) Tactile discrimination of curvature by humans using only cutaneous information from the fingerpads. *Experimental Brain Research* 86:663-672.

Goodwin,A.W., and H.E. Wheat (1992) Magnitude estimation of contact force when objects with different shapes are applied passively to the fingerpad. *Somatosensory and Motor Research* 9:339-344.

Johnson, K.O. and Lamb, G.D. (1981) Neural mechanisms of spatial tactile discrimination: neural patterns evoked by Braille-like dot patterns in the monkey. *Journal of Physiology* (London) 310, 117-144

LaMotte, R.H., and M.A. Srinivasan (1987a) Tactile discrimination of shape: Responses of slowly adapting mechanoreceptive afferents to a step stroked across the monkey fingerpad. *Journal of Neuroscience* 7:1655-1671.

LaMotte, R.H., and M.A. Srinivasan (1987b) Tactile discrimination of shape: Responses of rapidly adapting mechanoreceptive afferents to a step stroked across the monkey fingerpad. *Journal of Neuroscience* 7:1672-1681.

LaMotte, R.H., M.A. Srinivasan and A. Klusch-Petersen. (1992) Tactile discriminations and identifications of different shapes and orientations of ellipsoidal objects applied to the fingerpad. *Society for Neuroscience Abstracts* 18:830.

LaMotte, R.H., M.A. Srinivasan, C. Lu, and A. Klusch-Petersen (1994) Cutaneous neural codes for shape. *Canadian Journal of Physiology and Pharmacology* 72:498-505.

Morrison, D.F. (1967) The Structure of Multivariate Observations: I. Principal Components. In: *Multivariate Statistical Methods.* McGraw-Hill, New York. pp. 221-258.

Srinivasan, M.A., and R.H. LaMotte (1987) Tactile discrimination of shape: Responses of slowly and rapidly adapting mechanoreceptive afferents to a step indented into the monkey fingerpad. *Journal of Neuroscience* 7:1682-1697.

Srinivasan, M.A., and R.H. LaMotte (1991) Encoding of shape in the responses of cutaneous mechanorecptors. *In:* O. Franzen, and J. Westman (eds.): *Information Processing in the Somatosensory System*, MacMillan, London, pp.59-69.

Somesthesis and the Neurobiology of the Somatosensory Cortex
O. Franzén, R. Johansson and L. Terenius (eds.)

Tactual discrimination of softness: abilities and mechanisms

Mandayam A. Srinivasan and Robert H. LaMotte*

*Department of Mechanical Engineering and Research Laboratory of Electronics, Massachusetts Institute of Technology, Cambridge, MA 02139, USA. and *Department of Anesthesiology, Yale University School of Medicine, 333 Cedar St., New Haven, CT 06510, USA.*

Summary. The ability of human subjects to discriminate softness was investigated. In addition, the associated mechanistic cues and the peripheral neural codes on which the discrimination may be based were investigated using monkeys. Two sets of compliant objects were constructed: (1) those with a deformable surface, composed of transparent silicone rubber of variable softness; (2) rigid plates supported by springs of variable compliance in cylindrical sleeves. To assess the relative contributions of tactile and kinesthetic information, psychophysical experiments were performed on human subjects under both active and passive touch, with or without local anesthesia that blocked the tactile information from the fingerpads. The results for *objects with deformable surfaces* show that whereas the subjects are quite good at discriminating subtle differences in softness under both active and passive touch without anesthesia, they are unable to discriminate even large differences under active touch with anesthesia. Therefore, we conclude that the discrimination is based entirely on tactile information. The mechanistic data for rubber specimens indicates that the basis for the perception of softness of rubber-like objects is likely to be the spatio-temporal variation of pressure on the skin (or, equivalently, the skin displacement and its derivatives). Neurophysiological data shows that the resulting responses from slowly adapting type I afferent population within the skin might encode the compliance of the objects. For *compliant objects with rigid surfaces*, best discrimination was achieved only with active touch, due to the availability of both tactile and kinesthetic information; under passive conditions, the absence of kinesthetic information resulted in considerable deterioration of discriminability.

Introduction

Compliance, loosely defined as the amount of deformation of an object under a unit measure of applied force, is a fundamental physical property of objects that governs their mechanical behavior. To successfully perform any haptic task such as manual exploration or manipulation that involves physical contact with objects, it is necessary for us to know how compliant they are, since the object motions and deformations caused by imposed forces are directly dependent on the object compliance. Compliant objects can be classified into two major types: (1) deformable objects whose surface is also deformable (such as rubber, fruits, etc.) and (2) deformable objects whose surfaces are rigid (such as a key in a piano or typewriter). Because the mechanical stimuli imposed on the skin depend on the type of object compliance, it is to be expected that the representation and processing of tactual information (which includes both tactile and kinesthetic sensory information) needed to discriminate object compliance depend on the type of object compliance.

Very little is known about the human ability to sense softness, a subjective measure of the compliance of an object. In particular, no literature exists on the biomechanical and neural mechanisms that enable the human discrimination of softness. Early work was focused on the measurement of differential sensitivity of humans and the form of the psychophysical law (Katz, 1938; Harper, 1952; Harper and Stevens, 1964). Later work of Roland and Ladegaard-Pedersen (1977) examined the ability of subjects to discriminate the stiffness of springs held between the thumb and index finger. They showed that anesthetized skin and joints did not affect the discriminability. Based on a matching paradigm requiring the activation of the elbow joint, Jones and Hunter (1990) estimated the human compliance resolution to be about 23%. Recently, Tan, *et al.*, (1995) have measured the Just Noticeable Differences (JND) in compliance between two rigid plates (driven by computer controlled linear motors) grasped between the thumb and index finger. The JNDs ranged from about 8% to 99% depending on whether the subjects had cues stemming from terminal grasp force and/or total work done.

Our goals in this study are (1) to measure the ability of humans to tactually discriminate softness, (2) to clarify the roles played by tactile and kinesthetic information in the discrimination, and (3) to identify the nature of the peripheral neural codes, such as which fiber population enables the discrimination and whether the code is intensive, temporal, spatial or spatio-temporal. The psychophysical results on human discrimination of softness summarized here have been described in more detail in a recent paper (Srinivasan and LaMotte, 1995). In addition, we give here some of the results on the possible mechanistic cues and the associated peripheral neural codes from biomechanical and neurophysiological experiments performed on monkeys.

Materials and Methods

Compliant objects with deformable and rigid surfaces were fabricated so that they exhibited elastic behavior, that is, the deformed objects under the applied loads returned to their original unloaded state once the loads were removed. For objects with deformable surfaces, transparent rubber specimens that varied only in compliance (controlled by varying the amount of a diluent) were cast. For compliant objects with rigid surfaces, telescoping hollow cylinders with rigid end plates were constructed such that the inner cylinder was supported by several springs inside the outer cylinder. Objective compliance, defined as the average slope of the displacement-force trace for a 0.25 inch diameter rigid probe indenting the rubber specimen at constant velocity of 0.5 mm/s, was determined for each of the specimens using a computer controlled tactile stimulator. The traces for the rubber specimens are shown in Fig. 1.

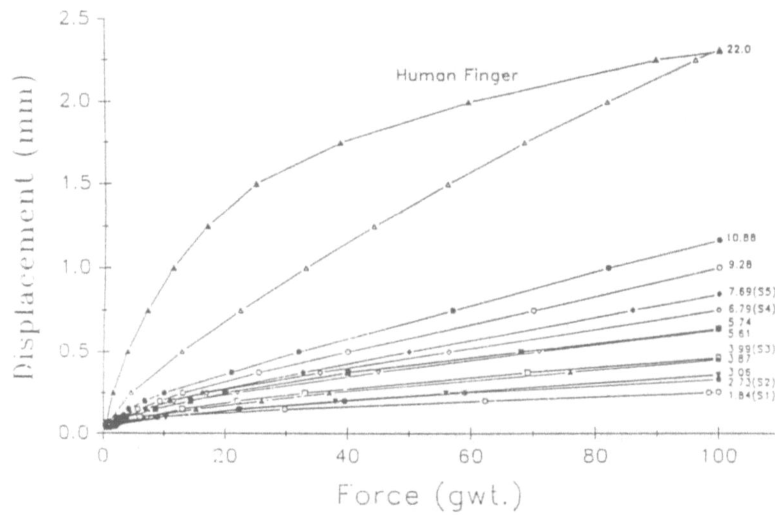

Figure 1. Traces of displacement vs. force for each of the rubber specimens and a human fingerpad indented by a flat-ended cylindrical probe (0.25 inch diameter).The velocity of indentation was maintained constant at 0.5 mm/s. Numbers at the *right end* of each trace: average slope, which represents an objective measure of the compliance (in microns/gwt) of the corresponding specimen. S1-S5: 5 specimens used for the pairwise discrimination experiments. Note that the fingerpad has a pronounced nonlinear force-displacement relationship, and is more compliant than all the rubber specimens (from Srinivasan and LaMotte, 1995).

To measure the human discriminability of softness and to isolate the associated information processing mechanisms, psychophysical experiments were performed under three conditions: (1) active touch with the normal finger, where both tactile and kinesthetic information was available to the subject; (2) active touch with local cutaneous anesthesia, so that only kinesthetic information was available; (3) passive touch, where a computer controlled mechanical stimulator brought down the compliant specimens onto the passive fingerpad of the subject, who therefore had only tactile information. The experiments consisted of ranking, where the subjects had to rearrange a set of 12 specimens in the order of increasing softness, and pairwise discriminations using a 2-interval, 2-alternative forced choice paradigm. These series of experiments are described in more detail in Srinivasan and LaMotte (1995).

To determine the possible mechanistic cues on which the softness discriminations and the associated neural codes could be based, biomechanical and neurophysiological data were simultaneously gathered for each of the specimens indenting the fingerpads of *Mecacca Fascicularis* monkeys under the same passive touch conditions as in the psychophysical experiments. The biomechanical data consisted of the variations over time of the net contact force (measured by using a load cell) and overall contact area (recorded with a videomicroscopy system), both of which enabled the calculation of the average pressure on the skin as a function of time. To infer the peripheral neural codes for softness, the responses of slowly adapting type I (SAI) and rapidly adapting type I (RAI) single afferent fibers in monkeys to the same stimulations used in psychophysical and biomechanical experiments were recorded electrophysiologically.

Results

1. Compliant objects with deformable surfaces

For rubber specimens, the ranking experiments with normal finger showed that the subjective sensation of softness correlates one-to-one with the objective determination of compliance - specimens with distinct compliances were discriminated, whereas those without, caused some confusion (Fig. 2). Compliance, as defined above, is therefore an appropriate objective measure relative to which the ability of the subjects can be judged. To distinguish the objective compliances of the specimens, high precision instruments needed to be operated under carefully controlled conditions. However, human subjects rank-ordered these specimens rather easily, and hence we conclude that humans have excellent softness discriminability. This ability was invariant across all subjects (14 in number) who used a variety of velocities and forces of indentation. Although our goal was not to measure the human JND (Just Noticeable Difference) for softness, based on the ranking experiments, we can estimate it to be about 12% or better.

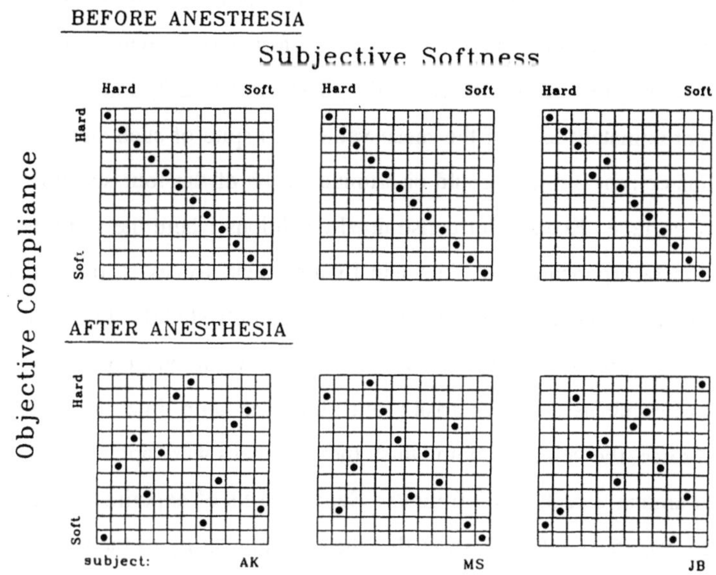

Figure 2. Results of ranking experiments before and after the administration of local anesthesia to the fingertips of 3 subjects. The confusion matrices plotted here show that the subjective judgment of the relative softness of the specimens coincided with objective compliance when tactile information was available to the subject. Under anesthetized conditions, when tactile information was absent, subjects could not even distinguish between the hardest and the softest specimens (from Srinivasan and LaMotte, 1995).

In both ranking and pairwise discrimination experiments, the difference in the performance of the subjects before and after the administration of local anesthesia to the fingertips was dramatic (Fig. 2). In contrast to the ranking results described above for active indentations with normal finger, under anesthetized conditions, the subjects could not even distinguish between the hardest and the softest specimens. In pairwise discriminations of the softness of rubber specimens S3 and S5 under active indentations, the subjects made 93% or greater correct judgments with normal finger. But after the administration of local anesthesia to the fingertips, the subjects could only achieve 55% correct or less, indicating that they were making random calls. Based on the results of both ranking and pairwise discrimination experiments with anesthetized fingerpads, we conclude that kinesthetic information alone is insufficient to discriminate the softness of rubber specimens and that tactile information is absolutely necessary.

Next, pairwise discrimination experiments were conducted under active and passive touch without anesthesia, with rubber specimen S3 as the standard and each of S1, S2, S4 and S5 as comparisons. Active touch experiments without any constraints on how the subjects indented the specimens enabled us to measure the velocities and forces used by subjects under almost normal conditions and to determine quantitatively their best performance among a variety of conditions. The standard deviations for both velocities (3.35 $\pm$ 0.85 mm/sec SD) and forces (98.71 $\pm$ 26.17 gwt SD) were approximately 25% of the mean values. In spite of such significant variations in indentation parameters, the subjects could discriminate each of the four specimen pairs at levels greater than 90% correct. Even under constrained conditions where the subjects were asked to employ only the metacarpophalangeal joint and to limit peak forces to pre-defined values, the performance of the subjects under active touch deteriorated only slightly (Fig. 3A).

In passive touch experiments, to mimic the velocity variations observed in active touch experiments and to eliminate any cues arising out of slower ramp times for softer specimens to attain the desired force, the velocities with which the specimens were applied to the passive finger were randomized among 2.4, 3, and 3.6 mm/s. Separate sets of experiments performed at peak force values of 25, 50, 75, and 90 gwt. show that the discriminability of the subjects under passive touch conditions, especially at higher forces, is about the same as that under active touch conditions (Fig. 3B). To further eliminate any possible temporal and intensive cues, the peak forces were also randomized among 60, 75, and 90 gwt., in addition to the three indentation velocities (Fig. 3C). The resulting subject performance was still almost the same as that under active touch at 50 or 75 gwt. peak force. Because almost no difference was observed in the discriminability of the subjects under active and passive touch, we conclude that tactile information alone is sufficient to discriminate the softness of the rubber specimens.

To investigate the biomechanical and peripheral neural basis for softness discrimination, each of the five rubber specimens S1 to S5 were applied to the passive monkey fingerpad by the tactile stimulator. For each indentation, the stimulator brought down a specimen at a constant vertical velocity until a desired force was reached, after which the specimen was withdrawn at the same speed. The same velocities (2.4, 3.0 and 3.6 mm/s) and forces (60 to 90 gwt) used in the psychophysical experiments were employed. Each stimulus was repeated twelve times to ensure the consistency of the fiber responses. Typical force vs. time traces and the associated responses of an SAI and an RAI to the application of specimen S1 (harder) and S3 (softer) are shown in Fig.

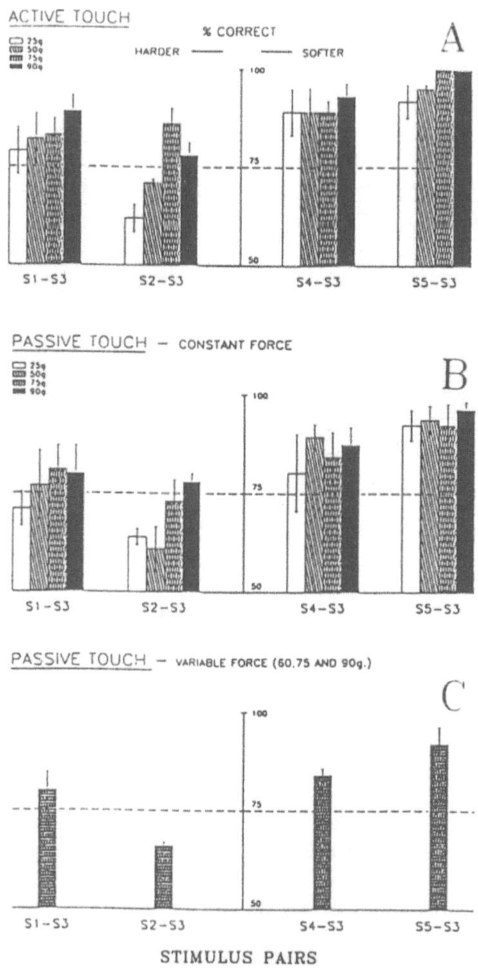

Figure 3. Results of pairwise discrimination experiments with rubber specimens under active and passive touch at various contact force values. In each experiment, specimen S3 was the standard and each of S1, S2, S4, and S5 were comparisons. Each bar represents the mean percent of correct calls for the corresponding specimen pair and force value, with the error bar indicating the SE. The labels 'HARDER' and 'SOFTER' indicate the relationship of the comparisons to the standard. (A) constrained active touch: The subjects were constrained to utilize 20° to 30° rotations of only the metacarpophalangeal joint, and peak forces within 20 gwt deviation from the mean values shown. Except for the pair S2-S3 at low forces, all the other pairs were discriminated by the subjects, with lower forces resulting in a slight deterioration in performance. (B) Passive touch under constant force: Separate sets of experiments were performed at each of the forces shown. For each set, the velocity of indentation was randomized among 2.4, 3, and 3.6 mm/s. The discriminability of the subjects in the absence of kinesthetic information deteriorated only slightly compared with active touch performance shown in (A), where both tactile and kinesthetic information were available to the subjects. (C) Passive touch under randomized forces: Both the forces (60, 75 and 90 gwt) as well as the velocities (2.4, 3, and 3.6 mm/s) were randomized. Despite the elimination of both temporal and intensive cues, together with the lack of kinesthetic information, the subjects performed almost as well as they did under active touch at 50 or 75 gwt (from Srinivasan and LaMotte, 1995).

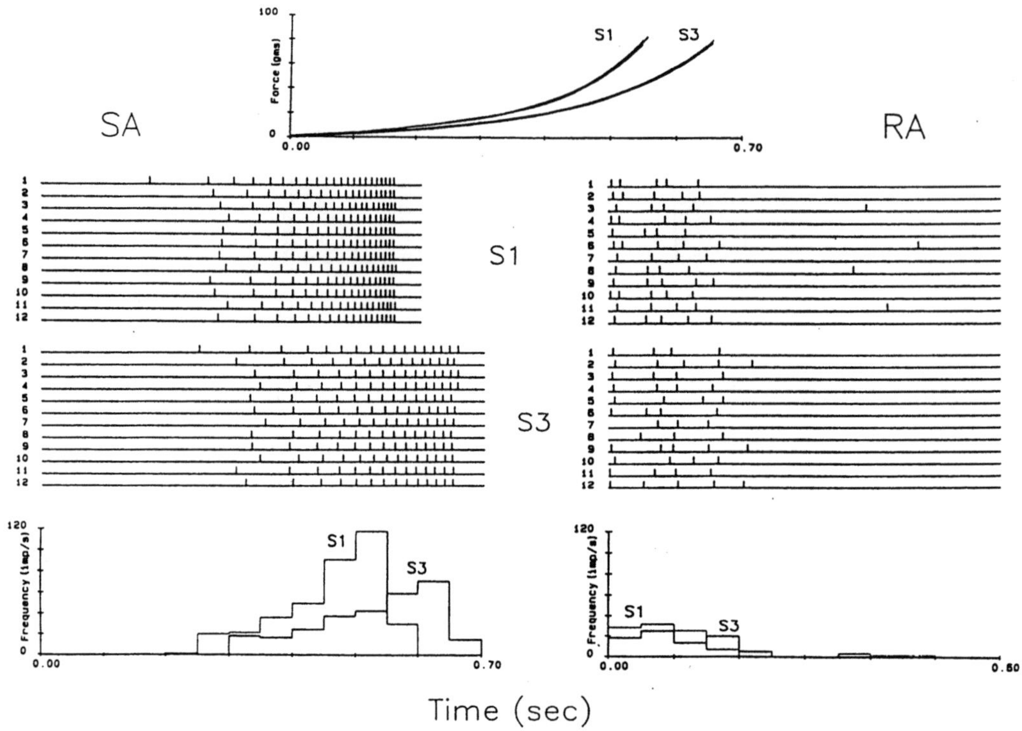

Figure 4. Responses of an SAI and an RAI to rubber specimens S1 and S3 indenting the monkey fingerpad at 3.0 mm/s. The top panel shows the corresponding force traces over time. Each horizontal line in the middle two panels represents one indentation, and each vertical tick represents the occurrence of an action potential at that time. The bottom panels show the corresponding discharge frequency histograms.

4. The rate of change of force is lower for the softer specimen. Correspondingly, the rate of change of the neural impulse frequency (impulses/sec) is lower in the SAI response. The RAI response, however, is about the same for both the specimens, which are quite easily discriminated by human subjects. Thus tactile discrimination of softness is likely to be based on SAI responses, and in particular, their discharge rates.

The plots in Fig. 5 show that when the specimens are applied to the fingerpad at the same velocity, the softer the specimen, the lower is the rate of change of net force and higher is the rate of change of overall contact area. Therefore, at a given instant of time during the indentation, the difference in the average pressures between two specimens is higher than the corresponding

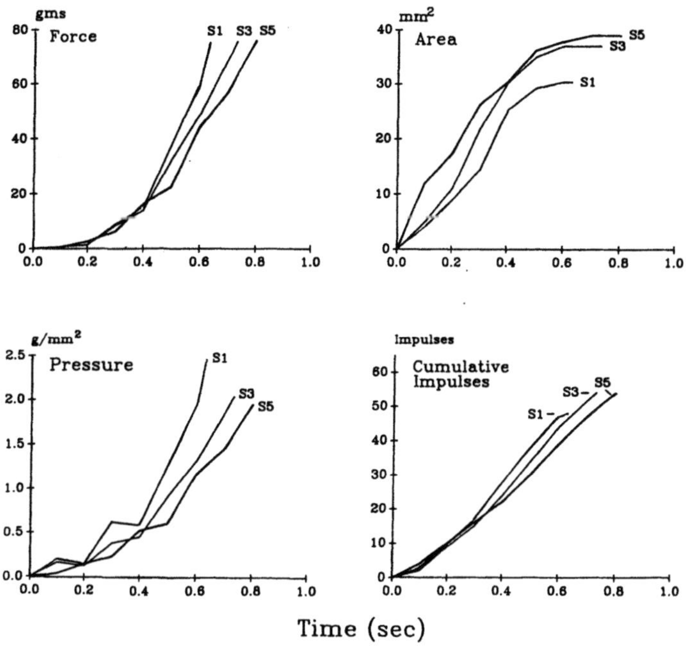

Figure 5. The variations over time of net force and overall contact area, together with the associated average pressure (net force/overall contact area) are shown for the standard specimen S3 and the two comparisons S1 and S5, all applied to the fingerpad at 3 mm/s. To illustrate the differences in the discharge rates of an SAI to the three specimens, its responses are plotted as cumulative impulses over time.

differences in either the forces or the contact areas. Just as the pressure increases more slowly for the softer specimen, the SAI discharge rate also increases more slowly (as shown in Fig. 4), resulting in a slower increase in cumulative impulses (Fig. 5).

However, the force, contact area, and discharge rate are affected by the velocity of indentation. For the same specimen, lower indentation velocity results in lower force and area rates, giving rise to a lower discharge rate (Fig. 6, top panel) at a given instant of time during the ramp. Since the discharge rate of a single fiber is affected by both the compliance of the specimen and the velocity of indentation, two specimens of differing compliances can be made to give rise to the same single fiber response by appropriately adjusting their velocities of indentation. The bottom panels of Fig. 6 show that the variation of cumulative impulses as a function of time is about the same for S3 applied at 3.0 mm/sand S1 applied at 2.4 mm/sec or for S3 applied at 2.4 mm/s and S5 applied at

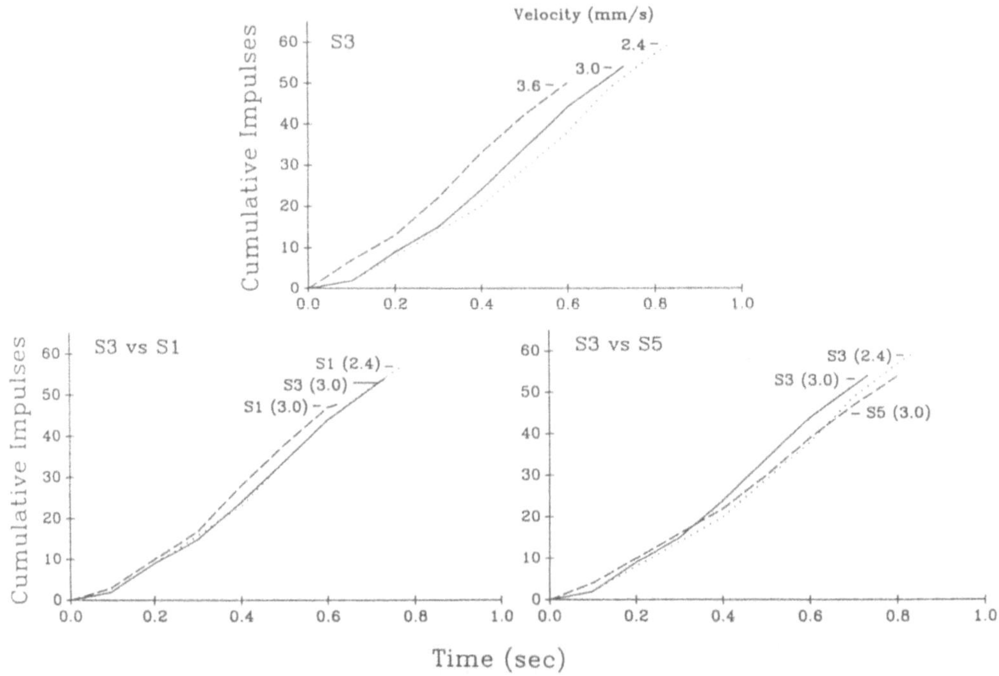

Figure 6. Cumulative impulses discharged by an SAI during the indentation ramp are shown as functions of time. Top panel shows the traces for the standard specimen S3 applied at indentation velocities of 2.4, 3.0, and 3.6 mm/s. Bottom left panel shows the traces for specimen S1 applied at 2.4 and 3 mm/s and S3 at 3 mm/s. Bottom right panel shows the traces for specimen S3 at 2.4 and 3 mm/s and S5 at 3 mm/s.

3.0 mm/s. Thus the discharge rate in a single SAI fiber does not unequivocally encode the compliance of the object.

2. Compliant objects with rigid surfaces

The ability of subjects to discriminate the compliance of the 'spring cells' was consistently poorer compared to that of the rubber specimens. In active touch experiments, the force control of the subjects was also poorer for spring cells. These results are to be attributed to the non-deformability of the surfaces of spring cells, which result in decreased tactile information regarding the compliance of the object: the contact area depends solely on the force applied and not the compliance of the spring cell, whereas it depends on both for the rubber specimens.

In the case of spring cells, in pairwise discrimination experiments involving active touch with normal fingers, the performance of the subjects was poorer compared to that for rubber specimen pairs with even closer compliances. Thus, even when both tactile and kinesthetic information was available, the subjects have less information with compliant objects with rigid surfaces as compared with those with deformable surfaces. In addition, in the passive touch experiments with spring cells, the subjects could not discriminate the specimens at all when irrelevant cues based on ramp times were eliminated. Therefore, unlike the case of objects with deformable surfaces, tactile information alone is insufficient and kinesthetic information is necessary to discriminate the compliance of objects with rigid surfaces.

Discussion

For rubber specimens, the one-to-one correlation between the subjective sensation of softness and the objective compliance indicates that the latter is a good standard relative to which the former can be judged. Both ranking and pairwise discrimination experiments with and without local anesthesia showed the insufficiency of the kinesthetic information and the necessity of tactile information. The active and passive pairwise discrimination experiments showed the sufficiency of tactile information for softness discrimination. Therefore, in identifying the mechanistic basis for the discriminability, we only need to examine the possible cues on the skin surface within the contact region.

It has been shown that subjects can discriminate softness based on the temporal signals resulting from tapping a specimen with a rigid probe (LaMotte, *et al.*, 1994). In the experiments described above, however, the ramp velocities are much lower and the specimens come into contact gradually with the soft fingerpad. Therefore the transient mechanical signals immediately after contact are not as significant to discrimination as the more gradual variation of net force and contact area over the indentation ramp. It should be noted that the contact area can be measured in two ways: (1) The overall or nominal contact area, which is the entire area within the boundaries of contact; (2) the actual contact area, which is the sum of the areas of all the islands of contact formed with the ridges on the fingerpad skin. The net force and overall area are dependent on indentation velocity whereas the subjects could discriminate the specimen presented even when the velocities were randomized. Therefore, the net force and overall area are not the relevant cues, but their ratio, the average pressure could be invariant with respect to indentation velocity (Srinivasan and LaMotte, 1995). Thus the temporal variations of average pressure, or the spatio-temporal variations of pressure distribution over the nominal or actual contact region might be the relevant mechanical signals for the discrimination of softness of rubber specimens.

It should be noted that the temporal variation of average displacement of the skin, or the spatio-temporal variation of skin displacement distribution are equivalent to the descriptions above in terms of pressure distributions. This raises the possibility that the sensitivity of SAIs to skin surface curvature changes (LaMotte and Srinivasan, 1993; Srinivasan and LaMotte, 1987, 1991) might play a major role in softness discrimination as well. Since single fiber response depends on indentation velocity in addition to the specimen compliance, the tactile neural code for softness must be based on population response, particularly in SAIs. Whether this response depends only on the temporal variation of average skin curvature or the spatio-temporal variation of skin curvature over either the nominal contact area or the actual contact region at the fine level of fingerprint ridges remains to be investigated.

In the case of spring cells, because their surface is not deformable, the area of contact and the spatial distribution of pressure as well as skin displacements at any instant of time during the ramp are completely governed by the net contact force, and are independent of the compliance of the cell. A consequence is that tactile information can only encode variation of net force over time, which is directly affected by indentation velocity. Therefore, to unequivocally determine the compliance, subjects need kinesthetic information to know the overall indentation velocity, in addition to the force rate information which may be obtained from either tactile or kinesthetic sources.

Acknowledgment

The research work reported here was supported by ONR (Grant Nos. N00014-91-J-1454 and N00014-92-J-1814) and NIH (Grant Nos. NS33778 and NS15888). We would like to thank Jim Whitehouse, Karen Heller, Terri Beattie, Jim Gregory, Andy Klusch-Petersen, John Beggs, and Terry O'Conner for their help in data analysis.

References

Harper, R. (1952) Psychological and psychophysical aspects of studies of craftsmanship in dairying. *British Journal of Psychology Monograph Supplement* 28:1-63.

Harper, R., and Stevens, S. S. (1964) Subjective hardness of compliant materials. *Quarterly Journal of Experimental Psychology* 16:204-215.

Jones, L. A., and Hunter, I. W. (1990) A perceptual analysis of stiffness. *Experimental Brain Research* 79:150-156.

Katz, D. (1938) The judgements of test bakers. A psychological study. *Occupational Psychology* 12:139-148.

LaMotte, R. H., and Srinivasan, M. A. (1993) Responses of cutaneous mechanoreceptors to the shape of objects applied to the primate fingerpad. *Acta Psychologica* 84:41-51.

LaMotte, R. H., Zhang, J. and Yan, J. (1994) Tactile signals generated by a tool used to discriminate the softness of objects. *Society for Neuroscience Abstracts* 20:1381.

Roland, P.E. and Ladegaard-Pedersen, H. (1977) A quantitative analysis of sensations of tension and of kinaesthesia in man. *Brain* 100:671-692.

Srinivasan, M. A., and LaMotte, R. H. (1987) Tactile discrimination of shape: Responses of slowly and rapidly adapting mechanoreceptive afferents to a step indented into the monkey fingerpad. *Journal of Neuroscience* 7:1682-1697.

Srinivasan, M. A., and LaMotte, R. H. (1991) Encoding of shape in the responses of cutaneous mechanoreceptors. In: O. Franzen and J. Westman (Eds.), *Information Processing in the Somatosensory System*, Wenner-Gren International Symposium Series, Macmillan Press, London, pp 59-69.

Srinivasan, M.A. and LaMotte, R. H. (1995) Tactual discrimination of softness. *Journal of Neurophysiology* 73:88-101.

Tan, H. Z., Durlach, N. I., Beauregard, G.L., and Srinivasan, M.A. (1995) Manual discrimination of compliance using active pinch grasp: The roles of force and work cues. *Perception and Psychophysics* 57:495-510.

O. Franzén, R. Johansson and L. Terenius (eds.)

Representation of the shape and contact force of handled objects in populations of cutaneous afferents

A.W. Goodwin, A.S. Browning and H.E. Wheat

Dept. of Anatomy & Cell Biology, University of Melbourne, Parkville Victoria 3052, Australia

Introduction

Everyday, humans grasp and manipulate objects with precision. In order to achieve this, the neural mechanisms controlling movements of the fingers must have continuous information about the relationships of the fingers to the object being handled. In particular, successful manipulations depend on a knowledge of the local shape of the object, its position on the fingers and the contact forces. Some of this information is relayed by receptors in the joints and in the muscles (Burgess et al., 1982), and joint angle is also signalled by cutaneous receptors around the joints and in the hairy skin of the hand (Edin and Abbs, 1991). The fingerpads are densely innervated with low threshold mechanoreceptors which signal a variety of precise information about grasp (Johansson and Vallbo, 1979). They are the only source of information about the region of contact with the fingers, an important consideration from the point of view of the dynamics of grasp (Fearing and Hollerbach, 1985). Receptors in the extrinsic and intrinsic hand muscles signal forces used by the fingers but information about contact forces between the fingers and the object is also signalled by the cutaneous receptors in the fingerpads (Cohen and Vierck, 1993).

While joint angles provide global information about the object's shape and size, the local shape of objects in contact with the fingers relies on signals from the fingerpads. There have been a number of psychophysical studies of tactile shape using active touch where all the above receptors may contribute (Lederman and Klatzky, 1987; Roland and Mortensen, 1987). The role of the cutaneous receptors alone can be studied by using a passive presentation paradigm. In a series of psychophysical and neurophysiological studies, LaMotte and Srinivasan applied steps of varying shapes to the fingerpads passively (LaMotte and Srinivasan, 1987a,b; Srinivasan and LaMotte, 1987). Their results highlight the importance of the local curvature of the stimulus.

Methods

We have chosen to use spheres as stimuli because the shape of a sphere is defined by a single parameter and it is simple to manufacture and specify such shapes with precision (Goodwin et al., 1991, 1995; Goodwin and Wheat, 1992). This single parameter is the radius of the sphere. Alternatively, it is often more convenient to describe the sphere by its curvature which is simply the reciprocal of the radius. There are also a number of theoretical reasons for using curvature to define local shape (Richards et al., 1986; Srinivasan and LaMotte, 1991). Figure 1 shows the range of spheres used in this study. A stimulator lowered the spheres onto the pad of an immobilized finger. With this passive presentation no information about the object is conveyed by receptors in the muscles or joints or in the hairy skin, and the cutaneous receptors in the fingerpads are the only source of information. The stimulator used was based on the principle of a balanced beam. Motion of the beam was damped so that objects were lowered onto the fingerpad with a constant velocity of about 20 mms^{-1}. By adjusting the counterweights on the beam the contact force could be controlled. The stimulator was mounted on a 2-dimensional stage driven by micrometers so that the position of the sphere on the finger could be controlled. Psychophysical experiments on humans (scaling and discrimination) were used to define the nature of the information relayed by the cutaneous receptors to the brain. The stimuli were in contact with the finger for about 1 second. Our primary interest is in how these signals are conveyed to the central nervous system by the primary afferent fibers in the digital nerves. To establish this, we recorded responses from single nerve fibers in the median nerves of anaesthetised monkeys. We used these single fiber data to synthesize the responses of populations of afferents. Since the same stimuli were used in the human psychophysical

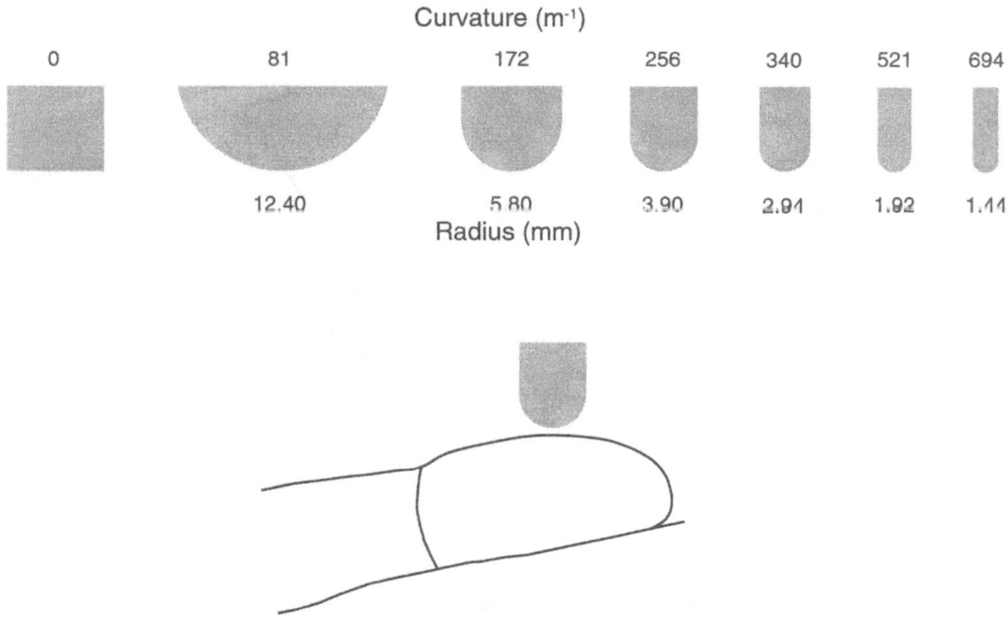

Figure 1. Stimuli consisted of spherically curved surfaces ranging from flat (curvature 0, radius ∞) to highly curved (curvature 694 m^{-1}, radius 1.44 mm). These were applied passively to the fingerpad with the direction of force normal to the plane tangential to the fingerpad at the point of contact.

experiments as in the neural recordings from monkeys, the results can be compared to elucidate the neural representations of the stimulus parameters.

Results and Discussion

While there have been many studies showing that humans can scale contact force (eg Greenspan, 1984), the ability to scale curvature has not been established. It is also important to know the extent to which these 2 parameters interact. Thus, two series of human scaling experiments were done. In the first series, subjects were presented with the 7 objects shown in Figure 1, at one of 3 contact forces. The stimuli were randomized and subjects were asked to scale for their perceived curvature. As shown in Figure 2 (left), an increase in the curvature of the stimulus resulted in an increase in perceived curvature, with contact force having only a small

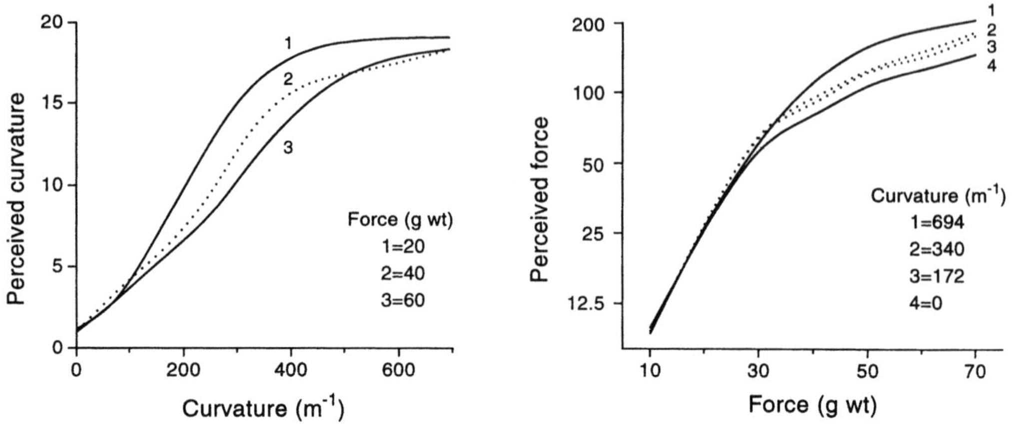

Figure 2. Psychophysical data for humans scaling objects with different curvatures applied at different contact forces. *Left.* Scaling for curvature (7 curvatures and 3 contact forces). *Right.* Scaling for contact force (7 forces and 4 curvatures).

effect. In the second series shown in Figure 2 (right), subjects were presented with objects at 7 different contact forces and one of 4 different curvatures in random order, and were asked to scale for their perceived contact force. An increase in contact force led to an increase in perceived force with little interaction from the curvature of the stimulus. These results show that the activity in the digital nerve fibers innervating the fingerpad contains information enabling the shape of the object to be extracted, even when contact force is varying, and enabling the contact force to be extracted, even when shape is varying. The precision of the representation is indicated by the results of discrimination experiments. For example, the ability of humans to discriminate the curvature of 2 spheres was measured using a forced choice paradigm analyzed by standard decision theory. For spheres with a curvature of 287m^{-1} (radius 3.48 mm) or 144 m^{-1} (radius 6.95 mm) the difference threshold was about 10%.

Responses of single fibers with receptive fields on the central region of a fingerpad were recorded in anesthetized monkeys. Slowly adapting type I afferents (SAIs) responded vigorously to our stimuli. Some of the rapidly adapting (Meissner) afferents did not respond at all and the remainder responded weakly. Most Pacinians did not respond. We will describe the responses of the SAIs. The measure of response used is the number of impulses evoked in the first second of response; this corresponds to the psychophysical data. In the first series of experiments we examined the responses of single afferents to stimuli applied at the receptive field center. Two

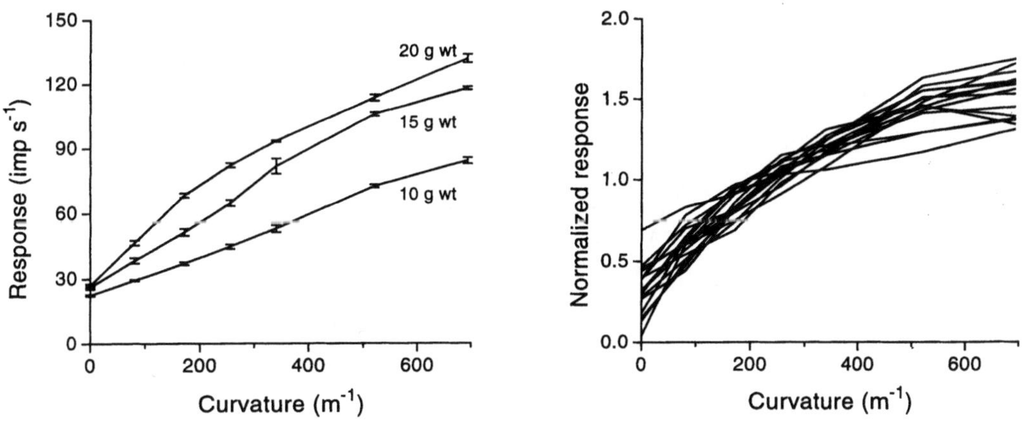

Figure 3. The effect of curvature and contact force on the responses of SAIs stimulated at the center of the receptive field. *Left.* Responses (mean±SEM) of a single fiber to stimuli repeated 8 times (over a period of 8.5 hours for 15 g wt). *Right.* Normalized responses of 16 SAIs at a contact force of 10 g wt. Each afferent's responses were normalized by dividing by the average response of that afferent to the seven curvatures.

parameters, the curvature and the contact force, were varied systematically in a random design. The responses of a single typical SAI are shown in Figure 3 (left). An increase in the curvature of the stimulus resulted in an increase in the response of the unit, and an increase in contact force also resulted in an increase in response. The low standard errors show the consistency of responses. All SAIs behaved in a similar way, but some fibers were more sensitive than others and as a consequence had greater responses magnitudes. The most sensitive fiber was 4.3 times as responsive as the least sensitive fiber. Differences in responses between fibers due to differences in sensitivity were removed by normalizing. The normalized responses were similar for all fibers regardless of their sensitivity as illustrated in Figure 3 (right). That is, the underlying function relating curvature and contact force to response was the same for all SAIs. This function can be expressed by:

$$y=sk[1.91-1.62\exp(-0.00243x)]$$

where y is the normalized response, x is the curvature, s is the sensitivity of the fiber, and k is the force factor (0.792, 1.01 and 1.21 for 10, 15 and 20 g wt respectively).

Single unit responses changed consistently with changes in curvature and also with changes in contact force, but there is clearly ambiguity in that an increase in a fiber's response may be due to an increase in curvature or due to an increase in contact force. This ambiguity can be resolved

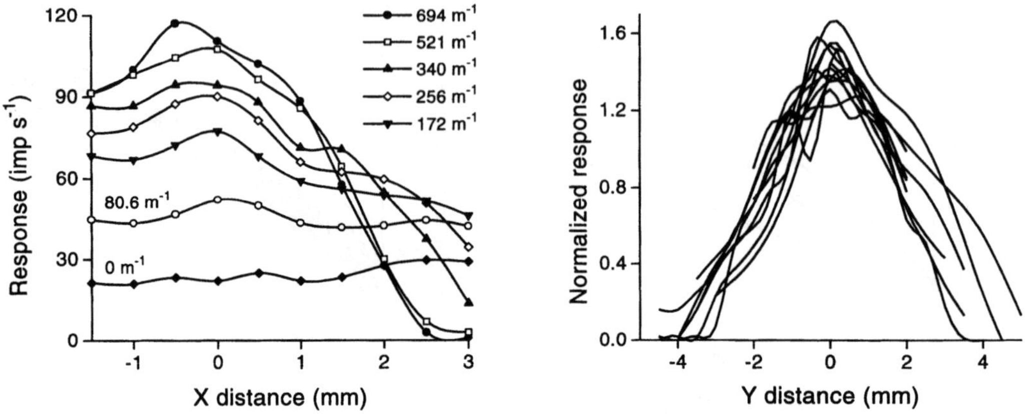

Figure 4. Response profiles. *Left.* Responses of a single afferent at points along a line through the receptive field center and orthogonal to the long axis of the finger. *Right.* Responses of 13 afferents normalized to remove the effect of the afferents' sensitivities. Profiles are shown along a line through the receptive field center and parallel to the long axis of the finger. The sphere had a curvature of 340 m^{-1}.

by examining the responses of the population of SAIs innervating the fingerpad. Since we were unable to record simultaneously from all the active SAIs, we synthesized the population response from the responses of single units.

Response profiles for single afferents were measured by systematically varying the position of the spheres with respect to the receptive field center of the fiber. These profiles are defined over the 2-dimensional skin surface; the convention used for coordinates is that the x and y axes pass through the center of the receptive field with the y axis parallel to the long axis of the finger. Profiles for stimuli positioned successively at 0.5 mm intervals along a line orthogonal to the long axis of the finger (x axis) are shown for a single SAI in Figure 4 (left). The line for the sphere with a curvature of 694 m^{-1} (filled circles) shows that as the sphere moved away from the center of the receptive field, the response declined. Comparing with the line for a less curved sphere (340 m^{-1} - up triangles) it can be seen that here too response declined as the sphere moved away from the receptive field center, but the decline was less rapid than for the more curved sphere. Also, as seen before, the response at the receptive field center was less for the sphere with a curvature of 340 m^{-1}. This effect of decreasing curvature is evident for all the spheres in Figure 4 (left); as the curvature of the sphere decreased, the profile decreased in height and became broader. Thus the shape of the profile corresponds to the shape of the stimulus.

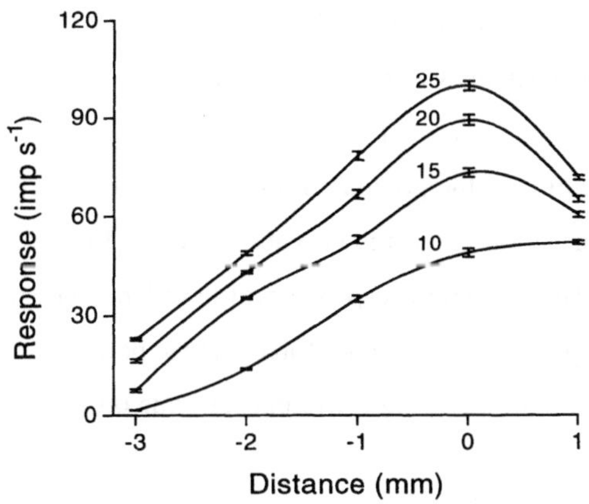

Figure 5. Response profiles (mean±SEM, n=8) for a single fiber responding to a sphere with a curvature of 521 m^{-1} positioned at points along a line through the receptive field center and parallel to the long axis of the finger. Contact forces were 10, 15, 20 or 25 g wt.

The shape of the profile was consistent from fiber to fiber as illustrated in Figure 4 (right) which shows normalized profiles for 13 fibers along a line parallel to the long axis of the finger (y axis). By positioning the stimuli at points on a 0.5 mm grid over the fingerpad, we defined the response profiles as a function of the 2-dimensional position on the fingerpad. Profiles were described well by the function:

$$z=a\exp(-bx^2-cy^2)$$

where z is the normalized response, and x and y are distances from the center of the receptive field in directions orthogonal to and parallel to the long axis of the finger respectively. The constants a, b and c are different for different curvatures. These profiles may be interpreted in 2 ways. They are a direct measure of the responses of a single afferent to stimuli positioned at different points in the receptive field. They may also be viewed as the profile of responses across a population of afferents innervating the fingerpad and responding to a sphere positioned at the center of the population (x=0, y=0). At this stage the simplifying assumption is made that all fibers have the same sensitivity. It should also be born in mind that these profiles are in fact not continuous, but are sampled by the responding afferents at a sampling rate determined by the innervation density.

The effect of a change in contact force on the response profiles is illustrated in Figure 5. As contact force increased the responses at each point increased. As a result, the profiles shifted upwards while maintaining their basic shape.

The above results explain how the population of slowly adapting type I afferents innervating the fingerpad provide information about the curvature and the force of a sphere in contact with the finger. The shape of the population response profile reflects the shape of the object and the height of the profile corresponds to the contact force. Thus, both parameters of the stimulus can be extracted with fidelity even though single afferents provide only ambiguous information.

Acknowledgments
This research was supported by a grant from the National Health and Medical Research Council of Australia.

References

Burgess, P.R., Wei, J.Y., Clark, F.J. and Simon, J. (1982) Signaling of kinesthetic information by peripheral sensory receptors. *Annual Review of Neuroscience* 5:171-187.

Cohen, R.H. and Vierck, C.J. Jr. (1993) Relationships between touch sensations and estimated population responses of peripheral afferent mechanoreceptors. *Experimental Brain Research* 94:120-130.

Edin, B.B. and Abbs, J.H. (1991) Finger movement responses of cutaneous mechanoreceptors in the dorsal skin of the human hand. *Journal of Neurophysiology* 65:657-670.

Fearing, R.S. and Hollerbach, J.M. (1985) Basic solid mechanics for tactile sensing. *International Journal of Robotics Research* 4:40-54.

Goodwin, A.W., John, K.T. and Marceglia, A.H. (1991) Tactile discrimination of curvature by humans using only cutaneous information from the fingerpads. *Experimental Brain Research* 86:663-672.

Goodwin, A.W. and Wheat, H.E. (1992) Magnitude estimation of force when objects with different shapes are applied passively to the fingerpad. *Somatosensory and Motor Research* 9:339-344

Goodwin, A.W., Browning, A.S. and Wheat, H.E. (1995) Representation of curved surfaces in responses of mechanoreceptive afferent fibers innervating the monkey's fingerpad. *Journal of Neuroscience* 15:798-810

Greenspan, J.D. (1984) A comparison of force and depth of skin indentation upon psychophysical functions of tactile intensity. *Somatosensory and Motor Research* 2:33-48.

Johansson, R.S. and Vallbo, A.B. (1979) Tactile sensibility in the human hand: relative and absolute densities of four types of mechanoreceptive units in glabrous skin. *Journal of Physiology* 286:283-300.

LaMotte, R.H. and Srinivasan, M.A. (1987a) Tactile discrimination of shape: Responses of slowly adapting mechanoreceptive afferents to a step stroked across the monkey fingerpad. *Journal of Neuroscience* 7:1655-1671.

LaMotte, R.H. and Srinivasan, M.A. (1987b) Tactile discrimination of shape: Responses of rapidly adapting mechanoreceptive afferents to a step stroked across the monkey fingerpad. *Journal of Neuroscience* 7:1672-1681.

Lederman, S.J. and Klatzky, R.L. (1987) Hand movements: A window into haptic object recognition. *Cognitive Psychology* 19:342-368.

Richards, W., Dawson, B. and Whittington, D. (1986) Encoding contour shape by curvature extrema. *Journal of the Optical Society of America* 3:1483-1491.

Roland, P.E. and Mortensen, E. (1987) Somatosensory detection of microgeometry, macrogeometry and kinesthesia in man. *Brain Research Reviews* 12:1-42.

Srinivasan, M.A. and LaMotte, R.H. (1987) Tactile discrimination of shape: Responses of slowly and rapidly adapting mechanoreceptive afferents to a step indented into the monkey fingerpad. *Journal of Neuroscience* 7:1682-1697.

Srinivasan, M.A. and LaMotte, R.H. (1991) Encoding of shape in the responses of cutaneous mechanoreceptors. *In*: O. Franzen and J. Westman (eds): *Information processing in the somatosensory system.* MacMillan Press, London, pp 59-69.

Somesthesis and the Neurobiology of the Somatosensory Cortex
O. Franzén, R. Johansson and L. Terenius (eds.)

Haptic object processing I: Early perceptual features

Roberta L. Klatzky[1], Susan J. Lederman[2] and Cheryl O'Neill[3]

[1]*Dept. of Psychology, Carnegi e Mellon University, Pittsburgh, PA 15213, USA*
[2]*Depts. of Psychology and Computer and Information Science, Queen's University, Kingston, Ontario, Canada K7L3N6*
[3]*Dept. of Psychology, Queen's University, Kingston, Ontario, Canada K7L3N6*

Summary. Early perceptual features are defined as values on dimensions that strongly activate one or more detector populations. This research is directed at identifying early perceptual features in the haptic modality, using a search task. Results of several experiments indicate that there are early perceptual features within the haptic system for material properties and spatial discontinuities, but not for spatial relations. Whether continuous surface contour constitutes a primitive remains an open question.

Introduction

This chapter describes a program of research that is directed at identifying perceptual features within the haptic modality that are available at early stages of processing. In this program we use a paradigm developed by Anne Treisman, which calls for a search for a target stimulus among a set of distractors. Targets and distractors differ from one another on some dimension; for example, the target may be rigid and the distractors compliant. Response time is measured as a function of the number of items in the search display. When this function is flat, the target is said to "pop out."

We use this task, as has Treisman, in an effort to identify early perceptual features. For our purposes, an early perceptual feature is a value on a dimension that strongly activates one or more populations of functional detectors following an initial brief contact. The term "functional" is meant to indicate that the detectors are presumed to reside at a relatively high level and to reflect the contribution of multiple levels still earlier in the haptic system, rather than being a peripheral neural population. As yet, little is known about the somatosensory encoding mechanisms that contribute to the representation of a perceptual feature at this level.

Treisman (1991; 1993) has made substantial progress in identifying functional detectors within the visual system. Such detectors appear to be selectively sensitive to a range of values on some quantitative dimension (e.g., line curvature, orientation, length). Values within this range of tuning collectively constitute the perceptual feature. When a target having a feature is presented among a set of non-feature distractor items, selective tuning to the target's value means that it tends to "pop out" to the perceiver. The result is that the number of distractor items does not affect the amount of time needed to detect the target. The mechanism used to explain perceptual pop-out is a high signal-to-noise ratio within the functional detctor population. That is, pop-out occurs when the target triggers a high level of detector activity relative to the pooled activity triggered by the set of non-feature distractors. In contrast, a non-feature target produces relatively little activity in the detector. When such a target is presented in a set of distractors within which the feature is present, the distractors produce a high level of detector activity and hence the signal-to-noise ratio is low within that detector population. Unless there is another detector population, selectively tuned to the value of the alternative feature value now serving as target, there will be no pop-out. When one value on a dimension pops out and the other does not, it produces an asymmetry in the slopes for visual search tasks (Treisman and Gormican, 1988).

In the haptic search task, search over the visual field becomes search over multiple fingers. On each trial in our studies, the subject is told about a designated target value on some dimension, for example, "rough". He or she is then presented with haptic stimulation to from one to six fingers and is asked to indicate whether the target value is present on any finger. On a given dimension, all distractors have the same value, which differs from that of the target; usually in our studies, the two values differed maximally.

We use a specialized apparatus for this task (Moore, Broekhoven, Lederman and Ulug, 1991). The subject's middle three fingers of each hand are placed on hand supports that leave the fingertip open to stimulation from below. That stimulation is accomplished by lifting from one to six surfaces toward the fingers. The surfaces are on a set of 6 stimulus drums, mounted on a platform that can be raised and lowered under software control. Each drum has 8 planar facets distributed around its circumference, to which various haptic stimuli such as textured surfaces and bars can be attached; a cutaway facet is used when no surface is to be presented to the corresponding finger. The display for a given trial is prepared by rotating the six drums under software control until the desired facets are facing upward. After a designated interval, the drums are simultaneously raised so as to to contact the selected finger combinations. Thumb switches are used to make the response.

Over a substantial number of studies, we have investigated four general types of features: (i) discriminations of material, such as finding a rough surface among smooth ones; (ii) discriminations involving surface discontinuities, such as looking for a surface with a raised bar among flat surfaces; (iii) discriminations between different continuous 3-dimensional surface contours, such as finding a slanted surface among flat distractors; and (iv) discriminations involving relative spatial positions of surfaces or surface discontinuities, such as looking for a surface with a horizontal bar among surfaces with vertical bars or looking for a raised dot to the right of a central indentation among distractors having the dot on the left. In most of our experiments, searches were done in both directions; that is, each feature value served as a target on some trials and a distractor on others. In all, we tested 13 dimensions, some along more than two values, as listed in Table 1.

Results and Discussion

The response times for these tasks were generally reasonably linear, but the slopes had a substantial range, as can be seen clearly in Figure 1, which contrasts two search conditions. When searching for a rough target among smooth distractors, slopes close to zero were obtained, indicating pop-out. When searching for a raised dot to the right of the center of a display, which was marked by a circular indentation, slopes averaged about 450 msec, indicating item by item search. We chose a 30-ms slope as the empirical cutoff generally used to define pop-out, based on a preliminary examination of the distribution of slopes and statistical tests of the slope values against zero. (Our classification is based on slopes averaged over target-present and target-absent trials, which were generally similar.)

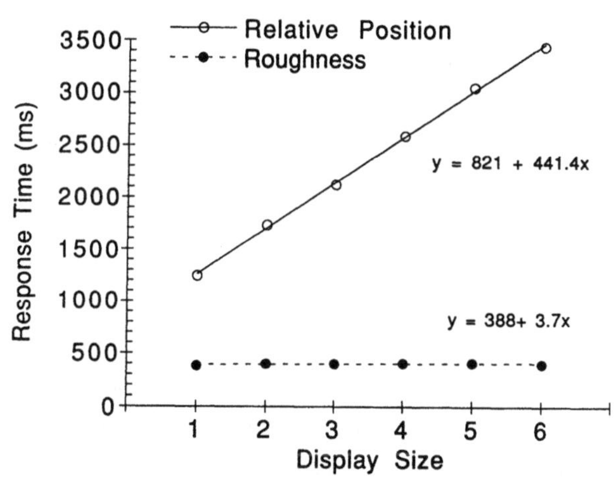

Figure 1. Response time to search for a target as a function of display size, for two conditions: search for a rough target in smooth distractors, and search for a raised dot to the right of an indentation among distractors having a raised dot to the left of an indentation. Data are averaged over target-present and target-absent conditions, and least-squares fit linear functions are also shown.

Table I. Search Outcome by Dimension

SEARCH DIMENSION	SEARCH OUTCOME	
	slope < 30 ms in 1 or more directions	slope > 30 ms in both directions
MATERIAL		
rough* vs. smooth	X	
hard* vs. soft*	X	
cool* vs. warm	X	
SURFACE DISCONTINUITIES		
vertical bar* vs. flat*	X	
horizontal bar* vs. flat*	X	
hole* vs. no hole	X	
deep* hole vs. shallow	X	
CONTINUOUS 3D SURFACE CONTOUR		
curved vs.flat		X
ramp† vs.flat	X†	
RELATIVE SPATIAL POSITION		
left planar position vs. right		X
horizontal bar vs.vertical		X
right oblique bar vs. left		X
right 3D ramp vs. left		X

*target value producing slope < 30 ms/item
†slope = 42 ms., but not significantly different from zero

The results can be seen in Table I. It indicates, for each search contrast, whether it produced a low slope, constituting pop-out, in at least one direction or a high slope in both directions. Clearly, the pattern is quite different across the dimension types.

To summarize the results, searches involving highly discriminable values on material dimensions generally produced the pattern called pop-out -- response times that were independent of the number of distractors in which the target was embedded -- at least in one direction. Searches involving surface discontinuities, when contrasted with flat surfaces, also tended to produce pop-out in at least one direction. However, judgments of relative spatial position apparently required item-by-item search. This pattern of results suggests a distinction between two broad classes of dimensions, those in which different features are defined by magnitude along the dimension (sometimes called intensive dimensions) and those in which different values have equal magnitude but different spatial location.

A number of searches in which one of the feature values produced a relatively low slope failed to do so with the other feature value, resulting in search asymmetries. The asymmetric pattern -- pop-out with one value of a dimension, but not the other -- was found among the material dimensions when searching for targets that were rough and when searching for those that were cool. However, searches involving the hard/soft distinction produced pop-out in both directions, indicating no preference for a functional detector population at one or another end of the dimension. With respect to the thermal contrast, apparent temperature was manipulated by the material presented to the skin, which would determine the rate of heat flow away from the body. Rapid heat flow, producing a cool sensation, is what appears to pop out.

Whereas contrasts between materials, and those between continuous and discontinuous surfaces, generally produced low search rates, distinctions between continuous surface contours are less clear. Two such contrasts were tested, with somewhat inconsistent results: Search for a ramped continuous 3D surface among flat surfaces produced a slope not significantly different from zero, but search for a curved one produced a significant slope. The ramped-vs.-flat discrimination may act not as a continuous contour, however, but as a surface discontinuity, because the finger may lose contact with the lower portion of the ramp. Thus we are uncertain whether there are early perceptual features defined by continuous 3D surface contours.

We believe that the lack of pop-out when targets and distractors are distinguished by spatial position, rather than magnitude along a dimension, is a general phenomenon. It was found within all searches of that type that we tested. The fact that error rates in spatial-position tasks were comparable to or better than those obtained from other tasks with much faster search rates suggests that it is not merely the difficulty of the discrimination that undermines rapid search. Our earlier work (see next chapter) indicates that the encoding of the layout of contours in space is difficult for the haptic system and often requires a specialized means of exploration -- following the surface contours with the fingers. We often observed subjects in these discrimination tasks using contour following. Spatial information that is obtained when each of several fingers follows a distinct contour is unlikely to be processed in parallel; indeed, motoric restrictions alone may prevent the fingers from simultaneous contour extraction in such circumstances.

Conclusions

These results indicate that there are early perceptual features within the haptic system for material properties and spatial discontinuities, but not for spatial relations. Whether continuous surface contour constitutes a primitive remains an open question. How these primitives are used in haptic object identification will be discussed in our companion chapter. A fuller discussion of the data reported here may be found in the paper by Lederman and Klatzky (1995).

References

Lederman, S. J., and Klatzky R.L. (1995) Early perceptual features for haptic object processing. *Submitted for publication.*

Moore, T., Broekhoven, M., Lederman, S.J. and Ulug, S. (1991) Q'Hand: A fully automated apparatus for studying haptic processing of spatially distributed inputs. *Behavior Research Methods, Instruments, & Computers* 23: 27-35.

Treisman, A. (1993) The perception of features and objects. *In*: A. Baddeley and L. Weiskrantz (eds.): *Attention: Selection, awareness, and control: A tribute to Donald Broadbent*, Clarendon Press, Oxford, pp 5-34.

Treisman, A. (1991) Search similarity, and integration of features between and within dimensions. *Journal of Experimental Psychology: Human Perception and Performance* 17: 652-676.

Treisman, A. and Gormican, S. (1988) Feature analysis in early vision: Evidence from search asymmetries. *Psychological Review* 95:15-48.

Somesthesis and the Neurobiology of the Somatosensory Cortex
O. Franzén, R. Johansson and L. Terenius (eds.)

Haptic object identification II: Purposive exploration

Susan J. Lederman[1] and Roberta L. Klatzky[2]

[1]*Dept. of Psychology and Dept. of Computing and Information Science, Queen's University, Kingston, Ontario, Canada K7L3N6*
[2]*Dept. of Psychology, Carnegie Mellon University, Pittsburgh, Pennsylvania, 15213 U.S.A.*

Summary. This paper reviews recent research on haptic exploration of multidimensional objects. We consider the nature of manual exploration in terms of the prevailing characteristics of various "exploratory procedures" ("EPs"), that is, stereotypical hand movement patterns; we also explore the impact of these characteristics on the sequence in which EPs are selected during a constrained object identification task.

Introduction

The first paper in this pair described our efforts to determine early perceptual features in the haptic system, using a haptic search task. By the term "haptic", we are referring to an information-processing perceptual system that uses inputs from receptors embedded in the skin, muscles, tendons and joints (Loomis and Lederman, 1986). The haptic search task requires subjects to search for a target stimulus among distractors, all presented to the fingertips. To infer the identity of early perceptual features, we used the slope of the function relating response time to the number of items presented. We have suggested a 30-msec slope as a reasonable cutoff point for identifying features. In this paper, we will consider the perceptual properties that are used in haptic object identification and how they are extracted, beginning with the early perceptual features available from brief contact and extending our focus to properties extracted by systematic exploration.

Some time ago, we (Klatzky, Lederman and Metzger, 1985) empirically demonstrated that people are very skilled at identifying common objects on the basis of touch alone. Blindfolded adults could name 100 common objects (e.g., cup) with almost perfect accuracy, usually in only a few seconds. We have proposed that the speed and accuracy of object identification with touch reflects parallel search of memory based on multiple object properties, both material and geometric. Subjects in our initial object identification study mentioned properties such as "smooth, hard, cold, round, and weighs little" when asked to explain the bases of their object recognition judgments, thus emphasizing the value of multidimensional sources of information. We have demonstrated that while shape is critical to identifying objects at the level of common naming, access to material information speeds identification (Klatzky, Loomis, Lederman, Wake, and Fujita, 1993).

From the haptic search task, we know that initial contact with an object provides at least coarse information about its perceptual properties (for details, see Lederman and Klatzky, 1995). Our focus here is on how perceived properties lead to object identification. First let us consider identification under conditions similar to those of the search task -- brief, confined to the fingertips, and with minimal movement. We call a brief fingertip exposure, on the order of 200 ms, a "haptic glance." An interval of 200 ms would allow search among 6 fingers at a 30-ms rate. Thus it should allow encoding of early perceptual features identified in the haptic search task, including roughness, compliance, local shape, and apparent coolness. The utility of this information for object identification was indicated in a recent study (Klatzky andLederman, 1995), in which we allowed subjects only 200 ms of fingertip contact with an informative local region on an object and found identification rates well above chance.

To reach virtually 100% accuracy on an arbitrary population of objects, however, subjects typically explore an object for a matter of seconds, not milliseconds, and their contact with an object is far more extensive than that of a haptic glance. In fact, we commonly observe subjects to use systematic hand movements to extract object properties. These hand movements have been the focus of our extensive and systematic research program on the "exploratory procedures" (EPs) used for haptic object perception and recognition.

Exploratory procedures and the extraction of associated properties

Our initial observations with unconstrained object identification led us to suspect that the explorer's hand movements might offer the key to understanding why touch alone can identify common objects so successfully. Accordingly, we investigated how people actually used their hands when asked to make haptic evaluations of various object properties (Lederman and Klatzky, 1987). On each trial, subjects were initially shown a multidimensional object (the "standard"), and asked to examine a particular property, for example, its roughness. Next, they were presented with a sequence of three other multidimensional objects (the "comparison" stimuli). Subjects had to select the comparison object that best matched the standard in terms of the named dimension (roughness). We videotaped and subsequently analyzed the hand movements performed during each trial.

One thing was immediately apparent. Although subjects were usually unaware of what they did with their hands, their unconstrained movements proved highly purposive and systematic.

Subjects performed very stereotypical movement patterns that possessed both necessary and typical features and that were linked to the target property. These patterns constitute the EPs. A selection of those we have emphasized in our work to date are shown in Figure 1, as they typically occurred; also shown is the linked property.

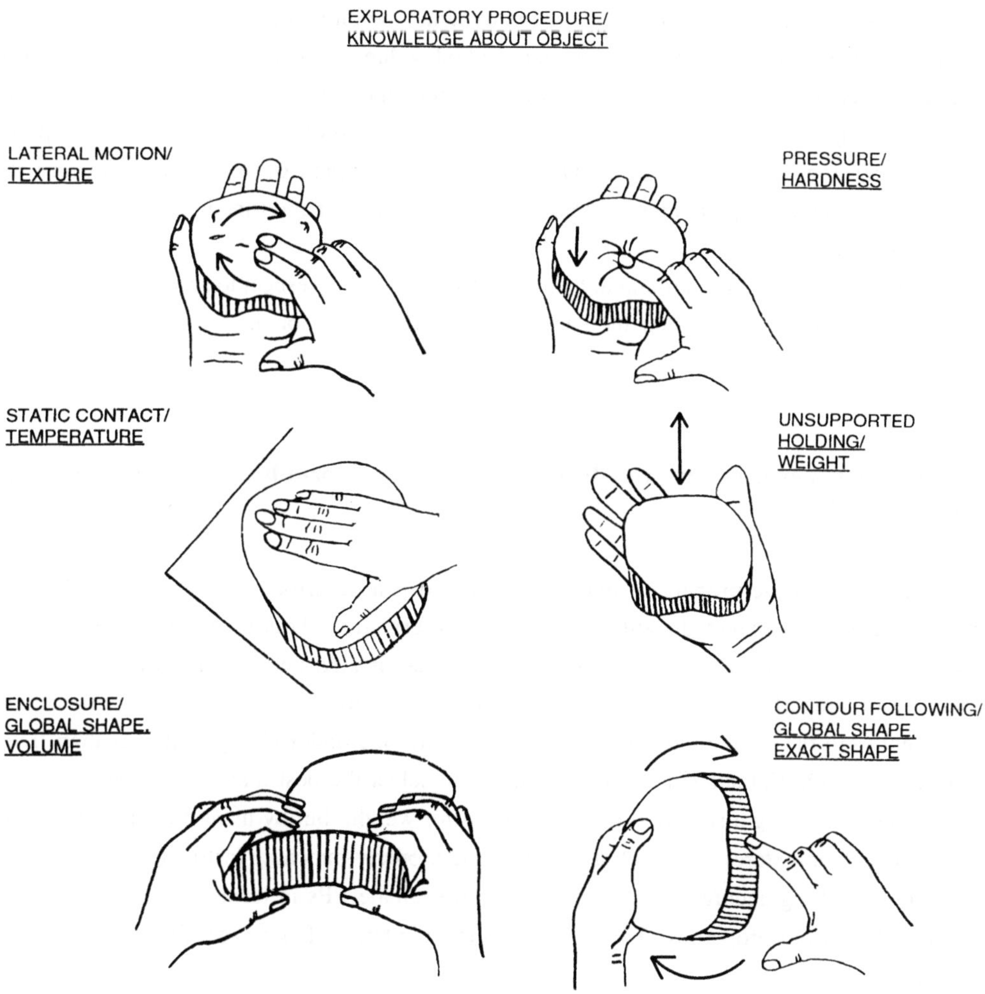

Figure 1. Exploratory Procedures and associated properties (reprinted with permission from Lederman 1991, as revised from Lederman & Klatzky, 1987).

The EP associated with texture judgments was *Lateral Motion,* a repetitive, lateral rubbing motion. *Pressure*, or opposing forces applied normally to a surface or a torque about some object axis, was used to extract information about hardness. *Static Contact*, involving stationary contact on a surface without contour molding, was used mainly to determine thermal properties. *Unsupported Holding*, or lifting an object away from a supporting surface, was used to extract information about weight. *Enclosure*, involving dynamic molding of the palm and/or finger(s) to the contours of an object, was chosen for extracting both volumetric cues and envelope information about an object's shape. *Contour Following*, which involved dynamic edge following or following along surfaces in the absence of edges, was used to obtain the more precise spatial details concerning an object's shape. Other EPs not mentioned or shown here concern the extraction of information pertaining to the motion of an object part and function, as determined by the object's structure.

At the level of analysis found to be most informative with respect to associations between EPs and object properties, neither the particular end effector(s) nor the area of skin contacted proved relevant.

Characteristics of exploratory procedures: Accuracy, speed, breadth of sufficiency, and EP-EP compatibility

A variant of the property-matching experiment just described (Lederman and Klatzky, 1987; Experiment 2) provided additional important information about the relative precision and speed with which each EP extracts various properties; it also enabled us to assess an EP's generality or breadth of sufficiency in simultaneously extracting multiple properties. The experiment used the same methodology as before, with one major change. Rather than freely exploring, on each trial subjects were constrained to select the best match on some designated property by performing a specified EP (e.g., match for weight, but by using Lateral Motion, the appropriate EP for texture). Across trials, all EPs and property-matching instructions were paired.

The resulting accuracy scores were used to evaluate EPs in terms of their precision in extracting the various properties. The results are shown in terms of an EP-property weight matrix, in Table I, top.

Table I. Characteristics of EPs: EP-to-property weights, average EP duration, and EP sufficiency (Adapted from Lederman and Klatzky, 1990a, and Klatzky and Lederman, 1993, reprinted with permission)

EP-to-property Weights

	Property						
EP	texture	hardness	temp.	weight	volume	global shape	exact shape
Lateral Motion	2	1	1	0	0	0	0
Pressure	1	2	1	0	0	0	0
Static Contact	1	0	2	0	1	1	0
Unsupported Holding	0	1	1	2	1	1	0
Enclosure	1	1	1	1	2	2	0
Contour Following	1	1	1	1	1	1	3

EP duration and breadth of sufficiency

EP	Duration (sec)	Breadth of Sufficiency
Lateral Motion	3.5	3
Pressure	2.2	3
Static Contact	0.1	4
Unsupported Holding	2.1	5
Enclosure	1.8	6
Contour Following	11.2	7

A cell has a "0" entered when the EP did not extract the property with above-chance accuracy. It has a "1" when the EP led to above-chance accuracy -- we say the EP is "sufficient" for extracting that property. It has a "2" when the EP led to better performance than any other -- in this case, the EP is considered "optimal" for extracting the designated property. Finally, the cell has a "3" when the EP was shown to be "necessary" for extracting the designated property, because no other EP yielded above-chance performance. This is the case for judgments of precise shape, where Contour Following is necessary.

In the bottom of Table I, we show the average duration (in seconds) for each EP, as determined from the first experiment, which used free exploration. Note that when full contour information is required, Contour Following is very much slower than the other EPs. Such temporal differences in exploration might generally bias a haptic observer against using Contour Following.

We also show in Table I, bottom, the relative breadth of sufficiency for each EP, based on the results of the second constrained-exploration experiment. The numeric entry for each EP represents the number of non-zero cells in the corresponding row above. This number indicates the number of properties for which that EP was sufficient. Thus, both Lateral Motion and Pressure are sufficient for extracting relatively few properties, whereas Enclosure and Contour Following are both broadly sufficient procedures -- each can provide fairly coarse information about almost all of the object properties shown here.

Finally, we have considered how compatible EPs are with one another, that is, the extent to which any two may be performed in tandem or very close in time. In assessing "EP-EP compatibility", we (Klatzky and Lederman, 1993) have determined four parameters to be necessary and sufficient for differentiating EPs from one another. The EPs are analyzed by these parameters in Table II.

Table II. Values of EPs on four parameters (from Klatzky and Lederman, 1993, reprinted with permission)

	Parameters			
	Movement	Direction	Region	Workspace Constraint?
Static contact	Static	Normal	Surface	No
Pressure	Dynamic	Normal	Surface	No
Lateral Motion	Dynamic	Tangential	Surface	No
Enclosure	Static	Normal	Surface & Edges	No
Contour Following	Dynamic	Tangential	Edges	No
Unsupported Holding	Static	Normal	Surface & Edges	Yes

The parameters are movement (whether the EP is static or dynamic), direction of force applied (normal; tangential), the object region explored (edges and/or central surface), and position relative to the workspace (whether or not the object must be removed from a supporting surface). Each EP has a value on the four parameters. These values act as constraints; for example, the value "normal" on the direction-of-force parameter is a constraint that the EP must be performed with normal force.

The relationships between the parameter values for any two EPS determines whether they can be performed in tandem, that is whether they are compatible. Based on an analysis of the parameters and a substantial body of observational data on EPs, we have developed a set of rules that indicate when two different values on a parameter may be satisfied simultaneously (for details, see Klatzky and Lederman, 1993). If joint satisfaction holds for all parameter values that differentiate two EPs, they can be performed together, or are "compatible." The EP-EP compatibilities are presented in Table III as a matrix of binary values. A "+" indicates compatibility, meaning that the differing parameter values of the two EPs do not prevent them from being performed together, whereas a "-" indicates incompatibility.

Table III: Compatibility relations between EPs (+ means compatible; - means incompatible) (from Klatzky and Lederman, 1993, reprinted with permission)

	EP				
	Pressure	Lateral Motion	Enclosure	Contour Following	Unsupported Holding
Static Contact	+	-	+	-	+
Pressure		+	+	-	+
Lateral Motion			-	+	-
Enclosure				-	+
Contour Following					-

A computational model of the manual-exploration sequence

From a connectionist point of view, we have treated the EP selection process as a constraint-satisfaction algorithm in which the EP-property weights (Table I) and EP-EP compatibility weights (Table III) function as constraints that are progressively relaxed until some element (an EP) is maximally activated. Our network successfully predicted the order in which subjects selected EPs in a cued object identification task, to which we now turn.

In the cued task, we (Lederman and Klatzky, 1990b) required subjects to decide whether or not an object placed in the hands was a member of a designated category, where the name was either at a relatively inclusive or specific level (e.g., a wine glass might be called a "drinking vessel" or a "glass"; see Rosch, 1978, for details on object classification taxonomy). Subjects were permitted to freely explore the objects; hand movements were videotaped on each trial. We had previously obtained judgments for each object category as to what properties were particularly diagnostic for identifying its members.

Exploratory activity during each trial was analyzed as a sequence of EPs. People consistently used EPs in a 2-stage sequence, shown in Figure 2. The graph shows schematically the percent cumulative occurrence of each EP as a function of its position in the EP sequence (i.e., the proportion of its total occurrences that occurred at each position). Stage 1 (represented by the thick, dark lines) consisted of a general grasp-and-lift (Enclosure and Unsupported Holding) routine, regardless of the object category targeted in the question. These two EPs, both of which are broadly sufficient and relatively fast to perform, can provide coarse information about almost all object properties. In contrast, Stage 2 (represented by the thinner, lighter lines) consisted of the more precise EPs that are optimal for extracting the property most diagnostic of the targeted object category.

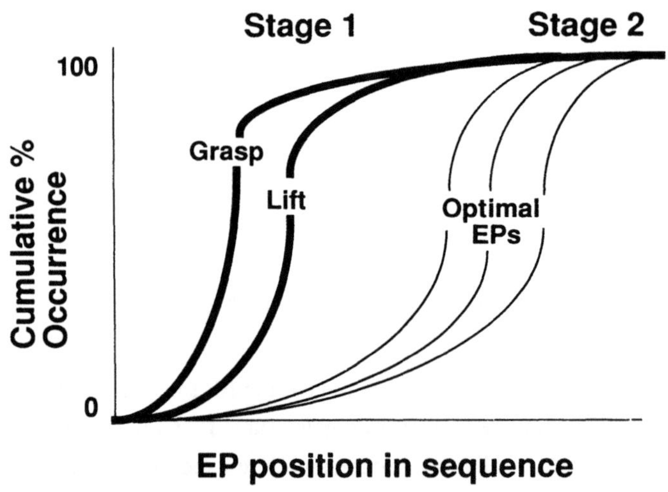

Figure 2. The two-stage sequence of haptic exploration during a constrained object identification task. Percent cumulative occurrence of each EP is plotted schematically as a function of its position in the EP sequence (reprinted with permission from Lederman and Klatzky, 1990b).

This 2-stage sequence was predicted by differentiating two versions of our network model. The first used the full EP-to-property weight matrix, thus giving positive weights to EPs that were merely sufficient for a desired property. This led to the grasp-and-lift EPs being selected for use, regardless of which property was diagnostic, reflecting their broad sufficiency to detect object properties. The second version of the model represented a need for greater precision and therefore used a reduced weight matrix, having only optimal EP-to-property weightings. This led to selection of the specific EP associated with a given targeted property. The model therefore indicates that optimal EPs will be used subsequent to an initial stage of exploration, which incorporates only grasping and lifting. That stage is itself preceded by brief initial contact, or a haptic glance, which we have found is in some cases sufficient by itself to effect object identification.

References

Klatzky, R.L. and Lederman, S.J. (1995) Identifying objects from a haptic glance. *Perception and Psychophysics; in press.*

Klatzky, R.L., and Lederman, S.J. (1993) Toward a computational model of constraint-driven exploration and haptic object identification. *Perception* 22: 597-621.

Klatzky, R.L., Lederman, S.J., and Metzger, V. (1985) Identifying objects by touch: An "expert system". *Perception and Psychophysics* 37: 299-302.

Klatzky, R.L., Loomis, J., Lederman, S.J., Wake, H., and Fujita, N. (1993) Haptic identification of objects and their depictions. *Perception & Psychophysics* 54: 170-178.

Lederman, S.J. (1991) Skin and touch. *Encyclopedia of human biology*, Vol. 7. Academic Press, San Diego, pp. 51-63.

Lederman, S.J. and Klatzky, R.L. (1995) Early perceptual features for haptic object processing. *Submitted for publication.*

Lederman, S.J. and Klatzky, R.L. (1990a) Flexible exploration by human and robotic haptic systems. *Proceedings of the Annual International Conference of the IEEE/Engineering in Medicine and Biology Society* 12: 1915-1916.

Lederman, S.J., and Klatzky, R.L. (1990b) Haptic classification of common objects: Knowledge-driven exploration. *Cognitive Psychology* 22: 421-459.

Lederman, S.J. and Klatzky, R.L. (1987) Hand movements: A window into haptic object recognition. *Cognitive Psychology* 19: 342-368.

Loomis, J.M. and Lederman, S.J. (1986) Tactual perception. *In:* K. Boff, L. Kaufman, and J. Thomas (eds.), *Handbook of perception and human performance.*, Wiley, New York, pp. 31-1 - 31-41.

Rosch, E. (1978) Principles of categorization. *In:* E. Rosch and B. Lloyd (eds.), *Cognition and categorization*, Erlbaum, Hillsdale, N.J., pp. 27-48.

Somesthesis and the Neurobiology of the Somatosensory Cortex
O. Franzén, R. Johansson and L. Terenius (eds.)

The somatosensory cortex

Jon H. Kaas, Ph.D.

Department of Psychology, Vanderbilt University, Nashville, TN, USA 37240

Summary. Mammals vary in the number of cortical areas involved in processing somatosensory information. Most mammals appear to have areas S1, S2, and PV, as well as strip-like somatosensory areas along the rostral and caudal borders of S1. In monkeys and other higher primates, cortex caudal to S1 is differentiated into areas 1 and 2, each also representing cutaneous receptors, and there is evidence for a ventral somatosensory area, VS, adjoining S2. Processing is more serial in monkeys so that the responsiveness of S2 depends on S1.

Introduction

The somatosensory system allows us to make accurate inferences about the outside world by using information from sensitive receptors in the skin that signal surface contact and from deeper receptors in muscles and joints that signal limb movements and positions. This information is modified and conducted by spinal cord, brain stem, and thalamic neurons to cortical areas where more complex processing occurs, and subdivisions of motor cortex are accessed. While all parts of the somatosensory system may vary according to the specializations and requirements of different mammalian species, the cortical stages of processing seem to be the most variable. Comparative studies suggest that most mammals have at least five cortical processing stations, and that higher primates have more. These areas and their cortical connections are considered here (also see, Kaas, 1988; 1990; 1993; Kaas and Pons, 1988).

The generalized plan

All or most mammals appear to have at least five somatosensory areas in common. These include the first or primary somatosensory area, S1 (see Kaas, 1983), the second somatosensory area, S2 (Woolsey, 1943), a parietal ventral area, PV, first described in squirrels (Krubitzer et al., 1986) but now identified in many mammals (Krubitzer, 1995), and zones of cortex rostral and caudal to S1 that would be identified as areas 3a and 1 in primates.

S1 has been described in a wide range of mammals (see Johnson, 1990) and it has a number of identifying characteristics (see Kaas, 1983). Architectonically, S1 is the most distinctive of somatosensory areas, with a more prominent layer IV, and dense myelination and cytochrome oxidase staining. S1 is the major target of neurons in the ventroposterior nucleus (VP) that are activated by slowly adapting (SA) or rapidly adapting (RA) cutaneous afferents (Dykes et al., 1981). Major cortical connections are with S2, PV, and cortex just rostral and just caudal to S1 (Beck et al., 1996; Krubitzer et al., 1993; Li et al., 1990). Typically, S1 is elongated with a mediolateral orientation so that the tail is represented medially, followed by the hindlimb, trunk, forelimb, head, and oral cavity in successively more lateral or rostrolateral cortex. Moles are exceptional in that S1 has a rostrocaudal orientation with the head rostral and hindlimb caudal (Catania and Kaas, 1995). Typically, the representations of the digits are rostral in S1, and the dorsal body and head midline are caudal. S1 in bats, with the forelimb specialized for flying, differs in having the digits of the wing represented caudally in S1 (Calford et al., 1985).

S2 adjoins S1 caudolaterally along the representation of the midline of the upper face in S1 (Nelson et al., 1979). The representation of the head in S2 is along the S1 border, with the representations of the digits displaced laterally from that border. The forelimb activates cortex rostral to that activated by the hindlimb. A major source of activation is from S1, but direct inputs from VP also exist in most mammals, where they form an effective alternative source of activation in most (Garraghty et al., 1991) but not all (Garraghty et al., 1990b) mammals.

PV is located along the lateral border of S1 just rostral to S2 (Krubitzer et al., 1986; Krubitzer and Kaas, 1990; Krubitzer and Calford, 1992; Beck et al., 1996). This representation of cutaneous receptors mirrors S2 so that the head adjoins S1 and the limbs are away from the border, with the hindlimb rostral to the forelimb. In some mammals, parts of PV are also responsive to auditory stimuli. Inputs to PV are from S1 and S2. Because of the position of PV along the rostral border of S2, early studies may have confounded observations on PV with those on S2.

Narrow bands of cortex along the rostral and caudal margins of S1 generally receive dense inputs from S1 (e.g., Beck et al., 1996; Akers and Killackey, 1978). In primates and cats, the rostral zone of cortex is identified as area 3a, and it receives thalamic inputs activated by muscle spindle receptors (see Kaas, 1983). 3a is less granular than S1 in that the fourth layer of granular cells is less developed. Cortex rostral to S1 has not been identified as area 3a in most other mammals, but this rostral zone of cortex is "dysgranular" in rats (Chapin and Lin, 1984) and it has a less pronounced layer IV in other mammals (see Beck et al., 1996). Although the rostral field in other mammals is likely to be homologous with area 3a of primates, we have used the more neutral term, the rostral somatic area or (SR) in opossums (Beck et al., 1996). Because SR is not very responsive to body stimuli in anesthetized animals, it is not yet certain if it receives muscle spindle activation from the thalamus. However, neurons in SR are sometimes responsive to taps on the body and movements that could activate muscle receptors. Neurons sometimes are also responsive to tactile stimuli, as are area 3a neurons in primates.

Cortex caudal to S1 is also generally unresponsive or poorly responsive to tactile stimuli in many anesthetized mammals, although there are dense inputs from S1. However, this caudal cortex may respond to moderately intense tactile stimulation (see Krubitzer and Calford, 1992) in a somatotopic pattern that parallels that of S1. This cortex is also less granular than S1. To be conservative, we have termed the field the caudal somatosensory area, SC (Beck et al., 1996), although SC resembles area 1 of monkeys in position, somatotopy, and connections with S1 (area 3b).

The more generalized decendents of early mammals may have additional somatosensory areas, but there seems room for only a few. Cortex immediately caudal to SC appears to be visual, but a bimodal region may exist as well. Cortex caudal to S2 and PV is auditory. A region ventral and rostral to PV along the rhinal sulcus appears to have connections with S2 and PV, and one or two further areas of predominately somatosensory processing may exist. Most mammals have a primary motor field, M1, as well as a supplimentary motor area, SMA or M2 (see Beck et al., 1996), and these fields have connections with somatosensory areas and probably have information processing as well as motor functions. If these motor fields are included, generalized mammals process tactile information over a network of 8-10 fields.

Somatosensory cortex in prosimian primates

The organization of somatosensory cortex in prosimian primates appears to be only somewhat progressive (Kaas, 1982). Prosimian primates, including galagos and slow lorises, have an S1 cortex that is somewhat narrower and more elongated than the more generalized form found in rats and opossums, but in other respects it is very similar. The somatotopic order is generalized, with digits pointing rostrally, the midline of the upper face and head caudolaterally, and a tail to tongue mediolateral progression of representation. Architectonically, S1 has a distinct layer IV of small granule cells, and has been termed somatic koniocortex (Sanides and Krishnamurti, 1967; Sur et al., 1980).

As in generalized mammals, cortex caudal and rostral to S1 is only weakly responsive to somatosensory stimuli in anesthetized animals, but these regions do receive topographic patterns of projections from S1 that indicate that each contains a representation in parallel with S1 (Wu et al., 1995). Thus, these fields are very much like SR and SC of generalized mammals. Areas 3a and 1 occupy these locations in monkeys, and thus SR and SC appear to be less differentiated versions of areas 3a and 1.

Cortex lateral to S1 in galagos includes areas S2 and PV (Burton and Carlson, 1986; Garraghty et al., 1991; Wu et al., 1995). As in generalized mammals, both S1 and S2 receive projections from the ventroposterior nucleus (Burton and Carlson, 1986), and lesions of S1 fail to abolish the responsiveness of S2 to tactile stimuli (Garraghty et al., 1991).

Little more is known about other somatosensory areas in galagos and other prosimians. They have a M1 and probably three premotor areas (Preuss et al., 1992; unpublished studies). Cortex caudal to SC (presumptive area 1) is enlarged and part of this cortex is not visual. Thus, it seems probable that part of posterior parietal cortex is involved in processing somatosensory stimuli. There may be several somatosensory areas in cortex lateral to S1 as well. Thus, some aspects of a dorsal pathway for spatial abilities and a ventral pathway for object identification (Mishkin, 1979) have emerged, but less so than in monkeys.

Somatosensory cortex in monkeys

Anterior parietal cortex of monkeys, apes and humans has long been divided into the traditional architectonic fields 3a, 3b, 1 and 2 of Brodmann (1909) and Vogt and Vogt (1919). Early investigators included all of these fields within "S1" (Marshall et al., 1937), since areas 3b, 1 and 2 and even parts of area 3a are highly responsive to tactile stimuli. However, we now know that each of these four architectonic fields contain a separate representation of body receptors, and that only area 3b has all of the identifying features of S1 (see Kaas, 1993). Thus, we equate area 3b with S1, and for consistency refer to the koniocortical field of S1 in non-primates as area 3b. Yet, architectonic terms continue to be used in an inconsistent manner. Thus, some investigators refer to only one field when they use the term S1, while others persist in including more than one architectonic field within S1 of cats, prosimians, and other mammals.

Features of somatosensory cortex vary somewhat across monkey taxa. Marmosets are small, smooth-brained monkeys that differ from other studied monkeys in that features of anterior parietal cortex organization seem intermediate between those of prosimian primates and other monkeys (Krubitzer and Kaas, 1990). Area 3b contains a representation that is characteristic of S1 in non-primates and of area 3b (S1) of other simians (Carlson et al., 1986). Cortex caudal to S1 in the position of area 1 usually is not very responsive to tactile stimuli in anesthetized marmosets, but under favorable conditions, recordings reveal a somatotopic representation much like that of area 1 in other monkeys. Major cortical inputs to area 1 are from area 3b, which also projects sparsely to more caudal cortex in the position of area 2. Yet, cortex in the position of area 2 has not been shown to be responsive to tactile stimuli. Architectonically, cortex along the rostral border of S1 resembles area 3a, and some responsiveness to tactile and muscle receptor stimulation seems to be present, but this cortex is much less responsive than area 3a in other monkeys. We interpret the organization of anterior parietal cortex of marmosets as having four fields (3a, 3b, 1 and 2), but some of these fields, especially area 2, are much less differentiated than in other monkeys.

Marmosets, as other mammals, also have areas S2 and PV. Major cortical inputs are from area 3b, and thalamic inputs are from the ventroposterior inferior nucleus, VPI, rather than VP (Krubitzer and Kaas, 1992). VPI seems to receive spinothalamic inputs (Gingold et al., 1991), rather than inputs relayed from mechanoreceptors, and lesions of S1 deactivate PV and S2 (Pons et al., 1987; Garraghty et al., 1990b). PV and S2 have connections with cortex caudal to S2, in the region of the ventral somatosensory area, and rostral to PV in a region we term parietal rostral, PR. Thus, marmoset cortex demonstrates some of the elaborations found in other arthropoid primates, but areas 1 and 2 are not comparably developed.

Other New World monkeys and especially Old World monkeys differ by having areas 1 and 2 that are highly responsive to cutaneous stimuli. Area 1 has a representation that is basically a mirror reversal of the somatotopic pattern in area 3b, although receptive fields are larger (Kaas et al., 1979). Area 1 depends on area 3b inputs for activation, at least in some monkeys (Garraghty et al., 1990a), although area 1 receives considerable direct input from VP (Lin et al., 1979; Cusick et al., 1985). Area 2 constitutes a more complex representation, that reverses somatotopy at the area 1 border, but also represents parts of the hand at least twice (Pons et al., 1985). Area 2 receives direct cortical inputs from both areas 3b and 1, and muscle spindle receptor information from the ventroposterior superior nucleus (Kaas et al., 1984; Cusick et al., 1985; Pons and Kaas, 1985).

The organization of posterior parietal cortex is not well understood, but the region is quite large in Old World monkeys. The rostral parts of posterior parietal cortex can be activated by somatosensory stimuli (Pons et al., 1985) and some regions receive inputs from area 2 (Pons and Kaas, 1985). Outputs include those to regions of motor and premotor cortex (Stepniewska et al., 1993; Preuss et al., 1993).

Areas S2 and probably PV (Krubitzer et al., 1995) receive inputs, not only from area 3b, but from areas 3a, 1, 2, and parts of posterior parietal cortex (area 5) as well (see Pons et al., 1987; Pons and Kaas, 1986). Another field in lateral parietal cortex, the ventral somatosensory area, VS, has not been extensively studied, but the area adjoins S2 and includes a systematic representation of cutaneous receptors (Cusick et al., 1989). Connections of VS have not been fully described, but S2 provides some inputs. Other fields with somatosensory functions also appear to exist in the cortex of the lateral fissure, but proposals differ in how to divide the region into areas, and other uncertainties exist.

Conclusions

In most mammals, three subdivisions of somatosensory cortex respond to tactile stimuli and systematically represent the body surface, the traditional S1 and S2 representations and the more recently defined PV area. Cortex rostral and caudal to S1 receives S1 projections, but is relatively unresponsive to tactile stimuli in most non-primate anesthetized mammals. S1, S2 and PV all receive inputs from neurons in the ventroposterior nucleus, and S2 and PV continue to respond to tactile stimuli after S1 lesions. Thus, the parallel thalamic input is an adequate source of activation.

Prosimian primates have retained major features of this generalized pattern so that cortex caudal and rostral to S1 remains relatively unresponsive to tactile stimuli, the ventroposterior nucleus provides a major parallel input to S2, and S2 remains responsive after S1 lesions. Marmoset monkeys present features of cortical organization that are intermediate between prosimian and other monkeys. Cortex rostral and caudal to S1 is less responsive to somatosensory stimuli than in other monkeys, but more responsive than in prosimians. As for other monkeys, major thalamic inputs to S2 and PV are from VPI rather than VP, and inputs from S1 are necessary for this activation of S2 and PV. Most monkeys have an area 1 caudal to S1 (3b) that is highly responsive to cutaneous stimuli, and a more caudal area 2 with responsiveness to tactile stimuli and the activation of deep receptors. Area 3a just rostral to S1 is also activated by both deep and cutaneous receptors, with the deep receptor input predominating. Processing is serial, with areas 1, PV and S2 depending on S1 for activation. Old World monkeys, and by implication of other higher primates, have greatly elaborated posterior parietal cortex, and some of this cortex is highly responsive to tactile stimuli. Thus, more cortical subdivisions probably exist in posterior parietal cortex of higher primates.

References

Akers, R.M. and Killackey, H.P. (1978) Organization of corticocortical connections in the parietal cortex of the rat. *Journal of Comparative Neurology 181:* 513-538.

Beck, P.D., Pospichal, M.W., and Kaas, J.H. (1996) Topography, architecture, and connections of somatosensory cortex in opossums: evidence for five somatosensory areas. *Journal of Comparative Neurology; in press.*

Brodmann K. (1909) Vergleichende Lokalisationslehre der Gro ßhirnrinde. Leipzig, Barth.

Burton, H. and Carlson, M. (1986) Second somatic sensory cortical area (S-II) in a prosimian primate, *Galago crassicaudatus. Journal of Comparative Neurology 247:* 200-220.

Carlson, M., Huerta, M.F., Cusick, C.G., and Kaas, J.H. (1986) Studies on the evolution of multiple somatosensory representations in primates: The organization of anterior parietal cortex in the New World Callitrichid, *Saguinus. Journal of Comparative Neurology 246:* 409-426.

Calford, M.B., Graydon, M.L., Huerta, M.F., Kaas, J.H., and Pettigrew, J.D. (1985) Altered somatotopy in the brain of a flying mammal. *Nature 313:* 477-479.

Catania, K.C. and Kaas, J.H. (1995) The organization of the somatosensory cortex of the star-nosed mole. *Journal of Comparative Neurology 351:* 549-567.

Chapin, J.K. and Lin, C.-S. (1984) Mapping the body representation in the SI cortex of anesthetized and awake rats. *Journal of Comparative Neurology 229:* 199-213.

Cusick, C.G., Steindler, D.A., and Kaas, J.H. (1985) Corticocortical and collateral thalamocortical connections of postcentral somatosensory cortical areas in squirrel monkeys: A double labeling study with wheat germ agglutinin (WGA) conjugated to horseradish peroxidase and radiolabeled WGA. *Somatosensory Research 3:* 1-31.

Cusick, C.G., Wall, J.T., Felleman, D.J. and Kaas, J.H. (1989) Somatotopic organization of the lateral sulcus of owl monkeys: area 3b, S-II, and a ventral somatosensory area. *Journal of Comparative Neurology 282:* 169-190.

Dykes, R.W., Sur, M. Merzenich, M.M., Kaas, J.H., and Nelson, R.J. (1981) Regional segregation of neurons responding to quickly adapting, slowly adapting, deep and pacinian receptors within thalamic ventroposterior lateral and ventroposterior inferior nuclei in the squirrel monkey. *Neuroscience 6:* 1687-1692.

Garraghty, P.E., Florence, S.L., and Kaas, J.H. (1990a) Ablations of areas 3a and 3b of monkey somatosensory cortex abolish cutaneous responsivity in area 1. *Brain Research 528:* 165-169.

Garraghty, P.E., Pons, T.P., and Kaas, J.H. (1990b) Ablations of areas 3b (S-I proper) and 3a of somatosensory cortex in marmosets deactivate the second and parietal ventral somatosensory areas. *Somatosensory and Motor Research 7:* 125-135.

Garraghty, P.E., Florence, S.L., Tenhala, W.N., and Kaas, J.H. (1991) Parallel thalamic activation of the first and second somatosensory areas in prosimian primates and tree shrews. *Journal of Comparative Neurology 310:* 1-11.

Gingold S-I, Greenspan, J.P., Apkavian A.V. (1991) Anatomical evidence of nociceptive inputs to primary somatosensory cortex: Relationships between spinothalamic terminals and thalamocortical cells in squirrel monkeys. *Journal of Comparative Neurology 308:* 467-490.

Johnson, J.I. (1990) Comparative development of somatic sensory cortex. *In:* E.G. Jones and A. Peters (eds): *Cerebral Cortex*. New York: Plenum, pp. 335-449.

Kaas, J.H. (1982) The somatosensory cortex and thalamus in Galago. *In:* E. E. Haines, (ed.), *The Lesser Bush Baby (Galago) as an Animal Model: Selected Topics*, CRC Press, Inc., Boca Raton, Fl., pp. 169-181.

Kaas, J.H. (1983) What, if anything, is S-I? The organization of the "first somatosensory area" of cortex. *Physiological Review 63:* 206-231.

Kaas, J.H. (1988) How the somatosensory thalamus is subdivided and interconnected with areas of somatosensory cortex in monkeys. *In:* M. Bentivoglio and R. Spreafico (eds.), *Cellular Thalamic Mechanisms*. Elsevier Science Publications, Amsterdam., pp. 143-150.

Kaas, J.H. (1990) The somatosensory system. *In:* G. Paxinos (ed.), *The Human Nervous System*. Academic Press, New York., pp. 813-844.

Kaas, J.H. (1993) The functional organization of somatosensory cortex in primates. *Annals of Anatomy 175:* 509-518.

Kaas, J.H. and Pons, T.P. (1988) The somatosensory system of primates. *In:* H. P. Steklis (ed.), *Comparative Primate Biology, Vol. 4: Neurosciences*. Alan R. Liss, Inc., New York, pp. 421-468.

Kaas, J.H., Nelson, R.J., Sur, M., Lin, C.,-S., and Merzenich, M.M. (1979) Multiple representations of the body within "S-I" of primates. *Science 204:* 521-523.

Kaas, J.H., Nelson, R.J., Sur, M., Dykes, R.W., and Merzenich, M.M. (1984) The organization of the ventroposterior thalamus of the squirrel monkey, Saimiri sciureus. *Journal of Comparative Neurology 226:* 111-140.

Krubitzer, L. (1995) The organization of neocortex in mammals: Are species differences really so different? *Trends in Neuroscience 18:* 408-417.

Krubitzer, L.A. and Calford, M.B. (1992) Five topographically organized fields in the somatosensory cortex of the flying fox. *Journal of Comparative Neurology 317:* 1-30.

Krubitzer, L.A. and Kaas, J.H. (1990) The organization and connections of somatosensory cortex in marmosets. *Journal of Neuroscience 10:* 952-974.

Krubitzer, L.A. and Kaas, J.H. (1992) The somatosensory thalamus of monkeys: Cortical connections and a redefinition of nuclei in marmosets. *Journal of Comparative Neurology 319:* 1-18.

Krubitzer, L.A., Sesma, M.A., and Kaas, J.H. (1986) Microelectrode maps, myeloarchitecture, and cortical connections of three somatotopically organized representations of the body surface in parietal cortex of squirrels. *Journal of Comparative Neurology 250:* 403-430.

Krubitzer, L.A., Clarey, J., Tweedale, R., Elston, G., and Calford, M. (1995) The organization of lateral somatosensory cortex in macaque monkey: Where is SII? *Journal of Neuroscience 15:* 3821-3839.

Krubitzer, L.A., Calford, M.B., and Schmid, L.M. (1993) Connections of somatosensory cortex in megachiropteran bats: The evolution of cortical fields in mammals. *Journal of Comparative Neurology 327:* 473-506.

Li, X.-G., Florence, S.L., and Kaas, J.H. (1990) Areal distributions of cortical neurons projecting to different levels of the caudal brain stem and spinal cord in rats. *Somatosensory and Motor Research 7:* 315-335.

Lin, C.-S., Merzenich, M.M., Sur, M., and Kaas, J.H. (1979) Connections of areas 3b and 1 of the parietal somatosensory strip with ventroposterior nucleus in the owl monkey (Aotus trivirgatus). *Journal of Comparative Neurology 185:* 355-372.

Marshall, W.H., Woolsey, C.N., Bard, P. (1937) Cortical representation of tactile sensibility as indicated by cortical potentials. *Science 85:* 388-390.

Mishkin M. (1979) Analogous neural models for tactile and visual learning. *Neuropsychologicia 17:* 139-151.

Nelson, R.J., Sur, M., and Kaas, J.H. (1979) The organization of the second somatosensory area (S-II) of the grey squirrel. *Journal of Comparative Neurology 184:* 473-490.

Pons, T.P. and Kaas, J.H. (1985) Connections of area 2 of somatosensory cortex with the anterior pulvinar and subdivisions of the ventroposterior complex in macaque monkeys. *Journal of Comparative Neurology 240:* 16-36.

Pons, T.P. and Kaas, J.H. (1986) Corticocortical connections of areas 2, 1, and 5 of somatosensory cortex in macaque monkeys: A correlative anatomical and electrophysiological study. *Journal of Comparative Neurology 248:* 313-335.

Pons, T.P., Garraghty, P.E., Cusick, C.G., and Kaas, J.H. (1985) The somatotopic organization of area 2 in macaque monkeys. *Journal of Comparative Neurology 241:* 445-466.

Pons, T.P., Garraghty, P.E., Mishkin M. (1987) Physiological evidence for serial processing in somatosensory cortex. *Science 237:* 417-420.

Preuss, T.M., Stepniewska, I. and Kaas, J.H. (1992) Microstimulation studies of motor cortical organization in lorisid primates. *European Journal of Neuroscience Suppl. 5:* 174.

Preuss, T.M., Stepniewska, I., and Kaas, J.H. (1993) Ipsilateral cortical connections of the dorsal and ventral premotor areas in owl monkeys. *Society of Neuroscience Abstracts 19:* 2211.

Sanides, F. and Krishnamurti, A. (1967) Cytoarchitectonic subdivisions of sensorimotor and prefrontal regions and bordering insular and limbic fields in slow loris (*Nycticebus coucang coucang*). *Journal für Hirnforschung 9:* 225-252.

Stepniewska, I., Preuss, T.M. and Kaas, J.H. (1993) Architectonics, somatotopic organization, and ipsilateral cortical connections of the primary motor area (M1) of owl monkeys. *Journal of Comparative Neurology 330:* 238-271.

Sur, M., Nelson, R.J., and Kaas, J.H. (1980) The representation of the body surface in somatic koniocortex in the prosimian (Galago senegalensis). *Journal of Comparative Neurology 180:* 381-402.

Vogt, C. and Vogt, O (1919) Allgemeinere Ergebnisse unserer Hirnforschung. *Journal of Physiological Neurology 25:* 279-462.

Woolsey, C.N. (1943) Second somatic receiving areas in the cerebral cortex of cat, dog, and monkey. *Federation Proceedings 2:* 55.

Wu, W.H. Beck, P.D. and Kaas, J.H. (1995) Ipsilateral cortical connections of S1 (3b) in prosimian primates: Evidence for five somatosensory areas. *Society of Neuroscience Abstracts 21:* 112.

Somesthesis and the Neurobiology of the Somatosensory Cortex
O. Franzén, R. Johansson and L. Terenius (eds.)

The organization of lateral somatosensory cortex in primates and other mammals

L. A. Krubitzer

Center for Neuroscience and Dept. of Psychology, University of California Davis, 1544 Newton Ct., Davis, CA. 95616, USA, and Vision, Touch, and Hearing Research Centre, Dept. Physiology and Pharmacology, University of Queensland, Australia

Summary: The organization of lateral somatosensory cortex has been described for several primates including Old World, and New World simians as well as prosimians. While descriptions of SII for New World monkeys and prosimians are consistent, those for Old World macaque monkeys vary considerably. However, all data conform to the proposal that at least two mirror symmetric representations reside in the area traditionally considered as a single field. These include the second somatosensory area, SII, and the parietal ventral area, PV. Discrepancies in the descriptions of SII in primates, as well as non primate mammals may be the result of the two fields being confused. Because of the ubiquity of SII and PV in all mammals, it is proposed that these areas are part of a basic plan of somatosensory processing. The addition of new fields, such as VS, Ri, and 7b in the line which led to extant primates, may account for the increase in dexterity, bilateral coordination of the hands, and tactile recognition abilities that characterize primates.

Introduction

Early in this century new techniques for examining the brain, particularly the cortex, not only allowed us to develop more accurate descriptions of cortical architecture and cell morphology, but revolutionized our ideas regarding brain construction and function. Architectonic studies of Brodmann (1909), Von Economo (1929), and Vogt and Vogt (1919), demonstrated that the brain is composed of a number of separate parts, and work in the middle of this century by Woolsey and colleagues (1946; 1958), established that those parts defined architectonically, were coextensive with maps of the sensory epithelium in visual, auditory, and somatosensory cortex. Since then, our efforts have been directed at uncovering smaller units of brain construction, and the properties of the separate components that compose the brain. Very recent studies have described the organization of separate fields in the cortex, have elucidated details of individual cell function, dendritic arborization, synaptic specialization, and neurotransmitter distribution across cortical regions. Not surprisingly, our level of analysis has again shifted to even smaller units of processing. However, while the techniques have become more refined, and our view of neurophysiological properties and neuroanatomical construction more acute, it is still unclear how all of these specific events or units of brain construction generate complex behaviors and perceptions. The following review on the organization of lateral somatosensory fields is an attempt to refocus our attention on more global aspects of somatosensory processing.

Early electrophysiological investigations in the somatosensory cortex reported that only a single primary area, SI, and a second somatosensory field, SII were present in parietal cortex (e.g. Woolsey and Fairman, 1946, Woolsey, 1958). Since then, SI has been further divided and found to contain separate representations coextensive with areas 3a, 3b, 1 and 2 of Brodmann (1909; e.g. Merzenich et al., 1978; Nelson et al., 1980; Pons et al., 1985), modules within SI related to segregated projections from slowly adapting and rapidly adapting inputs have been discovered (Sur et al., 1984), and specific patterns of connections for the different anterior parietal fields have been ascertained (e.g. Pons and Kaas, 1986). Recent work in which individual properties of neurons in areas 3b and 1 (e.g. Iwamura et al., 1983; Sinclair and Burton, 1991), area 2 and posterior parietal cortex (e.g. Iwamura et al., 1994) were investigated in awake monkeys allows us to appreciate the possible function of fields (see Kaas and Pons, 1988 for review). While our knowledge of anterior parietal and posterior parietal cortical organization and function has increased considerably in the last decade, only recently has cortex lateral to anterior parietal fields, in the SII region, been given attention. Work to date indicates that this region of cortex is involved in more complex discriminations, bilateral coordination of the hands, and discriminations made during active touch (e.g. Sinclair and Burton, 1993). Connections of this field with posterior parietal cortex (e.g. Cavada and Goldman-Rakic, 1989a; 1989b; Andersen et al., 1990), temporal lobe areas that are connected to the amygdala and hippocampus (e.g. Krubitzer and Kaas, 1990), and lesions along this pathway, indicate that areas here may be involved in tactile recognition and memory (Mishkin, 1979; Murray and Mishkin, 1984). Finally, this is the region of cortex where sensory modalities converge suggesting that fields in the lateral sulcus play a role in sensory integration across modalities. Thus, most data indicate that areas in this region are involved in more complex tactile discriminations than areas in anterior parietal cortex. Comparisons of this region in primates as well as non primate mammals demonstrate that this area is composed of at least two separate representations, the second somatosensory area, SII, and the parietal ventral area, PV, and that additional areas such as VS, Ri, and 7b may have evolved somewhat later, particularly in the lineage leading to extant primates. These areas may be responsible for the complex behavior associated with primates such as bilateral coordination of the hands, and precise tactile discriminations necessary for object identification and recognition.

The organization of lateral somatosensory areas in primates

The second somatosensory area, SII. The second somatosensory area, SII, has been described in a variety of primates including macaque monkeys (Whitsel et al., 1969; Robinson and Burton, 1980b; Pons et al., 1988; Krubitzer et al., 1995), marmosets (Krubitzer and Kaas, 1990); owl monkeys (Cusick et al., 1989), and galagos (Burton and Carlson, 1986). In Old world macaque monkeys, discrepancies exist in the size of SII, the position of SII, and the internal organization of SII (Fig. 1). However, recent electrophysiological as well as anatomical data have provided evidence for two separate fields in cortex traditionally defined as SII. It has been proposed that these discrepancies in the description of SII in Old World monkeys may be the result of the two fields being confused (Krubitzer et al., 1995). The internal organization, relative location, and architecture of SII in New World and prosimian primates is similar to recent descriptions of SII in macaque monkeys (Figs. 2 and 3). SII forms a complete representation of the body surface with the face, head and neck represented medially, the forelimb laterally or rostrolaterally, and the hindlimb most laterally. The trunk is represented caudally or caudolaterally in SII. Within the limb representation, the representation of the distal limbs, including glabrous digits and toes occupies a large region of cortex rostral to the representation of the proximal limbs. SII is also distinguished by a heavily myelinated appearance relative to surrounding cortex, as well as a dense granule cell layer in cortex that has been cut coronally or parasagittally. In some experiments, bilateral receptive fields have been described for neurons in the SII region (e.g. Whitsel et al., 1969; Robinson and Burton, 1980b). Another feature of SII is that it is proposed to be a major recipient zone of connections from SI (e.g. Friedman et al., 1980). However, at the time most investigations were done, SI included fields 3a, 3b, 1 and 2, and we now appreciate that only 3b should be considered as SI as described in other mammals (Kaas, 1983). Also, the definition of SII was vague and somewhat circular, i.e. neurons responsive to tactile simulation in cortex lateral to SI are in SII, and SII receives direct inputs from "SI". Recently, it has been shown that in marmoset monkeys (Krubitzer and Kaas, 1990) and macaque monkeys (Burton et al., 1995) there are a number of recipient zones of 3b projections in this region of cortex, several of which receive topographically matched inputs from 3b, and that a very large region of cortex lateral to anterior parietal fields, much larger than the traditional SII, contains neurons responsive to tactile stimulation. Thus, while neurons in SII respond to tactile stimulation, and while SII does indeed receive inputs from the primary somatosensory area, it is not the only field to do so. Finally, recent experiments in which the cortical connections of SII were directly investigated (Friedman and Murray, 1986; Friedman et al., 1986; Krubitzer and Kaas, 1990), support the contention that additional fields, densely interconnected with SII, are present in the lateral sulcus of primates.

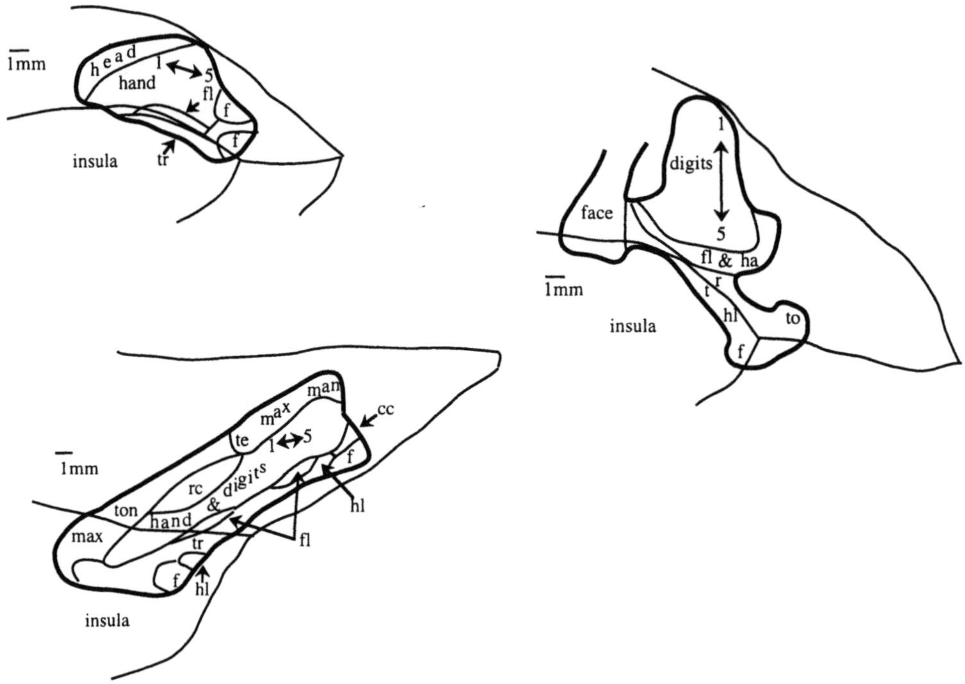

Figure 1. The organization of the second somatosensory area in the macaque monkey redrawn from Robinson and Burton, 1980 (top), Friedman et al., 1980 (middle), and Pons et al., 1988 (bottom). The top and bottom figures are based on results from electrophysiological mapping experiments and the middle figure is a summary based on results from connections from different body representations in SI. Note that the size, internal organization, and relative location of SII is different in the different studies. In this and the following figures, rostral is to the left and medial is to the top. Caudal complex, cc; foot, f; forelimb, fl; hand, ha; hindlimb, hl; mandibular, man; maxillary, max; teeth, te; tongue, ton; trunk, tr.

The parietal ventral area, PV. Single unit recording experiments in the early 1980's by Robinson and Burton, demonstrated that cortex surrounding the SII region contained neurons responsive to somatic stimulation (e.g. retroinsular, Ri, and granular insula, Ig), and were outside of the proposed boundaries of "SII". However, even within the region they defined as SII, multiple representations of similar body parts were described, suggesting that "SII" may contain separate fields. The notion that multiple representations exist in the lateral sulcus of primates has been further substantiated in more recent studies in marmoset and macaque monkeys. The details of how cortex in the lateral sulcus of primates should be subdivided is still contentious, but the data across studies are consistent with the proposition that more than a single field resides here. There is clear electrophysiological and anatomical evidence for two mirror symmetric representations. Both early and recent studies of connections of anterior parietal areas with cortex in the lateral sulcus (Friedman et al., 1980; Burton et al., 1995) demonstrate that multiple target areas exist in the lateral sulcus. One of which is SII, the other of which is similar to PV. In accordance with this interpretation, our laboratory has termed these fields SII and PV, because we propose that they are homologous to similar fields described in other mammals (Fig. 3, see below). Like, SII, PV contains a complete representation of the sensory epithelium. The mediolateral organization of PV is much like that of SII with the head represented most medially (superficially in the sulcus), and the forelimb and hindlimb more laterally (deeper in the sulcus). However, the rostrocaudal organization of PV is mirror symmetric to that of SII with the distal limbs represented caudally and proximal limbs and trunk represented rostrally (Fig. 2, marmoset, Fig. 3).

The ventral somatosensory area, VS. Areas in addition to SII and PV have also been identified in primates using several different, although not always overlapping criteria. Based on connections from anterior parietal fields combined with electrophysiological mapping in owl monkeys, Cusick et al. (1989) described two separate representations in cortex lateral to 3b and 1 (Fig. 2). These fields included SII, similar in organization, position, and connections from 3b, to SII described in macaque monkeys, and a ventral somatosensory, VS, just ventral to SII. Although complete maps for VS have not been obtained in owl monkeys, the representation in VS appears to reverse off of the SII representation (Fig. 3; see the description of VS in flying fox). A partial map, similar in organization and position to VS described in owl monkeys has been generated in macaque monkeys (Krubitzer et al., 1995), and as in owl monkeys, has been termed VS.

Area 7b. Studies of connections, architecture, and physiology indicate that there is a region of cortex just caudal to SII and the lateral portion of area 1, which is termed area 7b (e.g. Robinson and Burton, 1980a). This region has interconnections with some subdivisions of the somatosensory cortex, and is also densely interconnected with subdivisions of frontal and posterior parietal cortex (Cavada and Goldman-Rakic, 1989a; 1989b; Anderson et al., 1990). Recent electrophysiological recording results indicate that neurons in this region respond to somatic stimulation, but that the type of stimulation needed to elicit a response is more complex than for areas SII and PV (Krubitzer et al., 1995). Neurons in area 7b require a broader stimulus applied to a large region of the body, and often have a longer latency of response than neurons in other somatosensory regions. However, complete maps of this field have not been obtained, and studies of single neuron properties are needed to determine differences between 7b and surrounding fields.

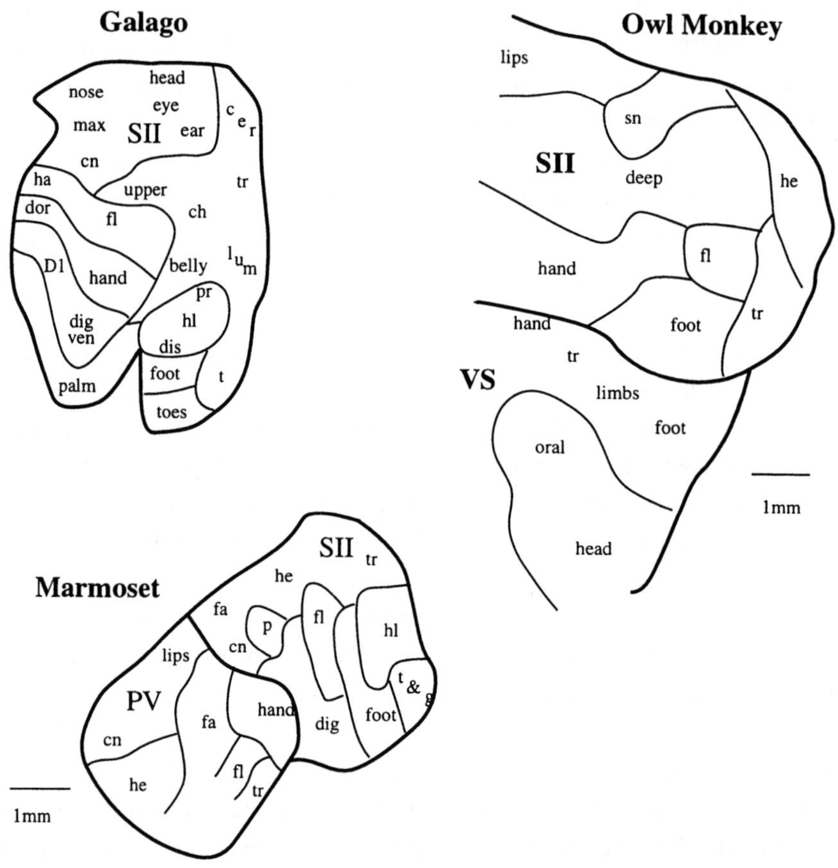

Figure 2. The organization of SII and surrounding fields redrawn from studies in galagos, marmosets, and owl monkeys. The internal organization of SII is very similar across these different species of primates. The pand representations. Cervical, cer; chest, ch; chin, cn; digits, dig; distal, dis; dorsal, dor; face, fa; genitals, g; head, he; palm,rietal ventral area, PV, shares a common border with SII at the representations of the lips and glabrous hands, and the ventral somatosensory area, VS, in owl monkeys shares a common boundary with SII at the foot and ha p; proximal, pr; tail, t; ventral, ven. Other abbreviations as in previous figures. (Galago: Burton and Carlson, 1986, Owl Monkey: Cusick et al., 1989; Marmoset: Krubitzer and Kaas, 1990).

Early studies in macaque monkeys identified regions outside of the area termed SII whose neurons were responsive to somatic stimulation, and neurons in these regions were related to architecture. These fields include the retroinsula, Ri, the granular insula, Ig and the post auditory Pa area (Robinson and Burton, 1980a). Similar subdivisions have been used in studies of connections of SII in macaque monkeys (Friedman et al., 1986). However, earlier studies in which these architectonically defined areas were related to electrophysiological recording results only sampled a small amount of cortex in any individual animal, and data was collapsed across animals to develop these subdivisions. Also, subdivisions such as Ig, Ri, and Pa do not appear to correspond with more recent electrophysiological recording results. Rather, it is likely that the relatively large region termed Ig is composed of more than a single field, and that the very small area Ri (approximately 1 mm x 1 mm) is part of a larger field.

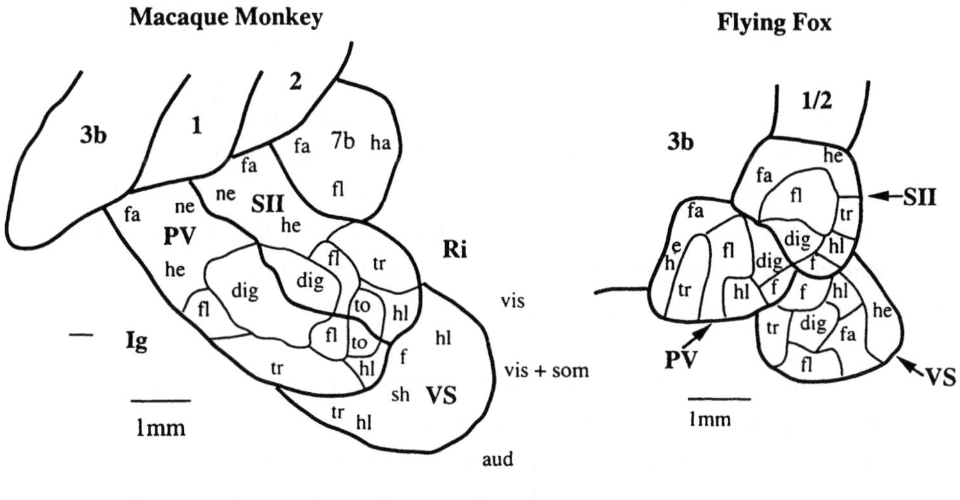

Figure 3. The organization of SII, PV, and VS, redrawn for the macaque monkey and flying fox. The internal organization, and relative location of these fields is similar in both species, although the location of SII and PV relative to anterior parietal fields is somewhat different. Neurons in cortex surrounding PV, SII and VS are also responsive to somatic stimulation and partial maps have been generated for at least one field, area 7b. Cortex in the region of Ri is responsive to visual and visual + somatosensory stimulation. Shoulder, sh; toe, t. Other abbreviations as in previous figures. (Macaque Monkey: Krubitzer et al., 1995; Flying Fox: Krubitzer et al., 1992).

Recently, studies in humans using imaging techniques such as positron emmission tomography (PET, Burton et al., 1993; Ledberg et al., 1995) have also demonstrated that there are multiple areas in the lateral sulcus of humans, and that these areas have different roles in discriminating microgeometric and macrogeometric stimuli (e.g. Ledberg et al., 1995). How the multiple representations in this region in humans relates to the separate fields identified in monkeys is not known.

The organization of lateral somatosensory areas in other mammals

Although early studies in non primates described the second somatosensory area as a large inverted representation of the body surface (e.g. Lende and Woolsey, 1956; Woolsey, 1967; Fig. 4), microelectrode mapping studies in cats demonstrated that SII is an upright representation of the body surface (Haight, 1972), similar to that described above for primates. Subsequent microelectrode mapping studies in other mammals (Fig. 4) also described SII as a non inverted representation. The discrepancy in the orientation and internal organization of SII in different mammals is likely to stem from two problems. First, early studies used evoked potentials to describe the organization of SII, and these types of techniques are not as sensitive to defining details of organization within the fields as electrophysiological recording techniques. The second possibility is that more than one field exists in the SII region, and that SII and other fields were being confused. Despite these differences in description, most recent observations in a variety of mammals (see Johnson, 1990 for review) demonstrate that all mammals contain a small, non inverted representation of the body surface lateral to SI and often sharing a common boundary with the face representation of SI. This field is the second somatosensory area, SII.

Recently, a field in addition to SII has been described in rodents (Krubitzer et al., 1986; Fabri and Burton, 1991), megachiropteran bats (Krubitzer and Calford, 1992), and marsupials (Beck et al., 1993; Elston et al., 1993) This field is similar in location, internal organization, architecture, and connections to PV described in primates. We proposed that SII and PV are common features of organization in mammals, and their ubiquity in the various lineages investigated, including two species of primates, suggests that these fields are part of a common plan of organization in mammals, and may be present in humans as well. However, these fields have been evolving independently in the different lineages for tens of millions of years, it unlikely that they are analogous (functionally similar).

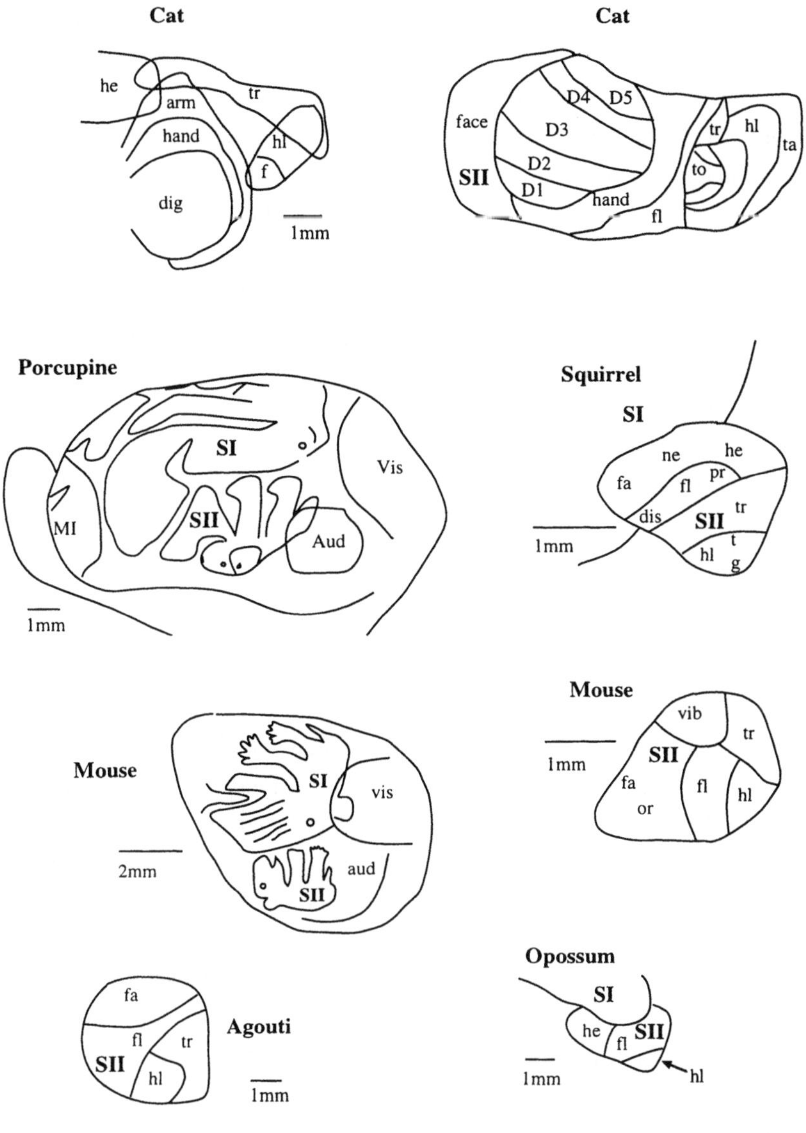

Figure 4. The organization of SII redrawn for a number of species including cats, porcupines, squirrels, mice, agoutis and opossums. The internal organization of SII differs in some investigations. In some investigations SII has been described as an inverted representation of the body surface and in other studies it has been described as a non inverted representation. Even in some studies in which the same species was examined (e.g. mouse), SII has been described as inverted (Woolsey, 1967) as subsequently as non inverted (Carvell and Simons, 1986). Abbreviations as in previous figures. (Cat (top, left): Haight, 1972; Cat (top, right): Burton et al., 1982; Porcupine: Lende and Woolsey, 1956; Squirrel: Nelson et al., 1979; Mouse (left): Woolsey, 1967; Mouse (right): Carvell and Simons, 1986; Agouti: Pimentel-Souza et al., 1980; Opossum: Pubols, 1977)

Work in cats demonstrates that at least two complete representations exist in this region, and that each of these separate regions is coextensive with a unique architectonic appearance and pattern of connections (Burton et al., 1982; Clemo and Stein, 1982; Burton and Kopf, 1984). The fields are termed SII, and SIV. The fourth somatosensory area, SIV, has some similarities with PV described in primates and other mammals (see below), but its relative location, internal organization, and patterns of connections, do not appear to be similar to PV.

In extensive electrophysiological mapping and studies of connections in the megachiropteran bat, at least four separate representations of the body surface have been identified and proposed to be involved in the processing of somatic inputs (Krubitzer and Calford, 1992; Krubitzer et al., 1993). These include, SII, PV, VS and the lateral somatosensory area, LS. These studies in the flying fox contain the first detailed description of the ventral somatosensory area in any mammal (Fig. 3). In the flying fox, VS shares a common boundary with both SII and PV. Although the organization of VS is less precise than that of SII and PV, VS clearly contains a complete representation of the body surface.

Conclusions

Based on comparisons of internal organization, patterns of connections, and architectonic distinctions, it is proposed that at least two complete representations of the body surface exist in all, or most mammals investigated, SII and PV. While other fields such as VS an SIV have been identified, their presence does not appear to be widely distributed across mammals, and is limited to only a few lineages. The presence of SII and PV in all mammals suggest that these areas form the basic building blocks of cortical organization of this regions of cortex, and that other fields were added independently in different lineages later in evolution. While some mammals appear to have an additional field, in primates, this region of cortex has expanded substantially, and contains at least four representations of the body. We propose that these additional fields compose networks that generate both perceptual and behavioral abilities directly associated with the extensive use of the glabrous, opposable digits in primates, and are involved in tactile exploration, object identification, and tactile recognition.

Acknowledgements

This work was supported by an Australian Research Council Fellowship (ARC), and by the Vision, Touch, and Hearing Research Centre.

References

Andersen, R.A., C. Asanuma, G. Essick, and R.M. Siegel (1990) Corticocortical connections of anatomically and physiologically defined subdivisions within the inferior parietal lobule. *Journal of Comparative Neurology* 296:65-113.

Beck, P.D., and J.H. Kaas (1993) Evidence for three somatosensory cortical areas in opossum. *Society for Neuroscience Abstracts* 19:764.

Brodmann, K. (1909) *Vergleichende Lokalisationsiehre der Grosshirnrinde in ihren Prinzipien Dargestellt auf Grund des Zellenbaues.* Leipzig: Barth:

Burton, H., and M. Carlson (1986) Second somatic sensory cortical area (SII) in a prosimian primate, *Galago crassicaudatus. Journal of Comparative Neurology.* 247:200-220.

Burton, H., M. Fabri, and K. Alloway (1995) Cortical areas within the lateral sulcus connected to cutaneous representations in areas 3b and 1: A revised interpretation of the second somatosensory area in macaque monkeys. *Journal of Comparative Neurology* 355:539-562.

Burton, H., and E.M. Kopf (1984) Ipsilateral cortical connections from the second and fourth somatic sensory areas in the cat. *Journal of Comparative Neurology* 225:527-553.

Burton, H., G. Mitchell, and D. Brent (1982) Second somatic sensory area in the cerebral cortex of cats: somatotopic organization and cytoarchitecture. *Journal of Comparative Neurology* 210:109-135.

Burton, H., T.O. Videen, and M.E. Raichle (1993) Tactile-vibration-activated foci in insular and parietal opercular cortex with positron emission tomography: Mapping the second somatosensory area in humans. *Somatosensory and Motor Research* 10:297-308.

Carvell, G.E., and D.J. Simons (1986) Somatotopic organization of the second somatosensory area (SII) in the cerebral cortex of the mouse. *Somatosensory Research.* 3:213-237.

Cavada, C., and P.S. Goldman-Rakic (1989a) Posterior parietal cortex in rhesus monkey: I. Parcellation of areas based on distinctive limbic and sensory corticocortical connections. *Journal of Comparative Neurology* 287:393-421.

Cavada, C., and P.S. Goldman-Rakic (1989b) Posterior parietal cortex in rhesus monkey: II. Evidence for segregated corticocortical networks linking sensory and limbic areas with the frontal lobe. *Journal of Comparative Neurology* 287:422-445.

Clemo, H.R., and B.E. Stein (1982) Somatosensory cortex: a 'new' somatotopic representation. *Brain Research.* 235:162-168.

Cusick, C.G., J.T. Wall, D.J. Felleman, and J.H. Kaas (1989) Somatotopic organization of the lateral sulcus of owl monkeys: area 3b, S-II, and a ventral somatosensory area. *Journal of Comparative Neurology.* 282:169-190.

Elston, G., L. Krubitzer, P. Manger, M. Calford, and T. Day (1993) The organization and connections of somatosensory cortex in the Australian marsupial, brush tailed possum (*Trichosurus vulpecula*). *Society for Neuroscience Abstracts* 19:764

Fabri, M., and H. Burton (1991) Ipsilateral cortical connections of primary somatic sensory cortex in rats. *Journal of Comparative Neurology.* 311:405-424.

Friedman, D.P., E.G. Jones, and H. Burton (1980) Representation pattern in the second somatic sensory area of the monkey cerebral cortex. *Journal of Comparative Neurology.* 192:21-41.

Friedman, D.P., and E.A. Murray (1986) Thalamic connectivity of the second somatosensory area and neighboring somatosensory fields of the lateral sulcus of the macaque. *Journal of Comparative Neurology.* 252:348-373.

Friedman, D.P., E.A. Murray, J.B. O'Neill, and M. Mishkin (1986) Cortical connections of the somatosensory fields of the lateral sulcus of macaques: Evidence for a corticolimbic pathway for touch. *Journal of Comparative Neurology.* 252:323-347.

Haight, J.R. (1972) The general organization of somatotopic projections to SII cerebral neocortex in the cat. *Brain Research* 44:483-502.

Iwamura, Y., A. Iriki, and M. Tanaka (1994) Bilateral hand representation in the postcentral somatosensory cortex.

Nature 369:554-556.

Iwamura, Y., M. Tanaka, M. Sakamoto, and O. Hikosaka (1983) Functional subdivisions representing different finger regions in area 3 of the first somatosensory cortex of the conscious monkey. *Experimental Brain Research* 51:315-326.

Johnson, J.I. (1990) Comparative development of somatic sensory cortex. *In* E. G. Jones and A. Peters (eds): *Cerebral Cortex* N. Y.: Plenum, pp 335-449

Kaas, J.H. (1983) What, if anything, is SI? organization of first somatosensory area of cortex. *Physiological Reviews*. 63:206-230.

Kaas, J.H., and T.P. Pons (1988) The somatosensory system of primates. *Comparative Primate Biology* 4:421-468.

Krubitzer, L., J. Clarey, R. Tweedale, G. Elston, and M. Calford (1995) A redefinition of somatosensory areas in the lateral sulcus of macaque monkeys. *Journal of Neuroscience* 15:3821-3839.

Krubitzer, L.A., and M.B. Calford (1992) Five topographically organized fields in the somatosensory cortex of the flying fox: Microelectrode maps, myeloarchitecture, and cortical modules. *Journal of Comparative Neurology* 317:1-30.

Krubitzer, L. A., M. B. Calford, and L. M. Schmid (1993) Connections of somatosensory cortex in megachiropteran bats: the evolution of cortical fields in mammals. *Journal of Comparative Neurology* 327:473-506.

Krubitzer, L.A., and J.H. Kaas (1990) The organization and connections of somatosensory cortex in marmosets. *Journal of Neuroscience*. 10:952-974.

Krubitzer, L.A., M.A. Sesma, and J.H. Kaas (1986) Microelectrode maps, myeloarchitecture, and cortical connections of three somatotopically organized representations of the body surface in the parietal cortex of squirrels. *Journal of Comparative Neurology* 250:403-430.

Ledberg, A., B.T. O'Sullivan, S. Kinomura, and P.E. Roland (1995) Somatosensory activations of the parietal operculum of man. A PET study. *Proceedings of the National Academy of Science, US in press:*

Lende, R.A., and C.N. Woolsey (1956) Sensory and motor localization in cerebral cortex of porcupine *(Erethizon dorsatum). Journal of Neurophysiology* 19:544-563.

Merzenich, M.M., J.H. Kaas, M. Sur, and C.-S. Lin (1978) Double representation of the body surface within cytoarchitectonic areas 3b and 1 in "SI" in the owl monkey *(Aotus trivirgatus). Journal of Comparative Neurology*. 181:41-74.

Mishkin, M. (1979) Analgous neural models for tactual and visual learning. *Neuropsychologia* 17:139-151.

Murray, E.A., and M. Mishkin (1984) Relative contributions of SII and area 5 to tactile discriminations in monkeys. *Behavioral Brain Research* 32:181-195.

Nelson, R.J., M. Sur, D.J. Felleman, and J.H. Kaas (1980) Representations of the body surface in postcentral parietal cortex of *Macaca fascicularis. Journal of Comparative Neurology* 192:611-643.

Nelson, R.J., M. Sur, and J.H. Kaas (1979) The organization of the second somatosensory area (SmII) of the grey squirrel. *Journal of Comparative Neurology* 184:473-490.

Pimentel-Souza, F., R.M. Cosenza, G.B. Campos, and J.I. Johnson (1980) Somatic sensory cortical regions of the agouti, *Dasyprocta aguti.. Brain Behavior and Evolution* 17:218-240.

Pons, T.P., P.E. Garraghty, C.G. Cusick, and J.H. Kaas (1985) The somatotopic organization of area 2 in macaque monkeys. *Journal of Comparative Neurology* 241:445-466.

Pons, T.P., P.E. Garraghty, and M. Mishkin (1988) Lesion-induced plasticity in the second somatosensory cortex of adult macaques. *Proceedings of the National Academy of Science, USA* 85:5279-5281.

Pons, T.P., and J.H. Kaas (1985) Connections of area 2 of somatosensory cortex with the anterior pulvinar and subdivisions of the ventroposterior complex in macaque monkeys. *Journal of Comparative Neurology* 240:16-36.

Pubols Jr, B.H. (1977) The second somatic sensory area (SmII) of opossum neocortex. *Journal of Comparative Neurology* 174:71-78.

Robinson, C.J., and H. Burton (1980a) Organization of somatosensory receptive fields in cortical areas 7b, retroinsula, postauditory, and granular insula of *M. fascicularis. Journal of Comparative Neurology* 192:69-92.

Robinson, C.J., and H. Burton (1980b) Somatotopographic organization in the second somatosensory area of *M. fascicularis. Journal of Comparative Neurology* 192:43-67.

Sinclair, R.J., and H. Burton (1991) Neuronal activity in the primary somatosensory cortex in monkeys (*Macaca mulatta*) during active touch of textured surface gratings: responses to groove width, applied force, and velocity of motion. *Journal of Neurophysiology* 66:153-169.

Sinclair, R.J., and H. Burton (1993) Neuronal activity in the second somatosensory cortex of monkeys (*Macaca*

mulatta) during active touch of gratings. *Journal of Neurophysiology* 70:331-350.
Sur, M., J.T. Wall, and J.H. Kaas (1984) Modular distribution of neurons with slowly adapting and rapidly adapting responses in area 3b of somatosensory cortex in monkeys. *Journal of Neurophysiology* 51:724-744.
Vogt, C., and O. Vogt (1919) Allgemeinere ergelnisse unserer hirnforschung. *Journal of Psychology and Neurology* (Leipzig) 25:279-462.
von Economo, C. (1929) *The cytoarchitectonics of the cerebral cortex.* Edited by H. Milford. London: Oxford University Press.
Welker, C., and M.M. Sinha (1972) Somatotopic organization of SmII cerebral neocortex in albino rat. *Brain Research* 37:132-136.
Whitsel, B.L., L.M. Pertrucelli, and G. Werner (1969) Symmetry and connectivity in the map of the body surface in somatosensory area II of primates. *Journal of Neurophysiology* 32:170-183.
Woolsey, C.N. (1958) Organization of somatic sensory and motor areas of the cerebral cortex. *In*: H. F. Harlow and C. N. Woolsey (eds). *The biological and biochemical bases of behavior.* Madison: University of Wisconsin Press.
Woolsey, C.N., and D. Fairman (1946) Contralateral, ipsilateral, and bilateral representation of cutaneous receptors in somatic areas I and II of the cerebral cortex of pig, sheep, and other mammals. *Surgery* 19:684-702.
Woolsey, T.A. (1967) Somatosensory, auditory and visual cortical areas of the mouse. *Johns Hopkins Medical Journal* 121:91-112.

Somesthesis and the Neurobiology of the Somatosensory Cortex
O. Franzén, R. Johansson and L. Terenius (eds.)

Serial processing in the somatosensory system of macaques

T.P. Pons

Department of Neurosurgery, Bowman Gray School of Medicine, Wake Forest University, Winston-Salem, North Carolina, USA, 27157

Summary. Electrophysiological, anatomical and immunocytochemical evidence for serial processing in the macaque somatosensory system is reviewed. The results from these studies are then compared with similar or dissimilar results obtained in other animal species. It is argued that the bulk of the current evidence strongly supports the view that the input supplying the major somatic drive to the second somatosensory area is from postcentral somatosensory cortex.

Introduction

Historically, the primary (SI) and secondary (SII) somatosensory cortex in all mammals was thought to have dense connections with the ventroposterior nucleus (VP) of the thalamus (Jones and Powell 1970; see Mountcastle 1984 for review). Based on combined ablation\recording studies in rabbits and cats (Woolsey and Wang 1945; Woolsey and Fairman 1946; Manzoni et al., 1984; Burton et al., 1987.), which left somatic activation of SII neurons intact after SI lesions, the major source of somatic activation in both SI and SII was thought to be from the dense direct projections from VP. Thus, the relay of somatic drive to SI and SII was theorized to occur via parallel projections from VP (see Mountcastle 1984 for review).

This view was called into question, at least for Macaque monkeys, by Pons et al., 1987, who found that selectively ablating body part representations in SI selectively deactivated body part representations in SII. They interpreted their findings, as well as additional anatomical and electrophysiological data, as indicating that the major source of somatic activation for neurons in SII is dependent upon a direct projection from SI. Recently, Rowe and colleagues (Murray et al., 1992; Turman et al., 1992; 1994) have performed cortical cooling experiments, mostly in rabbits and cats, in order to reversibly deactivate SI or SII cortex and study the effects of deactivating one cortical region on the other. As in the previous studies on these species (Woolsey 1945; Burton and Robinson 1987; Manzoni 1984), functional deactivation of portions of SI did not lead to deactivation of the corresponding portions of SII.

Interestingly, recent preliminary cortical cooling experiments of SI in Marmosets (a new world monkey) by Rowe and colleagues have yielded findings similar to those obtained in rabbits and cats, e.g. deactivation of SI does not deactivate SII. This is in contrast to the findings of Garraghty et al., 1990 that indicated ablation of SI in this same species does deactivate the second somatosensory area. Based on their reversible deactivation experiments Rowe suggested at this conference that the major source of somatic activation to SI and SII occurs over parallel projections from VP to SI and SII, and serial processing in the macaque should be reconsidered based on this evidence. The purpose of this chapter is to review current evidence regarding the major relay routes by which gross somatic responsivity to SI and SII cortex is conveyed in non-human primates.

Materials and Methods

A more detailed description of the Methods for the electrophysiological experiments can be found in Pons et al., 1987; 1992. Briefly, electrode penetrations were placed 0.5-1.0 mm apart in a rectangular grid across the entire extent of SII, and neuronal responses were sampled at 200um intervals through the depth of this cortex using standard recording techniques. The receptive fields of the neurons at the recording sites were determined by applying tactile stimulation at different locations on the contralateral body surface. Normal and operated hemispheres were studied, the operated hemispheres receiving various types of cortical ablations. Cortical ablations were made by aspiration and were verified later anatomically with Nissl staining. Both acute (recording performed immediately after the cortical ablation) and chronic (recording performed 6-8 weeks after cortical ablation) survival periods were examined in these experiments.

For the anatomical experiments, Cytochrome oxidase, Nissl and GABA staining were employed to study patterns of thalamic degeneration after cortical ablation of all or portions of SI and SII. The survival times were similar to those employed in the electrophysiological experiments described here and in some instances hemispheres that were used in electrophysiological experiments were also used in anatomical experiments. Cell counts were obtained to determine the pattern and degree of thalamic degeneration after cortical ablations. Our findings in VP after SI ablations were then compared with the findings in VP of Friedman and Murray 1986 after SII cortical ablations. For a full description of the anatomical techniques employed see Chmielowska and Pons (in press).

Results and Discussion

Electrophysiological Evidence for serial processing in macaques.

We previously reported (Pons et al., 1987) that in intact (or normal) hemispheres, receptive fields of neurons in SII were readily found for tactile stimulation of all contralateral body parts, with the majority of the fields representing loci on the glabrous and hairy surfaces of the hand. By contrast, in recording sites through SII of hemispheres in which the hand representations in SI had been removed (in each of cytoarchitectonic areas 3a, 3b, 1 and 2), no receptive fields could be found for tactile stimulation of the glabrous surface of the hand, and only a few were found that included the hand's hairy surface. Yet there was no difficulty in recording responses in the experimental hemispheres to stimulation of all other body parts, indicating that the near absence of a hand representation in SII was not due simply to a general postsurgical depression of the SII cortex. The functional dependency on postcentral cortex that was demonstrated for the SII representation of the hand held also for the SII representations of other body parts. For example, in recording sites distributed through the SII region in a case with a postcentral removal that spared only the hand representation, all of the receptive fields found were confined to the hand. Similarly, in recording sites through SII in a case with a total removal of the postcentral strip, no neuronal activation was observed from tactile stimulation of any body part. In short, the elimination of any body-part representation in the postcentral cortex eliminated it also in SII. The reverse was not true, however, as ablation of SII resulted in no obvious changes in the response properties of SI neurons, at least at the subjective level.

Burton et al., (1990) also functionally deactivated most of the SII hand representation in macaques when they ablated all of the SI hand region except for a portion of area 3a deep in the fundus of the central sulcus. Like Pons et al., (1987), they interpreted their results as supporting the view that the somatic activation of neurons in SII is dependent upon a serial relay of inputs from the SI region.

Of the four areas that comprise SI cortex in macaques, two (areas 3b and 1) process predominantly cutaneous inputs, whereas the other two (areas 3a and 2) process additional "deep" somatic inputs. In another study by Pons et al., (1992), selective removals of the hand representations in areas 3a, 3b, 1 and 2 in different combinations were made to determine each area's contribution to the responsivity and modality characteristics of neurons in SII. Ablations that left intact one or both postcentral areas that process "deep" inputs yielded SII recording sites mainly responsive to "deep" stimulation. Conversely, removals that left intact one or both postcentral areas that process predominantly cutaneous information yielded SII recording sites responsive to cutaneous stimulation, and none that was driven exclusively by "deep" stimulation.

When only the area 3b hand representation was ablated, leaving areas 3a, 1, and 2 fully intact, there was a sharp reduction (though not elimination) in the number of cutaneous recording sites in the SII hand representation, but not in the number of "deep" sites. The ratio of cutaneous and "deep" sites for body part representations other than the hand was not affected by any of the ablations, indicating that the above results were not due to a general postsurgical depression or to anesthetic effects. From these findings Pons et al., (1992) suggested that modality-specific information is relayed from postcentral cortex to SII along parallel channels, with cutaneous inputs being transmitted via areas 3b and to a lesser extent area 1, and "deep" inputs via areas 3a and 2. They further suggested that area 3b provides the major source of cutaneous inputs to SII, either directly or via area 1 and that somatic activation of neurons in SII is achieved by both serial and parallel cortical processing.

In other studies, Garraghty and his colleagues have shown differential effects of SI ablations on somatic activity in SII, depending on the species of animal studied (Garraghty et al., 1990;1991). In cats, tree shrews and galagos ablation of all of SI cortex appears **not** to deactivate SII. Interestingly, these animals have relatively dense projections from the ventroposterior lateral nucleus (VPL) to SII. In animals with a relatively sparse input from VPL to SII, such as macaques, neurons in SII are largely unresponsive to cutaneous stimulation after SI ablation (with the notable exception of pacinian receptor inputs). In a single study on marmosets, a New World primate, similar results were obtained (Garraghty et al., 1990).

Anatomical Evidence for serial processing in macaques.

In a recent study we examined the pattern of cytochrome oxidase (CO) and Nissl staining as well as gamma aminobutryic acid (GABA) immunoreactivity in the ventroposterior lateral nucleus (VPL) of macaque monkeys that were normal or had various combinations of ablations of the hand representations in the cytoarchitectonic areas comprising SI (Chmielowska and Pons, in press). The findings indicated that in the unoperated, or normal, animals, the VPL hand representation was characterized heavy but homogeneous CO staining throughout the entire nucleus. Counts of neurons in the normal VPL hand representation revealed 261.4 neurons/mm^2, 78.4/mm^2 of which stained positive for GABA.

In stark contrast, macaque monkeys in which the total SI or postcentral hand representations had been ablated showed a dramatic reduction in CO staining in the VPL, this reduction in CO staining was confined to the expected location of the VPL hand representation. Counts of neurons in VPL from Nissl stained material, as well as sections that underwent GABA immunostaining, revealed an average of 46.9/mm^2 Nissl stained neurons, a dramatic and statistically significant reduction from normal. There was also a statistically significant reduction in the number of neurons staining positive for GABA. Most importantly, large to medium sized (>180um^2) GABA(-) neurons, generally thought to be projection neurons (Rausell and Jones 1992) were almost completely eliminated in the VPL hand representation.

The above results of dramatic degeneration of thalamic neurons following SI cortical ablations contrast sharply with the findings of Friedman and Murray (1986) who showed that there was **no** statistical reduction in the number of neurons in VP after SII cortical ablations. Even though it appears that the overwhelming output of cells in VP is to SI, there may still be some direct inputs from VP to SII in macaques. All previous anatomical studies that have reported that VPI supplies the major thalamic input to SII have recognized a relatively sparse input from VP to SII as well (Manzoni et al., 84; Friedman et al., 1986; Krubitzer and Kaas 1990). Nevertheless, the virtual elimination of projection-sized neurons in VP after SI (but not SII) cortical ablations lends overwhelming support to the view that in Old World macaque monkeys the major output of VP is to the SI region (areas 3a, 3b, 1 and 2). The above anatomical findings, taken in conjunction with our previous physiological findings that complete ablation of SI cortex deactivates tactile input in SII, and that SII ablation does not deactivate tactile activation of SI, makes a compelling case for serial processing in Old World macaque monkeys.

Having stated the case for the major source of somatic activation in SII being conveyed serially from SI, there is room for disagreement. Perhaps the best evidence for a parallel relay of tactual information from VP to SI and SII is electrophysiological collision data showing that in Marmosets at least some individual VPL neurons project to both SI and SII. Such elegant evidence presented by Dr. Rowe at this conference is not easily refuted and demonstrates that such neurons certainly exist in Marmoset VPL and suggests they may exist in macaque VPL as well. Given the dramatic degeneration of neurons in VPL in macaques following the SI ablations, however, and the apparent inability of projections from VP to SII to sustain tactual activity after SI ablations, the weight of the evidence, in my view, still supports serial processing as the means by which the vast majority of neurons in SII receive their major somatic drive.

It could be argued that SI cortical ablations in macaques involve the underlying white matter and inadvertently affect VPL projections to SII. This seems highly unlikely, however, for several reasons. 1) There is no evidence that any projections from VPL or VPI to SII course under SI cortex. 2) Even assuming that VPL or VPI projections to SII take such a tortuous route under SI, as a control we have ablated area 4 and a portion of area 3a across the fundus of the central sulcus. This ablation was as likely to affect any fibers projecting to SII as the SI ablations, but no decrement in SII responsiveness was observed in these cases. 3) The precise correspondence between the body part representation ablated in SI and that deactivated in SII would not be possible if projection fibers to SII were severed. 4) The sharp mediolateral border of degeneration in VP of animals receiving various types of SI ablations seems implausible if our ablations extended extensively into the white matter.

Differences in SII responsivity of Marmosets, a New World monkey, after deactivation of SI cortex by cortical cooling or ablations are harder to explain. According to Rowe and colleagues, SII is largely responsive to tactile inputs following cooling of SI cortex in Marmosets. In contrast, in a single study, Garraghty et al., 1990 reported that ablation of SI cortex resulted in complete deactivation of the SII region in this same species. Why these groups obtained such different findings is not entirely clear. One possibility is that with additional sampling of electrophysiological activity in SII that Garraghty et al., (1990) may have seen tactual responses. Another possibility, however, is that the cortical cooling of SI cortex by Rowe and colleagues was not complete. In a Pons et al., study (1992) if any portion of SI was left intact, including a relatively small portion of area 3a, tactual responsiveness in SII was not abolished. Burton et al., (1990) reported similar findings. Whether area 3a was defined as a part of "SI" cortex was not clear in the Marmoset study. With cortical cooling, if any portion of SI was not affected, then responsiveness in SII cortex might not have been abolished.

Besides methodological differences in using ablations and cooling techniques to deactivate the cortex one should be aware that major species differences can exist as well. Certainly the pattern of projections from VPL to SI and SII varies in different species (Garraghty 1991 for review), as well as the effects of SII responsiveness after SI cortical ablations. It is probably premature to state that a general "primate plan" for serial or parallel processing exists. It is noteworthy, however, that virtually all of the anatomical and electrophysiological work in macaques clearly supports a serial relay of tactual inputs from SI to SII as providing the major source of somatic activation in SII in this species, though parallel inputs from the thalamus probably also exist. The extent to which a serial processing scheme exists in other primates, including humans, remains to be seen. Of cours the precise functional role of parallel inputs from VP to SII, as well as serial inputs from SI to SII, remains an open field of inquiry.

Acknowledgments

I thank Drs. Mortimer Mishkin, Preston Garraghty and Jolanta Chmielowska for their helpful discussion of issues related to serial processing in sensory cortex. This work was supported in part by grant 1 RO1 MH53369-01 from the National Institute of Mental Health, the A. Tab Williams, Jr. Family Neuroscience Research and Program Development Endowment and an Alfred P. Sloan Research Foundation Fellowship.

References

Burton, H. (1984) Corticothalamic connections from the second somatosensory area and neighboring regions in the lateral sulcus of macaque monkeys. *Brain Research. 309*:367-372.

Burton, H. and Robinson, C.J. Responses in the first or second somatosensory cortical areas in cats during transient inactivation of the other ipsilateral area with lidocaine hydrochloride. *Somatosensory and Motor Research.* 4: 215-236, 1987.

Burton, H., Sathian, K. and Dian-hua, S. Altered responses to cutaneous stimuli in the second somatosensory cortex following lesions of the postcentral gyrus in infant and juvenile monkeys. *Journal of Comparative Neurology.* 291: 395-414, 1990.

Chmielowska, J. and Pons, T. P. Patterns of Thalamocortical Degeneration after Ablation of Somatosensory Cortex in Monkeys. *Journal of Comparative Neurology.* (in press).

Friedman, D.P., and E.A. Murray (1986) Thalamic connectivity of the second somatosensory area and neighboring somatosensory fields of the lateral sulcus of the macaque. *Journal of Comparative Neurology. 252* :348-373.

Garraghty, P.E., Florence, S.L., Tenhula, W.N. and Kaas, J.H. Parallel thalamic activation of the first and second somatosensory areas in prosimian primates and tree shrews. *Journal of Comparative Neurology.* 311; 289-299, 1991.

Garraghty, PE., Pons, T.P. and Kaas, J.H. Ablations of areas 3b (SI proper) and 3a of somatosensory cortex in marmosets deactivate the second and parietal ventral somatosensory areas. *Somatosensory and Motor Research.* 7: 125-135, 1990.

Jones, E.G. (1985) *The Thalamus.* Plenum Press, New York.

Jones, E.G. and T.P.S. Powell. (1970) Connexions of the somatic sensory cortex of the rhesus monkey, III. Thalamic connexions, *Brain 93*:37-56.

Kaas, J.H. and Pons, T.P. The somatosensory system of primates. In: *Comparative Primate Biology* Vol 4, Neurosciences, edited by H. Steklis. New York: Alan R. Liss, Inc., 1988, p. 421-468.

Krubitzer, L.X. and Kaas, J.H. The organization and connections of thesomatosensory cortex in marmosets. *Journal of Neuroscience.* 3: 952-974, 1990.

Manzoni, T.P., Barbaresi, R. and Conti, F. Callosal mechanism for the interhemispheric transfer of hand somatosensory information in the monkey. *Behavioral Brain Research.* 11: 155-170, 1984.

Mountcastle, V.B. Central nervous mechanisms in mechanoreceptivesensibility. In: *Handbook of physiology-The nervous system III*, edited by V.B. Mountcastle. Bethesda, Maryland: American Physiological Association, 1984, p. 789-878.

Murray, G.M., Zhang, H.Q., Kaye, A.N., Campbell, D.H. and Rowe, M.J. Parallel processing in rabbit first (SI) and second (SII) somatosensory cortical areas: effects of reversible inactivation by cooling of SI responses on responses in SII. (1992) *Journal of Neurophysiology..* 68; 3; 703-710.

Olszewski, J. (1952) *The Thalamus of Macaca mulatta:* An Atlas for Use with the Stereotaxic Instrument. Karger, Basel.

Pons, T.P., and J.H. Kaas (1985) Connections of area 2 of somatosensory cortex with the anterior pulvinar and subdivisions of the ventroposterior complex in macaque monkeys.*Journal of Comparative Neurology. 240:* 16-36.

Pons, T.P., Garraghty, P.E., Friedman, D.P. and Mishkin, M. (1987) Physiological evidence for serial processing in the somatosensory system. *Science.* 237: 417-420.

Pons, T.P., Garraghty, P.E., and Mishkin, M. (1992) Serial and parallel processing in somatosensory cortex of rhesus monkeys. *Journal of Neurophysiology.* 68: 518-527.

Pons, T.P., Garraghty, P.E. and Mishkin, M. Lesion-induced plasticity in the second somatosensory cortex of adult macaques. *Proceedings of the National Academy of Sciences.* 85: 5279-5281, 1988a.

Pons, T.P., Garraghty, P.E. and Mishkin, M. Plasticity in nonprimary somatosensory cortex of adult monkeys. *In Post Lesion Neural Plasticity.* edited by H. Flohr. Berlin, Heidelberg, Springer-Verlag, 1988b, p. 511-517.

Rausell, E., and E.G. Jones (1991a) Histochemical and immunocytochemical compartments of the thalamic VPM nucleus in monkeys and their relationship to the representational map. *Journal of Neuroscience. 11:* 210-225.

Rausell, E., and E.G. Jones (1991b) Chemically distinct compartments of the thalamic VPM nucleus in monkeys relay principal and spinal trigeminal pathways to different layers of the somatosensory cortex. *Journal of Neuroscience.* 11*:* 226-237.

Rausell, E., Bae, C.S., Vinuela, A., Huntley, G.W., and Jones, E.G. (1992) Calbindin and parvalbumin cells in monkey VPL thalamic nucleus: distribution, laminar cortical projections, and relations to spinothalamic terminations. *Journal of Neuroscience. 12(10)*:4088-4111.

Turman, A.B., Morley, J.W. and Rowe, M.J. (1983) Parallel distributed processing in somatosensory cortex: Effects of SII inactivation on SI responses in cat. *Neuroscience Abstracts.* 18. 1543.

Turman, A.B., Ferrington, D.G., Ghosh, S. and Morley, J.W. and Rowe, M.J. (1992) *Journal of Neurophysiology.* 67; 2: 411-429.

Woolsey, C.N. 'Second' somatic receiving areas in the cerebral cortex of cat, dog and monkey. *Federation Proceedings.* 2; 55, 1943.

Woolsey, C.N. and Fairman, D. Contralateral, ipsilateral and bilateral representation of cutaneous receptors in somatic areas I and II of the cerebral cortex of pig, sheep and other mammals. *Surgery*, 19: 684-702, 1946.

Woolsey, C.N. and Wang, G.H. Somatic areas 1 and 2 of the cerebral cortex of the rabbit. *Federation Proceedings..* 4: 79, 1945.

Somesthesis and the Neurobiology of the Somatosensory Cortex
O. Franzén, R. Johansson and L. Terenius (eds.)

Parallel processing in somatosensory areas I and II of the cerebral cortex

M.J. Rowe, A.B. Turman, G.M. Murray and H.Q. Zhang

School of Physiology and Pharmacology, The University of New South Wales, Sydney, N.S.W. 2052 Australia

Summary. Somatosensory areas I and II (SI and SII) constitute two of the principal tactile processing areas in the cerebral cortex. They each receive direct projections from the thalamus and, as well, are linked through intracortical reciprocal connections. Tactile inputs may therefore be conveyed to SII, for example, over *either* a direct path from the thalamus or an indirect, or *serial*, path from the thalamus via SI. We have examined in the cat, rabbit and marmoset monkey the behaviour of SII neurons before, during and after the selective, rapidly-reversible inactivation of SI by means of localized cooling. The results demonstrate that in cat and rabbit, SII responsiveness is never abolished and infrequently affected by SI inactivation and that tactile inputs to SII therefore traverse a direct path from thalamus, organized in parallel with that to SI. In the marmoset (*Callithrix jacchus*), in contrast to earlier studies based on ablation of SI we found that with reversible inactivation of SI, SII responsiveness was unaffected in 25% of neurons and, although reduced in the remainder, was rarely abolished (<10% of SII neurons). The results indicate that there is substantial direct thalamic input to SII, even in this simian primate, and therefore necessitate revision of the hypothesis that tactile processing at the thalamocortical level in simian primates is based on a strict serial scheme in which tactile information is conveyed from the thalamus to SI and thence to SII.

Multiple representation of sensory inputs within the cerebral cortex

Each of the major sensory systems in placental mammals is characterized by multiple representation within the cerebral cortex. Therefore, whenever sensory inputs are generated by stimulation of a given point on the peripheral receptor array, whether on the retina, along the cochlea, or on the skin surface, several spatially separate processing regions within the cerebral cortex are activated (reviews by Brugge and Reale, 1985; Jones, 1986; Rowe, 1990; Felleman and van Essen, 1991).

The first demonstration that dual or multiple cortical sites might be involved in processing a given source of sensory input arose with electrophysiological studies in the 1930s and 40s, that revealed in a variety of species, the existence of two spatially separate cortical areas of somatosensory representation identified as the primary and secondary somatosensory areas, SI and SII (Marshall et al., 1941; Adrian, 1941; Woolsey and Fairman, 1946). These initial studies of sensorimotor organization within the mammalian neocortex relied for the most part upon evoked potential recording for the identification of sensory areas, and cortical surface stimulation for the identification of motor areas. However, with the more sensitive and detailed analysis of neocortical organization that is possible with microelectrode recording techniques and newer fiber tracing methods, it has become clear that larger proportions of neocortex are identifiable as visual, auditory or somatosensory, as multiple representations have been identified for a given sensory modality within areas that were previously designated association cortex (Burton, 1986; Jones, 1986; Pandya and Yeterian, 1985; Felleman and van Essen, 1991). The most striking example of this has taken place for the visual representations in the macaque monkey which, it is now argued, occupy the posterior half of the hemisphere and make up approximately 30 distinctive areas of cortex that are largely or exclusively visual in function. Their identification as visual areas is based upon the connections, topographic organization, cytoarchitecture and functional properties of the different areas (Felleman and van Essen, 1991). As these multiple visual areas are estimated to occupy ~55% of neocortical space in the macaque monkey (Felleman and van Essen, 1991), it is hardly surprising that fewer somatosensory and auditory areas have been identified. Nevertheless, a more modest population explosion of cortical representation areas has also taken place for these systems, with estimates for the macaque of about a dozen somatosensory-motor areas and about half that number of auditory areas (Friedman et al. 1986; Felleman and van Essen, 1991).

In the case of the somatosensory system, the proliferation of cortical areas of body representation has been based on both a fractionation of the originally identified SI and SII areas, in particular in the studies of primate somatosensory cortex, and upon the identification of additional areas beyond these. The areas now identified as somatosensory in the macaque in particular, are 3a, 3b, 1, 2, SII, the retroinsular (RI), granular insula (Ig), dysgranular insular (Id), the postauditory (Pa) and areas 5 and 7b (Burton, 1986; Friedman et al., 1986). In non-primates there are fewer areas, and no general agreement that SI warrants the fragmentation accorded it in primates. In the cat, in particular, the designations are based largely on the chronology of identification, with SI, SII, SIII, SIV and SV areas reported (Burton, 1986 for review, and Mori et al., 1991).

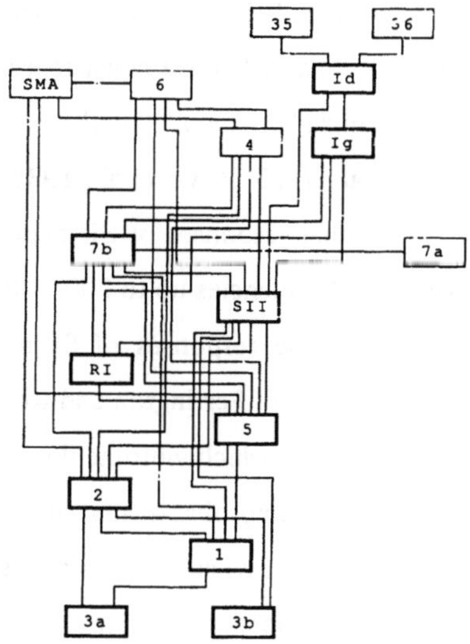

Figure 1. A proposed hierarchical scheme of organization of cortical somatosensory and motor areas in the macaque monkey, taken from Felleman and van Essen (1991). Somatosensory areas are outlined with heavy lines. Motor (areas 4, 6 and supplementary motor area, SMA), associational (areas 35 and 36 of the perirhinal cortex) and visual (area 7a) areas are indicated with finer outlines. RI, Ig and Id represent the retroinsular, granular insula and dysgranular insula areas respectively.

Hierarchical and parallel streams in cortical sensory processing

For the approximately 30 visual cortical areas identified for the macaque monkey, Felleman and van Essen (1991) have proposed an elaborate scheme of parallel and hierarchical (or serial) interconnections that form a highly distributed network for visual cortical processing. In its hierarchical dimension the network incorporates 10 levels of cortical processing. In their equivalent scheme for somatosensory and motor processing for the macaque cerebral cortex, Felleman and van Essen (1991) have proposed (based on several studies cited in their paper) a similar number of hierarchical levels (Fig. 1) as was proposed in their scheme for visual cortical processing. According to their scheme, the four cytoarchitectural areas, 3a, 3b, 1 and 2, that form the postcentral somatosensory area in the macaque and which, in other species, are usually grouped together as the primary somatosensory or SI area, occupy the lowest levels of the hierarchy (Fig. 1). In contrast, the second somatosensory area, SII, is placed at a higher level in the processing hierarchy (Fig. 1) despite a direct thalamic input to SII known to be present in the macaque and other primate and non-primate mammals that have been investigated (Friedman and

Murray, 1986; and reviews by Jones, 1985; Burton, 1986). Furthermore, in the cat at least, the functional properties of the tactile neurons of SII are consistent with these neurons receiving a direct input from the thalamus, organized in parallel with that to SI (Bennett et al., 1980; Ferrington and Rowe, 1980; Fisher et al., 1983; Rowe et al., 1985). Part of the evidence for this conclusion came from the observation that responses of SII neurons to vibrotactile stimuli may be phaselocked (see below) at vibration frequencies up to 200-300 Hz, in contrast to the absence of phaselocking at these frequencies in the responses of SI neurons of the equivalent class (Mountcastle et al., 1969; Bennett et al., 1980; Ferrington and Rowe, 1980).

In a system with strict serial processing in which vibrotactile information is conveyed from the thalamus to SI and thence to SII, it is improbable that neurons within SII could display phaselocked responses to cutaneous vibration when neurons at the preceding level of processing (SI) have poorer phaselocking in their responses. For this reason, and because of the direct anatomical projection from the thalamus to both SI and SII, it had been assumed by many investigators that SI and SII occupied an equivalent hierarchical level and were organized in parallel for tactile processing (Fisher et al., 1983; Rowe, 1990). However, one of the complications for this *parallel processing* model, is the presence of reciprocal, cortico-cortical connections between SI and SII which could mean that tactile processing in one area, SII for example, is not just dependent upon direct inputs from the thalamus, but may be determined in part at least, by inputs that come over a *serial*, or *hierarchical* path from the thalamus to SI and are then conveyed to SII at a higher hierarchical level of tactile processing.

Organization of SI and SII areas for tactile processing: experimental investigation of parallel and serial schemes

The relative importance of the direct and indirect paths to SII, and therefore of the *parallel* or *serial* processing schemes, may be evaluated by examining tactile processing within SII when SI is inactivated. This was addressed directly for the macaque monkey by Pons et al. (1987) in a study in which they observed an abolition of tactile responsiveness within the hand area of SII subsequent to inactivation of the hand area of the postcentral, or SI, area of cortex by means of surgical ablation. Similar observations in the marmoset monkey Garraghty et al., 1990) indicated that in *New-*, as well as *Old-World* simian primates there was a strict serial scheme of

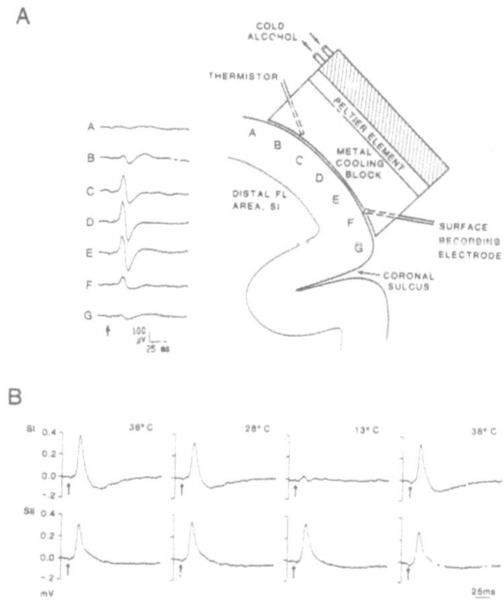

Figure 2. Experimental arrangement for reversible inactivation of the SI forelimb area in the cat. A: transverse view from the caudal side of the right hemisphere showing the position of the cooling block over the distal forelimb area of SI on the postcruciate region of cortex adjacent to the coronal sulcus. B: effect of cooling the distal forelimb area in SI on evoked potentials recorded from the distal forelimb areas of SI (upper part) and SII (lower part). Each trace is an average of 20 sweeps. Simultaneous recordings were made when the temperature of the block overlying SI was at 38°C, 28°C, 13°C and again at 38°C. Arrows indicate the onset of the 5 ms duration tap stimulus (amplitude 400 μm) delivered to the central footpad (from Turman et al., 1992).

tactile processing. In contrast, when the SI area in the cat was inactivated by local anesthetic blockade, SII responsiveness was not abolished (Burton and Robinson, 1987). However, neither of these procedures employed for SI inactivation permits proper analysis of individual SII neurons in both *control* and *test* circumstances. The ablation procedure is clearly irreversible and the local anesthetic blockade only slowly and somewhat unpredictably reversible. We have therefore employed a rapidly reversible inactivation procedure, based on cooling, to determine whether the *parallel* or *serial* hypothesis more closely describes the functional organization of the SI and SII tactile areas in non-primate and primate species. With this technique we can inactivate SI within 1-2 minutes by lowering the temperature at the cortical surface beneath a silver cooling block (Fig.2A) from 38 to ~13°C and can reactivate SI over a similar time course. The speed of the procedure allows several repetitions of the inactivation while recording from the one SII neuron and thereby enables any uncertainty about the effect of SI inactivation to be resolved. The cooling method also avoids some possible risks associated with ablation, namely, damage to underlying thalamocortical fibers that project to SII and vascular impairment that may extend beyond SI itself.

Parallel organization of the SI and SII areas for tactile processing in the cat

Our initial studies have been conducted in the cat and rabbit (Turman et al., 1992, 1995; Murray et al., 1992) and more recent ones in the marmoset (Zhang et al. 1993a, b; Murray et al. 1993). The present report deals principally with data from the cat (Turman et al., 1992).

The SI and SII areas for the distal forelimb were first identified by evoked potential (EP) recording from the cortical surface (above positions A-G for SI in Fig.2A) in response to a brief, 5 ms tap stimulus to the central foot pad. Once the SI area was identified, a 7 mm diameter silver cooling block fitted with a Peltier thermoelectric device was placed over the area. The block was equipped with a thermistor to record the temperature at the cortical surface and an electrode to record surface EPs (Fig.2A). Inactivation of the cortex beneath the block was achieved with surface cooling to 10°-15°C, as reflected in the abolition of the SI EP at 13°C in the upper row of Fig.2B. The small residual SI response reflects the thalamocortical input volley, unmasked by the abolition of the cortical response (Turman et al., 1992). Inactivation of SI was also verified at the single neuron level by microelectrode recording around the margins of the block.

The simultaneously recorded SII EP was unaffected in four out of eight experiments in the cat (lower row in Fig.2B) and underwent a small reduction in the remainder. In no experiment was it abolished. As the SII EPs survive SI inactivation it is clear that SII responsiveness is not dependent upon inputs that traverse a serial path from the thalamus via SI. However, the SII EP is generated by a brief afferent volley, perhaps before there is time for the indirect path through SI to contribute to the response. Therefore, in order to test properly whether the serial route contributes to SII responsiveness, we examined, in association with SI inactivation, the responsiveness of single SII neurons to maintained and controlled forms of tactile stimulation. This was done for 29 SII neurons, with quantitative single neuron analysis. In 23, or 80% of the sample, responsiveness was unchanged, as illustrated in Fig.3 for an SII neuron sensitive to high-frequency vibrotactile stimuli, whose input appeared to come selectively from Pacinian corpuscle (PC) receptor sources. The responses to a 1s train of vibration at 300 Hz (1, 5 and 25 μm) appear to be unchanged when SI was inactivated (middle column in Fig.3; SI at 8°C) incomparison with the pre- and post-cooling controls (left and right columns, Fig.3; SI at 38°C). However, in order to evaluate reliably any possible effect, the SII responses were quantified precisely by constructing stimulus-response (S-R) relations (Fig.4) in

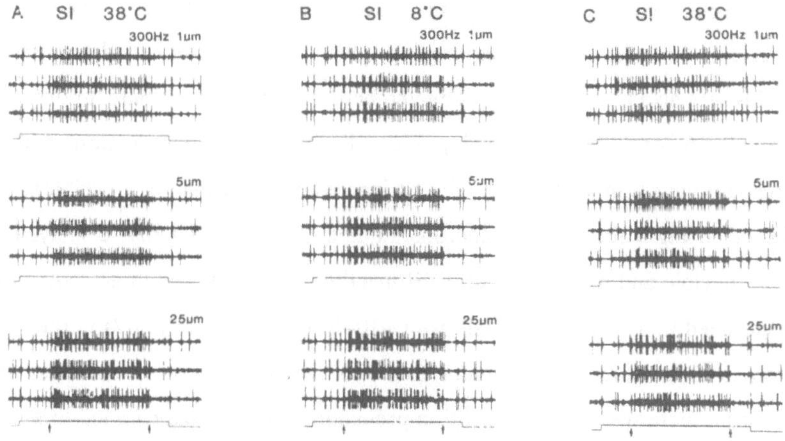

Figure 3. Absence of effect of SI inactivation on responsiveness of an SII neuron in the cat to a 1s train of 300 Hz vibrotactile stimulation delivered at three intensities (1, 5 and 25 μm) in the period indicated by the arrows in the lowest traces of the stimulus waveforms in A, B and C. The vibration was superimposed on a 1.5 s step indentation of the skin (400 μm amplitude). A: control responses obtained when SI was held at 38°C; B: responses obtained during SI inactivation induced by cooling the block over the forelimb area of SI to 8°C; C: responses obtained on rewarming SI to 38°C (from Turman et al., 1992).

which the mean response (± SD) was plotted as a function of stimulus intensity. Relations A and C in Fig. 4 were the controls obtained *before* and *after* SI inactivation. Relation B was obtained when SI was inactivated, and is no different from the control S-R relations in its overall form, or in details such as the plateau level of response, the slope of the initial phase of the relation, the threshold for responsiveness, or in the dynamic range for the S-R relation, defined as the range

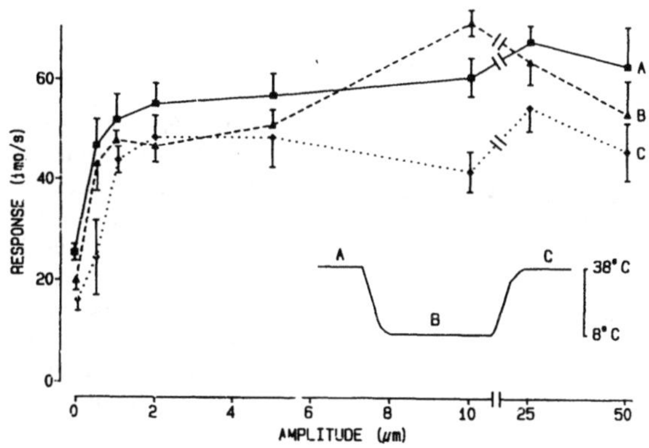

Figure 4. Stimulus-response (S-R) relations for an SII neuron in cat, constructed before, during and after SI inactivation by cooling, showing the absence of effect of the SI inactivation. The graphs plot the mean response ± S.D. (impulses/s; n = 5) obtained at a series of different amplitudes (μm) of the 300 Hz vibration. S-R relation A was obtained with SI at 38°C, B with SI at 8°C and C with SI restored to 38°C (from Turman et al., 1992).

of stimulus intensities over which response increments were obtained. Several different functional classes were represented among the SII neurons whose responsiveness was unchanged. This was also the case among affected neurons. The latter made up only a minority of the SII sample (6, or 20%) and displayed usually only a small reduction, and never an abolition of responsiveness in association with SI inactivation (Turman et al., 1992). Response levels are plotted in Fig.5 for the two neurons studied in the cat SII that displayed the most consistent, reproducible and marked effect of SI inactivation. In A, the three repetitions of the cooling procedure led to a reproducible fall from ~30-40 imp./s to ~25 imp./s in the response of the SII neuron to a 1s vibrotactile stimulus that was repeated every 10 seconds. In B, for another SII neuron, there were two inactivation procedures that, once again, led to a clear fall in responsiveness.

Effect of SI inactivation on impulse patterning in the responses of SII neurons to vibrotactile stimuli

As the response level was affected in only 20% of SII neurons in the cat in association with SI inactivation, other response features were examined for more subtle contributions that might be made to SII responsiveness by inputs that traversed the proposed indirect, *serial* path via SI. In particular, the phaselocking and temporal patterning of impulse activity evoked by vibrotactile

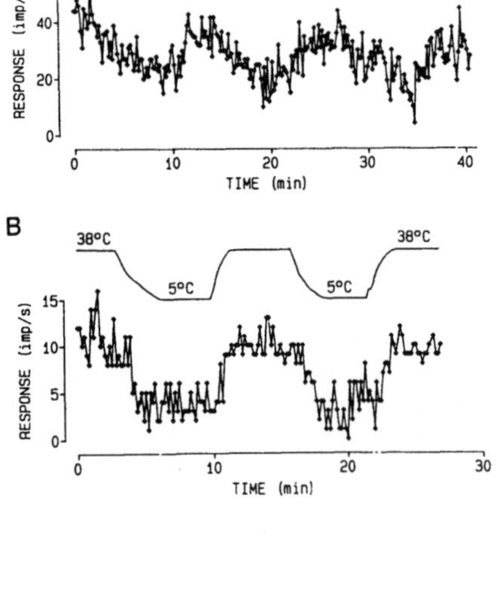

Figure 5. Altered responsiveness to tactile stimuli in SII neurons of the cat in association with SI inactivation. A and B: response level (impulses/s) to successive deliveries (1 per 10s) of a 1s tactile stimulus plotted for two SII neurons against time. During the 40 min. sampling period in A, SI transmission was reversibly inactivated three times as a result of cooling the overlying block from 38°C to 13°C. In B, the SI inactivation was repeated twice over a 30 min. period for a different SII neuron. The tactile stimulus in A was a 1s train of vibration (300 Hz, 30 µm) and in B was a 1s train of rectangular pulses (5 Hz, 125 µm). (from Turman et al., 1992, where full statistical analysis of data is presented).

stimuli were examined. The reason for examining these features was that if the input were coming to the SII neuron over *both* a *direct* path from the thalamus and an *indirect* intracortical path from the thalamus, via SI, one might expect some temporal discrepancy in the arrival of vibrotactile-induced signals over the two paths. Therefore, if the indirect path via SI were to be inactivated, some tightening of the phaselocking in SII responses might result during that time. The impulse trace replicas of Fig.6 for responses of an SII neuron to a 300 Hz train of vibration show that most impulse occurrences fall between the peak and subsequent trough of the illustrated waveform whether one considers the *control* traces in A and C *before* and *after* SI inactivation (SI at 38°C) or those in B, obtained when SI was inactivated (SI at 18°C). In order to quantify the tightness of phaselocking in vibrotactile-induced responses, cycle histograms (CHs) were constructed (Fig.7) to plot the probability of impulse occurrence throughout successive cycles of the vibration train. A rectangular distribution is obtained in the cycle histogram if the impulses occur randomly in relation to the vibration cycle, that is, if there is no phaselocking in the response. In each of the CHs in Fig.7 there is a clear preferential grouping of impulses indicative of tight phaselocking in the responses of two SII neurons that were responding to 30 Hz vibration. The middle distributions, obtained when SI was inactivated, (SI at 18° or 13°C) show no systematic change from the controls on the left and right (SI at 38°C), either in the general form of the distribution or in quantitative measures of phaselocking that

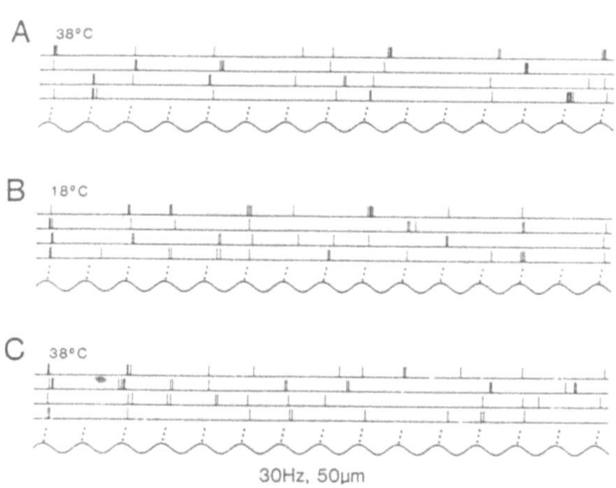

Figure 6. Absence of effect of SI inactivation on response phase for an SII neuron activated by 30 Hz vibration (50 µm amplitude) applied to the forelimb central footpad. The four traces in A, B and C represent replicas of the impulse sequence from records stored in a laboratory computer. The dotted lines extending from the peak of each vibration cycle indicate the preferred phase of response of the neuron in relation to the plotted waveform. The duration of each trace was 500 ms as indicated by the presence of the 15 cycles of vibration at 30 Hz (from Turman et al. 1992).

were based on the *percentage entrainment* value indicated in the histograms (Turman et al., 1992). Among SII neurons examined in this way at different vibration frequencies in the range from 30 to 300 Hz, there was no evidence for an enhancement, or indeed, any change, in phaselocking in association with SI inactivation (Turman et al., 1992).

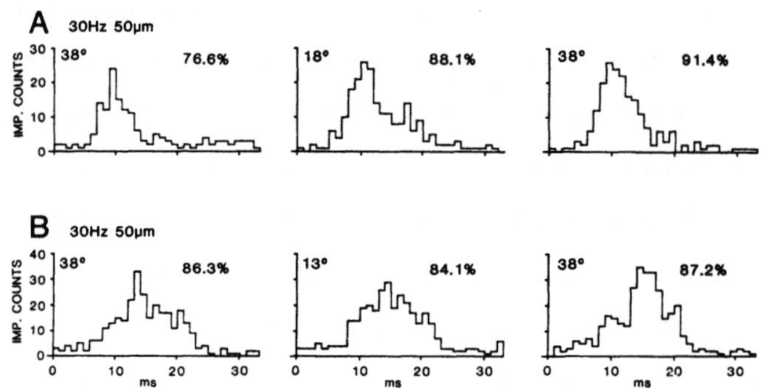

Figure 7. Effect of SI inactivation on phaselocking of SII responses in the cat to cutaneous vibration. A and B show cycle histograms constructed from the responses of two SII neurons to 30 Hz vibration (amplitude, 50 μm). The histograms, whose analysis times correspond to the vibration cycle period of 33.3 ms at 30 Hz, show the distribution of impulse activity within the vibration cycle period (from Turman et al., 1992).

The observations in the cat, and similar observations in the rabbit in which only 2/28 SII neurons were affected (Murray et al., 1992) when SI was reversibly inactivated with the cooling method, provide unequivocal evidence that in these non-primate species, the SII area is driven predominantly or exclusively by direct inputs from the thalamus and is therefore organized in parallel with SI for the processing of tactile information.

Are there fundamental differences between non-primates and primates in the organization of SI and SII for tactile processing?

Our results therefore suggest that there may be fundamental differences in thalamocortical organization between these non-primate species, with their parallel organization of SI and SII tactile processing networks (Fig.8A), and simian primates with their reported serial functional schemes (Figs.1 and 8B) for SI and SII processing (Pons et al., 1987, 1988; Garraghty et al., 1990). However, because of the obvious advantages of the rapid, reversible cooling procedure for SI inactivation, we have re-investigated the issue of *serial* versus *parallel* processing in SI

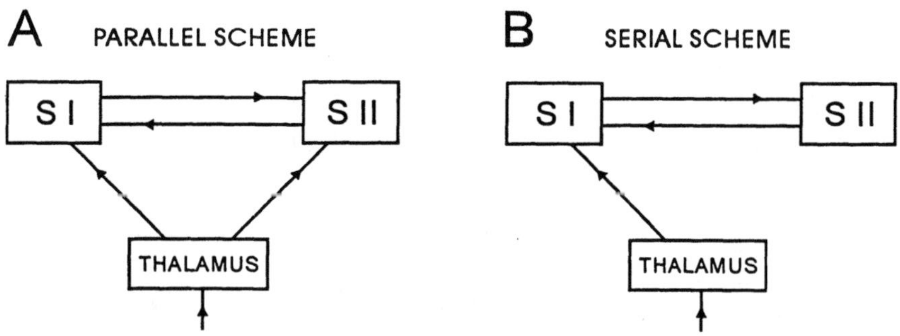

Figure 8. Schematic representation of hypotheses for functional organization of thalamocortical tactile systems in non-primate mammals (A) and simian primates (B). A shows the *parallel* scheme for cat, rabbit and other non-primates in which tactile sensory signals are conveyed directly from thalamus to both the SI and SII cortical areas. B shows the *serial* functional scheme that appeared to operate in simian primates (based on studies involving ablation of SI) in which tactile information was thought to travel from thalamus to SI before being conveyed to SII.

and SII in one of the primate species, the marmoset monkey (*Callithrix jacchus*), that was the subject of the surgical ablation studies (Garraghty et al., 1990). In these experiments, on account of the small size of the marmoset, the hand area of SI could be covered by a 5 mm diameter cooling block. Both evoked potential and single neuron analyses have been carried out in studies that have been reported in abstract form (Zhang et al., 1993a,b; Murray et al., 1993) and are now prepared for journal submission. In brief, the results show that SI inactivation never abolished the SII evoked potential and, in the majority (12/20) of experiments the SII EP was unchanged. At the single neuron level, responsiveness was unchanged in almost 30% of a final sample of 47 SII neurons examined quantitatively in association with SI inactivation, while in the remainder, there were reductions in response level that ranged from minor to more substantial, but in only <10% of the sample, was responsiveness essentially abolished. As responsiveness in >90% of SII neurons survived SI inactivation or underwent only partial reduction, the result is very different from the findings in the study based on surgical ablation of SI in which SII responsiveness was abolished.

Even in the case of SII neurons that underwent some fall in responsiveness in association with SI inactivation, and this was found not just in the marmoset but also in a proportion of SII neurons in both cat (Turman et al., 1992) and rabbit (Murray et al., 1992), it is not clear that these neurons received part of their peripheral tactile input via an indirect, *serial* pathway from

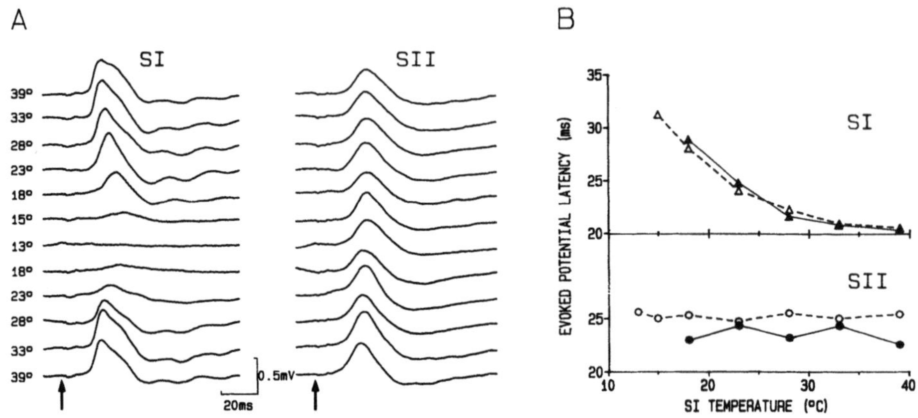

Figure 9. Effect of progressive cooling and rewarming of SI on the averaged EPs (n=20 sweeps) recorded from the distal forelimb areas of SI and SII in the rabbit. The face of the SI block was cooled to 13°C and rewarmed to 39°C in 5 C° steps with a waiting period of 1 minute at each temperature. Arrows: onset of stimulus delivery. B: measurement of onset latency to the peak amplitude of the SI EP (upper graph) and SII EP (lower graph) shown in A, and plotted against temperature (°C) over SI. (from Murray et al., 1992).

the thalamus via SI. An alternative possibility, for which there are several lines of evidence, is that the reduction in response in association with SI inactivation reflects a loss of background facilitation that operates as a tonic modulatory influence mediated from SI to SII over the intracortical connections known to link the two areas. One reason for this interpretation is that, in both the cat and marmoset, there was no change, in particular, no tightening of phaselocking in the responses of SII neurons to vibrotactile stimuli during SI inactivation (Figs.6 and 7). This was confirmed with quantitative measurements of phaselocking in 11 vibrotactile-sensitive SII neurons in the cat (Figs.15, 16 and 17 in Turman et al., 1992) and in 7 SII neurons studied in the marmoset (Zhang et al., 1993b).

A second argument in support of the interpretation that reductions in SII EPs or single neuron responsiveness in association with SI inactivation reflects a loss of background facilitation from SI is that, in cases where SII responsiveness was reduced by SI inactivation there was also usually a fall in the background activity of the SII neuron (Turman et al., 1992). As the background activity may be endogenous to SII, its reduction with SI inactivation may also suggest a disfacilitation. The third argument comes from observations on the time course of SII responses as SI was cooled. Preceding the actual inactivation of SI, there was a marked delay or slowing in the SI response whether in the rabbit (Fig.9), the cat (Figs.2 and 11 in Turman et al., 1992), or the marmoset. If the SII responses were dependent upon a path through SI there should be a comparable delay in the SII response, or in that component attributable to the serial path. As there was not, one has to argue that any effects on SII were attributable to a removal of facilitation.

The observations made in the cat, rabbit and marmoset on SII responsiveness in association with the reversible inactivation of SI by cooling demonstrate that in all three species there is a very substantial, perhaps exclusive, parallel organization of the SI and SII areas for tactile processing. The recent results in the marmoset (Zhang et al., 1993a,b; Murray et al., 1993) necessitate a revision of any hypothesis that tactile processing at thalamocortical levels in simian primates, is based on a strict serial scheme in which signals are conveyed from the thalamus to SI, and thence to SII. It appears therefore that there is no fundamental difference in terms of serial and parallel organization of the SI and SII areas, between simian primates in general, and other mammals. Schemes such as those represented in Fig.1 and Fig.8B that place SII above SI, or its component cytoarchitectural zones 3a, 3b, 1 and 2, in a hierarchical network for tactile processing, do not appear to apply to the marmoset or the non-primate species that have been examined, such as cat and rabbit. Whether the hierarchical scheme of Figs.1 and 8B can be confirmed for the macaque monkey, will depend upon the application of refined methods, such as cooling, for SI inactivation, and quantitative evaluation of SII responsiveness in association with the reversible inactivation procedures.

The explanation for the differences between our observations, based on SI inactivation by localized cooling and earlier results in the marmoset monkey (Garraghty et al., 1990) in which SII responsiveness was abolished as a consequence of SI ablation, is unclear. However, it is possible that the surgical ablation may have created an injury discharge in the cortico-cortical neurons that project to the topographically-related hand area of SII, and that this could have led to changes in this region of SII, for example, potassium ion accumulation leading to accommodation and impaired responsiveness in the constituent SII neurons.

References

Adrian, E.D. (1941) Afferent discharges to the cerebral cortex from peripheral sense organs. *Journal of Physiology (London)* 100: 159-191.

Bennett, R.E., Ferrington, D.G. and Rowe, M.J. (1980) Tactile neuron classes within the second somatosensory area (SII) of the cat cerebral cortex. *Journal of Neurophysiology* 43: 292-309.

Brugge, J.F. and Reale, R.A. (1985) Auditory cortex. *In*: A. Peters and E.G. Jones (eds): *Cerebral cortex. Vol 4: Association and Auditory Cortices.* Plenum, New York, pp. 229-271.

Burton, H. (1986) Second somatosensory cortex and related areas. *In*: E.G. Jones and A. Peters (eds): *Cerebral Cortex. Vol 5:Sensory-Motor Areas and Aspects of Cortical Connectivity.* Plenum, New York, pp. 31-98.

Burton, H. and Robinson, C.J. (1987) Responses in the first or second somatosensory cortical area in cats during transient inactivation of the other ipsilateral area with lidocaine hydrochloride. *Somatosensory Research* 4: 215-236.

Felleman, D.J. and Van Essen, D.C. (1991) Distributed hierarchical processing in the primate cerebral cortex. *Cerebral Cortex* 1: 1-47.

Ferrington, D.G. and Rowe, M.J. (1980) Differential contributions to the coding of cutaneous vibratory information by cortical somatosensory areas I and II. *Journal of Neurophysiology* 43: 310-331.

Fisher, G.R., Freeman, B. and Rowe, M.J. (1983) Organization of parallel projections from Pacinian afferent fibers to somatosensory cortical areas I and II in the cat. *Journal of Neurophysiology* 49: 75-97.

Friedman, D.P. and Murray, E.A. (1986) Thalamic connectivity of the second somatosensory area and neighboring somatosensory fields of the lateral sulcus of the macaque. *Journal of Comparative Neurology* 252: 348-373.

Friedman, D.P., Murray, E.A., O'Neill, J.B. and Mishkin, M. (1986) Cortical connections of the somatosensory fields of the lateral sulcus of macaques: evidence for a corticolimbic pathway for touch. *Journal of Comparative Neurology* 252: 323-347.

Garraghty, P.E., Pons, T.P. and Kaas, J.H. (1990) Ablations of areas 3b (SI proper) and 3a of somatosensory cortex in marmosets deactivate the second and parietal ventral somatosensory areas. *Somatosensory and Motor Research* 7: 125-135.

Jones, E. G. (1985) *The Thalamus.* Plenum, New York.

Jones, E.G. (1986) Connectivity of the primate sensory-motor cortex. *In*: E.G. Jones and A. Peters (eds): *Sensory-motor areas and aspects of cortical connectivity.* Plenum, New York, pp. 113-183.

Marshall, W.H., Woolsey, C.N. and Bard, P. (1941) Observations on cortical somatic sensory mechanisms of cat and monkey. *Journal of Neurophysiology* 4: 1-24.

Mori, A., Hanashima, N., Tsuboi, Y., Hiraba, H., Goto, N. and Sumino, R. (1991) Fifth somatosensory cortex (SV) representation of the whole body surface in the medial bank of the anterior suprasylvian sulcus of the cat. *Neuroscience Research* 11: 198-208.

Mountcastle, V.B., Talbot, W.H., Sakata, H. and Hyvärinen, J. (1969) Cortical neuronal mechanisms in flutter-vibration studied in unanesthetized monkeys. Neuronal periodicity and frequency discrimination. *Journal of Neurophysiology* 32: 452-484.

Murray, G.M., Zhang, H.Q., Kaye, A.N., Sinnadurai, T., Campbell, D.H. and Rowe, M.J. (1992) Parallel processing in rabbit first (SI) and second (SII) somatosensory cortical areas: effects of reversible inactivation by cooling of SI on responses in SII. *Journal of Neurophysiology* 68: 703-710.

Murray, G.M., Zhang, H.Q., Turman, A.B., Mackie, P.D. and Rowe, M.J. (1993) Parallel or serial processing in cortical somatosensory areas I and II of the primate: Effects on unitary responses in SII of SI reversible inactivation. *Proceedings of the International Union of Physiological Sciences* 87: 3/P

Pandya, D.N. and Yeterian, E.H. (1985) Architecture and connections of cortical association areas. *In:* A. Peters and E.G. Jones (eds): *Cerebral Cortex. Vol. 4: Association and Auditory Cortices.* Plenum, New York, pp. 3-61.

Pons, T.P., Garraghty, P.E., Friedman, D.P. and Mishkin, M. (1987) Physiological evidence for serial processing in somatosensory cortex. *Science* 237: 417-420.

Pons, T.P., Garraghty, P.E. and Mishkin, M. (1988) Lesion-induced plasticity in the second somatosensory cortex of adult macaques. *Proceedings of the National Academy of Science* 85: 5279-5281.

Rowe, M.J., Ferrington, D.G., Fisher, G.R. and Freeman, B. (1985) Parallel processing and distributed coding for tactile vibratory information within the sensory cortex. *In*: M.J. Rowe and W.D. Willis (eds): *Development, Organization and Processing in Somatosensory Pathways.* Alan R. Liss Inc., New York, pp. 247-258.

Rowe, M.J. (1990) Organization of the cerebral cortex in monotremes and marsupials. *In*: E.G. Jones and A. Peters (eds): *Cerebral Cortex. Volume 8: Comparative Structure and Evolution of Cerebral Cortex.* Plenum, New York, pp. 263-334.

Turman, A.B., Ferrington, D.G., Ghosh, S., Morley, J.W. and Rowe, M.J. (1992) Parallel processing of tactile information in the cerebral cortex of the cat: effect of reversible inactivation of SI on responsiveness of SII neurons. *Journal of Neurophysiology* 67: 411-429.

Turman, A.B., Morley, J.W., Zhang, H.Q. and Rowe, M.J. (1995) Parallel processing of tactile information in cat cerebral cortex: effect of reversible inactivation of SII on SI responses. *Journal of Neurophysiology* 73: 1063-1075.

Woolsey, C.N. and Fairman, D. (1946) Contralateral, ipsilateral and bilateral representation of cutaneous receptors in somatic areas I and II of the cerebral cortex of pig, sheep and other mammals. *Surgery* 19: 684-702.

Zhang, H.Q., Turman, A.B., Murray, G.M., Mackie, P., Coleman, G. and Rowe, M.J. (1993a) Parallel or serial processing in cortical somatosensory areas I and II of the primate: Effects on SII evoked potentials of reversible inactivation of SI. *Proceedings of the International Union of Physiological Sciences*. 87:2/P

Zhang, H.Q., Murray, G.M., Turman, A.B., Mackie, P.D. and Rowe, M.J. (1993b) Are there direct and indirect paths for tactile inputs to SII from the thalamus in the primates? *Proceedings of the Australian Neuroscience Society* 4: 83.

Linearity as the basic law of psychophysics: Evidence from studies of the neural mechanisms of roughness magnitude estimation

Kenneth O. Johnson, Steven S. Hsiao and David T. Blake

Krieger Mind/Brain Institute and Neuroscience Department, Johns Hopkins University, Baltimore, MD 21218, USA

Summary. The search for general laws governing our perceptual representation of the external world goes back at least to Fechner and is no less important now than it was then. Fechner used human discrimination performance to infer that the basic relationship between subjective intensity and the intensity of a physical stimulus was logarithmic. A hundred years later, Stevens demonstrated that direct subjective report frequently results in a power law rather than a logarithmic law. However, it was soon appreciated that power laws and logarithmic laws are indistinguishable without peering inside the mechanisms leading to a subject's responses. After doing so, Mountcastle and his colleagues proposed that the basic law is linearity between the neural representation of an intensive continuum and its subjective magnitude. We were drawn to the same conclusion in a study of the neural coding mechanisms of roughness perception that was designed to be independent of any assumptions about the basic laws of psychophysics. We adopted consistency as the test of neural coding hypotheses. The single, consistent neural coding hypothesis that survived all challenges resulted in a linear relationship between the neural coding measure and subjects' reports of intensity. That hypothesis was that roughness perception depends on spatial variation in the SAI afferent population response. This study and other similar studies support the idea that once the internal neural representation of intensity is established, the fundamental laws are linear, as Mountcastle proposed over thirty years ago.

Introduction

The quest for the metrics that govern our internal representation of the external world has a long history. The original formulation, which persisted for nearly a hundred years, was Fechner's (1966), who showed that internal, intensive scales could be reconstructed if there is a relationship between just noticeable differences, jnd's, and increments in subjective intensity. He hypothesized that jnd's correspond to equal increments in subjective intensity and that the constancy of Weber fractions implies a logarithmic relationship between physical and subjective intensity. Stevens (1957) showed, however, that when subjects are asked to report their sense of subjective intensity, either directly by numerical report or indirectly by selecting an equivalent intensity in another modality, the relationships are described better by power law functions than by a logarithmic function. On the basis of the widespread occurrence of power law relationships, Stevens suggested that the basic law is a power rather than a logarithmic law. However, MacKay (1963) showed that unless we peer inside the mechanisms leading to magnitude estimation we cannot distinguish between power and logarithmic laws: the expected result in magnitude estimation and cross-modality matching is a power law even if the fundamental relationship between external stimuli (including numbers) and subjective intensity is logarithmic. After observing neural responses similar to human reports of subjective magnitude, Mountcastle and his colleagues (Mountcastle et al., 1963; Werner and Mountcastle, 1965) concluded that "The implication [of matching neural and psychophysical response functions] is, we believe, that the neural transforms intervening between input and the final verbal description of an introspective magnitude estimation must be linear for the intensive continuum. This does not imply, of course, that the intervening neural transforms must all be linear, but that the sum of their serial superpositions must be so." This view and Stevens' view are not inconsistent. Although Stevens advocated the power law as the basic law of perceptual intensity, his search for corresponding power functions in the relationship between primary, sensory neural responses and physical intensity (Stevens, 1970; Stevens, 1975) implies a belief in linearity. While there is little reason to believe that the neural responses of sensory systems are governed by any single class of mathematical functions, the evidence is strong that once the neural basis for a perceptual, intensive category is established, the remaining mechanisms are linear. Franzén has repeatedly called attention to aspects of his own work that support Mountcastle's hypothesis of linearity (Franzén and Offenloch, 1969; Franzén et al., 1991).

We were drawn to this conclusion by our recent studies of the neural mechanisms underlying roughness magnitude estimation, which were specifically designed to be free of assumptions about the form of the relationship between subjective magnitude and the neural measure on which it is based. The motivation was to understand the neural coding mechanisms underlying texture perception. We were reminded of the earlier quest for universal psychophysical laws only when near perfect linearity appeared in the correlation of subjects' responses with hypothetical neural codes. Whenever there was a consistent relationship between roughness magnitude judgments and a putative neural code the relationship was linear. We pursue this aspect of our studies in some detail because of the importance of establishing the metric basis of perceptual operations.

Consistency as the fundamental test of a neural coding hypothesis

The general method for investigating neural coding mechanisms, which is well known and widely used in sensory physiology, uses identical stimuli in psychophysical and neurophysiological experiments. The psychophysical experiments provide an estimate of the psychometric function

$$r = p(S) + {}_{p} \qquad (1)$$

where r is a subject's response to a single stimulus presentation, S, and $_p$ represents the trial-to-trial variation in the subject's responses. The convention here is that upper and lowercase letters represent multidimensional and unidimensional quantities, respectively. The neurophysiological experiments provide estimates of the neural population responses underlying those psychophysical responses:

$$N = F(S) + E_N \quad (2)$$

where E_N represents the trial-to-trial variation in the population response, N. The assumption is, of course, that the relevant neural responses have been sampled and that the psychophysical responses, r, depends on the neural activity, N. The problem of finding the relationship between r and N is akin to multiple regression where the hypotheses are regression models, which may or may not be linear (Draper and Smith, 1969).

If nothing more was known about the problem the next step would be to formulate multivariate linear and nonlinear models as hypothetical links between subjects' responses and the neural population responses, selecting between the models with some measure of goodness of fit. However, this confounds several steps in the process leading to the subjects' responses. We assume that subjects' responses are related to an internal sense of subjective magnitude, m, in a consistent manner and that subjective magnitude, in turn, is related to some unidimensional measure, c, of the neural activity sampled in the neurophysiological experiments; that is,

$$c = c(N) + e_c \quad (3)$$

$$m = m(c) + e_m \quad (4)$$

$$r = r(m) + e_r \quad (5)$$

where c(N) is the unidimensional neural measure (neural code) that is the basis of the subjective magnitude being probed in the magnitude estimation experiment; m(c) expresses the relationship between subjective magnitude and the neural code on which it is based; and r(m) expresses the relationship between subjective magnitude and the subjects' responses. As before, e_c, e_m, and e_r represent trial-to-trial variation in these quantities.

The principal object of a neural coding study is the function c(N). A typical hypothesis is that magnitude judgments depend on the mean firing rate in the entire population or a subpopulation of the neurons activated by the stimulus. However, much more complex bases for magnitude estimation are possible and even likely considering the wide range of complex intensive continua that allow magnitude judgments (Stevens, 1970). A problem confronting any neural coding study is that the only test of any putative neural code, c(N), is its ability to predict subjects' responses, which involves the unknown functions m(c) and r(m). How are these linking functions to be handled in the hypothesis testing phase?

One approach to the linking functions is to try various general mathematical functions like power or logarithmic functions and assess their efficacy based on some measure of goodness of fit. This is not a very satisfying approach because of the large number of possibilities. A lack of fit between the predicted and observed responses may as easily result from selection of the wrong linking functions as the wrong neural coding hypothesis. To circumvent these difficulties, we developed a different approach that is independent of specific assumptions about the form of the linking functions; that is simply to require *consistency* between the putative neural code and the subjects' responses. Consistency in a functional relationship between, for example, x and y means that except for measurable trial-to-trial variation in y there is a single value of y for each value of x. We assume also a reasonable degree of smoothness but beyond that nothing about the form of the function, not even that it is monotonic. The only basis for rejecting a coding hypothesis is lack of consistency between the hypothetical neural coding parameter and the subjects' responses. Since the hypothetical coding measure, c, and the psychophysical responses, r, associated with each stimulus in the experimental design are univariate, consistency or its lack can be assessed in a simple scatter plot of r versus c.

The neural code underlying roughness magnitude judgments

Three studies employing consistency as the basic test of neural coding hypotheses have been completed. Each study employed the same experimental design: identical textured surfaces were used in psychophysical and neurophysiological studies to obtain estimates of the neural population responses underlying the psychophysical roughness magnitude estimates. Then, all of the viable coding hypotheses, c(N), that we could devise were tested for consistency and rejected when they yielded a grossly inconsistent relationship between subjects' responses and the putative neural code (e.g., see Fig. 2). The differences between the three studies lay entirely in the surfaces used to probe roughness magnitude perception. The surfaces in the first study were designed to span a wide range of roughness magnitudes from smooth to very rough and to include a wide range of surface conditions. The surfaces in the second study were designed to distinguish between temporal and spatial coding mechanisms. The outcome of the second study was more general than intended, allowing discrimination between whole classes of coding mechanisms. The surfaces in the third study were designed to distinguish between codes based on SAI and RA responses.

The surfaces in the first study (Connor et al., 1990) were tetragonal arrays of raised dots designed to challenge a wide range of possible neural coding hypotheses. Preliminary experiments showed that roughness magnitude estimates for such patterns are related to the element spacing by an inverted U-shaped function that peaks at center-to-center spacings near 3 mm and that these estimates are affected by the diameter of the raised dots. Eighteen surfaces with three dot diameters and six spacings were designed to obtain the result illustrated in Fig. 1, three inverted U-shaped psychometric functions displaced by small amounts because of the differences in dot diameter.

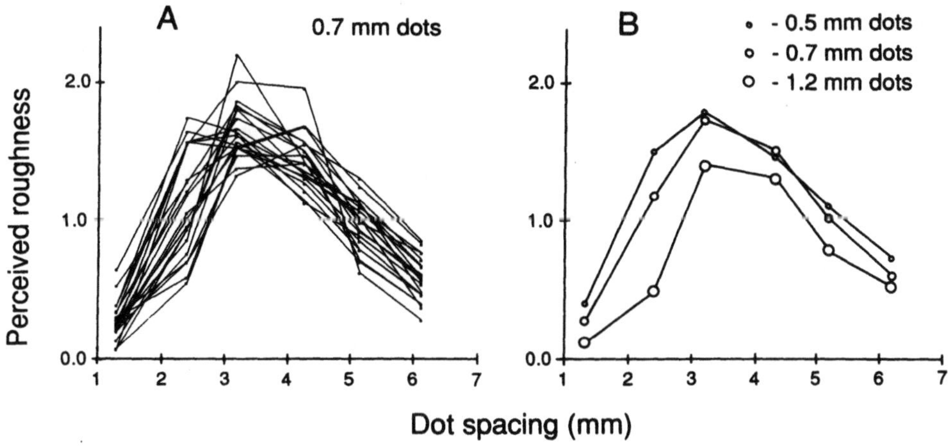

Figure 1. Normalized subjective roughness estimates for 18 surfaces with tetragonal raised-dot patterns. The 18 surfaces were constructed from all combinations of 6 dot spacings and 3 dot diameters. A. Responses of 21 subjects to the surfaces with 0.7 mm dot diameters. Roughness magnitude estimates were normalized within subjects to eliminate differences in individual scaling ranges. Each point is the mean of five responses. B. Grand average responses for all 18 surfaces. The roughest surface (0.5 mm diameter dots, 3.2 mm center-to-center spacing) was about as rough as 36 grit sandpaper. The smoothest surface (1.2 mm diameter, flat-topped dots, 1.3 mm center-to-center spacing) felt only slightly rougher than smooth plastic. From Connor et al. (1990).

The object of the design was to challenge any coding hypothesis severely. The result was that while there was a unique, consistent relationship between the stimuli and the magnitude judgments there was no unique inverse relationship between the judgments and the stimuli. Over a wide range of roughness magnitudes, horizontal lines in Fig. 1 (single values of subjective magnitude) intersect the psychophysical response functions at six different locations; that is, the experiments show that six different surfaces with very different neural responses evoke the same psychophysical response. Consistency requires that the coding value be the same for all of those different neural responses. An invalid neural coding hypothesis is likely to fail that test.

A wide range of neural coding hypotheses was tested in the first study. There was no consistent relationship between subjects' responses and any measure of mean impulse rate as illustrated in Fig. 2. Each point in each graph in Fig. 2 represents the mean impulse rate and the mean roughness report evoked by one of the 18 surfaces used in the study. Some smooth and some very rough surfaces evoked nearly identical mean impulse rates in the SAI and RA afferent population responses; thus, neither SAI nor RA mean impulse rate provided a consistent basis for the psychophysical responses. Codes based on the Pacinian population response were rejected as a class because Pacinian afferents responded to all of the surfaces with nearly identical impulse rates and with no consistent spatial or temporal structure that we could find. In contrast to measures based on mean impulse rate, the first, simple measure of firing rate variation that we tried (firing rate variance) was consistent with subjects' responses. More physiologically plausible measures of spatial and temporal variation in SAI and RA firing rates yielded even more consistent bases for the psychophysical responses as illustrated in Fig. 2. Temporal variation was defined as the mean absolute difference in firing rates at time separated by 130 msec. Spatial variation was defined as the mean absolute difference in firing rates between regions of the population response whose somatotopic loci were separated by 2 mm.

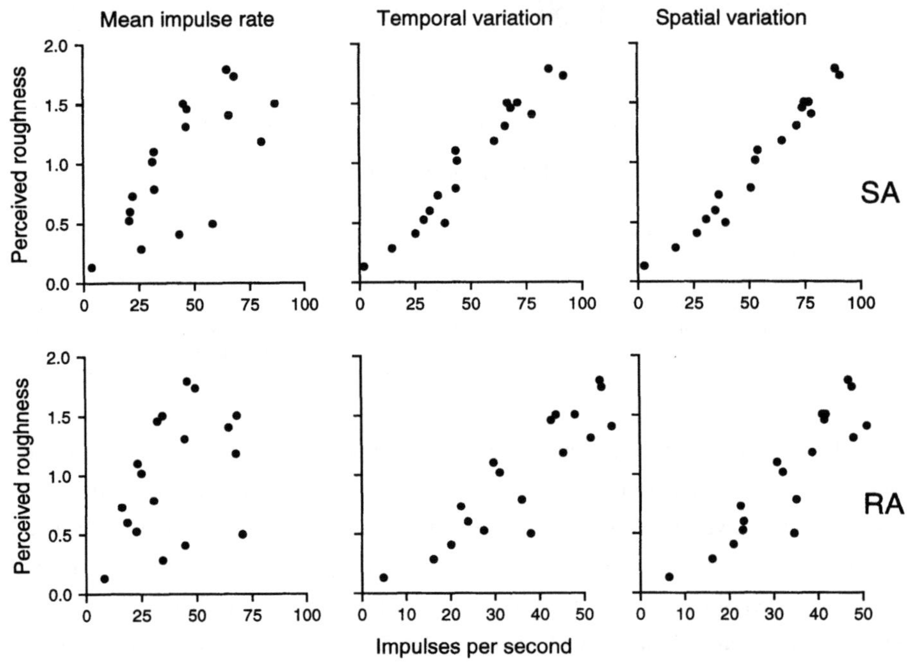

Figure 2. Consistency plots of subjective roughness estimates versus several measures of the SAI and RA population responses to the same 18 surfaces as in the psychophysical study illustrated in Fig. 1. Each of the measures was estimated from the responses of 11 SA and 10 RA fibers. Each point in each graph represents a single neural population response measure and the mean roughness judgment evoked by one of the 18 stimulus surfaces. Spatial and temporal variation is defined in the text. The correlation coefficients for SAI mean impulse rate, temporal variation and spatial variation were 0.710, 0.973, and 0.983 respectively. The comparable RA correlation coefficients were 0.429, 0.922, and 0.901. From Connor et al. (1990).

Remarkable features of the graphs of temporal and spatial variation illustrated in Fig. 2 are their consistency and linearity, particularly for the SAI afferents. Nothing in the analyses presumed linearity or biased the search for consistency toward linearity. Pearson's product-moment correlation coefficient was adopted as the measure of consistency only after determining that there were no significant quadratic or cubic trends. If a significant nonlinearity was found a different measure of consistency would have to be used.

Surfaces in the second study (Connor and Johnson, 1992) were designed to address questions that remained after the first study. First, the dot spacings beyond 3-4 mm were well beyond any used in previous studies, raising the possibility that subjects cannot judge roughness for surfaces with such wide spacings. None of the subjects expressed any difficulty and the results displayed in Fig. 2 suggest that subjects were judging a single, consistent subjective continuum. Nonetheless, the concern warranted further experimentation. Consequently, spacings were restricted to 4 mm and less. Second, we wanted to reexamine our rejection of rate codes since mean impulse rate is such a common basis for subjective magnitude. Third, the first study provided only circumstantial evidence that the code is spatial rather than temporal. It was argued that roughness perception was more likely to rely on a spatial neural code because the spatial structure of the afferent neural discharge (Johnson and Lamb, 1981; Phillips et al., 1990), like roughness perception itself (Lederman, 1974) is affected minimally by changes in scanning velocity (4-16 cm/sec, see Johnson Lamb, 1981) whereas temporal neural codes are very velocity dependent. The surfaces in the second study were designed to address all three concerns.

One way to divide putative neural coding measures into two broad, general classes is into within-fiber and between-fiber codes. A within-fiber code is defined here as any measure of the firing patterns within single afferents averaged or summed to obtain a population code. Measures of mean rate within single afferents summed over any part or the whole of the active afferent population are examples of within-fiber codes. Codes based on ratios or differences of impulse rates between populations are between-fiber codes. Measures of neural activity that are sensitive to exact spike timing such as temporal order codes, entrainment to repetitive stimulus features of the stimulus, and measures of spike rate variation within single afferents are also examples of within-fiber codes. Between-fiber codes could be very complex, but we consider here only local measures of spatial variation in firing rates. The relative engagement of the three afferent fiber populations is an example of a between-fiber code that has been suggested as a possible basis for

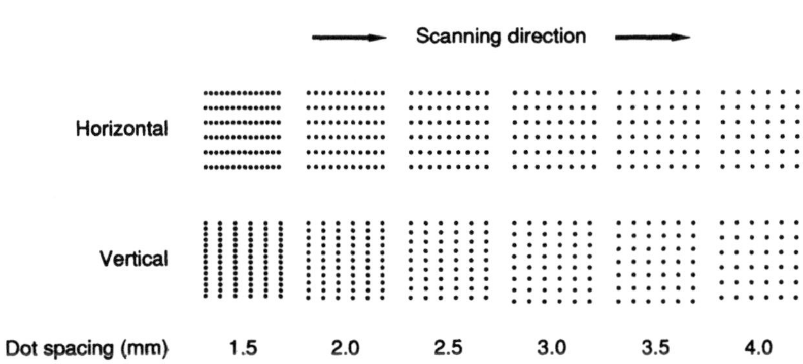

Figure 3. Stimulus patterns employed by Connor and Johnson (1992). Each embossed dot had the shape of a truncated cone 0.5 mm in height with a flat surface 0.5 mm in diameter. Columns in the "Vertical" patterns contained raised dots with center-to-center spacings ranging from 1.5 to 4.0 mm; the columns were separated horizontally by 4.0 mm. The top row represents the same surfaces rotated by 90 ; that is, "Vertical" patterns were rotated 90 before presentation to obtain the "Horizontal" patterns. From Connor and Johnson (1992).

texture perception (Johnson, 1983) but that was meant to account for the multidimensional character of texture perception (Hollins et al., 1993) in the same way that the relative engagement of three retinal cone types accounts for the multidimensional character of color. No measure of relative engagement (differences and ratios of firing rates in different populations) was related to roughness with a consistency approaching measures of spatial and temporal variation in firing rate.

The surfaces in the second study (see Fig. 3) were designed to discriminate between within- and between-fiber codes. In particular, the surfaces with constant spacing in the horizontal direction but differing spacing in the vertical direction (bottom row in Fig. 3) were designed to have opposite effects on within- and between-fiber codes. As the dots move closer together vertically, codes based on within-fiber measures should increase in magnitude while those based on between-fiber measures such as spatial variation should decrease. The psychophysical roughness magnitude estimates decrease in value over this range, suggesting that the code is not a within-fiber code.

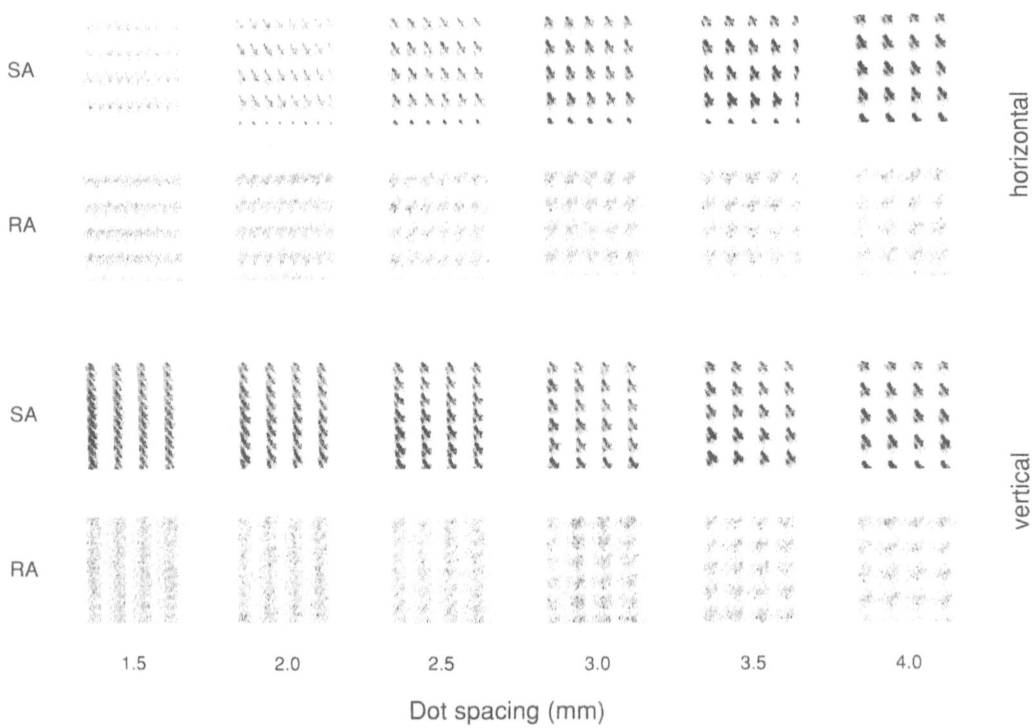

Figure 4. Spatial event plots of the responses of typical SAI and RA afferent fibers to the dot patterns shown in Fig. 3. Each plot comprises a set of nerve spike rasters resulting from repetitive parallel scans of the stimulus pattern with 0.2 mm lateral shifts between scans. From Connor and Johnson (1992).

To explain the predicted increase of within-fiber measures with increasing vertical dot density consider the effect of vertical dot density on the responses of individual fibers. When the dots are closely spaced in the vertical direction (left surface, bottom row, Fig. 3) every afferent fiber is stimulated vigorously by each column as it passes by (left lower SAI and RA plots in Fig. 4). When the dots are more widely spaced in the vertical direction, afferents whose receptive fields ride over the tops of dots will still be stimulated vigorously (right upper and lower plots in Fig. 4) and the within-fiber coding contribution from those fibers will be undiminished. However, fibers whose receptive fields fall between the dots will be activated minimally or not at all, as illustrated in Fig. 4, and their contribution to the within-fiber measure will be diminished. The overall mean within-fiber measure, which is based on the mean or sum across fibers, must decrease but the psychophysical responses increased, as illustrated in Fig. 5. The opposite prediction, a prediction matching the psychophysical outcome, is obtained if roughness depends on spatial variation (a between-fiber code). When the dots are close together they produce a nearly continuous band of activity in the vertical direction (see Fig. 4), reducing spatial variation in that direction. As the dots move apart, regions of diminished activity appear in the vertical direction, thus contributing to an increasing overall spatial variation.

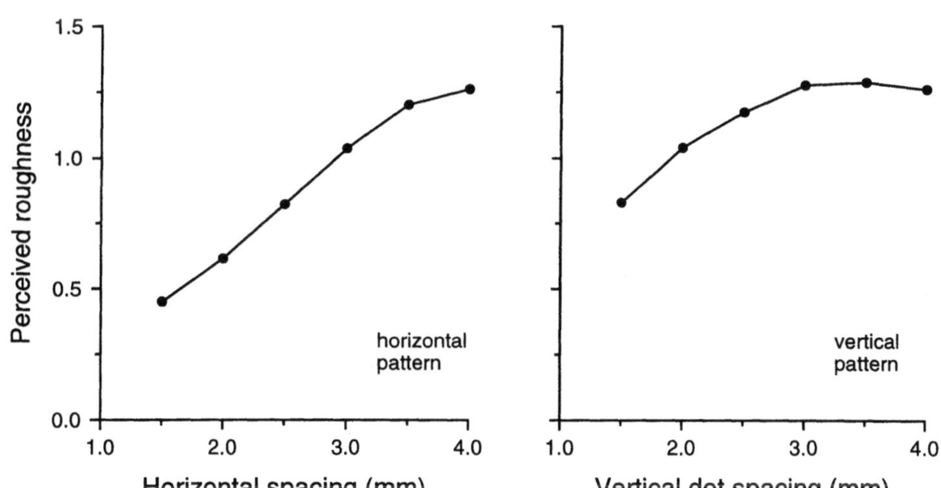

Figure 5. Subjective roughness estimates evoked by the surfaces shown in Fig. 3. Roughness magnitude estimates were averaged across 36 subjects after normalization within subjects to eliminate differences in individual scaling ranges. The responses evoked by patterns with horizontal rows and vertical columns of dots are shown in the left and right panels respectively. From Connor and Johnson (1992).

Further, if roughness perception is based on spatial variation then roughness should be similar whether the lines of closely spaced dots are oriented along (Fig. 3, top row) or across (Fig. 3, bottom row) the scanning direction. The psychophysical results illustrated in Fig. 5 refute hypotheses based on within-fiber codes and favor hypotheses based on between-fiber codes.

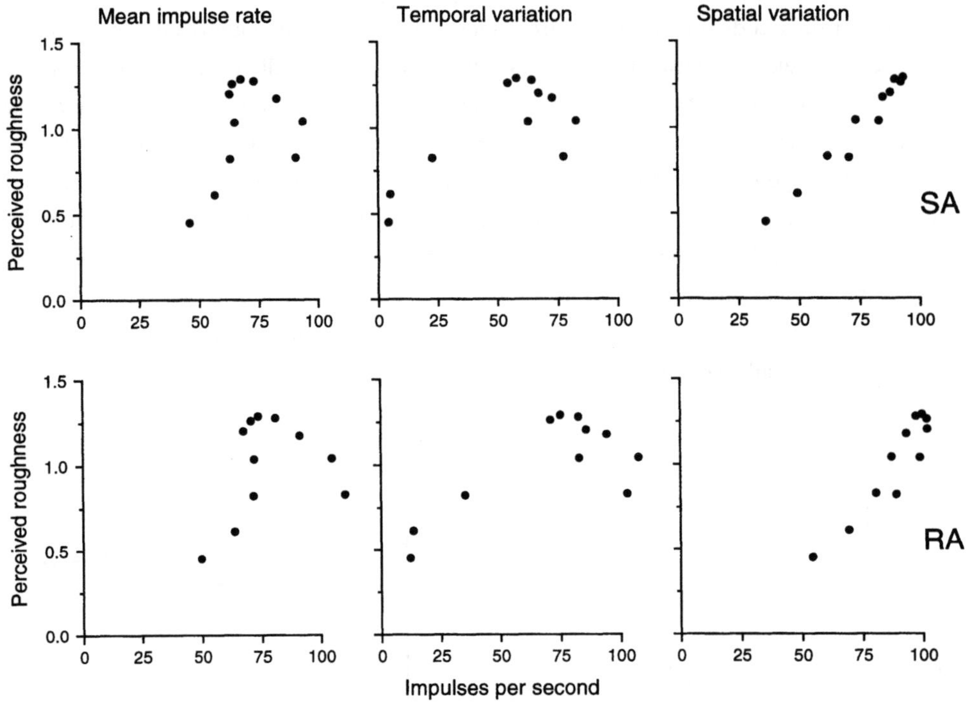

Figure 6. Consistency plots of subjective roughness estimates versus mean impulse rate, temporal variation, and spatial variation in the SAI and RA population responses evoked by the surfaces illustrated in Fig. 3. The display format is the same as in Fig. 2. The correlation coefficients for the roughness estimates versus SAI and RA neural coding measures were 0.404 and 0.301 for mean impulse rate, 0.741 and 0.714 for temporal variation, and 0.984 and 0.938 for spatial variation. Adapted from Connor and Johnson (1992).

Consistency plots of psychophysical magnitude judgments versus mean impulse rate, temporal variation, and spatial variation in the SAI and RA afferent population responses are illustrated in Fig. 6. Figure 7 displays the same data in a more conventional way, showing that lack of consistency in within-fiber codes is qualitative as well as quantitative; mean impulse rate and temporal variation decline with increasing vertical dot spacing, as predicted, whereas the subjects' magnitude estimates rise. The opposite occurs when the dot spacing is varied in the horizontal direction. In that case, mean impulse rate, temporal variation, and subjects' reports rise together as the dot spacing increases. Only SAI and RA spatial variation changes in the same way as the subjects' psychophysical responses for both sets of surfaces. Thus, the first and second studies show unequivocally, we believe, that no within-fiber code can account for roughness magnitude perception and that spatial variation in either the SAI or RA afferent population responses accounts for roughness perception.

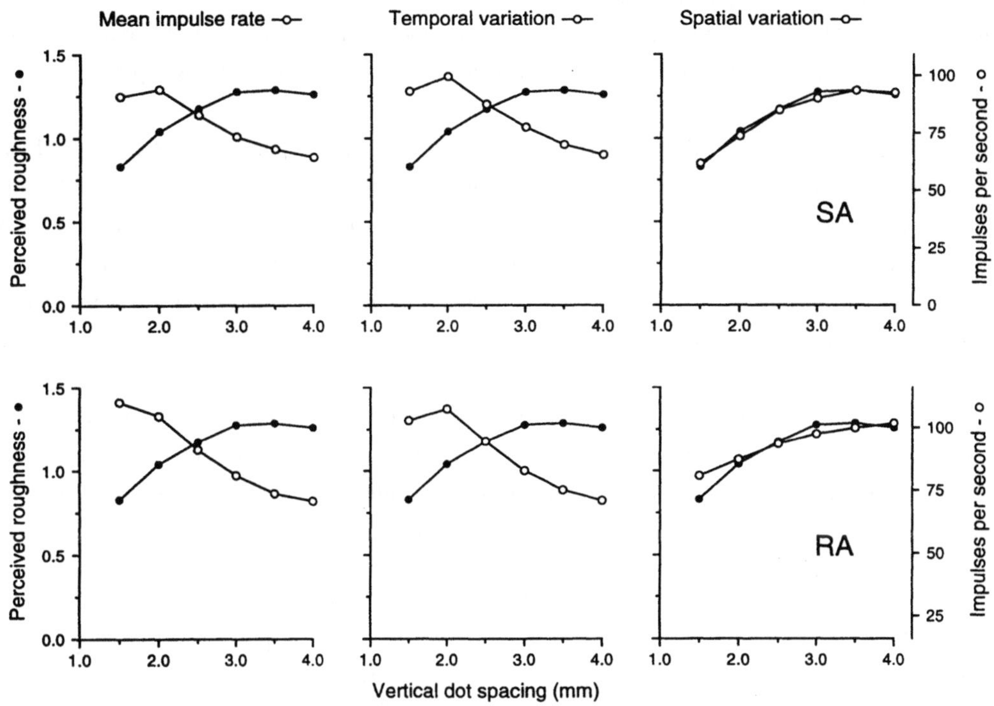

Figure 7. Direct comparison of psychophysical responses and neural responses to the vertical stimulus patterns illustrated in Fig. 3. Closed circles represent psychophysical responses. Open circles represent measures of the neural population responses evoked by the patterns. The left ordinates represent subjective magnitude estimates. The right ordinates, which represent neural measures, all have units of impulses per second. The position and scale of the right ordinates are selected to produce the least mean squared difference between the neural and psychophysical data over all stimulus surfaces. Data associated with the vertical patterns are illustrated to show not just a lack of consistency but that the within-fiber measures (mean rate and temporal variation) vary in the wrong way with vertical dot spacing. From Connor and Johnson (1992).

Both the first and second studies provided some evidence that roughness perception is based on spatial variation in the SAI rather than the RA population but none of it was conclusive. In both studies the consistency plots for both afferent types were linear (Figs. 2 and 6) but the SAI correlation was higher and the relationship between roughness magnitude and SAI spatial variation was nearly proportional, rising linearly from a threshold near zero. If the RA afferents were shown to be responsible for roughness perception, the very high RA spatial variation threshold would have to be explained. Because the RA afferents are more sensitive to fine spatial variation (Johnson, 1983) the RA spatial variation measures were very high even for surfaces with very low roughness ratings. Surfaces judged to differ by a factor of 10:1 in roughness differed by less than 2:1 in RA spatial variation. Perhaps more compelling is the fact that decreasing dot diameters in the first study (Connor et al., 1990) produced parallel increases in roughness magnitude judgments and SAI spatial variation but the opposite effect on RA spatial variation (Johnson and Hsiao, 1994). The opposing effects of dot diameters on the SAI and RA responses are primarily responsible for the difference in consistency between the SAI and RA plots shown in Fig. 2.

Textured surfaces in the third study (Blake et al., 1994;) were designed to affect the SAI and RA afferent population responses differently making it unlikely that coding measures in both populations would be consistent with the psychophysical responses. The first study had shown that dot diameter affects the SAI and RA population responses differently. Also, preliminary studies had shown that changes in dot height within the range from 0.3-0.6 mm have a large effect on SAI responses but little, if any, effect on RA responses; RA responses may be saturated at 0.3 mm. Consequently, surfaces comprising tetragonal arrays of dots that varied in height and diameter but not spacing were constructed. Both dot height and diameter had major effects on roughness perception. As expected, a surface comprising narrow, tall dots felt very rough whereas a surface comprising broad, low dots felt quite smooth. SAI responses, like the roughness judgments, were strongly affected by both dot height and width.

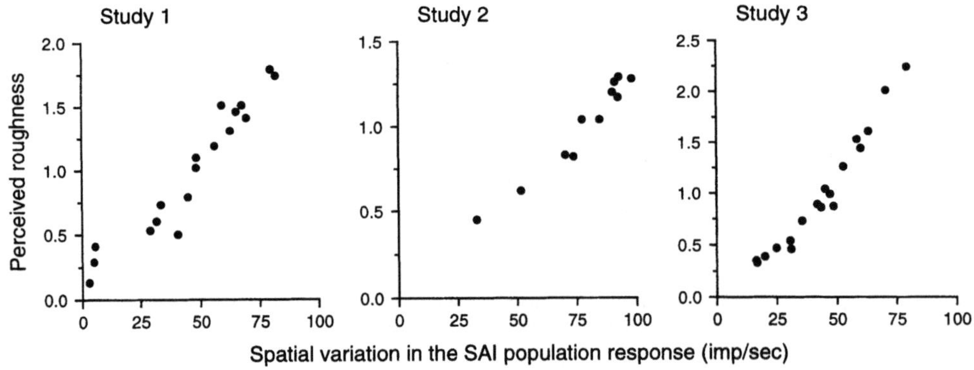

Figure 8. Consistency between subjects' roughness judgments and a single, uniform measure of SAI spatial variation from the two studies illustrated in Figs. 2 and 6 and from a third study. The SAI neurophysiological data from all three studies were reanalyzed using a single measure of spatial variation, which was the mean impulse rate in a hypothetical population of central neurons with Gabor-like receptive fields. The Gabor receptive fields had excitatory and inhibitory subregions separated by 1.5 mm and had orientations distributed uniformly in all directions. The correlation coefficients are 0.959, 0.972, and 0.974 for the data from the three studies.

RA responses, in contrast, were affected by dot diameter but were unaffected by changes in dot height, suggesting that the roughness judgments were not based on activity in the RA population response.

Consistency plots of roughness judgments versus SAI spatial variation from all three studies suggest that roughness magnitude judgments are proportional to spatial variation in the SAI afferent discharge. If this is so, a single hypothesis should account for all of the data from all three studies. Consequently, the data from all three studies were reanalyzed using a single measure of spatial variation, which was the hypothetical firing rate of a population of central neurons whose receptive fields are described by Gabor filters with excitatory and inhibitory lobes separated by 1.5 mm (Connor and Johnson, 1992). Consistency plots between subjects' roughness judgments and the same measure of spatial variation from all three studies are shown in Fig. 8. As emphasized before, nothing in the analysis biased the outcome either toward spatial variation or linearity as the appropriate coding relationship.

Physiological interpretation of spatial variation

The hypothesis that roughness magnitude is based on spatial variation in SAI afferents implies that a central mechanism computes differences in firing rates between afferents, and that the perception of roughness is based on this measure. All central neurons that have receptive fields with spatially separated regions of excitation and inhibition are sensitive to some form of spatial variation. Connor and Johnson (1992) tested whether two commonly observed receptive field structures involving spatially separated regions of excitation and inhibition could account for the perception of roughness: center-surround fields, which are modeled well by two-dimensional difference of Gaussian functions (DOG functions, Rodieck, 1965), and simple-cell-like receptive fields with two or more oriented bands of excitation and inhibition, which are modeled well by two-dimensional Gabor functions (Jones and Palmer, 1987; Worgotter and Holt, 1991). Neurons with subregions of excitation and inhibition have been described in areas 3b and 1 of primate SI cortex (Mountcastle and Powell, 1959; Hyvärinen and Poranen, 1978; Sur, 1981; Gardner and Costanzo, 1980; Phillips et al., 1988; Bankman et al., 1990). The idea that we favor is that roughness perception is based on the mean impulse rate in a subpopulation of neurons in one of the cortical areas leading to the perception of roughness (e.g. areas 3b or 1). Spatial variation measures based on Gabor functions fit the data much better than did measures based on DOG functions; the consistency plots for SAI spatial variation in Figs. 6, and 8 are all based on Gabor measures. The analysis yielding the data in Figs. 6 and 8 is mathematically equivalent to the simulated mean firing rates of a population of central neurons with Gabor-like receptive fields driven by the SAI firing patterns that we recorded in our neurophysiological experiments.

The psychophysical law

The principal issue addressed by Fechner was whether there are simple, universal laws governing our subjective sense of the external world in the same way that there are simple, universal laws governing the external world itself. Fechner and, later, Stevens believed that there are and that they can be expressed in terms of relationships between physical and subjective intensity although they had different conceptions of the laws (Stevens, 1961). Fechner (1860) called this quest for the relationship between physical and subjective intensity 'outer psychophysics' to distinguish it from 'inner psychophysics', which assumes the existence of an inner stimulus, E, as the source of subjective intensity. Fechner was aware that any attempt to link subjective intensity to physical intensity would have limited generality but he was much more confident that general laws would prevail in inner psychophysics (Fechner, 1860, page 56): "While [the logarithmic law] is of limited validity with respect to stimulus and sensation in the area of outer psychophysics, it probably has unlimited validity in the area of inner psychophysics." The neural coding measure, c, (Equation 3 in this paper) is equivalent to Fechner's inner stimulus, E. Fechner even considered linearity as the possible inner law; that is, he considered the possibility that his logarithmic law might result from a logarithmic relationship between the inner and outer stimuli ($c = k \cdot \log S$) and a linear relationship between subjective intensity and the inner stimulus ($m = k \cdot c$ in Equation 4) but rejected that idea in favor of a linear transducer function followed by a logarithmic law for the important, inner psychophysical relationships (Boring, 1950). Fechner seemed to have a clear notion of what had to be done to translate the study of outer psychophysics to the study of inner psychophysics (Fechner, 1860, page 56): "Quantitative dependence of sensation on the [outer] stimulus can eventually be translated into dependence on the [neural activity] that directly underlies sensation — in short, the psychophysical processes — and the measurement of sensation will be changed to one depending on the strength of these processes." It was Mountcastle and his colleagues (Mountcastle et al., 1963; Werner and Mountcastle, 1965) who eventually achieved this translation and suggested linearity as the basic law of psychophysics based on their observation of matching psychophysical and neural response functions. We were drawn to the same conclusion by the spontaneous appearance of linearity in the neural coding studies described in this paper.

The occurrence of linearity in our studies of the neural coding mechanisms of roughness perception and the many previous observations of matching neural and psychophysical responses suggest that the central mechanisms of psychophysics are linear. The strongest hypothesis is, we believe, Mountcastle's original conjecture that subjective intensive continua are served by linear mechanisms. Linearity is important because additive and proportional relationships are preserved in successive operations on the neural signals for intensity. An intriguing possibility is that linearity applies more broadly to all or a large part of the metrical structure of perception, which may account for the success of multidimensional scaling, and multidimensional linear models (Ashby, 1992).

Acknowledgments. This research described here was supported by NIH grant NS18787 and the W.M. Keck Foundation.

References

Ashby, F.G. (1992) *Multidimensional Models of Perception and Cognition.* Lawrence Erlbaum, Hillsdale.

Bankman, I.N., Johnson, K.O. and Hsiao, S.S. (1990) A neural network model of transformation in the somatosensory system. *Cold Spring Harbor Symposia on Quantitative Biology,* 55: 611-620.

Blake, D.T., Hsiao, S.S. and Johnson, K.O. (1994) Neural basis for tactile roughness perception: the relative contributions of slowly adapting and rapidly adapting afferents, Society for Neuroscience Abstract 20: 1387.

Boring, E.G. (1950) *A History of Experimental Psychology.* Second edition, Appleton-Century-Crofts, New York.

Connor, C.E., Hsiao, S.S., Phillips, J.R. and Johnson, K.O. (1990) Tactile roughness: Neural codes that account for psychophysical magnitude estimates. *Journal of Neuroscience,* 10: 3823-3836.

Connor, C.E. and Johnson, K.O. (1992) Neural coding of tactile texture: Comparisons of spatial and temporal mechanisms for roughness perception. *Journal of Neuroscience,* 12: 3414-3426.

Draper, N.R. and Smith, H. (1969) *Applied Regression Analysis.* John Wiley & Sons, New York.

Fechner, G.T. (1860) *Elemente der Psychophysik.* Breitkopf and Hartel, Leipzig, translated by H.E. Adler. Holt, Rinehart and Winston, New York, 1966.

Franzén, O., Kenshalo, D.R. and Essick, G.K. (1991) Neural population encoding of touch intensity. In: O. Franzén and J. Westman (eds.): *Information Processing in the Somatosensory System, Macmillan,* London, pp 71–80.

Franzén, O. and Offenloch, K. (1969) Evoked response correlates of psychophysical magnitude estimates for tactile stimulation in man. *Experimental Brain Research,* 8: 1–18.

Gardner, E.P. and Costanzo, R.M. (1980) Spatial integration of multiple-point stimuli in primary somatosensory cortical receptive fields of alert monkeys. *Journal of Neurophysiology,* 43: 420-443.

Hollins, M., Rao, S., Young, F. and Faldowski, R. (1993) Perceptual dimensions of tactile surface texture: A multidimensional-scaling analysis. *Perception and Psychophysics,* 54: 697-705.

Hyvärinen, J. and Poranen, A. (1978) Receptive field integration and submodality convergence in the hand area of the post-central gyrus of the alert monkey. *Journal of Physiology* (London*),* 283: 539-556.

Johnson, K.O. (1983) Neural mechanisms of tactual form and texture discrimination. *Federation Proceedings,* 42: 2542-2547.

Johnson, K.O. and Hsiao, S.S. (1994) Evalutation of the relative roles of slowly and rapidly adapting afferent fibers in roughness perception. *Canadian Journal of Physiology and Pharmacology,* 72: 488–497.

Johnson, K.O. and Lamb, G.D. (1981) Neural mechanisms of spatial tactile discrimination: Neural patterns evoked by braille-like dot patterns in the monkey. *Journal of Physiology* (London), 310:117–144.

Jones, J.P. and Palmer, L.A. (1987) An evaluation of the two-dimensional Gabor filter model of simple receptive fields in cat striate cortex. *Journal of Neurophysiology,* 58: 1233-1258.

Lederman, S.J. (1974) Tactile roughness of grooved surfaces: The touching process and the effects of macro- and microsurface structure. *Perception and Psychophysics,* 16: 385-395.

MacKay, D.M. (1963) Psychophysics of perceived intensity: a theoretical basis for Fechner's and Stevens' laws. *Science,* 139: 1213-1216.

Mountcastle, V.B., Poggio, G.F. and Werner, G. (1963) The relation of thalamic cell response to peripheral stimuli varied over an intensive continuum. *Journal of Neurophysiology,* 26: 807-834.

Mountcastle, V.B. and Powell, T.P.S. (1959) Neural mechanisms subserving cutaneous sensibility, with special reference to the role of afferent inhibition in sensory perception and discrimination. *Bulletin of the Johns Hopkins Hospital,* 105: 201-232.

Phillips, J.R., Johnson, K.O. and Hsiao, S.S. (1988) Spatial pattern representation and transformation in monkey somatosensory cortex. *Proceedings of the National Academy of Science, U.S.A.* 85: 1317-1321.

Phillips, J.R., Johansson, R.S. and Johnson, K.O. (1990) Representation of Braille characters in human nerve fibers. *Experimental Brain Research,* 81:589–592.

Rodieck, R.W. (1965) Quantitative analysis of cat retinal ganglion cell response to visual stimuli. *Vision Research,* 5: 583-601.

Stevens, S.S. (1957) On the psychophysical law. *Psychological Review,* 64: 153-181.

Stevens, S.S. (1961) To honor and repeal his law. *Science,* 133: 80–86.

Stevens, S.S. (1970) Neural events and the psychophysical law. *Science,* 170: 1043-1050.

Stevens, S.S. (1975) *Psychophysics: Introduction to its perceptual, neural and social aspects.* John Wiley and Sons, New York.

Sur, M. (1981) Receptive fields of neurons in areas 3b and 1 of somatosensory cortex in monkeys. *Brain Research,* 198: 465-471.

Werner, G. and Mountcastle, V.B. (1965) Neural activity in mechanoreceptive cutaneous afferents: stimulus-response relations, Weber functions, and information transmission. *Journal of Neurophysiology,* 28: 359-397.

Worgotter, F. and Holt, G. (1991) Spatiotemporal mechanisms in receptive fields of visual cortical simple cells: A model. *Journal of Neurophysiology,* 65: 494-510.

Somesthesis and the Neurobiology of the Somatosensory Cortex
O. Franzén, R. Johansson and L. Terenius (eds.)

Form processing and attention effects in the somatosensory system

Steven S. Hsiao, Kenneth O. Johnson, Alexander Twombly and James DiCarlo

The Krieger Mind/Brain Institute and the Department of Neuroscience, The Johns Hopkins University, Krieger Hall, 3400 N. Charles St. Baltimore, MD 21218, USA

Summary. Tactile spatial form is initially encoded as an isomorphic representation across the peripheral SAI afferent population. This representation appears to be preserved in the dorsal column nuclei and thalamus since neurons in the input granular layers of area 3b of SI cortex are also isomorphic. The information progresses from the granular to the supra granular layers where it is partially transformed by neurons that are sensitive to local spatial gradients. Neurons in SII cortex show complex responses suggesting that neurons in SII lie at a higher processing stage than SI. The behavioral relevance of the stimuli has a large effect on the responses of neurons in both SI and SII cortex. Neurons show both enhanced (SI and SII) and suppressed (SII only) responses to stimuli that are relevant for the animal's perceptual goals.

Introduction

Present evidence suggests that three of the four primary mechanoreceptive afferent systems subserve separate roles in tactile perception (see Johnson and Hsiao, (1992) for a review). The slowly adapting type I system (SAI) which consists of the primary SAI afferents and all of the ascending pathway(s) related to these afferents is involved in the perception of tactile spatial form and texture. The rapidly adapting system (RA) is involved in processing movement on the skin surface and is important for detecting slip when grasping and lifting objects. The Pacinian system (PC) is involved in processing temporal information such as high frequency vibratory stimuli transmitted to the skin when we handle tools.

Our present working hypothesis is that the different tactile systems utilize a common plan for processing information. Tactile stimuli are initially encoded as an isomorphic representation in the peripheral sensory sheet of receptors. This information then flows centrally where it is transformed in successive processing stages to a representation that is matched against stored memories. The transformation is highly dynamic and is susceptible to descending influences such as selective attention. In this paper we focus on tactile form processing and discuss experimental and practical considerations of how spatial form is processed in the central nervous system.

The primary question being asked is, how are two-dimensional spatial forms, like Braille characters or embossed patterns, represented deep in the brain where information is matched against stored memories? Although this central representation could be isomorphic, it is unlikely since experimental data (discussed below) and practical considerations suggest that information is coded in a form suitable for rapid pattern recognition. An example of our high capacity to perform rapid tactile pattern recognition is the ability of subjects to read Braille characters. Normal high school aged subjects can read Braille at rates around 80 to 90 words per minute (wpm) with some subjects capable of reading at rates up to 200 wpm (Foulke, 1991). The mean recognition rates for individual characters is about 90 msec per character (range of 60 to 130 msec) (Nolan and Kederis,1969). A rate of about one hundred milliseconds per character is very rapid since it includes the time necessary to encode the information within the peripheral afferents, transmit the information to the central nervous system (which takes about 25 msec) and match the pattern against all stored representations regardless of size, orientation, or position on the finger tip. Considering that neurons and synaptic junctions are slow processing elements (1-2 msec/event) it is highly likely that pattern matching within the central nervous system is based on a massively parallel search mechanism where a nonisomorphic representation (i.e., size, position, and orientation invariant) is matched against previously stored representations.

Perception involves a complex interaction between incoming or "bottom-up" signals and descending or "top-down" influences that shape the neural responses of incoming signals to accommodate the animal's perceptual goals. While one view is that these descending influences are simply gating signals that allow some inputs to reach perception and disallow others, psychophysical and neurophysiological evidence suggest that the interaction is much more complex. Oldfield and Phillips (1983) have shown that a small letter b pressed into the fingertip is perceived as a b when the hand is pronated and is perceived as a d when the hand is supinated. Perception of tactile stimuli is also greatly affected by the attentive state of the animal. Experiments on the effects of selective attention show that attention is specific to body location and that attended stimuli are perceived more rapidly and accurately than unattended stimuli. Thus, the perception of tactile stimuli such as clothes touching the skin are normally suppressed unless attention is selectively focused on a particular body site. When attention is focused, stimuli such as vibrations or textured surfaces are perceived more rapidly and accurately (Craig, 1985; Whang et al., 1991; Post and Chapman, 1991). Neurophysiological evidence supporting these psychophysical findings will be discussed in detail later.

In this paper we present psychophysical and neurophysiological data and computational results that support the notion that, in the somatosensory system, central processes transform spatial information to a nonisomorphic form and that this transformation is modifiable depending on the animals perceptual goals. These studies use embossed san-serif capital letters of the alphabet as test stimuli for a number of reasons. Letters are commonly used in psychophysical experiments (Phillips et al., 1983; Johnson and Phillips, 1981; Loomis, 1982) to investigate spatial acuity and spatial form perception because they constitute a diverse and well-known set of complex, spatially distinct patterns composed of simple geometric features (i.e., "B" is composed of a single vertical line connected to two half circles with three short horizontal line segments). In the neurophysiological experiments described below, embossed letters were used for three main reasons. First, letters are composed of a variety of simple geometric shapes that test whether neurons are selectively responsive to individual features. Second, embossed letters span a wide range of spatial frequencies which provide neurophysiological data that are well suited for neural modeling studies. Third, and most significantly, the combined results from the psychophysical and neurophysiological experiments provide a consistent set of data that can be used to investigate the neural mechanisms underlying the perception of complex spatial form.

The primary objectives of the research described here are to understand the neural mechanisms underlying spatial form perception. Specifically the aims are to 1) demonstrate which neurons and neural pathways are involved in spatial form perception. 2) determine how information is represented and transformed along those pathways, and 3) show how selective attention affects the responses of neurons at each stage of processing.

Psychophysics of tactile letter recognition

In both vision and touch, psychophysical experiments show that the threshold for perception (heights yielding 50% correct recognition) is determined by the receptor spacings in the retina and fingertips (about one afferent per mm in the fingertip and 0.45 min of arc (cone spacing) in the retina) (Johnson and Hsiao, 1992; Vega-Bermudez et al., 1991; Phillips et al., 1983; O'Brien, 1951). In touch, the threshold letter height (height that subjects respond correctly 50% of the time) at the fingertip is about 5.0 mm which corresponds to about five receptor spacings (Phillips et al., 1983; Johnson and Phillips, 1981). A threshold near 5.0 mm makes intuitive sense since to discern the details of a typical letter, like the letter E, requires an innervation density high enough to resolve the individual features of the letters. For the letter E the threshold should be around five receptor spacings which in the vertical direction corresponds to the three horizontal bars separated by two spaces. In psychophysical studies, employing letters 6.0 mm high, the letter E is recognized at a rate of about 58% with letters composed of simpler features recognized at higher rates (e.g., I and L) and letters with more complex features (e.g., S) recognized more poorly (Vega-Bermudez et al., 1991). The thresholds for tactile letter recognition are unaffected by the mode of presentation (stationary or scanned at velocities from 10 to 40 mm/sec) or whether the letters are actively (allowed to scan the letters at any velocity or force that they wanted) or passively (scanned at 20mm/sec across their stationary fingertips) presented to the subject's fingertips (Vega-Bermudez et al., 1991). Repeated presentation of letters over multiple days, however, appears to have a small but significant effect illustrating that pattern recognition is dynamic and affected by memory and learning. Figure 1 shows a confusion matrix from 64 subjects who were presented letters (6.0 mm high) both actively and passively. The diagonal elements in the matrix represent correct responses while the off-diagonal elements represent confusions. About 50% of the subjects' responses are found in the off-diagonal elements with a majority of the responses concentrated at specific cells (22 out of 325 possible confusions) and, in many cases, subjects respond to the off-diagonal letter even more frequently than they did to the stimulus letter (e.g., B-D, Q-O). The concentrated responses at individual cells imply that the subjects were not making random guesses but were truly mistaken as to the identity of the letter. As will be shown later, many of the confusions can be explained by the primary representation of letters in the peripheral responses (see figure 2).

Response

Stimulus	A	B	C	D	E	F	G	H		J	K	L	M	N	O	P	Q	R	S	T	U	V	W	X	Y	Z	n
A	71	1				1				1	3	1	2	8				5				3	1	2		1	324
B	2	18	2	[27]	1		[5]	2			1		2	6	6	4	3	6	[3]		6		1				324
C		1	46	3			[15]								[15]		[15]	2	1		2						324
D	1	[2]	2	49	1		2	1					1		[22]	5	4	1	1		7		1				324
E	2	6	1	1	60	2	1	2			[8]		1	2			1	8	1		1	1	1	1		1	324
F	1	1			7	57	1	1	1		4			2		[11]		3	3	2		2		1	2		324
G	1	[11]	[2]	3	1		41						4	1	6	1	[18]	6	[5]								324
H	1	1		1	1	1	1	49			4		9	[15]		1	1	[5]			6	1	4	1			324
									98	1																	324
J		1		1	1	1	1		1	84	1								3		1	1		1	1	2	324
K	1	1	2		[10]	1					57		3	1		1		9			1		2	[7]		4	324
L	1		1							1	1	93						1								1	324
M	1	4			2	1	2	2			8		34	[4]			1	5	1		6	1	[30]				324
N		5		1	1	1		[12]			7		[20]	15		1		[9]	2		4		[19]	2			324
O		1	[2]	[6]			4								83	1	[3]										324
P		2	1	1	1	[23]	2	1			3			2		44	1	2	4	5		3	1		4	1	324
Q		2	[2]	8			[7]							1	[49]	1	28	1	1								324
R	6	4	1	4	1		2	[17]			3		8	[14]	1	2		31			4		1				324
S		[21]	2	6	2		[14]	1			1	1	2	1	5	3	6	8	25		1		2			1	177
T	1	1	1			2			1	1	1	1	1			1			3	61		2		4	[15]	3	324
U		1		2			1	1					1	1	2	1	1	1	1		80	4	2			1	324
V				1							3			2		2			1		3	62	3	1	[21]	1	324
W	1	1			1	1	1	1			3		[11]	[2]		1		1	2		1	1	71			1	324
X	1	1	1		5	1				1	[11]		2	1			1	3	1	1			2	57	1	[12]	324
Y						2	1			2	5		1	2		1		1	1	[6]		[15]		3	58	3	324
Z	1				3	3	1	1	1	2	7			1				2	1	2			1	[31]	2	41	324
	92	86	66	114	98	97	102	91	102	93	131	96	102	81	189	81	83	110	60	77	123	96	142	111	104	73	8,278

Figure 1. Confusion matrix from 64 subjects performing a tactile letter identification task. The test patterns were capital san-serif (Helvetica) letters, 6.0 mm high, raised 500 microns above the surface. Subjects identified the letters by either actively scanning the letters or by having the letter scan across their passively restrained fingers. Each entry in the matrix represents the percentage of trials that a stimulus letter was reported as the letter within the column. The total number of times that each letter was presented across all subjects is listed in the right hand column and boxes represent pairs where the combined confusion rate was greater than 16%. From Vega-Bermudez et al., 1991).

An important finding of these psychophysical studies is that the subjects performed at a rate near 50% even though they had no previous experience in this task and received no training or feedback during the experiments. Therefore subjects must have based their responses on visual memories of letters for comparison. This suggests either that both touch and vision have access to the same central representations or that there are two separate representations that are linked by some associative mechanism. The fact that the subjects perform at a high rate without training, however, suggests that the first explanation is correct and that the transformation occurring within the tactile system must be similar to the one used by the visual system. One possibility is that the nervous system employs a common mechanism for processing spatial form and, if so, studies in the tactile system are valid for understanding the neural mechanisms underlying the processing of spatial form within the visual system and vise-versa.

Neurophysiological studies of tactile spatial form: Peripheral representation

Among the four kinds of peripheral afferents that innervate the glabrous skin of the hand, the slowly adapting type I afferents (SAI) appear to be critical for processing spatial form (Phillips et al., 1988). Two afferent types (PC and SAII) can not be involved in processing form because they have little spatial acuity and do not innervate the skin with sufficient density to convey a coherent spatial image (Phillips et al., 1990). The other two afferents types (SAI and RA) however, have higher spatial acuity and innervate the skin densely. In particular, SAI afferents have small receptive fields (about 2 mm in diameter) and exhibit a kind of surround suppression by responding more vigorously to edges and discontinuities than to flat surfaces (Phillips and Johnson, 1981). The response of typical SAI and RA afferents to scanned embossed letters are shown in figure 2. This figure, which is termed a spatial event plot (SEP), was constructed by continuously scanning the letters from right to left across the neuron's receptive field. Two things are immediately evident in the SEP's. First, the neural responses from the SAI, and less so from the RA afferents preserve much of the spatial details of the stimulus letters. Because the responses of the peripheral SAI and RA afferents are homogeneous within their respective classes, it can be inferred that the image conveyed to the central nervous system across the population of peripheral afferents are like the ones illustrated in figure 2. Second, it is evident that the peripheral images are not perfect renditions of the stimulus letters. Not only are the responses blurred but some features are well represented and others are either poorly represented or missing. For all the letters, features that lie along the leading edge (w.r.t. the direction of scanning across the finger) are better represented in the SEP than are trailing features and internal horizontal features. For example, the leading edge of the letter E is crisp and well formed while the training features are poorly represented and the internal horizontal bars are missing.

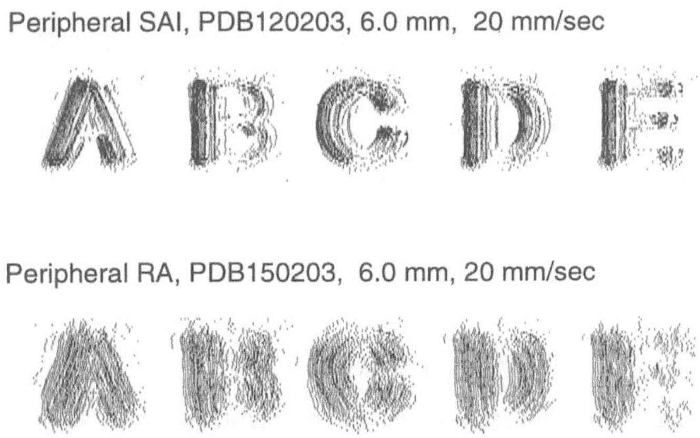

Figure 2. Response of typical monkey SAI and RA afferent fibers to scanned embossed letters of the alphabet. The stimulus letters were 6.0 mm high and were raised 0.5 mm above the background. The letters were mounted on a drum stimulator and were repeatedly scanned (20mm/sec) across the animal's restrained fingertip. Initially no part of the letter was in contact with the neurons receptive field, however on every rotation of the drum the letters were shifted 200 microns so that after about 75 scans the entire letter scanned across the receptive field.

These observations suggest that although the general form of the peripheral responses are much like the stimulus letters, the skin and receptor transduction mechanisms impose distortions on the information available to the central perceptual processes. Thus, it is not surprising that subjects performing the tactile letter discrimination task interpret the letter B as the letter D since the peripheral neural input of the letter B appears more like the letter D (see fig 2). Our working hypothesis, is that all of the confusions observed in the confusion matrix are explainable by the distortions imposed at the level of the peripheral nerve.

Although evidence based on the responses to scanned letters does not clearly show that SAI afferents are solely responsible for tactile form perception, other studies clearly do (see Johnson et al. (1995) for a review). Briefly, those studies show that RA afferents are unable to resolve gaps in gratings that are easily resolved in psychophysical experiments and that spatial acuity measured using the optacon, which activates RA and not SAI afferents, is 5 times poorer (>20mm) than when measured using scanned embossed letters. In contrast, SAI afferent responses account for all of the psychophysical data.

In summary, the primary representation of fine spatial form is conveyed by the population response of peripheral SAI afferents. The representation, which appears to account for the psychophysical behavior, is a blurred form of the stimulus with some features being well represented and other features being either missing or poorly represented.

Central representation

Area 3b in SI cortex and area SII have been shown to be critical for processing tactile form. Both of these areas receive ascending projections from the ventroposterior lateral nucleus of the thalamus (VPL) and ablation of either area severely disrupts the ability of animals to do all tactile tasks, including both form and texture discrimination tasks (Murray and Mishkin, 1984; Carlson, 1981; Semmes and Turner, 1977; Ridley and Ettlinger, 1978). Although there is an ongoing controversy whether SI and SII cortex process information in series or in parallel, the evidence from studies using embossed letters suggest that the information is processed in series. The controversy arises because anatomically, both areas receive inputs from VPL. However, ablating SI tends to silence the responses of neurons in SII but ablation of SII does not silence the responses of neurons in SI, but, activity in SII remains after cooling of SI (Pons et al., 1992; Rowe et al., 1992; Garraghty et al., 1990; Pons et al., 1987).

Two lines of neurophysiological evidence suggest that SI lies at a processing stage earlier than SII. The first is that neurons in SI cortex (especially area 3b) have small localized receptive fields that are often restricted to single phalanges on individual digits while neurons in SII have complex receptive fields that encompass multiple digits or even the entire hand with some neurons (about 20%) having bilateral receptive fields (Hsiao et al., 1993; Robinson and Burton, 1980). The loss of fine somatotopy suggests that SII neurons lie at a higher stage of processing and are receiving convergent inputs from neurons in SI cortex. The second line of evidence is that a greater percentage of neurons in SII are affected by selective attention than in SI. Previous studies and the ones described below have shown that a large fraction of neurons in SII cortex (about 80%) are affected by the attentional state of the animal which suggests that neurons in SII are closely related to higher integrative and perceptual functions (Hsiao et al., 1993; Poranen and Hyvarinen, 1982; Hyvarinen et al., 1980). In contrast, the reported percentages of neurons in SI affected by attention are smaller (between 16 and 50%). Thus the evidence suggests that neurons in SI cortex are performing primary functions (i.e., "bottom-up processing") like extracting information about different attributes of stimuli.

Three principal findings have emerged from our studies on the responses of neurons in area 3b of SI cortex. The first finding is that neurons in area 3b have heterogeneous responses that we have grouped into four classes based on their similarity to the input stimuli. The first class, which comprises about 19% of the cells, shows responses that are isomorphic to the ones observed in the periphery (see figure 3). SEP's from these neurons are readily identified as letters and are untransformed versions of the peripheral inputs. The large number of isomorphic neurons in area 3b imply that the ascending nuclei (i.e., dorsal column nuclei and VPL) do not transform spatial information in any significant way and that these subcortical nuclei are most likely relay stations for tactile information. If this observation is valid then neural recordings from these nuclei should be dominated by neurons with isomorphic responses. The second class of neurons are those that show nonisomorphic but highly structured responses (see figure 3). In area 3b approximately 28% of the neurons that we have studied fall within this class. These responses differ from isomorphic responses in that neurons respond in some complex way to the letters with individual letters not being easily discernable in the responses. In figure 3, the two nonisomorphic, highly structured neurons show clear bursts of neural activity when the letters are scanned across the fingertip. The pattern of activity is not related solely to the spatial structure of the letter but to some complex interaction between the features of the letter with the neuron's receptive field. Neurons within this class represent a transformed version of the stimulus letters where the representation has changed from an isomorphic to a locally nonisomorphic form. The third class of neurons are those that are nonisomorphic but weakly structured. An example of this kind of neuron is shown in figure 3. Neurons grouped within this class (about 31% of the neurons) responded to the letters but did not show the high spatio-temporal structure that is evident in the isomorphic and nonisomorphic highly structured responses. These neurons are most likely not involved in the transformation since they lack high spatial information. One possibility is that these neurons are playing some modulatory role. The fourth class of neurons (about 22% of the neurons) are termed unstructured. Although this last class of neurons could be activated by a blunt probe on the fingertip, they did not respond at all to scanned letters which suggests that they are not directly involved in spatial form processing and must play another role in processing tactile information.

The second finding, which is presently being investigated in detail, is that the neural responses in area 3b are not uniformly distributed across the different laminae (see figure 4). In these studies, simultaneous recordings are made into the post-central gyrus using a dense array of microelectrodes spaced 400 microns apart. The electrode array is oriented to form a linear array perpendicular to the pial surface with the electrodes driven down and positioned so that the electrodes record simultaneously from four neurons in different laminae in area 3b with overlapping receptive fields. The preliminary results from this study, shown in figure 4, are that neurons with isomorphic responses are found in the central layers, neurons with nonisomorphic but highly structured responses in the upper layers, neurons with nonisomorphic but weakly structured responses uniformly spread across all layers, and neurons with unstructured responses in the deeper layers

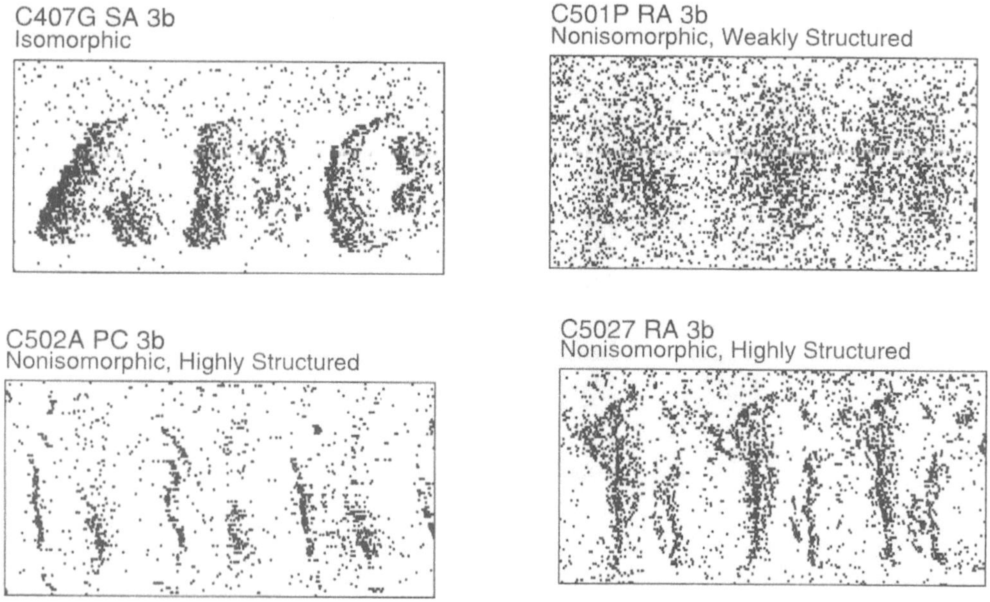

Figure 3. Cortical SEPs from 4 neurons in area 3b. The stimulus letters were 8mm high and scanned at 50 mm/sec across the monkey's finger. The method of scanning was identical to figure 2 except that the animals were awake and performing a light dimming detection task that was unrelated to the stimulus letters.

(DiCarlo et al., 1994). This finding suggests that the granular layer (IV) receives an untransformed version of the stimulus and that there is a significant transformation of spatial form between the granular and supragranular layers. The lack of spatial structure in the infragranular layers, which project to subcortical structures, implies that these neurons are not directly involved in processing spatial information. These results are now being confirmed using more quantitative methods (DiCarlo et al., 1995).

Isomorphic responses can be most easily explained by neurons having simple receptive fields consisting of a single excitatory region. For these neurons the neural response is predictable by the response to a single embossed dot. Explanations for nonisomorphic responses, however, are not

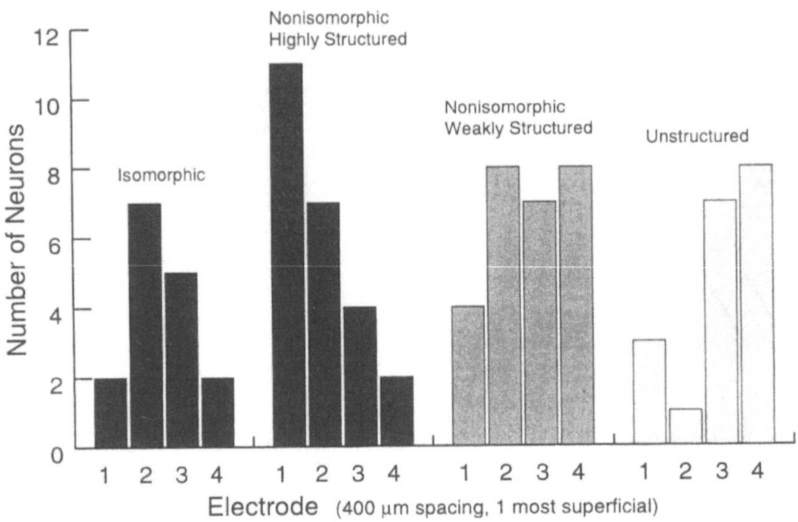

Figure 4. Histograms showing the classifications of 86 neurons that were recorded from four electrodes labeled 1-4 along the bottom axes. The electrodes (spaced 400 microns apart) were driven down into area 3b and positioned so that electrode 1 was in the supra granular layers, electrode 4 was in the infragranular layers and electrodes 2 and 3 were near the middle layers of cortex. Neurons were placed in one of four categories based on the subjective similarity of the SEPs to the stimulus letters (see figure 3).

so simple. The responses of three SA neurons in area 3b to a scanned embossed dot are shown in figure 5. The responses of the upper and lower neurons to a dot are relatively simple which accounts for the isomorphic neural responses to the letters. However the response of the middle neuron is much more complex and appears to have multiple subregions. We estimated the spatial structure of of the receptive fields of these neurons by employing two and three-layer neural network models that use the response from a peripheral SAI afferent as their input and the response of a cortical neuron as the output response to be modeled. The network is then trained on the letters A-M and tested using the letters N-Z. An estimate of the neurons receptive field is then made by "probing" the receptive field of the trained network. The result from these neural network studies is that, as expected, neurons with isomorphic responses have receptive fields dominated by a single excitatory region and that neurons with nonisomorphic responses have receptive fields composed of multiple excitatory and inhibitory subregions separated by 2-3 mm (see figure 5). The neuron illustrated in the bottom panel is isomorphic, however it is responding in an on-off-on fashion to each of the bars in the letters. Although the response to the dot appears to be composed of a simple excitatory region, the figure illustrated in the right hand panel shows that the receptive field is composed of an excitatory and inhibitory region. Neurons that have receptive fields with excitatory and inhibitory subregions are predicted by Connor et. al. (Connor and Johnson, 1992) in their studies of tactile roughness which implies that roughness may be encoded by a subset of neurons in area 3b of SI cortex.

The third principal finding is that a significant fraction of neurons in area 3b of SI cortex are affected by the animal's focus of attention. Monkeys were trained to do two behavioral tasks: a tactile task that required the animal to discriminate embossed letters scanning across its fingertip and a visual task in which the letters continued to scan across the animals fingertip but the animal was required to do a visual detection task. In the tactile task, a letter of the alphabet appeared on the screen in front of the monkey and the animal twisted a switch when a letter scanning across its finger matched the pattern on the screen. The results from this study showed that 50% of the neurons had increased activity in the tactile task when the letters scanning across the fingertip were relevant (Hsiao et al., 1993). These results illustrate that neurons even at the earliest stages of cortical processing are under behavioral control.

Thus far, the studies in SII cortex are consistent with the notion that SII lies at a higher processing stage than SI. Neurons in SII respond in more complex ways to embossed letters than do neurons in SI cortex in several ways. First, none of the neurons studied in SII show isomorphic responses. Instead, neurons have large receptive fields and show complex, nonisomorphic responses that are less directly related to the stimulus. For example, the neuron illustrated in the top panel in figure 6 responded moderately well the trailing parts of letters except for the letter E to which it does not respond at all. In the bottom panel, the neuron again responded well to the trailing features of letters

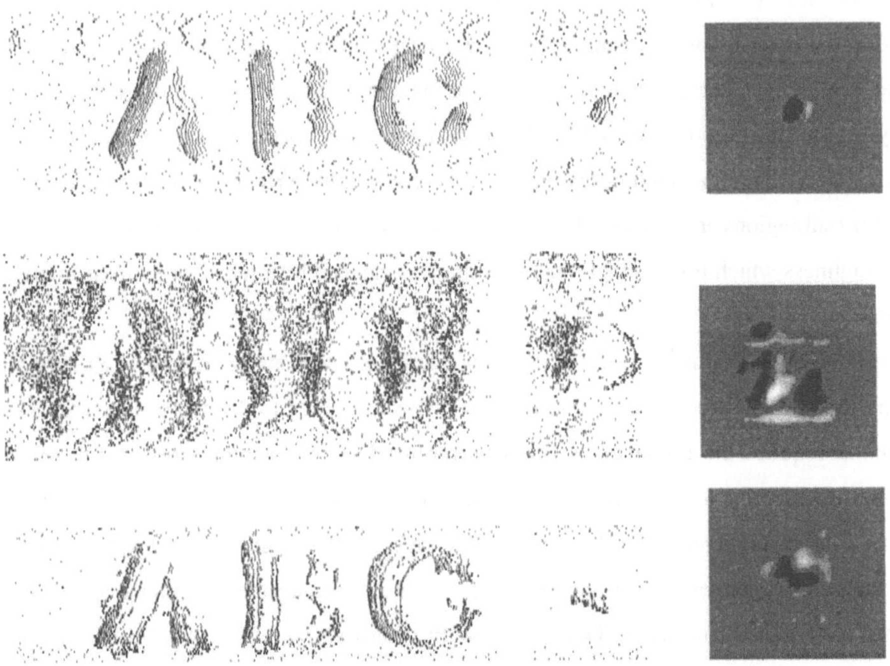

Figure 5. Responses of three cortical, area 3b, SA neurons to the letters ABC and to a scanned dot. The "receptive fields" shown to the right were derived by probing the 3-level neural network models with a one millimeter dot. The network consisted of an input layer into which were fed the peripheral response of an SAI afferent, an intermediate layer with adjustable weight connections and an output layer that contained a single node. The network were adjusted by back propagation to achieve the least mean square error between the actual cortical response and the model output. The black regions represent excitatory (positive) and white regions inhibitory (negative) subregions within the receptive field.

and responded well to the leading arc in the C and to the middle portion of the leading bar in the D. No neurons in SI cortex respond in such complex ways. In the same attention task described earlier, 80% of the neural responses were affected by the animal's focus of attention with 58% yielding enhanced responses and 22% yielding suppressed responses during the letter recognition task. In addition, the effects observed were not simply increases or decreases in firing rates but were

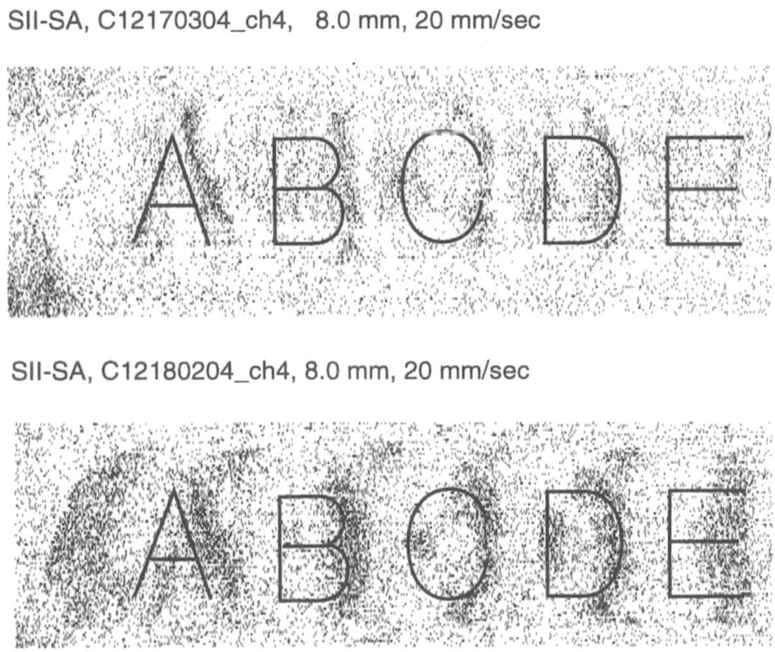

Figure 6. Responses from two cortical SII, SA neuronal responses to letters, 8 mm high, scanned at 20 mm/sec.

significant modifications in the form of the response. Examples of two neurons that were affected by the animals focus of attention are illustrated in figure 7. In the top figure, the impulse rates evoked by the leading and trailing parts of the letter are significantly enhanced and in the bottom figure the entire response is greatly enhanced when the animal performed the tactile task. One possibility is that the modified responses that are observed during the tactile task are critical for the animal's behavior and as such the decreased activity recorded during missed trials (trials in which the animal failed to respond to the target letter) could be reflecting a decrease in attentional effort that resulted in the animal not correctly identifying the letter.

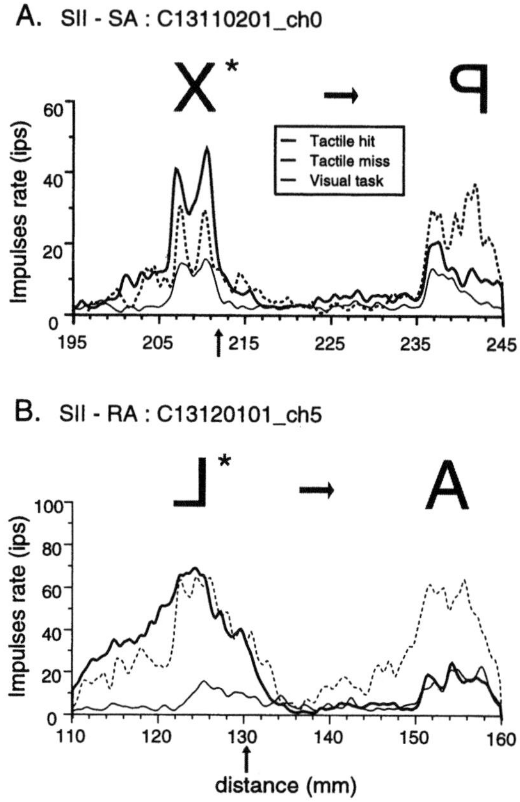

Figure 7. Response of two neurons in SII cortex to letters (6.0 mm high, 20 mm/sec) scanning across the finger pad while the animal's focus of attention was directed alternately every 15-20 revolutions of the drum between the tactile stimuli or an unrelated visual stimuli. X, L, A, and P are embossed letters mounted on a rotating drum. In the tactile task a target letter (denoted by *) appeared on the screen in front of the monkey and the animal twisted a switch with the opposite hand when the letter on the screen matched the letter on the animals finger. *Tactile hits* (correct responses) were rewarded by a liquid reward, *tactile misses* (animal failed to respond) were ignored. During the *visual task*, the letter on the screen was replaced by three illuminated squares. The animals task was to turn the switch when one of the squares dimmed. The abscissa represents distance in millimeters from a fixed point on the drum surface and the arrows along the abscissa represent the mean location of when the animal correctly identified the letter. The letters P and L are plotted backwards because that is the direction that the letters were scanned across the finger pad (from left to right in the figure). Adapted from Hsiao et. al., 1993.

Whether or not attentional effects were observed in SI and SII cortex depended on the behavioral relevance of the embossed letters. During the tactile task (see above) there were two periods when no attentional effects were observed. First, there were times when the animal was being punished for responding to an incorrect letter (false-positives). During these periods, called time-out periods, which lasted for the next letter following the incorrect response, the screen went blank and the animal could not receive a reward. Second, were periods following "hits" (i.e. monkey correctly responded to the target letter) when the animal was being rewarded. During time-out periods, the tactile letters were irrelevant since the animal could not receive a reward and we believe that during these periods the animals attention was withdrawn from the tactile letters. The reason why the animal's focus of attention was withdrawn during the reward period is not as easily explained. One possibility is that the animal's attentional focus was redirected during the reward period to it's liquid reward. The other possibility is that the animal learned that (in this experiment) the target letters never occurred in succession and thus it ignored letters following target letters. The effect of behavioral relevance during the reward period is illustrated in figure 7. In this figure, the response to the following letter is dramatically affected by the animals behavioral response to the target letter. The response to letters following hits show minimal attentional effects while the response following misses show large attentional effects.

In summary, spatial form is initially encoded as an isomorphic representation across the peripheral SAI afferent population. Neurons in the central granular layers of SI cortex also have isomorphic responses suggesting that the primary representation is preserved in the ascending pathways. In SI however, neurons in the supragranular layers have transformed responses. Neural modeling studies indicate that neurons in these upper layers have receptive fields that are sensitive to local spatial gradients. The outputs from these neurons then project to SII cortex where the next transformation step occurs. The behavioral relevance of the stimuli has a large effect on the responses of neurons in both SI and SII cortex.

References

Carlson M. (1981) Characteristics of sensory deficits following lesions of Brodmann's areas 1 and 2 in the postcentral gyrus of Macaca mulatta. *Brain Research* 204:424-30.

Connor CE, Johnson KO. (1992) Neural coding of tactile texture: Comparisons of spatial and temporal mechanisms for roughness perception. *Journal of Neuroscience* 12:3414-26.

Craig JC. (1985) Attending to two fingers: two hands are better than one. *Perception and Psychophysics* 38:496-511.

DiCarlo JJ, Hsiao SS, Johnson KO. (1994) Transformation of tactile spatial form within a cortical column in area 3b of the macaque. [Abstract] *Society for Neuroscience Abstracts* 20:1387

DiCarlo JJ, Lane JW, et al. (1995) Marking Microelectrode Penetrations with Fluorescent Dyes.. *Journal of Neuroscience Methods* In press

Foulke E. (1991) Braille. ; Heller MA, Schiff W, editors.*The Psychology of Touch.* Lawrence Erlbaum associates, publishers, Hillsdale, New Jersey: 9, p. 219-34.

Garraghty PE, Pons TP, Kaas JH. (1990) Ablations of areas 3b (SI proper) and 3a of somatosensory cortex in marmosets deactivate the second and parietal ventral somatosensory areas. *Somatosensory and Motor Research* 7:125-35.

Hsiao SS, O'Shaughnessy DM, Johnson KO. (1993) Effects of selective attention on spatial form processing in monkey primary and secondary somatosensory cortex. *Journal of Neurophysiology* 70, no 1:444-7.

Hyvarinen J, Poranen A, Jokinen Y. (1980) Influence of attentive behavior on neuronal responses to vibration in primary somatosensory cortex of the monkey. *Journal of Neurophysiology* 43:870-82.

Johnson KO, Hsiao SS. (1992) Neural mechanisms of tactual form and texture perception. *Annual Review of Neuroscience* 15:227-50.

Johnson KO, Hsiao SS, Twombly IA. (1995) Neural mechanisms of tactile form recognition. ; Gazzaniga M, editors.*The cognitive sciences.* MIT press, Cambridge, Ma: p. 253-68.

Johnson KO, Phillips JR. (1981) Tactile spatial resolution: I. Two-point discrimination, gap detection, grating resolution, and letter recognition. *Journal of Neurophysiology* 46:1177-91.

Loomis JM. (1982) Analysis of tactile and visual confusion matrices. *Perception and Psychophysics* :41-52.

Murray EA, Mishkin M. (1984) Relative contributions of SII and area 5 to tactile discrimination in monkeys. *Behavioral Brain Research* 11:67-83.

Nolan CY, Kederis CJ. (1969) *Perceptual factors in braille word recognition.* New York: American Foundation for the Blind.

O'Brien B. (1951) Vision and resolution in the central retina. *Journal of the Optical Society of America* :882-94.

Oldfield SR, Phillips JR. (1983) Spatial characteristics of tactile form perception. *Perception* :615-26.

Phillips JR, Johansson RS, Johnson KO. (1990) Representation of Braille characters in human nerve fibers. *Experimental Brain Research* 81:589-92.

Phillips JR, Johnson KO. (1981) Tactile spatial resolution: II. Neural representation of bars, edges, and gratings in monkey afferents. *Journal of Neurophysiology* 46:1192-203.

Phillips JR, Johnson KO, Browne HM. (1983) A comparison of visual and two modes of tactual letter resolution. *Perception and Psychophysics* 34:243-9.

Phillips JR, Johnson KO, Hsiao SS. (1988) Spatial pattern representation and transformation in monkey somatosensory cortex. *Proceedings of the National Academy of Sciences USA* 85:1317-21.

Pons TP, Garraghty PE, et al. (1987) Physiological evidence for serial processing in somatosensory cortex. *Science* 237:417-20.

Pons TP, Garraghty PE, Mishkin M. (1992) Serial and parallel processing of tactual information in somatosensory cortex of rhesus monkey. *Journal of Neurophysiology* 68:518-27.

Poranen A, Hyvarinen J. (1982) Effects of attention on multiunit responses to vibration in the somatosensory regions of the monkey's brain. *Electroencephalography and Clinical Neurophysiology* 53:525-37.

Post LJ, Chapman CE. (1991) The effects of cross-modal manipulations of attention on the detection of vibrotactile stimuli in humans. *Somatosens Mot Res* 8(2):149-57.

Ridley RM, Ettlinger G. (1978) Further evidence of impaired tactile learning after removal of the second somatosensory projection cortex (SII) in monkey. *Experimental Brain Research* :475-88.

Robinson CJ, Burton HA. (1980) Somatotopographic organization in the second somatosensory area of M. fascicularis. *Journal of Comparitive Neurology* 192:43-67.

Rowe MJ, Morley JW, et al. (1992) Parallel processing of tactile information in the cerebral cortex of the cat: effect of reversible inactivation of SI on responsiveness of SII neurons. *Journal of Neurophysiology* 67:411-29.

Semmes J, Turner BH. (1977) Effects of cortical lesions on somatosensory tasks. *Journal of Investigative Dermatology* 69:181-9.

Vega-Bermudez F, Johnson KO, Hsiao SS. (1991) Human tactile pattern recognition: Active versus passive touch, velocity effects, and patterns of confusion. *Journal of Neurophysiology* 65:531-46.

Whang KC, Burton HA, Shulman GL. (1991) Selective attention in vibrotactile tasks: Detecting the presence and absence of amplitude change. *Perception and Psychophysics* 50:157-65.

Somesthesis and the Neurobiology of the Somatosensory Cortex
O. Franzén, R. Johansson and L. Terenius (eds.)

Functional plasticity of cortical representations of the hand

Michael M. Merzenich, Xiaoqin Wang[1], Christian Xerri[2] and Randolph Nudo[3]

Keck Center and Coleman Laboratory, University of California at San Francisco, San Francisco, CA 94143-0732, U.S.A.; [1]Department of Biomedical Engineering and Mind/Brain Institute, Johns Hopkins University School of Medicine, 710 Traylor Building, 720 Rutland Avenue, Baltimore, MD 21205, U.S.A.; [2]Lab. de Psychophysiologie, Faculté de St Jerome, B.P. 362, Av. Esc. Normandie-Niemen, 13397 Marseille Cedex 13, France; [3]Dept. of Neurobiology and Anatomy, University of Texas Medical School, Rm 7.046, P.O. Box 20708, Houston, TX 77225, U.S.A.

In this brief review, we summarize some of the main features of the representations of hand movements and of the skin of the hand defined in the primary motor cortex (area 4) and in the principal fields of the anterior parietal somatosensory strip ("S1" areas 3a, 3b, 1 and 2). We shall then describe how representations of the hand are remodeled in detail in a monkey performing a simple small-object retrieval task that stereotypically engages specific skin surfaces and involves specific hand movements. Some of the implications of this cortical representational plasticity for understanding a) the nature, and b) the variability of representations of skin afferents, kinesthetic afferents, and movements of the hand in different individual primates and in different primate species will then be briefly discussed.

Observations

1. *A note on hand use differences in primates*

Our principal models for studying the cortical representations of hand skin and hand movements have been the New World owl monkey (*Aotus*) and squirrel monkey (*Saimiri*). These primate models were chosen in large part because their central sulci, which are deep in most primates, do not usually extend into the hand representational zone in the anterior parietal (S1) or posterior frontal (M1) cortical fields. A main objective of our experiments has been to reconstruct, in entirety, the S1 and M1 representations of hand movements and hand afferent sensory inputs, a task that is simplified in these species by their being represented on the exposed lateral hemisphere surface.

It should be noted at the outset that there are significant differences in hand use in these small primates, as compared with humans or the more widely studied macaque monkeys. In general: a) Digital movements are relatively poorly differentiated in owl and squirrel monkeys. While thumb movements are differentiated from those of the long fingers, the latter most commonly move in concert, and palmar grasps are heavily and preferentially employed in many object retrieval behaviors. b) Natural behavioral schedules for fine finger movements and object manipulation differ across primate species. This difference is especially marked between arboreal and ground-dwelling monkeys, and between monkeys and contemporary humans, in which the heavy, daily use of fine tools (eating utensils, writing implements, occupational and household tools, etc.) has evolved to a high level. c) Hand sizes of New World monkey models also differ substantially from those of macaque monkeys, apes and humans. As will be noted below, that difference can have substantial impacts, given an input coincidence-based behavioral shaping of central nervous representations of afferent inputs and movements. In general, heavier schedules of temporally differentiated inputs can arise from larger hands.

2. *Cortical representations of the hand in primates*

a) CORTICAL "MAPS" OF CUTANEOUS AFFERENTS. Early maps of the cortical representations of the skin surfaces of the hand were derived with surface recording of evoked responses with response sampling at sites separated by a millimeter or more (e.g., see Woolsey et al., 1943; Pubols and Pubols, 1971). By contrast, contemporary maps are commonly derived by much more discrete intracortical microelectrode recording, and with response sampling at densities of about 20 to 50 samples/mm^2 (e.g., see Merzenich et al., 1987; Xerri et al., 1995). This difference in recording precision, along with the experimental combination of recording data from response mapping studies conducted in different individual animals, were apparent main contributors to the early erroneous conclusion that there was a single representation of the body surface within the four cytoarchitectonic fields of the S1 cortex (e.g., Penfield and Boldrey, 1937; Woolsey et al., 1943; Powell and Mountcastle, 1968; Werner and Whitsel, 1971; Dreyer et al., 1977; McKenna et al., 1982; Iwamura et al., 1983). We now appreciate that there are at least four different cortical areas within the classical S1 zone in monkeys, distinguished by field-specific sources and destinations of anatomical connections, and each containing a relatively complete representation of the body surface (cortical areas 3b and 1; see Paul et al., 1972; Merzenich et al.,

1978; Nelson et al., 1980; Sur et al., 1982), of muscle and joint afferents ordinarily masking anatomically convergent cutaneous inputs (cortical area 3a; see Recanzone et al., 1992c; Phillips et al., 1972; Zarzecki and Herman, 1983), or of mixed cutaneous and proprioceptive inputs (cortical area 2; see Pons et al., 1985; Iwamura et al., 1990). Representations of the hands in these cortical areas have a roughly mirror-image topographic relationship with one another, with representational reversals occurring along the lines of representation of the fingertips between areas 3a and 3b, and 1 and 2, or along a line separating duplicative palmar representations between areas 3b and 1 (Merzenich et al., 1978; Nelson et al., 1980; Sur et al., 1982). Cortical area 2 in macaque monkeys may be further subdivided, i.e., a) can have sharply segregated, exclusively deep-afferent or predominantly cutaneous afferent zones; and b) may encompass more than one functional cortical area (Pons et al., 1985).

Cortical representations of the surfaces of the hand have been described for "S1" or for one or more of its component fields in humans (Penfield and Boldrey, 1937; Penfield and Rasmusson, 1950; Yang et al, 1993), great apes (Woolsey et al., 1951; 1958; Welt, 1963), macaque monkeys (Woolsey et al., 1943; Powell and Moutncastle, 1968; Paul et al., 1972; Dreyer et al., 1977; Nelson et al., 1980; McKenna et al., 1982; Pons et al., 1985; 1987), New World owl (Merzenich et al., 1978; 1987), squirrel (Benjamin and Welker, 1957; Sur et al., 1982; Merzenich et al., 1987), spider (Pubols and Pubols, 1971), cebus (Hursch and Coxe, 1958; Felleman et al., 1989) and marmoset (Carlson et al., 1986; Krubitzer and Kaas, 1990) monkeys, and in several prosimian primates (Krisnamurti et al., 1976; Carlson and Fitzpatrick, 1982; Fitzpatrick et al., 1982). Given wide differences in response sampling densities in mapping studies, it is difficult to draw many definite conclusions about differences between the hand representations of different primate species because so few have been studied in adequate detail. Representations of the hands relative to the rest of the body surface and of individual glabrous digits relative to palmar or dorsal surfaces are proportionally large in all studied primates. Representations of digits are relatively discrete in area 3b in most primates, less so in cortical areas 1 and 2, and rarely so in area 3a. Some New World monkeys have exaggerated representations of their fingernails, which are used on a heavy behavioral schedule in grasping; and the representation of the thumb of humans and great apes would appear to be proportionally large, in comparison with other species. In general, all primates have a similar general layout for the representation of the hand in these fields. The radial margin of the hand is represented lateralward, there discontinuously related to the face representation. The digits and palmar pads are represented in order proceeding medialward in the cortex, in a radial to ulnar sequence across the hand. The wrist and forearm are represented medial to the hand representation.

Maps have been derived in different individuals of the same species in several studies of the representations of hand surfaces in these S1 cortical areas, and substantial ideosyncratic variability in representational detail has been described. Documentation of this variability has been most complete in contemporary studies in which maps have been derived with higher sampling densities (Merzenich, 1986; Merzenich et al., 1987; Pons et al., 1987). They show that in cortical area 3b: a) territories of representation of the entire hand may vary more than 2-fold in their extents in different adult monkeys of approximately equivalent ages and weights; b) representations of individual digits or segments of digits may vary more than 3-fold; c) representations of the dorsal surfaces may be substantially complete and detailed, or may be completely absent, representing many-fold differences in their 3b representational territories; d) the representation of the back of the hand can occupy nearly half of the hand zone of area 3b in one individual, and not more than a few percent in another; e) entirely different skin surfaces can be represented along the border with the face, or along the anterior border with cortical fields 3a in different individual monkeys; f) among many other examples of intraspecies representational differences. These studies have led to the conclusion that representational differences between individuals of a given species can be as great or greater than are the differences recorded for studied individuals of different primate species (Merzenich, 1986; Merzenich et al., 1987).

Ideosyncratic differences in hand representations are even greater in cortical areas 3a, 1, 2 and 5 than in the true "primary somatosensory field", area 3b. In cortical area 1, for example, up to more than 7-fold differences in the territories of representations of the glabrous surfaces of specific digits have been recorded (Merzenich et al., 1987), and variability in the extents of territories representing different digit segments, dorsal hand surfaces or specific palmar pads are as great or greater. In cortical area 3a, we have studied monkeys with very large representations of palmar movements and proportionally small digital-movement muscle afferent representations, contrasted with other monkeys with almost no such palm-muscle afferent representations; and monkeys in which over half of field 3a is dominated by cutaneous inputs, contrasted with other monkeys in which few or no cutaneous response are recorded in the hand zone of this field. A limited number of maps derived in cortical areas SII (e.g., Robinson and Burton, 1980a; Burton and Robinson, 1981; Friedman, 1981) and cortical area 7b (Robinson and Burton, 1980b) also show striking ideosyncratic representational feature differences.

b) MOVEMENT MAPS IN MOTOR CORTICAL AREA 4. Cortical representations of evoked movements of the hand in "primary motor cortex" (area 4) have been defined for a number of primate species by the use of surface stimulation mapping, and over the past 20 years by the use of intracortical microstimulation mapping in many species including humans (Penfield and Boldrey, 1937; Penfield and Rasmusson, 1950;), great apes (Leyton and Sherrington, 1917; Woolsey, 1951; 1958; Welt, 1963), baboons (Hern et al., 1962; Phillips, 1967; Phillips et al., 1971; Waters et al., 1990), macaque monkeys (e.g., Ferrier, 1975; Sherrington, 1906; Franz, 1915; Chung, 1947; Woolsey, 1951; 1958; Asanuma and Rosen, 1972; Humphrey, 1986), and New World owl (Gould et al., 1986) and squirrel (Kwan et al., 1978; Strick and Preston, 1982; Nudo et al., 1992; 1995) monkeys. While studies in humans and apes have been limited to surface stimulation mapping or even cruder non-invasive imaging or magnetic stimulation mapping (e.g., Levy et al., 1991; Yang et al., 1993), in general, they reveal more refined digital and hand movements than are evoked by either surface or intracortical stimulation in monkeys. Movements evoked by low-level intracortical microstimulation in New World monkeys appear to be less differentiated than are those evoked in macaques or baboons, at least roughly paralleling differences in their hand dexterities. In New World monkeys, evoked movements of the thumb are substantially separately represented from those of the other digits, but the representations of the four fingers broadly overlap, and independent long-finger movements are rarely if ever evocable.

Maps derived in M1 (area 4) in different species again differ greatly in evoked-movement sampling densities, and their commonalities and differences are therefore incompletely determined. Most early surface-stimulation maps were interpreted as being consistent with single, simple continuous representations of body movements (for an important exception, see Franz (1915). Later studies described a double hand representation (Strick and Preston, 1982), or a bullseye representation (Kwan et al., 1978) of digit, hand, arm and shoulder movements. By contrast, contemporary intracortical microstimulation maps conducted with a sampling density of 20 or more samples/mm^2 have shown that movement representations are internally much more complex than was formerly believed, with any given wrist or hand or finger movement represented in a series of discrete islands or patches, and with a very rich variety of neighborhood relationships for different zones of representation of any specific movements (Gould et al., 1986; Nudo et al., 1992; 1995; see Wise, 1993, for review). Moreover, a number of surrounding cortical fields from which movements can be evoked and which, like cortical area 4, also project directly to the spinal cord have now been been identified, indicating that "area 6", "supplementary motor cortex" and other zones bordering cortical area 4 are subdividable into 7 or 8 or more input- and output-specific functional cortical motor areas (e.g., Nudo and Masterton, 1990; Dum and Strick, 1992; He et al., 1993).

Early investigators had noted that the relatively crudely defined movement maps derived in the same species by the use of surface-stimulation methods were highly variable in detail. Sherrington and colleagues provided a historic explanation for that variability, in studies that showed that movement maps could be altered by a short schedule of conditioning surface electrical stimulation, or by stimulating a nerve innervating a specific related muscle, e.g., an antagonist for the movement originally evoked (Graham-Brown and Sherrington, 1912). These landmark but long-forgotten experiments, supported by similarly motivated, more contemporary intracortical microstimulation mapping studies (Nudo et al., 1990; Recanzone et al., 1992a), indicated that movement representations are shaped in their details by motor system use, with that functional dynamism likely accounting for recorded representational variability. Variability in movement maps between individuals and between hemispheres in the same species (macaques) was also recorded in the important early cortical area 4 surface mapping experiments of Ivory Franz (1915), and led him to conclude that they probably reflected the motor experiences and abilities of different studied monkeys. Franz' studies were a starting point for historic experiments conducted by his former colleague, Karl Lashley (1923), in which Lashley derived repeated movement maps in the same macaque monkeys. He found that each of a series of successively recorded maps differed substantially in detail from the preceding map derived several days earlier, and still more from a map derived in the same monkey at an earlier time; while the overall probability of evoking any given movement was unchanged over time. This evident representational dynamism led Lashley to conclude that movement representations in cortical area 4 are continually changing constructs, altered in their representational details by intervening movement behaviors --- and conversely, that any point in motor cortex was "equipotential" for representing a broad range of movements.

Ideosyncratic variability in movement maps was also a main finding of the human motor cortex studies of Penfield and colleagues (1937; 1950). Indeed, Penfield justified his cortical mapping as a part of neurosurgical procedures on the basis of this variability, which, prior to direct mapping, provided him with only a limited ability for predicting the locations and extents of the representation(s) of any given movement in any given individual.

This individual variability in area 4 movement maps has been defined at another level in contemporary, highly detailed intracortical microstimulation mapping experiments conducted in New World monkeys (Nudo et al., 1992; 1995; also see Gould et al, 1986). These studies have quantified very marked variability between movement representations between individuals, and for the representation of the two hands in the same individual. They have further shown that the representation of specific movements of the hand preferred for use in small-object retrieval are often more complex, more elaborate, have more complex relationships with the representations of other movements, and are represented over larger total areas than are those same movements in the M1 cortical representation for the non-preferred hand. The finding that there are more representations of a given movement in much more elaborate relationships to the representations of other movements in the area 4 zone contralateral to a preferred hand is consistent with a general view of motor cortex organization originally posited by Hughlings Jackson (see Taylor, 1958). Jackson argued that what is represented in the cortex is a specific movement product; specific muscles and simple component movements must be represented in combination with other movements that contribute to a movement product. Those combinations are much more varied and richly expressed in the representation of a preferred than a non-preferred hand.

Again, these studies documenting the ideosyncratic variability of movement representations of the hand have suggested to investigators for nearly a hundred years that variability at least substantially arises from an individual primate's hand use, and reflects a given individual primate's manual capacities. Had earlier investigators appreciated that representational variability in sensory cortices paralleled that in the motor cortex, they would doubtless have extended their arguments and experiments into that broader realm.

3. *Cortical representations of the skin surfaces of the hand, of deep afferents in muscles controlling hand movements, and of hand movements evoked by microstimulation in cortical area 4 are remodeled by behaviorally important hand use, throughout life.*

On the basis of early evidence of cortical motor map variability, as noted earlier, Sherrington and colleagues (1906) and later Franz (1015) and Lashley (1923) hypothesized that representational variability at least partly manifested changes induced by motor skill learning in adult monkeys. Many more recent experiments conducted using Pavlovian or operant conditioning behaviors and unit recording in the cerebral cortex support that conclusion, by demonstrating that changes in motor (e.g., Woody and Engel, 1972; Aou et al., 1992; Nudo et al., 1995) and sensory (e.g., Olds et al., 1972; Disterhoft and Stuart, 1976; Diamond and Weinberger, 1986; Weinberger and Diamond, 1986; Merzenich et al., 1988; Weinberger et al., 1990; Jenkins et al., 1990; Merzenich et al., 1990; Recanzone et al., 1992c-d; 1993; Merzenich and Jenkins, 1993; Merzenich and Sameshima, 1994; Merzenich and DeCharms, 1995) cortices specific for conditioned responses and unconditioned and conditioned stimuli can be generated by behavioral training, and reversed when the associative conditioning is extinguished (e.g., Disterhoft and Stuart, 1976; Buchhalter et al., 1978; Diamond and Weinberger, 1986; Ahissar et al., 1992), or when the operantly conditioned behavior is unpracticed (Jenkins et al., 1990; Nudo et al., 1995). In general, with associative (Pavlovian) conditioning, in an appropriate cortical area: a) a specific conditioned stimulus ("S+", e.g., a specific auditory or tactile or visual stimulus) comes to be represented over a larger cortical area, b) with higher representational specificity (e.g., smaller receptive fields, and with neuronal responses more specific to conditioning stimuli); c) with greater sensitivity; d) with higher strengths of connections between neurons in the engaged cortical sectors that conjointly represent the conditioned stimulus. Opposite changes, i.e., less specific response selectivity, less sensitive responses and weaker coupling of nearby cortical neurons are recorded for equally-heavily-presented, non-associated ("S-") stimuli (see Woody, 1986; Weinberger et al., 1990; Weinberger, 1994; Merzenich and Jenkins, 1993; Merzenich and Sameshima, 1994; Merzenich and DeCharms, 1995, for reviews).

Equally great representational changes have been documented in area 4 studies documenting learning-induced changes in responses to a unconditioned or conditioned stimulus (e.g., a click, or a puff of air to the nose or eyelids), as well as in the area 4 zone from which the unconditioned response (e.g., a twitch of the nose, or an eyeblink) can be evoked by intracortical microstimulation (see Woody, 1986, for review).

In operant conditioning studies, in which animals progressively improve their abilities to detect or make distinctions about, or categorize sensory stimuli, in an appropriate cortical field: a) The cortical zone directly engaged by behaviorally important stimuli usually enlarges progressively through the period of progressive discriminative learning (Jenkins et al., 1990; Recanzone et al., 1992b,c, 1993). b) Representational changes are also induced in a broad surround that is not directly engaged by behaviorally important stimuli (Jenkins et al., 1990; Recanzone et al., 1992 b,c; Wang et al., 1995). c) Response thresholds for behaviorally important stimuli can be progressively reduced (Recanzone et al., 1992 b,c; 1993; Jenkins et al., 1993). d) Receptive fields can progressively shrink (e.g., Jenkins et al., 1990; Recanzone et al., 1993) or enlarge (e.g., Recanzone et al., 1992b,c; Wang et al., 1995), depending upon whether or not highly spatially restricted and positionally variable inputs, or substantially nonspecific or spatially invariant inputs are delivered into the cortical net, respectively. e) Again, coupling strengths between neurons in the engaged cortical network region(s) appear to progressively increase, generating a progressively more sharply coherent neuronal population response to individual, behaviorally important stimulus events (Recanzone et al., 1992b,d; Wang et al., 1995). f) Input effectiveness changes underlying learning-induced representational plasticity only occur when inputs are behaviorally attended (Jenkins et al., 1990; Recanzone et al., 1992b-d; 1993).

Similar changes in cortical representations have also been induced by engaging an adult mammal in a new experience, or in an instrumental conditioning regime (e.g., Xerri, 1992; Xerri et al., 1995; Nudo et al., 1995). Such studies continue a long history of experiments in which physical changes in the cortex (i.e., in its thickness, layers, compartments, neurons, dendritic branchlets, spines, synapses) have been induced by new or "enriched" experiences (see Greenough and Chang, 1988, for review). Consider, for example, the changes in cortical maps that result from an animal practicing a small-object retrieval task. To document such changes, we have trained adult owl and squirrel monkeys to pick up 3 mm diameter food pellets from small food wells (Nudo et al., 1995; Xerri et al., 1995). In this task, under the conditions presented to these monkeys, they initially made multiple attempts before a successful object retrieval. However, as they practiced this behavior day by day, they ultimately successfully retrieved a pellet with nearly one attempt. From initially clumsy attempts in which movements of the four fingers were poorly differentiated, monkeys came to be adept at a) introducing two adjacent digits into the small well, b) palpating the pellet and drawing it to the top edge of the well with the digit extended and stiffened, c) then grasping it between the two finger tips or capturing it in the distal interphalangeal joint by a rapid finger flexion, d) then bringing it up to the mouth. When this behavior was perfected, the monkeys performed it in a highly stereotyped way, always involving two (in one monkey, three) fingers, with movements and pellet palpation and grasp achieved very fluidly. How does 8 to 30 days of practice at this task resulting in its perfection --- as measured by successful pellet retrieval for virtually every attempt --- change somatosensory and motor cortex representations? How do those plastic changes in the cortex relate to the acquisition of this simple skill?

a) CORTICAL AREA 3b. In picking up small food pellets, again, a trained monkey brought the tip of two digits into a well while extending the other long digits out of the way, palpated the round pellet, drew it up the side of the well, then grasped it between the fingertips and/or at the distal interphalangeal joint before bringing it to the mouth. In this behavior, the critical, tactile information for successfully performing this task came principally from the digit tips that the monkey used to come in contact with the well surfaces and the pellet. To successfully locate, drag out and grasp these small round pellets that are about 1/3rd the diameter of a monkey's finger tip, the monkey has to know how the pellets were located in relation to its finger surfaces to accomplish errorless retrieval.

In one simple investigation of this behavior, we determined how these critical surfaces of the hand are represented in the cortex after an owl monkey or squirrel monkey had mastered this pellet retrieval task. The representations of the tips of the behaviorally engaged digits were consistently enlarged by nearly 2X when compared with the representations of other long fingers, or of the same fingers in control monkeys or hemispheres (Xerri et al., 1995). The representation of the skin of the middle digit segment on these behaviorally engaged digits was also significantly larger for these exercised fingers, albeit changes were smaller. Receptive fields recorded in the territory of representation of the fingertip surfaces that the monkey used were less than 1/2 as large as were receptive fields on control digits. That is, as an apparent consequence of a monkey performing this behavior retrieving 100 pellets each day for A TOTAL OF A FEW HOURS OVER 8-30 sessions, the representation of the skin surfaces that the monkey engaged differentially in the behavior almost doubled in size, with this behaviorally important skin surface now represented in correspondingly finer spatial grain.

In this behavior, monkeys initially adopted different strategies in their attempts at pellet retrieval. They might first try to drop one finger or any combination of two or three or even four fingers into the well in their initial object-retrieval attempts. A given strategy usually predominated during each training session. Given the conditions of this small-object retrieval, a monkey usually first succeeded at this behavior when it used two adjacent, long digits, which were needed for capturing the pellet, and which can be reliably introduced together into these small wells. It may take 5 or 15 or 25 or more days before the monkey arrives, through trial and error, at this successful strategy. Once it becomes predominant, the monkey rapidly perfects the retrieval behavior, i.e., comes to be successful in nearly every retrieval attempt. Thus, when we map the cutaneous representations of skin surfaces engaged in this task after its perfection and find them to be represented in much greater detail than are all of the other digits, that indicates that these differential changes have been driven by only a few hundred practice trials at this task; and that similar effects that must have applied earlier to other digital surfaces in the training are no longer evident, i.e., have faded over the time course of those few hundred two-digit practice trials.

b) CORTICAL AREA 4. We have conducted the same basic experiment in New World squirrel monkeys, but in this case reconstructing changes in evoked-movement representations in cortical area 4. Those changes have been documented by deriving intracortical microstimulation maps of evoked movements on a grain of 15-20 samples/mm^2. Maps were determined in monkeys before the small object retrieval task was initiated, under sterile surgical conditions (Nudo et al., 1995). After the monkey had "perfected" the task, the same cortical zone was mapped again, in equal detail.

These intracortical microstimulation maps derived before and after this behavioral training showed that major changes in evoked movement representations in cortical area 4 are induced by this practiced motor behavior. Motor cortex changes were specific to the movements involved in the skill learning. Overall, the representations of digital movements expanded about 1.5X in the motor cortex, and about 2X in the anterior half of cortical area 4. The representation of specific movements, e.g., of digit extension, increased several fold in extent. Movements that co-concurred in the behavior were evoked with higher probability at equal electrical stimulation thresholds in area 4. In a repeatedly mapped monkey, changes in cortical maps directly paralleled the monkey's task performance.

Thus, as a consequence of the monkey performing this simple behavior, the representations of movements within cortical area 4 were dramatically remodelled, with evoked movement map changes that were related to the emergent movement dexterity.

c) OTHER SENSORY AND MOTOR AREAS. While we have looked only to a limited extent in other cortical areas with this particular behavior, we have studied the plasticity induced in other similar behaviors in two other S1 areas, area 1 and area 3a. Other investigators have recorded plastic changes in a cued retrieval task that have produced behaviorally specific neuronal response changes on a major scale in different sectors of premotor area 6 (e.g., Mitz et al., 1992), and of supplementary (or bordering) motor cortex (e.g., Aizawa et al., 1991).

Cortical area 3a is normally dominated by inputs from muscle spindles and other deep inputs and is thought to provide inputs to area 4 contributing to the initiation of movements. In another behavior involving a reaction-time-limited hand movement from a resting position, we have seen dramatic representational remodeling in this area that generated a manyfold exaggeration of the representation of the specific sensory inputs that cued movement initiation (Recanzone et al., 1992c; and unpublished observations). In this pellet retrieval behavior, the movement was usually initiated initially from the hand or wrist, but as the behavior was perfected, movement initiation came to involve all of the joints of the arm, wrist, hand and fingers, i.e., was initiated fluidly from the entire limb. We hypothesize that we would record changes in area 3a consistent with these behavior changes in the small-object retrieval behavior, but have not yet documented them. Consistent with their presence, one major change in cortical area 4 is a progressive increase in cortical territory over which extension of the fingers, hand, wrist and elbow are evoked together as the behavior is perfected. Such movement combinations are relatively infrequently recorded in monkeys that are naive to this behavior, while they were common, and increased progressively, in trained monkeys.

The hand representational zone of cortical area 1 does not appear to be activated during behavioral periods in which the digits are in motion (Nelson et al., 1991; see Merzenich et al., 1988; 1990); inputs are believed to be inhibited by gating inputs originating from cortical area 4. On the other hands, maps derived in several of the monkeys trained to retrieve pellets showed representations of behaviorally engaged digits that were significantly larger in representation than were other digits. Those changes less purely represented fingers involved only in the last days of training, when the monkey restricted pellet palpation and manipulation to just two fingers. On the other hand, digits that these monkeys never used to pick up pellets, like digits 1 and 5, appeared to be decidedly under-represented when compared to control hand representations. We hypothesize that when a monkey introduced its fingers into the food well and shifted from extension to flexion and withdrawal, afferent inputs might be momentarily effective in engaging cortical areas 1. In any event, area 1 also appears to be remodeled by this behavior, and induced changes in this field presumably contribute to it.

There are a number of other somatosensory cortical fields in the parietal and insular cortex that are engaged by a behavior like this one, including what are probably multiple cortical areas within "area 2", "area 5" and "area 7b", "S2" and the "insular cortex". With high likelihood, <u>all</u> of these sensory cortical fields will be significantly remodeled by this behavior. Similarly, changes must be effected on an important scale in the area 7 or 8 or more spinal cord-projecting motor areas that surround cortical area 4. In macaque monkeys performing a visually or auditory cued movement behavior, nearly half of neurons sampled over a broad zone of cortical area 6 developed more selective responses related to cueing stimuli or cued movements in the behavior (Mitz et al., 1992). Changes on a major scale also emerge with behavioral training in the supplementary motor cortex (Aizawa et al., 1991). Indeed they occur in *every* relatively directly engaged sensory and motor cortical area that has been studied in a learning context like this one (see Merzenich et al., 1990; Merzenich and Jenkins, 1993; Merzenich and DeCharms, 1995, for review).

Discussion and some general conclusions

1. *Summary of plasticity effects induced by motor skill learning.*

How is skill acquisition accounted for by these representational changes in cerebral cortical fields? When a monkey takes on a simple, new, behaviorally important task, like picking up small food pellets from cylindrical food wells, its attended practice drives plastic representational changes in sensory and motor cortical areas that are engaged by the object contact and object manipulation, and by the kinesthetic inputs from muscles and from other deep receptors that are differentially excited in the behavior. These changes progress with practice repetition to differentially represent these cutaneous and deep inputs when the monkey performs a rewarded trial. A connectional effectiveness change/reward progression commonly applies to several movement strategies that are successively, temporarily adopted before the monkey initiates a highly effective one. When that highly effective strategy is identified, it is rapidly stereotyped. The emergent representational changes that track this behavioral evolution *are* the cortical parts of skill learning, with each of a number of engaged cortical areas making field-specific contributions to its cutaneous sensory or kinesthetic or movement planning or movement initiation or movement performance aspects.

2. *Changes in cortical plasticity with the development of "automatic" motor behaviors.*

Why are behaviors like small-object retrieval not improved indefinitely? Why don't repetitive behaviors like this practiced small-object retrieval overwhelm the hand movement zone of cortical area 4 representations at the expense of other manual skills? A large body of evidence has shown that the creation and maintenance of these representations is a function of the animal or human's level of attention at the task. Cortical representational plasticity in skill acquisition is self-limiting, in the sense that as the behavior comes to be more "automatic", i.e., less closely attended, representational changes induced in the cortex fade, and ultimately disappear or (because of unlearning effects) reverse (Haier et al., 1992; Pascual-Leone et al., 1994; R. Nudo, W. Jenkins and M. Merzenich, unpublished observations). The facet of behavioral performance that enables maintenance of the behavior with minimum involvement of the cortical learning machinery is probably stereotypic movement sequence repetition. As a movement behavior destined to evolve to a stereotypic form is practiced, ultimately an effective, highly statistically predictable movement sequence is adopted that enables the storage of the learned behavior in a permanent form that requires only minimal --- or no --- behavioral attention. Note that the brain creates a kind of behavioral homeostasis for a well-learned behavior: If behavioral performance declines, or behavioral or brain conditions change to again render a task more difficult, attention to the behavior will again increase, which will produce an invigorated cortical response to the new learning challenge.

By this view, *the cerebral cortex is a learning machine*. William James (1890) was perhaps the first to point out the great practical advantages for a self-organizing cortex of the development of what he called "habits". Once a skill is overlearned, he argued, it will engage pathways that are so reliably tracked that they can be followed without attention. The learning machine, no longer necessary for this habitual reiteration, is freed to focus on new enterprises. The great classical physiologists Leyton and Sherrington (1917) echoed a similar theme from another perspective, arguing that although cortical plasticity probably accounts for or reflects changes accounting for skill learning, nonetheless there must be a basic and relatively reliable cortical map of movements that the forebrain never strays too far from.

How is sensorimotor learning sustained when the adaptive representations of learned behavior "fade" in the cerebral cortex? There are two general possibilities; neither has been assessed at a level of experimentation at which this important issue can be entirely resolved. One view is that habits come to be represented in an enduring form extracortically, whose creation might be contributed to greatly by a "teaching" cortex, e.g., in the spinal cord, basal ganglia, red nucleus and/or cerebellum. In support of this view, while the learning of manual skills requires a motor cortex, some over-learned motor skills are not very significantly reduced by the induction of wide area 4 lesions.

A second possibility is that behaviorally induced changes in the cortex endure in a highly efficient representational form that can sustain the representation of its key features in the cortex itself, ultimately engaging only limited distributed populations of cortical neurons to represent the behavior with high fidelity. The general finding that the responses of cortical neurons in a very heavily practiced behavior can be strongly correlated with a monkey's discriminative abilities or movement performance abilities support this alternative view. Of course many behaviors require constant attended practice at a highly cognitively reinforced level, and may be expected to reflected by enduring, territorially exaggerated representations in highly manually skillful individuals

3. *Origin of topographic sensory and motor cortical "maps".*

If skin surfaces or movements can be represented in many alternative detailed forms in the cortex, what accounts for their orderliness and their specific detailed structures? A large body of fundamental (e.g., see Merzenich et al., 1988; 1990; Merzenich and Jenkins, 1993; Merzenich and DeCharms, 1995; for review) and representational plasticity evidence (e.g., Allard et al., 1991; Wang et al., 1995) has shown that cortical plasticity mechanisms are coincident input-dependent. That is, connectional strengths for behaviorally important inputs that are delivered nearly simultaneously into cortical networks are strengthened together, to achieve their co-selection. By the operation of these basic "Hebb-like" mechanisms (Hebb, 1949), *cortical representations map input time continua* (Merzenich et al., 1984; 1988; 1990). Temporal input structure is topographic. Inputs from adjacent skin surfaces on the hand are excited with high probability nearly coincidentally in time. In an adaptive cortical network operating with Hebb-like synapses, they will be represented in adjacent cortical locations. On the other hand, temporally non-coincidental skin inputs will create discontinuities in their cortical representations (Wang et al., 1995).

It should be noted that only a limited subset of inputs that are acted upon by adaptive cortical connection-strengthening mechanisms are delivered from the mainline somatosensory or kinesthetic afferent systems. Extensive inputs also come from other thalamic, cortical, basal ganglion and modulatory control system sources. Still more are delivered from within the cortical network itself. *All* of these input strengths are presumably subject to plastic modification in attended behaviors on the bases of their nearly-coincident activation, with those changes presumably all contributing to evolving representational plasticity. It should be noted that any spatially limited stimulation of the cortex will confer strong inputs into a broad surround, both from intrinsic network sources and from backward, top-down projections; and b) conversely, that inputs from a cortical network surround delivered from intrinsic and from more topographically dispersed extrinsic sources must contribute significantly to neuronal response specificity and its behavioral remodeling.

Motor cortical areas are almost certainly organized on the same coincident-input connectional-strengthening basis, which is hypothesized to apply generally for the neocortex. Coincident input-based plasticity is a prominent feature of the adaptive physiology of area 4 (Keller et al., 1990; Iriki et al., 1991; Baranyi, et al., 1991; Woody, et al., 1991; Keller et al., 1992), and leads to major functional and physical remodeling within this zone. The demonstration that movements that co-concur in behavior are evoked together by threshold stimulation over growing regions in trained monkeys is consistent with the hypothesis that coincident afferent inputs evoked by the small-object retrieval behavior in motor cortex are also co-selected in this field (Nudo et al., 1995). Neurons that participate together in the distributed representations of specific learned target-capture reaches are selectively positively coupled. These behaviorally induced representational features of M1 parallel those recorded in the sensory cortical areas of S1.

The representations of any given specific movement in a few to many small cortical "islands" is a probable consequence of the relatively wide anatomical convergence and divergence of inputs projecting into the motor cortices. It presumably also reflects the involvement of that given specific movement as a component of a number of different movement behavior sequences, which would result in the generation of a variety of different neighborhood relationships for that movement in cortical area 4.

4. *Origins of individual differences in cortical representations. Implications for strategies employed for investigating neocortex physiology and function.*

Studies conducted principally over the past decade have shown that learning-induced cortical plasticity constitutes a main source of cortical representational variability in adult mammals. They show that cortical neuronal response specificity is remodeled in detail throughout life by attended behaviors. Unfortunately, in most single unit cortical experiments directed toward defining the cortical origins of neuronal response specificity, relevant experiential histories and behavioral capabilities are unknown.

While many primate studies are conducted in behaviorally conditioned monkeys, experimental animals are commonly overtrained. Overtraining can result in the evolution of cortical representational changes to an asymptotic representational state. In that state, representation will be only minimally altered by further learning. This representational rigidity can mask the fundamentally adaptive nature of learning and memory processes in cortical fields. Moreover, when behaviors are performed under easily-performed and more automatic conditions, as must often be the case in waking primate studies, at least some of the consequences of earlier, closely attended learning may have significantly faded in the cortex. Finally, behavioral training paradigms are commonly designed to minimize the influences of stimulus context. In fact, projections from other cortical and extracortical sources providing cueing, context and other cognitive influences, which play a major and crucial role in the dynamic life of real cortex. That role is, again, largely hidden from view by the design of most cortical physiological/behavioral studies.

5. *Origins of species differences in cortical response specificity and in the details of cortical maps.*

Primates have species-specific differences in their hands that must contribute to differences in hand representations in sensory and motor cortical areas. The thumb extends from a more proximal position on the hand of New World owl and squirrel monkeys than in macaques, apes or man, generating less temporally correlated inputs from the thumb vs fingers that must contribute to their cortical representational separation, given coincident input-based cortical plasticity mechanisms. Hands of New World monkeys have a projecting pisiform bone at the base of the hypothenar eminence that provides a platform for heavily employed palmar grasp behaviors.

Hands of primates also differ greatly in size, which contribute substantial differences in the temporal schedules of inputs delivered from their sensory surfaces and controlling muscles. In general, afferent inputs from skin and muscles are more temporally differentiated in larger than in smaller hands and fingers.

The schedules of behaviorally generated inputs differ greatly in different species because of the different natures and schedules of manual experiences that apply for each primate. The generation of heavier schedules of nearly coincident inputs from different fingers plausibly contributes to the much greater evident coupled finger movements that mark New World species and human infants, and to a lesser extent, old world monkeys. It is interesting to note that a brief period of training requiring separate digit control in a New World monkey results in more sharply differentiated representations of digit movements in the motor cortex. Thus, their brains have the capacity to separate digit representations to at least some extent, but afferent input schedules hypothetically normally limit the development of independent finger movement behaviors and M1 cortical representations in these small monkeys.

For well-studied primates, differences in hand use are perhaps greatest between humans and monkeys, given the extraordinary schedules of tool use involving highly differentiated hand movements and highly specific skin stimulation in humans. Collectively, these differences in hand structure, hand size and especially differently weighted, behaviorally important hand stimulation and movements are presumably a main contributor to the cortical representational differences recorded between studied primate species.

Acknowledgments
This research was supported by NIH grant NS-10414, Hearing Research, Inc., the Sooy Fund, and the Coleman Fund. The authors thank the many collaborators who participated in these cortical plasticity experiments.

References

Ahissar, E., Vaadia, E., Ahissar, M., Bergman, H., Arieli, A., and M. Abeles (1992) Dependence of cortical plasticity on correlated activity of single neurons and on behavioral context. *Science* 257:1412-1415.
Aizawa, H., Inase, M., Mushiake, H., Shima, K., and J. Tanji (1991) Reorganization of activity in the supplementary motor area associated with motor learning and funcitonal recovery. *Experimental Brain Research* 84:668-671.
Allard, T.A., S. A. Clark, W. M. Jenkins and M. M. Merzenich (1990) Reorganization of somatosensory area 3b representations in adult owl monkeys following digital syndactyly. *Journal of Neurophysiology* 66:1048-1058.
Aou, S., Woody, C.D., and D. Birt (1992) Increases in excitability of neurons of the motor cortex of cats after rapid acquisition of eye blink conditioning. *Journal of Neuroscience* 12:560-569.
Asanuma, H., and Rosén, I. (1972) Topographic organization of cortical efferent zones projecting to distal

forelimb muscles in the monkey. *Experimental Brain Research* 14:243-256.
Baranyi, A., Szent, M.B., and C.D. Woody (1991) Properties of associative long-lasting potentiation induced by cellular conditioning in the motor cortex of conscious cats. *Neuroscience* 42:321-334.
Benjamin, R.M., and W. I. Welker (1957) Somatic receiving areas of cerebral cortex in the squirrel monkey (*Saimiri sciureus*) *Journal of Neurophysiology* 20:286-299.
Burton, H., and C. J. Robinson (1981) Organization of the S2 parietal cortex. Multiple sensory representations within and near the second somatic area of cynomologous monkeys. *In:* C. N. Woolsey, ed., *Cortical Sensory Organization: Multiple Somatic Areas.* Humana: Clifton, New Jersey, pp. 1-27
Carlson, M. and K.A. Fitzpatrick. (1982) Organization of the hand area in the primary somatic sensory area (SmI) of the prosimian primate, *Nycticebus cougang. Journal of Comparative Neurology* 209:280-295.
Carlson, M., Heuerta, M.F., Cusick, C.G., and J.H. Kaas (1986) Studies on the evolution of multiple somatosensory representations in primates: The organization of anterior parietal cortex in the New World callitrichid, *Saguinus. Journal of Comparative Neurology* 246:409-426.
Chung, H.-T., Ruch, T.C., and A.A. Ward (1947) Topographic representation of muscles in the motor cortex of monkeys. *Journal of Neurophysiology* 10:39-56.
Diamond, D.M. and N.M. Weinberger (1986) Classical-conditioning rapidly induces specific changes in frequency receptive fields of single neurons in secondary and ventral ectosylvian auditory cortical fields. *Brain Research* 372:357-360.
Dreyer D.A., Schneider, R.J., Metz, C.B., and B.L. Whitsel (1977) Differential contributions of spinal pathways to body representation in postcentral gyrus of *Macaca mulatta. Journal of Neurophysiology* 36:119-145.
Dum, R.P., and P. L. Strick (1992) Medial wall motor areas and skeletomotor control. *Current Opinion in Neurobiology* 2:836-839.
Disterhoft, J.F. and D.K. Stuart (1976) Trial sequence of changed unit activity in auditory system of alert rat during \conditioned response acquisition and extinction. *Journal of Neurophysiology* 39:266-281.
Felleman, D.J., Nelson, R.J., Sur, M., and J. H. Kaas (1989) Representations of the body surfac e in areas 3b and 1 of postcentral parietal cortex of cebus monkeys. *Brain Research* 268:15-26.
Ferrier, D. (1975) Experiments in the brain of monkeys. *Proceedings of the Royal Society of London, Series B.* 23:409-430.
Fitzpatrick, K.M., Carlson, M., aqnd J. Charlton (1982) Topography, cytoarchiture and sulcal patterns in primary somatic sensory cortex (SmI) of the prosimian primate, *Perodicticus potto. Journal of Comparative Neurology* 204:196-210.
Franz, W. (1915) *Psychology Review* 19:80.
Friedman D.P. (1981) Body topography in second somatic sensory area. Monkey S2 somatotopy. *In:* C.N. Woolsey, ed., *Cortical Sensory Organization: Multiple Somatic Areas.* Humana: Clifton, NJ, pp. 121-165.
Gould, H.J., C. G. Cusick, T. P. Pons and J. H. Kaas (1986) The relationship of corpus callosum connections to electrical stimulation maps of motor, supplementary motor, and frontal eye fields in owl monkeys. *Journal of Comparative Neurology* 247: 297.
Graham Brown, T., and C.S. Sherrington (1912) On the instability of a cortical point. *Proceedings of the Royal Society of London, Series B.* 85:250-249.
Greenough, W.T., and F. F. Chang (1988) *In:* A. Peters and E. G. Jones, eds., *Cerebral Cortex. Vol. 7. Development and Maturation of Cerebral Cortex.* Plenum:New York, pp. 335-392.
Haier R. J., Siegel, B.V., Jr., MacLachlan, E., Soderling, E., Lotenberg, S., and M. S. Buchsbaum (1992) Regional glucose moetabolic changes after learning a complex visuospatial/motor task: A Positron emission tomographic study. *Brain Research* 570:134-143.
He, S.Q., Dum, R.P., and P. C. Strick (1993) Topographic organization of corticospinal projections from the frontal lobe: Motor areas on the lateral surface of the hemisphere. *Journal of Neuroscience* 13:952-980.
Hebb, D.O. (1949) *The Organization of Behavior: A Neuropsychological Theory.* New York: Wiley.
Hern, J., Landgren, S., Phillips, C., and R. Porter (1962) Selective excitation of corticofugal neurones by surface-anodal stimulation of the baboon's motor cortex. *Journal of Physiology* 161:73-90.
Hirsch, J.F., and W.S. Coxe (1958) Representation of cutaneous tactile sensibility in cerebral cortex of *Cebus. Journal of Neurophysiology* 21:481-498.
Humphrey, D.R. (1986) Representation of movements and muscles within the primate precentral motor cortex: Historical and current perspectives. *Federation Proceedings* 45:2687-2699.
Iriki, A., Pavlides C., Keller, A., and H. Asanuma (1991) Long-term potentiation of thalamic input to the motor

cortex induced by coactivation of thalamocortical and corticocortical afferents. *Journal of Neurophysiology* 65:1435-1441.
Iwamura Y, M. Tanaka, O. and O. Hikosaka (1990) Overlapping representation of fingers in area 2 of the somatosensory cortex of the conscious monkey. *Brain Research* 197:516-520.
Iwamura, Y., Tanaka M., Sakamoto M., and O. Hikosaka (1983) Functional subdivisions representing different finger regions in area 3 of the first somatosensory cortex of the conscious monkey. *Experimental Brain Research* 51:315-326.
James, W. (1890) *The Principles of Psychology.* Vol. 1. New York: Dover Publications, Inc.
Jenkins W.M., Beitel, R., Xerri, C., Peterson, B., Wang, X., and M. M. Merzenich (1993) Behaviroal measurements of vibrotactile frequency response and temporal response properties of neurons in SI cortex of trained owl monkeys: Threshold shifts and duration effects. *Society for Neuroscience Abstract.* 19:162.
Jenkins, W.M., M.M. Merzenich, M. Ochs, T. Allard, T. and E. Guic-Robles (1990) Functional reorganization of primary somatosensory cortex in adult owl monkeys after behaviorally controlled tactile stimulation. *Journal of Neurophysiology* 63:82-104.
Keller A., Iriki, A. Ned H. Asanuma (1990) Identification of neurons producing long=-term potentiation in the cat motor cortex: Intracellular recordings and labeling. *Journal of Comparative Neurology* 300:47-60.
Keller A., Arissian C., and H. Asanuma (1992) Synaptic proliferation in the motor cortex of adult cats after long-term thalamic stimulation. *Journal of Neurophysiology* 68:295-308.
Krishnamurti, A., Sanides, F., and W. I. Welker (1976) Microelectrode mapping of modality-specific somatic sensory cerebral neocortex in slow loris. *Brain Behav. Evol.* 13:267-283.
Krubitzer, L.A., and J. H. Kaas (1990) The organization of connections of somatosensory cortex in marmosets. *Journal of Neuroscience* 10:952-974.
Kwan, H.C., MacKay, W.A., Murphy, J. T., and Y. C. Wong (1978) Spatial organization of precentral cortex in awake primates. II. Motor outputs. *Journal of Neurophysiology* 41:1120-1131.
Lashley, K. (1923) Temporal variation in the function of the gyrus precentralis in primates. *American Journal of Physiology* 65:585.
Levy, W.J., Amassian, V.E., Schmid, U.D., and C. Jungreis (1991) Mapping of motor cortex gyral sites non-invasively by transcranial magnetic stimulation in normal subjects and patients. Electroencephalography & Clinical Neurophysiology 43:51-75.
Leyton, A.S.F., and C. S. Sherrington (1917) Observations on the excitable cortex of the chimpanzee, orang-utan, and gorilla. *Quarterly Journal of Experimental Physiology* 11:135.
McKenna, T.M., Whitsel, B.L., and D.A. Dreyer (1982) Anterior parietal cortical topographic organization in macaque monkey: A reevaluation. *Journal of Neurophysiology* 48:389-417.
Merzenich, M.M. (1986) Sources of intraspecies and interspecies cortical map variability in mammals: conclusions and hypotheses. *In:* M.J. Cohen and F. Strumwasser, eds., *Comparative Neurobiology: Modes of Communication in the Nervous System.* Wiley:New York, pp. 105-116.
Merzenich, M.M., and R. C. DeCharms (1995) Neural representations, experience and change. In: R. Llinas and P. Chruchland, eds., *Mind and Brain,* MIT Press: Boston.
Merzenich, M.M., and W. M. Jenkins (1993) Cortical representation of learned behaviors. In: P. Andersen, O. Hvalby, O. Paulsen and B. Hokfelt, eds., *Memory Concepts.* Elsevier:Amsterdam.
Merzenich, M.M., and K. Sameshima (1993) Cortical plasticity and memory. *Current Opinion in Neurobiology* 3:187-196.
Merzenich, M.M., W.M. Jenkins, and J.C. Middlebrooks (1984) Observation and hypotheses on special organizational features of the central auditory nervous system. *In:* G.M. Edelman, W.E. Gall, and W.M. Cowan, eds., *Dynamic Aspects of Neocortical Function,* Wiley:New York, pp 397-424.
Merzenich, M.M., Kaas, J.H., Sur, M. and C.-S. Lin (1978) Double representation of the body surface within cytoarchitectonic areas 3b and 1 in "SI" in the owl monkey (*Aotus trivirgatus*). *Journal of Comparative Neurology* 181:41-73.
Merzenich, M.M., R.J. Nelson, J.H. Kaas, M.P. Stryker, W.M. Jenkins, J.M. Zook, M.S. Cynader, and A. Schoppmann (1987) Variability in hand surface representations in areas 3b and 1 in adult owl and squirrel monkeys. *Journal of Comparative Neurology* 258:281-296.
Merzenich, M.M., G.H. Recanzone, W.M. Jenkins, T. Allard and R. J. Nudo (1988) Cortical representational plasticity. *In:* P. Rakic and W. Singer, eds., *Neurobiology of Neocortex.* Wiley:New York, pp. 41-67.
Merzenich, M.M., G.H. Recanzone, and W.M. Jenkins (1990) How the brain functionally rewires itself. *In:* M. Arbib and J. A. Robinson, eds., *Natural and Artificial Parallel Computations.* MIT Press:New York, pp. 177-210.

Mitz, A.R., Godschalk, M., and S. P. Wise (1992) Learning-dependent neuronal activitiy in the premotor cortex: Activity during acquisition of conditioned motor associations. *Journal of Neuroscience* 11:1855-1872.
Nelson, R.J., Smith, B.N., and V. D. Douglas (1991) Relationships between sensory responsiveness and premovement activity of quickly adapting neurons in areas 3b and 1 of monkey primary somatosensory cortex. *Experimental Brain Research* 84:75-90,
Nelson, R.J., Sur, M., Felleman, D.J., and J. H. Kaas. (1980) Representations of the body surface in postcentral parietal cortex of *Macaca fascicularis*. *Journal of Comparative Neurology* 192:611-643.
Nudo, R.J., and R. B. Masterton (1990) Descending pathways to the spinal cord. III. Sites of origin of the corticospinal tract. *Journal of Comparative Neurology* 286:559-583.
Nudo, R.J., W.M. Jenkins, and M.M. Merzenich (1990) Repetitive microstimulation alters the cortical representation of movements in adult rats. *Somatosensory & Motor Research* 7:463-483.
Nudo, R.J., Jenkins, W.M., Merzenich, M.M., Prejean, T., and R. Grenda (1992) Neurophysiological correlates of hand preference in primary motor cortex of adult squirrel monkeys. *Journal of Neuroscience* 12:2918-1947.
Nudo, R.M., G. W. Milliken, W. M. Jenkins, and M. M. Merzenich (1995) Use-dependent alterations of movement representations in primary motor cortex of adult squirrel monkeys. *Journal of Neuroscience* (submitted)
Olds, J., J.F. Disterhoft, M. Segal, C.L. Kornblith and R. Hirsh (1972) Learning centers of rat brain mapped by measuring latencies of conditioned unit responses. *Journal of Neurophysiology* 35:202-219.
Paul, R.L., Merzenich, M.M., and H. Goodman (1972) Representation of slowly and rapidly adapting cutaneous mechanoreceptors of the hand in Brodmann's areas 3 and 1 of *Macaca mulatta*. *Brain Research* 36:229-249.
Penfield, W., and E. Boldrey (1937) Somatic motor and sensory representations in the cerebral cortex of man as studied by electrical stimulation. *Brain* 60:389-443.
Penfield, W., and T. Rasmussen (1950) *The Cerebral Cortex of Man: A Clinical Study of Localization of Function.* MacMillan, New York.
Phillips, C.G. (1967) Corticomotorneuronal organization. Projections from the arm area of the baboon's motor cortex. *Archives of Neurology* 17:188-195.
Phillips, C.G., T. P.S. Powell and M. Wiesendanger (1971). Projection from low-threshold muscle afferents of hand and forearm to area 3a of baboon's cortex. *Journal of Physiology* 217:419-446.
Pons, T.P., Garraghty, P.E., Cusick, C.G., and J.H. Kaas (1985) The somatotopic organization of area 2 in macaque monkeys. *Journal of Comparative Neurology* 241:445-466.
Pons, T.P., Wall, J.H., Garraghty, P.E., Cusick, C.G., and J. H. Kaas (1987) Consistent features of the representation of the hand in area 3b of macaque monkeys. *Somatosensory & Motor Research* 4:309-331.
Powell, T.P.S., and V.B. Mountcastle. (1968) Some aspects of the functional organization of the cortex of the postcentral gyrus of the monkey: A correlation of findings obtrained in a single unit analysis with cytoarchitecture. *Bulletin of Johns Hopkins Hospital* 105:301-334.
Pubols, B.H., and L. M. Pubols (1971) Somatotopic organization of spider monkey somatic sensory cerebral cortex. *Journal of Comparative Neurology* 141:63-76.
Recanzone, G.H., H.A. Dinse and M. M. Merzenich (1992a) Expansion of the cortical representation of a specific skin field in primary somatosensory cortex by intracortical microstimulation. *Cerebral Cortex* 2:181-196.
Recanzone, G.H., M.M. Merzenich, W. M. Jenkins, Grajski, K.A., and H. A. Dinse (1992b) Topographic reorganization of the hand representational zone in cortical area 3b paralleling improvements in frequency discrimination performance. *Journal of Neurophysiology* 67:1031-1056.
Recanzone, G.H., M.M. Merzenich and W.M. Jenkins (1992c) Frequency discrimination training engaging a restricted skin surface results in an emergence of a cutaneous response zone in cortical area 3a. *Journal of Neurophysiology* 67:1057-1070.
Recanzone G.H., M.M. Merzenich and C. E. Schreiner (1992d) Changes in the distributed temporal response properties of SI cortical neurons reflect improvements in performance on a temporally-based tactile discrimination task. *Journal of Neurophysiology* 67:1071-1091.
Recanzone, G. H., C. E. Schreiner and M. M. Merzenich (1993) Plasticity in the primary auditory cortex following discrimination training in adult owl monkeys. *Journal of Neuroscience* 13:87-104.
Robinson, C.J., H. Burton (1980) Somatotopographic organization in the second somatosensory area of *M. fascicularis*. *Journal of Comparative Neurology* 192:43-67.
Robinson, C.J., and H. Burton (1980) Organization of somatosensory receptive fields in cortical areas 6b,

retroinsular, postauditory and granular insula of *M. fascicularis*. *Journal of Comparative Neurology* 192:265-292.

Rosenzweig, M.R., Bennett, E.L., and M. C. Diamond (1978) *In:* G. Gottlieb , ed., *Studies on the Development of Behavior and the Nervous System. Vol. 4. Early Influences.* Academic Press:New York, pp. 289-327.

Sherrington, C.S. (1906) *The Integrative Action of the Nervous System.* Yale Univ. Press:New Haven.

Snow, P.J., R.J. Nudo, W. Rivers, W.M. Jenkins and M. M. Merzenich (1988) Somatotopically inappropriate projections from thalamocortical neurons to the S1cortex in the cat. *Somatosensory & Motor Research* 5:349-372.

Strick, P.L., and J. B. Preston (1982) Two representations of the hand in area 4 of a primate. I. Motor output organization. *Journal of Neurophysiology* 48:139-149

Sur, M., Nelson, R. J., and J. H. Kaas (1982) Representations of the body surface in cortical areas 3b and 1 of squirrel monkeys: Comparisons with other primates. *Journal of Comparative Neurology* 211:177-192.

Sur, M., Nelson, R.J., and J.H. Kaas (1980) The representation of the body surface in somatic koniocortex in the prosimian (*Galago senegalensis*) *Journal of Comparative Neurology* 180:381-402.

Taylor, J., ed. (1958) Selected Writings of John Hughlings Jackson. Basic Books:New York.

Wang, X., M.M. Merzenich, K. Sameshima and W.M. Jenkins (1995). Remodeling of Hand Representation in Adult Cortex Determined by Timing of Tactile Stimulation. *Nature* (in press).

Waters, R.S., Samlack D.D., Dykes, R.W., and P.A. McKinley (1990) Topographic organization of baboon primary motor cortex: Face, hand, forelimb, and shoulder representation. *Somatosensory & Motor Research* 7:485-514.

Weinberger, N.M., and D.M. Diamond (1988) Dynamic modulation of the auditory system by associative learning. *In:* G. M. Edelman, W. E. Gall and W. M. Cowan, eds., *Auditory Function: Neurobiological Bases of Hearing.* Wiley:New York, pp. 485.

Weinberger, N.M., J.H. Ashe, R. Metherate, T.M. McKenna, D.M. Diamond and J. Bakin (1990) Retuning auditory cortex by learning: A preliminary model of receptive field plasticity. *Concepts in Neuroscience* 1:91.

Welker W. I., Benjamin, R.M., Miles, R.C., and Woolsey, C.N. (1957) Motor effets of stimulation of cerebral cortex of squirrel monkey (*Saimiri sciureus*) *Journal of Neurophysiology* 20:347-364.

Welt, C. (1963) *Topographical Organization of the Somatic Sensory and Motor Areas of the Cerebral Cortex of The Gibbon (Hylobates) and Chimpanzee (Pan)* Ph.D. Thesis, University of Chicago.

Werner, G., and B.C. Whitsel (1971) The functional organization of the somatosensory cortex. *In:* A. Iggo, ed., *Handbook of Sensory Physiology.* New York: Springer, pp. 612-700.

Wise, S.P. (1993) Monkey motor cortex: Movements, muscles, motoneurons and metrics. *Trends in Neuroscience* 16:46-49.

Woolsey, C.N., (1958) Organization of somtic sensory and motor areas in the cerebral cordtex. *In:* H. F. Harlow, C. N. Woolsey, eds., *Biological and Biochemical Bases of Behavior.* U. Wisconsin Press:Madison, pp. 63-81.

Woolsey, C.N. (1951) Patterns of localization in sensory and motor areas of the cereral cortex. *In:* *Biology of Mental Health and Disease.* Hoeber: New York, pp. 193-206.

Woolsey, C.N., Marshall, W.H., and Bard, P. (1943) Representation of cutaneous tract sensibility in the cerebral cortex of the monkey as indicated by evoked potentials. *Bulletin of Johns Hopkins Hospital* 60:399-441.

Woody, C.D. (1986) Understanding the cellular basis of memory and learning. *Annual Review of Psychology* 37:433-493.

Woody, C.D., and J. Engel (1972) Changes in unit activity and thresholds to electrical microstimulation at coronal-pericruciate cortex of cat with classical conditioning of different facial movements. *Journal of Neurophysiology* 36:230.

Xerri, C., M. M. Merzenich, W. M. Jenkins and S. Santucci (1995) Representational plasticity in cortical area 3b contributing to tactual -motor skill acquisition in adult monkeys. *Journal of Neurophysiology* (submitted)

Xerri, C., Stern, J., and Merzenich, M.M. (1994) Alterations of the cortical representation of the rat ventrum induced by nursing behavior. *Journal of Neuroscience* 14:1710-1721.

Yang, T.T., Gallen, C.C., Schwartz, B.J., and F. E. Bloom (1993) Noninvasive somatosensory homunculus mapping in humans by using a large-array biomagnetometer. *Proceedings of the National Academy of Sciences of the United States of America* 90:3098-3102.

Zarzecki, P. and D. Herman (1983) Convergence of sensory inputs upon projection neurons of somatosensory cortex: vestibular, neck, head, and forelimb inputs. *Experimental Brain Research* 50:408-14.

Somesthesis and the Neurobiology of the Somatosensory Cortex
O. Franzén, R. Johansson and L. Terenius (eds.)

Somatosensory signals and sensorimotor transformations in reactive control of grasp

R.S. Johansson

Department of Physiology, Umeå University, S-901 87 Umeå, Sweden

Summary. While restraining an object subjected to unpredictable destabilizing load forces, subjects use sensory information related to load changes to automatically generate grip responses via a task-specific 'load-to-grip' sensorimotor transformation. Tactile receptors of the skin in contact with the object are the only type of mechanoreceptors capable of triggering and scaling appropriate grip responses, but muscle and joint afferents may provide information related to the reactive forces produced by the subject. The employed sensorimotor transformation appears to exploit short term predictions of the rate of load changes. Furthermore, the friction in the object-digit interface modifies the gain of the entire load-to-grip force transformation based on frictional information obtained during initial skin-object contact. The initiation, as well as the regulation of the motor responses, seem to depend on supraspinal control mechanisms. Evidence is provided that the human motor cortex is involved in the regulation of already initiated grip responses, but that their initiation most likely depends on brain stem networks. Afferent inflow may also contribute in the regulation of already initiated responses via fast spinal networks.

Introduction

Most manipulative tasks require us to stabilize an object within our grasp as we move or restrain the object or use it as a tool. To prevent slips and accidental loss of the object, adequately large forces normal to the grip surfaces ('grip' forces) are applied in relation to destabilizing forces tangential to the grip surfaces ('load' forces). At the same time, exceedingly large grip forces are avoided because they cause unnecessary fatigue and may crush fragile objects or injure the hand. Hence, the term 'grasp stability' entails preventing accidental slips as well as exceedingly large fingertip forces.

Based on the properties of the object, manipulative tasks may be divided into two principal classes: One involves the handling of mechanically predictable, 'passive', objects whose relevant physical properties are stable over time (i.e., objects subjected only to inertial and gravitational forces). Because the motor output is self-paced and the objects are predictable, the spatio-temporal control of the muscle commands can rely largely on anticipatory mechanisms based on memory information from previous manipulation. The second class of tasks involves the handling of objects which are subjected to unpredictable loading forces, such as when one holds a dog's leash, restrains a child by holding her arm or when operating power tools. The role of anticipatory control strategies is obviously of limited use in such tasks: Somatosensory information related to the load must be used to control the forces of the grasp in order not to loose the object. In the present account I will discuss sensory afferent signals and the sensorimotor transformation underlying the reactive control of the precision grip when humans restrain 'active' objects from moving. For further information on the sensorimotor control during handling of 'passive' objects, the reader may refer to overviews by Johansson (1991), Johansson and Westling (1991) and Johansson and Edin (1993).

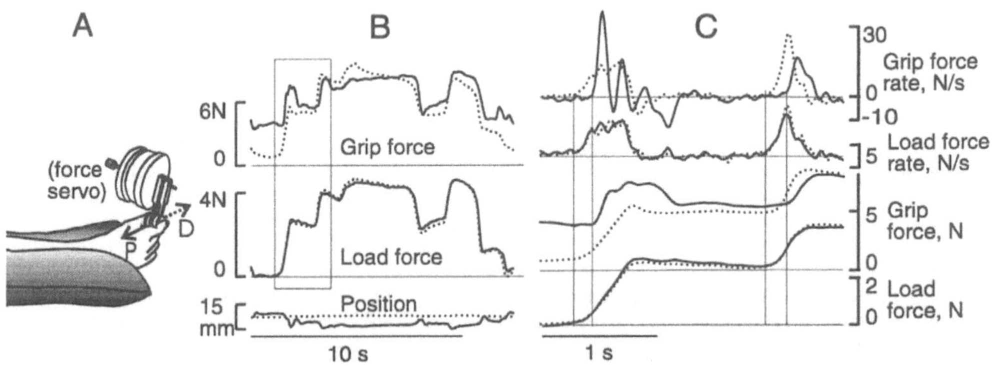

Figure 1. Comparison of the grip forces employed during load changes applied by the subject to a passive manipulandum with identical load profiles imposed on the grasp through the same manipulandum that was loaded by a servo controlled force motor ('active' object). **A**. Schematic drawing of the apparatus. The manipulandum consists of two parallel grip surfaces (30 mm diameter; spaced 25 mm apart). The servo-controlled motor can deliver load forces in two opposite directions and the arrows indicate the directions of loading (D—distal direction, P—proximal direction). The grip and load forces, perpendicular and tangential to the grip surfaces, respectively, were recorded, as well as the position of the object in the loading directions. **B** — **C** Load and grip forces during manipulation of a 'passive' object (dashed curves); the subject only pulls manipulandum that is immovable to produce various periods of force changes mixed with hold periods. The resulting load pattern is recorded and later played back through the force servo while the subject is asked to restrain the 'active' manipulandum from moving (solid curves). In both tasks the blindfolded subject grasped the object by the tips of the thumb and index finger. The rectangle in **B** marks the period represented at a faster time base in **C** together with the grip and load force rate profiles. Pairs of vertical lines in **C** indicate onset of grip force responses during the passive and active conditions. Single subject. Adapted from Johansson *et al.* 1992c.

General features of the employed sensorimotor transformation

Figures 1 compares the grip forces employed by a subject while generating load forces on a 'passive' manipulandum (dashed curves) with those employed while identical load force profiles were imposed on the grasp through the same manipulandum (active object; solid curves).

(Johansson et al., 1992c). Note that in both tasks the load and grip forces increase and decrease in parallel, ensuring that adequately strong grip forces are used to prevent frictional slippage at any load force. This parallel coordination, supporting grasp stability, emerges automatically and proceeds even if no instructions to respond with grip changes are given. However, there are differences in the behavior: (i) The change in the grip and load forces are nearly synchronous when the manipulandum is 'passive' and the grip force changes are programmed in parallel with (rather than as a reaction to) the self-induced load changes. In contrast, with the 'active' manipulandum the grip force changes lag behind those of the load force because they are generated reactively (Fig. 1 C). Subjects also automatically produce reaction forces that prevents excessive displacement of the manipulandum in the direction of the loading. (ii) Higher grip forces are generally used with the 'active' manipulandum rather than with the "passive", especially at low load forces. This strategy prevents frictional slips during the delay between the onset of the increases in load forces and the onset of the grip force response. (iii) Finally, with the 'passive' manipulandum, the force rate profiles are generally unimodal, indicating that the target forces are determined in advance (see 'anticipatory parameter control' in Johansson and Edin 1993). In contrast, with the active object, the grip force responses to extended periods of load force increases consist of an initial strong grip force rate pulse followed by a period of slower, often stepwise, grip force increases (Fig. 1 C).

We have further characterized the somatosensory control of grasp stability while handling 'active' objects, using an apparatus similar to that schematically illustrated in Fig. 1A. Trapezoidal load-force profiles with unpredictable onsets, amplitudes and rates of loading and unloading were delivered tangential to the grip surfaces. Figure 2 A shows basic features of the observed grasp response pattern. The sensory signals from the loading initially trigger a prompt grip force increase lasting for 0.2—0.3 s. This force increase has been termed the 'catch-up' response and is initiated after a short delay (≥70 ms) that varies with the load force rate (see Fig. 3 B). The catch-up response (C in Fig. 2 A) compensates for the response delay in the sense that the grip force adequately matches the load demands (i.e. it 'catches up') and reestablishes an adequate safety margin against frictional slips by the end of this response, i.e., the grip:load force ratio slightly exceeds the required minimum determined by the coefficient ('slip ratio' in Fig. 2). The grip force increases in parallel with further changes in the load force, and maintains an approximately constant safety margin. This 'tracking' response (T in Fig. 2 A) continues until the load ceases to increase; accordingly, at the end of the load plateau phase the grip force is approximately proportional to the load amplitude (Figs. 2 A and 3 B).

Interestingly, the employed sensorimotor transformation is tailored to the diverse consequences of frictional slippage in different loading directions (Häger-Ross et al., 1995). For instance, the grip response onset latencies are some 10 ms shorter for loads in the direction away from the palm and in the direction of gravity than for loads in the opposite directions. These directional effects reflect different central delays for triggering grip responses, because there are no directional dependent difference in the latency of the afferent signals triggering the grip responses (see Häger-Ross et al., 1995 for details). We interpret the shorter grip response latency for loads in certain directions to reflect a default response preparation by the CNS for rapidly issuing a response to loads in these directions (cf. Favilla et al., 1990). The response latency for loads in other directions would be longer, since these loads would require implementation of a different grip force scaling to the frictional conditions (see below) because the finger tips show frictional anisotropies (Häger-Ross et al., 1995), and the implementation of a different muscle activation pattern to produce the adequate reactive forces in the axis of loading. Such influences by intrinsic task variables indicate that the input-output relationships of sensorimotor transformations employed by manipulative tasks cannot be captured by traditional concepts of stereotyped reflexes; rather, they depend on the task and its intrinsic properties.

Parametric adjustments of grip responses to rate and amplitude of load force and to frictional conditions in the digit-object interface

Grasp stability during this task requires adapting the force output to the rate and amplitude of the load force increase. These requirements are met automatically, rapidly and parametrically (Figs. 2 B and 3 B). Sensory information reflecting the initial load force rate during the grip response latent period specifies the rate of the grip-force response in a feed-forward predictive manner (Johansson et al., 1992b). From its onset, the amplitude of the catch-up response varies linearly with the load force rate, but the duration of the response stays constant (Fig. 2 C). Hence, the catch-up response is a centrally programmed default response, whose size is controlled by sensory information according to a 'pulse height control policy'. This type of control policy may simplify a rapid scaling of the force output and has also been described for other types of isometric motor tasks (e.g., Freund et al., 1978; Ghez and Vicario 1978; Gordon and Ghez 1987).

Although the tracking response may rely on sensory data sampled more or less continuously, grip tracking at an insufficient rate is corrected by actions best described as 'additional catch-up responses' to offset accumulated tracking error (Cole and Johansson 1993). These are probably driven by tactile afferent responses to localized slips in the object-digit interface, since they occur when the grip:load force ratio falls too quickly (Johansson and Westling 1987). Accordingly, the tracking and the catch-up responses may share a common control mechanism; there is also a critical drop in the grip:load force ratio during the grip response latent period (Fig. 2 A). Furthermore, the idea that the grip tracking exploits short term prediction mechanisms similar to those of the catch-up responses is also suggested by the fact that it goes on for some 0.2 s after ramp termination, i.e., a period similar to the duration of the catch-up response.

The friction between the skin and the contact surface modifies the 'gain' of the entire load-to-grip force sensorimotor transformation and this provides an adequate safety margin to prevent frictional slips during different frictional conditions (Figs. 2 D and E) (Cole and Johansson 1993). The background grip force and the grip responses to the load increase (catch-up and tracking responses) are all scaled in approximate proportion to the inverse friction ('slip ratio' in Fig. 2 D) resulting in a small but adequate safety margin against frictional slip regardless of surface material (cf. Johansson and Westling 1984). Hence, information about friction apparently adjusts a central scaling factor that is applied uniformly, or 'globally', to motor commands

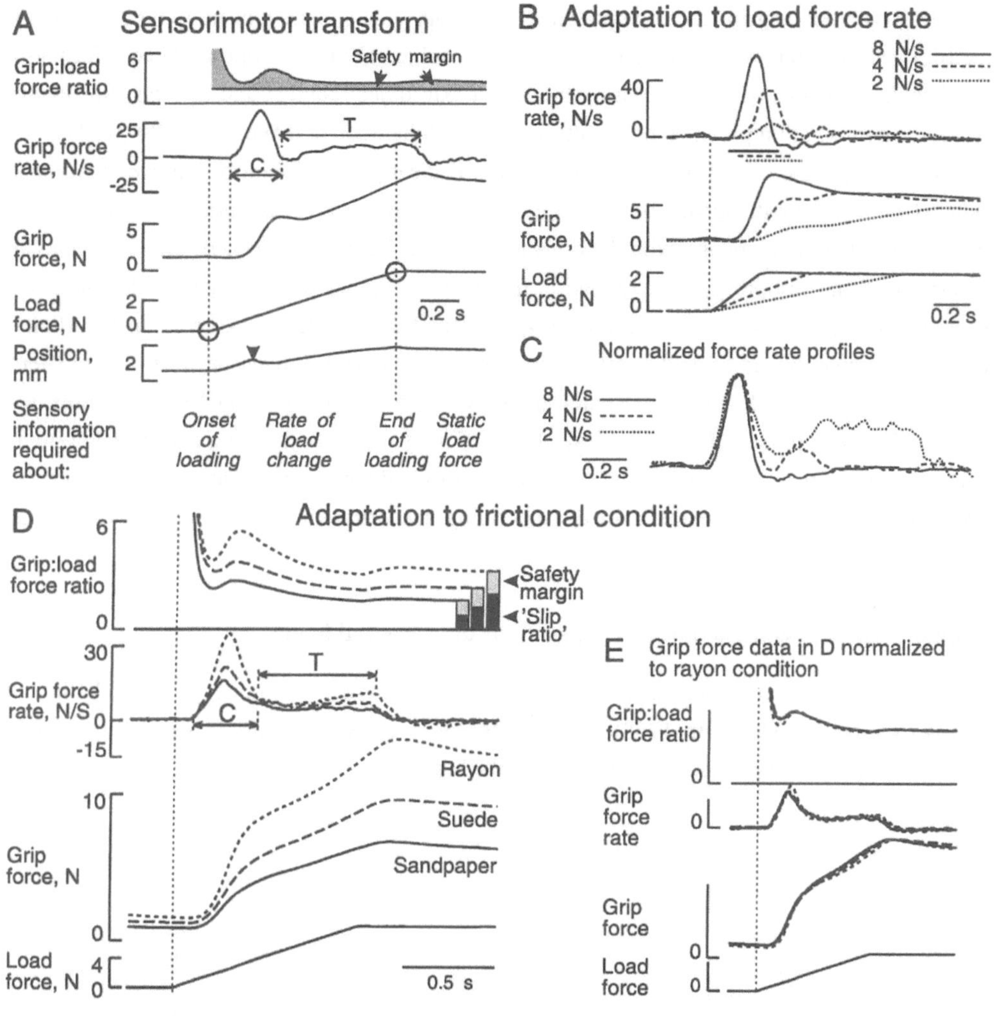

Figure 2. A Basic features of the employed 'load-to-grip' force sensorimotor transformation. The precision grip was loaded by distal pull to 4 N amplitude at 4 N/s. Top traces show the grip:load force ratio. Horizontal solid line indicates the minimum ratio to prevent slips ('slip ratio') and the hatched zone the safety margin. 'C' and 'T' indicate 'catch-up' and 'tracking' responses. Vertical lines indicate start and end of the of loading phase, and the onsets of the reactive grip force changes. Adapted from Johansson *et al.*, 1992c. **B — E** Sensory based adaptations of grip force responses. **B** Loading at three different force rates (2 N/s, 4 N/s and 8 N/s) to a constant force amplitude (2 N). Horizontal bars indicate the period of the catch-up responses. **C** Grip force rate profiles in **B** normalized in amplitude and horizontally aligned to peak rate. Note the similar catch-up response profiles and the brief and lengthy tracking responses at 4 N/s and 2 N/s loading, respectively. **D — E** Scaling by the frictional condition of the overall sensory-motor transform. Data from trials with sandpaper (least slippery), suede, and rayon (most slippery) surfaces are indicated by different lines. **D** The solid portions of the bars imposed upon the grip:load force ratio signals indicate the force ratio at which the finger slipped; the stippled portions represent the 'safety margin'. **E** Data shown in **D** for the sandpaper and suede conditions have been normalized in amplitude to the rayon condition using the peak grip forces. Note that the traces nearly superimpose indicating that the various response components maintained proportional relationships under the different frictional conditions. Adapted from Johansson *et al.*, 1992c (**A**), Johansson *et al.*, 1992b (**B** and **C**) and from Cole and Johansson, 1993 (**D** and **E**).

controlling the various grip force components. This 'gain' adjustment appears to be controlled in a feed-forward manner: when lifting 'passive' objects, subjects extract friction related information from signals in cutaneous sensors during the initial skin-object contact and use this information to set force coordination parameters for future actions (Johansson and Westling 1984, 1987).

Interestingly, the overall grip response pattern is clearly reminiscent of the eye movements produced during smooth pursuit of a moving target when the head remains stationary (Fuchs 1967). The catch-up response would correspond to the 'catch-up' saccade, and the 'tracking response' to visual smooth pursuit. Likewise, in a manner similar to the way saccades punctuate visual smooth pursuit to offset accumulated gaze retinal errors, grip tracking of an insufficient rate is interrupted by additional catch-up responses (Cole and Johansson 1993). Short-term prediction may play a role not only in catch-up responses but also in grip tracking responses, just as in visual smooth pursuit (Lisberger et al., 1987). Finally, for both the hand and visual systems, changes in the relationship between the sensory stimulus and the required motor response are met parametrically by changes in the gain of underlying sensorimotor transformations (Robinson, 1981; Cole and Johansson 1993).

Peripheral afferent control of the grip reactions to tangential loads

Using tungsten micro-electrodes inserted percutaneously into the median nerve of awake human subjects (Vallbo and Hagbarth 1968), we have recently recorded signals in various potentially useful mechanoreceptive sensory channels from the hand.

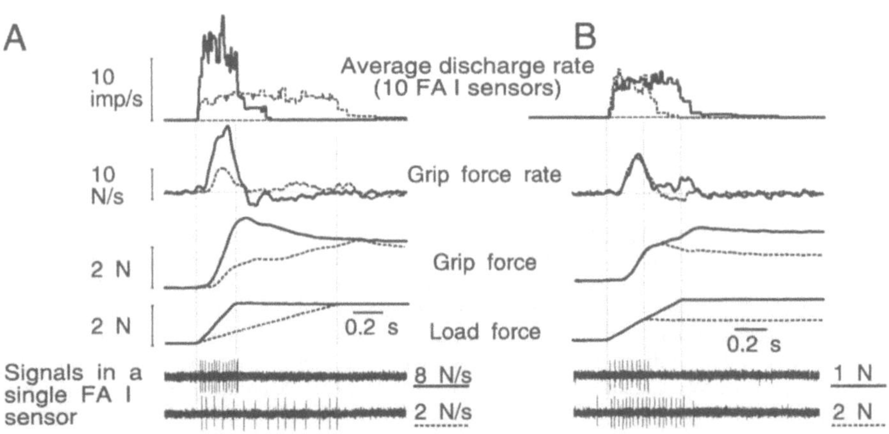

Figure 3. A Average instantaneous frequency of the impulse response in 10 FA I sensors to 2 N loads delivered to the receptor-bearing digit at 2 N/s (dashed lines) and 8 N/s (solid lines). **B** Average instantaneous frequency response of the same sensors to 1 N (dashed lines) and 2 N (solid lines) loads delivered to the receptor-bearing digit at 4 N/s. **A** and **B** Below the load force signals are shown example of impulse responses of a single FA I sensor.

Tactile afferents. Signals from tactile afferents with receptive fields in the glabrous skin areas in contact with the manipulandum, provide rapid and accurate information about load force changes and could therefore be responsible for both the initiation and the scaling of the grip responses (cf. list of 'required sensory information' in Fig. 2 A) (Macefield et al., 1995). The rapidly adapting FA I afferents respond during the loading (and unloading) phase and they faithfully encode the rate of load force change (Fig. 5). In contrast, these afferents are surprisingly insensitive to the subject's grip force responses. The FA II (Pacinian) afferents are largely insensitive to the type load trials delivered: only with rapid 'step-loads' do some afferents respond. Both classes of slowly-adapting (SA) afferents are sensitive to the load and grip forces and their changes. The discharges of the SA I afferents appeared to be relatively more influenced by the subject's grip force response than those of the SA II afferents, which are more influenced by the load-force stimulus. During the hold phase, slowly adapting afferents exhibit tonic discharges that probably provide information about the load force during the hold phase; at the level of individual afferents the tonic discharge is differentially influenced by the grip force and by the load force. In addition, information about the maintained load may have been indirectly provided by the FA I afferents, which by virtue of their silence during the hold phase would have indicated that changes in tangential forces had not occurred. The direction in which the tangential load force is applied to the skin influences the firing of most afferents and, in particular, the SA II afferents (Johansson and Westling 1987); also see (Goodwin and Morley 1987; Srinivasan et al., 1990; Edin et al., 1995). However, as a population the tactile afferents would efficiently encode load changes in any direction.

Individual afferents within each class (except for the FA II afferents) responds to the loading ramp before the onset of the subject's grip response, and may thus be responsible for initiating the automatic increase in grip force. However, nearly half of the FA I afferents recruited by the load trials respond to the loading phase early enough to trigger the subject's grip force response (cf. Fig. 3), whereas only about one fifth of the SA I's and SA II's do so. These observations, together with the high density of FA I receptors in the digits (Johansson and Vallbo 1979) and their encoding of the load force rate, might place the FA I afferents in a unique position to convey the information required to initiate and scale the reactive grip force responses to the imposed load forces.

Muscle and joint afferents. In contrast to the tactile afferents, the afferents from the intrinsic and extrinsic hand muscles do not respond to load increases early enough to allow them to contribute to the initiation and the initial scaling of the grip responses during normal digital sensibility; these respond during load trials, essentially after the onset of the subject's grip response to the imposed load (Macefield and Johansson 1995). Likewise, the ensemble pattern from the joint afferents revealed no significant encoding of changes in finger-tip forces before the onset of the increase in grip force. Furthermore; joint receptors are much more sparse in the digits than cutaneous mechanoreceptors (Johansson and Vallbo 1979; Burke et al., 1988; Edin 1990), and are certainly less sensitive to mechanical changes at the object-digit interface.

The fact that mechanoreceptors in the flexors of the digits and in the interphalangeal joints cannot be allotted a significant role in triggering the automatic changes in grip force agrees with our finding that subjects with experimentally-induced, impaired digital sensibility often do not respond with grip force changes to object loading (Johansson et al., 1992a). If present, the grip force responses are weak and markedly delayed compared to normal sensibility, and the normal pattern of catch-up and tracking responses is disrupted, as is the scaling of the responses to the load force rates and amplitudes. Higher grip forces than normal are used during the periods between load trials as a necessary strategy to prevent slips during the prolonged grip response latencies. Furthermore, during digital anesthesia much of the automatic character of the control is lost and an enhanced mental attention is required to complete the restrain task. Interestingly, grip control may improve somewhat during postural conditions which afford large wrist and arm movements caused by the imposed loads (Häger-Ross and Johansson 1995). Thus, signals from different kinds of non-digital mechanoreceptors could have contributed to some grip control during impaired digital sensibility, e.g., muscle and joint receptors (Macefield and Johansson 1995), as well as skin-stretch sensitive cutaneous mechanoreceptors innervating the dorsum of the hand (Edin and Abbs 1991). Presently we cannot determine their relative importance. Nevertheless, the increased grip response latencies during digital anesthesia would be partly explained by the higher thresholds of receptors for load changes compared to those of tactile afferents from the digits.

Observations on central neural mechanisms

In general terms, the control of grasping and manipulation appears to rely on distributed processes in the CNS, engaging most areas known to be involved in sensorimotor control. These parallel processes may, to some extent, be responsible for specific control functions such as selection of task-related sensorimotor transformations, initiation of motor actions, and specification of various motor command parameters based on memory systems.

Recently we have used transcranial magnetic brain stimulation (TMS) directed at the hand area of the human motor cortex to assess the influences by the motor cortex over various hand and arm muscles during our restrain task (Johansson et al., 1994). In this experiment, subjects gripped an object that was subjected to an unexpected brisk pulling load of 1N, delivered at a high rate (ca. 80 N/s) (Fig. 4). This 'step-load' increase triggered an increase in EMG activity in the gripping muscles at a minimum onset latency around 51 ms (see 'background' EMG signals referring to control trials in Fig. 4) followed shortly afterwards (10-20 ms) by an increase in grip force. Because the applied load generates a distinct cutaneous input ('Tactile afferent signal' in

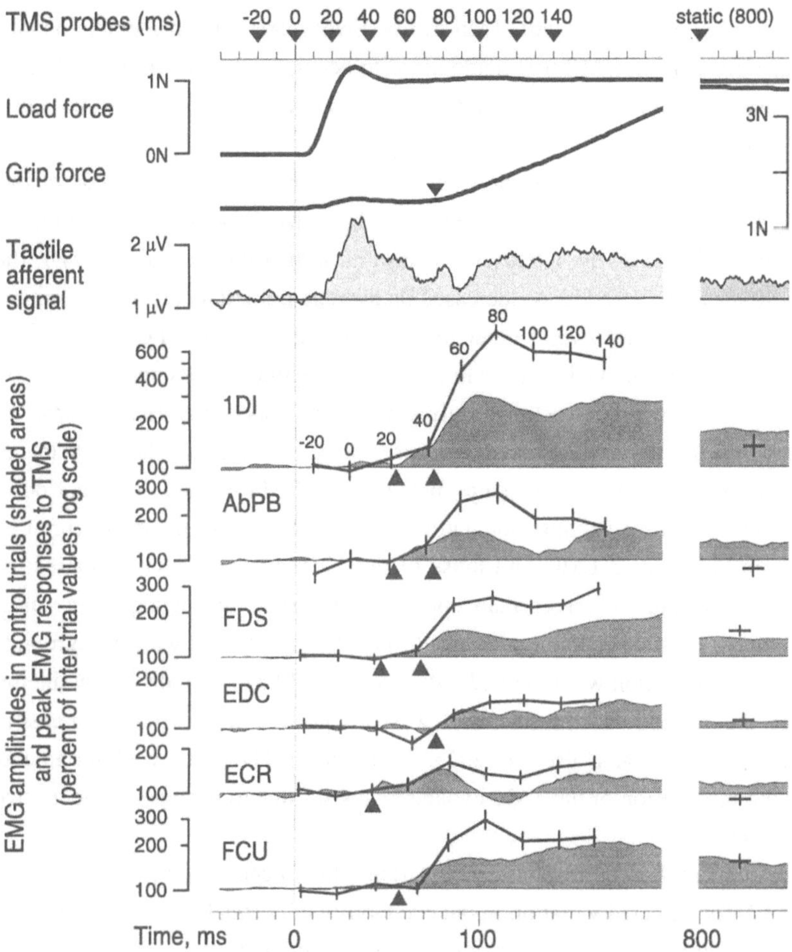

Figure 4. Grip force and EMG changes produced by an unexpected pulling force of 1N (with a force ramp of ca. 80 N/s) generated away from the subject. In control trials 'background' EMG was recorded (r.m.s. processed) from the first dorsal interosseous (1DI), abductor pollicis brevis (AbPB), flexor digitorum superficialis (FDS), extensor digitorum communis (EDC), extensor carpi radiales (ECR) and flexor carpi ulnaris (FCU) muscles. The 'tactile afferent signal' refers to rectified multi-unit nerve activity recorded from cutaneous fascicles in the median nerve at the midlevel of the upper arm. In separate trials, TMS was delivered at different times before and after the command to generate the force pulse (see top line) referenced to the computer-generated command to the force servo, which actually preceded the start of the load increase by ca. 5 ms. The arrow on the grip force trace indicates the onset of the reflex grip response during trials in which no TMS was given. Onset of reflex EMG changes are indicated by arrows; 1DI, AbPB and FDS showed a small early response, followed by a second, larger response (second arrow). The figure also shows the variation in the size of the peak amplitude of the TMS-evoked EMG responses in the different muscles (line with vertical bars, which represents the mean +/- SEM). The amplitude of this response and 'background' EMG as a percentage of the amplitude of the intertrial values (note the logarithmic scale). The peaks are represented on the time scale by the time of their occurrence rather than by the point in time of the corresponding TMS probe, which occurred 21-28 ms earlier, depending on the muscle; the numbers above the vertical bars of the curve referring to 1DI indicate the time of delivery of these TMS probes. Adapted from (Johansson et al., 1994).

Fig. 4), the timing of its interaction with TMS can be investigated in considerable detail. In our experiments, the timing of the TMS was randomized to occur at intervals from ca. 20 ms before the onset of the unexpected load to 140 ms after it, or during the steady intertrial periods of grip ('TMS probes' in Fig. 4 A).

Figure 4 shows, for different hand muscles, the variations in peak amplitude of the EMG responses evoked by the TMS (solid line with error bars) by depicting their amplitudes as a percentage of the amplitude of the intertrial response on a logarithmic scale. In 1DI, the TMS response to the 80 ms probe (see Fig. 4) increased on average more than eight times compared to the intertrial responses, and in several other muscles (AbPB, FDS and FCU) the change was more than two times. These marked changes in response amplitude should not be taken to simply reflect the level of motoneuronal activity during the grip response. First, in most muscles, changes in amplitude of the TMS-evoked responses were substantially larger than could be explained by a proportional relationship between a TMS-evoked constant-descending corticospinal volley and a time varying modulation in the excitability of the motoneuron pool (cf. relative changes in background EMG and those of the TMS-evoked EMG responses; note the logarithmic ordinates), and far larger than could be explained by an algebraic summation. Second, modulation of response amplitude did not necessarily co-vary with background EMG. Third, the optimum modulation was obtained with intensities (0.8T) of TMS below the threshold (T) for exciting the relaxed 1DI muscle, and was not present with stronger intensities (see Johansson et al., 1994 for details; also see Lemon et al., 1995). Therefore, we infer an additional contribution to the modulation of the TMS-evoked responses due to changes in the corticospinal volley elicited by TMS.

Indeed, we suggest that the response modulation which occurred beyond that observed in the background EMGs reflected variation in the excitability of cortical neurons during the course of the loading trial, and that it was these changes in excitability that were probed by low intensity TMS. Thus, a particularly large corticospinal volley would be generated if corticospinal cells were themselves active at the time of delivery of TMS, thereby increasing the probability of their discharge by magnetic stimulation (see Flament et al., 1992). Hence, the results are consistent with a central interaction between the effects of tactile afferent inputs and those of TMS. A likely site of interaction is the primary motor cortex. Interestingly, in monkeys performing a similar restraint task, an equal proportion of neurons in the motor cortex responsive to the perturbations had proprioceptive and cutaneous receptive fields, although those with cutaneous fields showed stronger excitation (Picard and Smith 1992).

Interestingly, the very earliest increase in the TMS-evoked EMG response over that of the background EMG was observed for the 60 ms TMS probe (see Fig. 4). This is more than 20 ms after the predicted first cortical arrival of tactile signals based on somatosensory-evoked potential measurements (Macefield and Johansson 1994). Thus, the cortex would have received, and probably somehow processed, most of the initial components of the afferent volley before emitting a descending command. These findings, together with various observations on response latencies etc. (Johansson *et al.*, 1992b), suggest a subcortical but supraspinal origin of response initiation. However, motor cortical mechanisms would powerfully contribute to the regulation of the already initiated response. Similar conclusions have been drawn from recordings of electroencephalographic activity over the scalp overlying the motor area of the hand during the same task (Macefield and Johansson 1994). Furthermore, there is evidence that fast spinal regulatory mechanisms also contribute to the regulation of the grip force response after initiation (Johansson et al., 1992b). Thus, in parallel with the cortical mechanisms, spinal networks provide response regulation after being adequately preset to the task by supraspinal mechanisms. Indeed, there is evidence to show that in man, that different sensory-motor processes may account for initiation of motor actions versus specification of response parameters (Cordo and Flanders 1989; Johansson et al., 1992b) and that these mechanisms may be organized in parallel (Ghez et al., 1990).

Acknowledgment

This work was supported by the Swedish Medical Research Council (project 08667) and the Office of Naval Research, Arlington, VA (grant no N00014-92-J-1919).

References

Burke, D., Gandevia, S.C. and Macefield, G. (1988) Responses to passive movement of receptors in joint, skin and muscle of the human hand. *Journal of Physiology* 402:347-361.

Cole, K.J. and Johansson, R.S. (1993) Friction in the digit-object interface scales the sensorimotor transformation for grip responses to pulling loads. *Experimental Brain Research* 95:523-532.

Cordo, P.J. and Flanders, M. (1989) Sensory control of target acquisition. *Trends in Neurosciences* 12:110-116.

Edin, B.B. (1990) Finger joint movement sensitivity of non-cutaneous mechanoreceptor afferents in the human radial nerve. *Experimental Brain Research* 82:417-422.

Edin, B.B. and Abbs, J.H. (1991) Finger movement responses of cutaneous mechanoreceptors in the dorsal skin of the human hand. *Journal of Neurophysiology* 65:657-670.

Edin, B.B., Essick, G.K., Trulsson, M. and Olsson, K.A. (1995) Receptor Encoding of Moving Tactile Stimuli in Humans .1. Temporal Pattern of Discharge of Individual Low-Threshold Mechanoreceptors. *Journal of Neuroscience* 15:830-847.

Favilla, M., Gordon, J., Hening, W. and Ghez, C. (1990) Trajectory control in targeted force impulses. VII. Independent setting of amplitude and direction in response preparation. *Experimental Brain Research* 79:530-538.

Flament, D., Hall, E.J. and Lemon, R.N. (1992) The development of cortico-motoneuronal projections investigated using magnetic brain stimulation in the infant macaque. *Journal of Physiology* 447:755-768.

Freund, H.-J., Büdingen, H. and J. (1978) The relationship between speed and amplitude of the fastest voluntary contractions of human arm muscles. *Experimental Brain Research* 31:1-12.

Fuchs, A. (1967) Saccadic and smooth pursuit eye movements in the monkey. *Journal of Physiology* 191:609-631.

Ghez, C., Hening, W. and Favilla, M. Parallel interacting channels in the initiation and specification of motor response features. *In:* M. Jeannerod (ed.): *Attention and Performance XIII.* Lawrence Erlbaum associates,, Hillsdale, NJ, pp 265-293.

Ghez, C. and Vicario, D. (1978) The control of rapid limb movement in the cat. II. Scaling of isometric force adjustments. *Experimental Brain Research* 33:191-202.

Goodwin, A.W. and Morley, J.W. (1987) Sinusoidal movement of a grating across the monkey's fingerpad: Effect of contact angle and force of the grating on afferent fiber responses. *Journal of Neuroscience* 7:2192-2202.

Gordon, J. and Ghez, C. (1987) Trajectory control in targeted force impulses. II. Pulse height control. *Experimental Brain Research* 67:241-252.

Häger-Ross, C., Cole, K.J. and Johansson, R.S. (1995) Grip force responses to unanticipated object loading: Load direction reveals body- and gravity-referenced intrinsic task variables. *Experimental Brain Research* (In press).

Häger-Ross, C. and Johansson, R.S. (1995) Non-digital afferent input in reactive control of fingertip forces during precision grip. *Experimental Brain Research* (In press)..

Johansson, R.S. How is grasping modified by somatosensory input? *In:* D.R. Humphrey and H.-J. Freund (eds.): *Motor Control: Concepts and Issues. Dahlem Konferenzen.* John Wiley & Sons Ltd, Chichester, pp 331-355.

Johansson, R.S., Lemon, R.N. and Westling, G. (1994) Time varying enhancement of human cortical excitability mediated by cutaneous inputs during precision grip. *Journal of Physiology* 481:761-775.

Johansson, R.S. and Edin, B.B. (1993) Predictive Feedforward Sensory Control During Grasping and Manipulation in Man. *Biomedical Research* 14:95-106.

Johansson, R.S., Häger, C. and Bäckström, L. (1992) Somatosensory control of precision grip during unpredictable pulling loads. III. Impairments during digital anesthesia. *Experimental Brain Research* 89:204-213.

Johansson, R.S., Häger, C. and Riso, R. (1992) Somatosensory control of precision grip during unpredictable pulling loads. II. Changes in load force rate. *Experimental Brain Research* 89:192-203.

Johansson, R.S., Riso, R., Häger, C. and Bäckström, L. (1992) Somatosensory control of precision grip during unpredictable pulling loads. I. Changes in load force amplitude. *Experimental Brain Research* 89:181-191.

Johansson, R.S. and Vallbo, A.B. (1979) Tactile sensibility in the human hand: relative and absolute densities of four types of mechanoreceptive units in glabrous skin. *Journal of Physiology* 286:283-300.

Johansson, R.S. and Westling, G. (1984) Roles of glabrous skin receptors and sensorimotor memory in automatic control of precision grip when lifting rougher or more slippery objects. *Experimental Brain Research* 56:550-564.

Johansson, R.S. and Westling, G. (1987) Signals in tactile afferents from the fingers eliciting adaptive motor responses during precision grip. *Experimental Brain Research* 66:141-154.

Johansson, R.S. and Westling, G. Afferent signals during manipulative tasks in man. *In:* O. Franzen and J. Westman (eds.): *Somatosensory Mechanisms.* Macmillan Press, London, pp 25-48.

Lemon, R.N., Johansson, R.S. and Westling, G. (1995) Corticospinal control during reach, grasp and precision lift in man. *Journal of Neuroscience* 15: 6145-6156.

Lisberger, S.G., Morris, E.J. and Tychsen, L. (1987) Visual motor processing and sensory-motor integration for smooth pursuit eye movements. *Annual Review of Neuroscience* 10:97-129.

Macefield, V.G., Häger-Ross, C. and Johansson, R.S. (1995) Control of grip force during restraint of an object held between finger and thumb: responses of cutaneous afferents from the digits. *Experimental Brain Research* (In press).

Macefield, V.G. and Johansson, R.S. (1994) Electrical Signs of Cortical Involvement in the Automatic-Control of Grip Force. *Neuroreport* 5:2229-2232.

Macefield, V.G. and Johansson, R.S. (1995) Control of grip force during restraint of an object held between finger and thumb: responses of muscle and joint afferents from the digits. *Experimental Brain Research* (In press).

Picard, N. and Smith, A.M. (1992) Primary motor cortical responses to perturbations of prehension in the monkey. *Journal of Neurophysiology* 68:1882-1894.

Robinson, D.A. The control of eye movements. *In:* V.B. Brooks (ed.): *Handbook of Physiology: The Nervous System.* Am. Physiology Soc., Bethesda, MD, pp 1275--1320.

Srinivasan, M.A., Whitehouse, J.M. and LaMotte, R.H. (1990) Tactile detection of slip: surface microgeometry and peripheral neural codes. *Journal of Neurophysiology* 63:1323-1332.

Vallbo, Å.B. and Hagbarth, K.E. (1968) Activity from skin mechanoreceptors recorded percutaneously in awake human subjects. *Experimental Neurology* 21:270-289.

Somesthesis and the Neurobiology of the Somatosensory Cortex
O. Franzén, R. Johansson and L. Terenius (eds.)

Strain-sensitive mechanoreceptors in the human skin provide kinaesthetic information

B.B. Edin

Department of Physiology, Umeå University, S-901 87 Umeå, Sweden

Summary. Sensory sources that are believed to be particularly important for sensorimotor control and kinaesthesia include vision, vestibular organs, afferents from muscles and joints, and skin mechanoreceptors. The functional relationship between these sources is only partially known. That vision effectively suppress illusions from vibratory stimulations of limb muscles indicate, for instance, an hierarchical organisation of the sensory systems. In the past it has generally been supposed that among the somatosensory systems, afferents from muscle spindles play a particularly important role in kinaesthesia. Recent studies challenge this view: skin mechanoreceptors in the hairy skin have been found to be exquisitely sensitive to movements in nearby joints, and display static position sensitivities at least on par with those reported for muscle spindle afferents. Moreover, at least under some experimental conditions, information from skin mechanoreceptors have precedence over that provided by muscle spindle afferents with regard to kinaesthesia and sensorimotor control.

Movement associated skin deformations

Every human movement is associated with skin deformation changes. These skin changes consist of particular patterns of increased or decreased strains in skin areas influenced directly or indirectly by the activated muscles. Whereas many facial muscles attach directly in the skin, limb muscles primarily cause skin deformations by their action on one or several joints but also by bulging and dimpling of the skin. This paper is primarily concerned with limb movements or specifically finger movements but most of the reasoning would also apply to the face.

Movement associated skin deformations can be assessed qualitatively by eye. During finger movements, for instance, the skin on the back of the hand clearly show different strain patterns with different postures at the metacarpophalangeal (MCP) joint of the fingers. These patterns are easily modified by changing the posture at the thumb or the wrist. Moreover, again from visual inspection, each joint configuration seems to be associated with a particular skin strain pattern and, conversely, each pattern of skin deformations seems to be related to a certain joint configuration. Accordingly, a system that is able to (i) encode movement associated strain patterns and (ii) map the strain patterns to joint configuration would be able to serve as a proprioceptive system, i.e., a sensory or sensorimotor system that process information related to joint configurations.

Responses in hairy skin mechanoreceptors to hand and finger movements

The starting point for a series of microneurographic studies (Edin, 1990, 1992; Edin and Abbs, 1991), was some remarkable observations made by Abbs and co-workers on complex lip and finger movements (Abbs and Gracco, 1984; Cole, Gracco and Abbs, 1984; Abbs and Cole, 1987). Both of these motor tasks involve the co-operation of two moveable parts and are successful only if the movement of the involved parts are spatially and temporally well co-ordinated. For instance, when humans produce the 'p' sound, they move the upper and lower lips together to close the mouth, quickly increase the intraoral pressure and then again open the mouth. Similarly, to produce a specific pinch force between the index finger and the thumb, subjects must first bring the pulps of these digits together and this again require spatial and temporal co-ordination. Importantly, with both of these motor tasks powerful so-called 'non-autogenic' responses to perturbations were observed implicating feed-forward control of the moving parts involved in the task. When the *lower* lip, for instance, was mechanically depressed just prior to the production of the 'p' sound, significant compensatory force increases were observed in the *upper* lip (Abbs and Gracco, 1984). Similarly, perturbations to the thumb prior to the moment in time when the index finger and thumb would make contact, caused powerful phase-dependent responses in the index finger (Cole, Gracco and Abbs, 1984).

The term 'non-autogenic' may seem to implicate muscle receptors but that was not necessarily the case. For the lip control, for instance, cutaneous and mucosal mechanoreceptors seem to be the only available sensory source (Folkins and Larson, 1978; Halata and Munger, 1983; Munger and Halata, 1983; Abbs and Gracco, 1984; Johansson *et al.*, 1988). In contrast, sensorimotor mechanisms involved in digit control may depend on not only skin but also joint and muscle afferents. The skin deformed during index finger and thumb movements include not only the palmar and dorsal aspects of the digits themselves but also the radial parts of the palm and dorsum of the hand. Although partial knowledge was available at this time about the response properties of the glabrous skin mechanoreceptors to movements (Hulliger *et al.*, 1979), exploring the capacity of hairy skin receptors to encode movements in nearby joints seemed fruitful for several reasons (Edin and Abbs, 1991). The hairy skin is, for instance, rarely engaged directly in tactile explorative tasks. On the other hand, it seemed unlikely that afferents from the glabrous skin simultaneously play an important role in tactile exploration and grasp control as well as

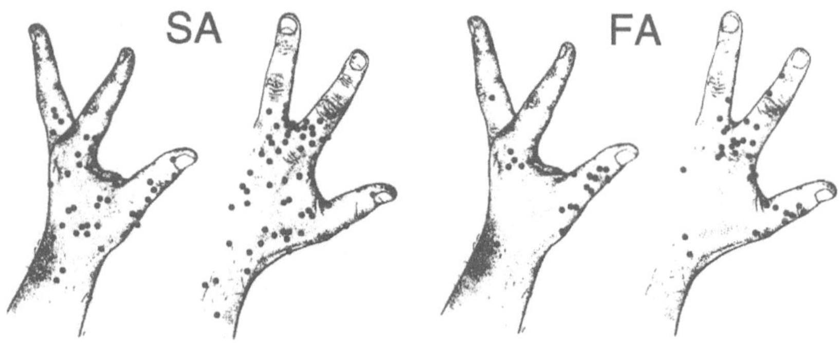

Figure 1. Receptor location of skin mechanoreceptor units (n=121). Note that FA units are found primarily close to joint whereas SA units are more uniformly distributed (from Edin & Abbs, 1991).

kinaesthesia. Indeed, the discharge rates even for the 'best' glabrous skin units were much lower during finger movements than those observed during tactile stimulation (Hulliger *et al.*, 1979). Moreover, the mechanical properties of the hairy and the glabrous skin are markedly different: the glabrous skin is tightly fixed to the subjacent tissues whereas the hairy skin can easily be translated in all directions.

The nerve chosen for exploration was the radial nerve or more specifically, those nerve fascicles that supply the radial aspect of the dorsum of the hand and the proximal parts of the thumb, the index finger and middle finger (the supply area rarely extends beyond the proximal interphalangeal (PIP) joints). Cutaneous afferents were identified to originate from receptors within the whole supply area of the radial nerve but the distribution within this area was different for slowly adapting (SA) and fast adapting (FA) afferents (Fig. 1): whereas SA receptors were more or less evenly distributed, FA receptors seemed to aggregate close to joints.

The responses of each of 121 afferents were recorded while the experimenter produced movements at the MCP joints, interphalangeal joints, and the wrist joint. A large majority (92 %) responded to experimenter induced movements at one or more joints and for a substantial number of these unit, their discharge clearly reflected the position and movements in individual joints. Common to all responsive units, the unit in Figure 2 displayed a pronounced dynamic responsivity as indicated by an unproportional increase in discharge frequency when the finger movements increased in velocity (Fig. 2, upper right panel).

Importantly, many SA receptors responded to joint movements that took place at significant distances from their receptive fields. In contrast, responses from FA receptors were typically only elicited if the moving joint was fairly close to its receptive field. Accordingly, among the receptors that did not respond quite a few were FA receptors located at some distance from any joint. This observation paired with the tendency of FA units to be located close to joints made sense teleologically if they do in fact play a role in encoding the dynamics of joint movements. Conversely, the distribution of SA units seems to fit their ability to respond to movements also at remote joints.

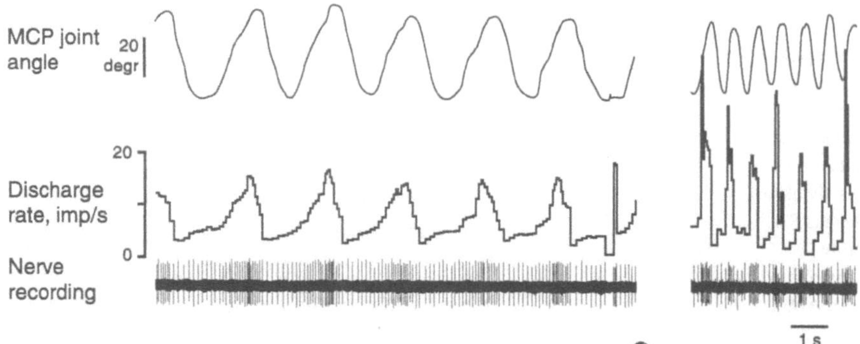

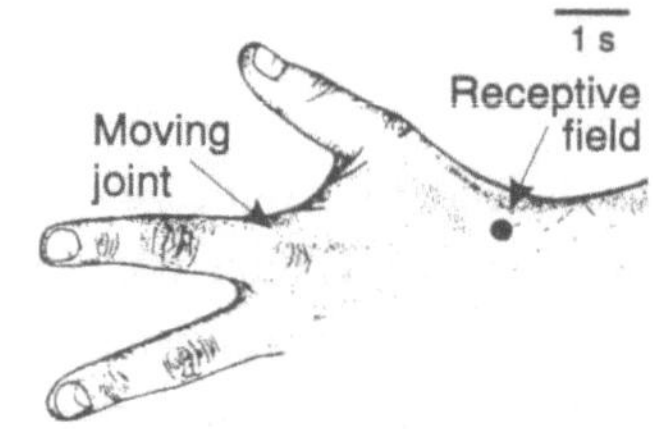

Figure 2. Response of an SAII unit to slow (left panel) and rapid flexion-extension movements (right upper panel). Note that the receptor was located close to the wrist joint but responded readily to movements at the metacarpophalangeal (MCP) joint of the index finger. (Data from Edin and Abbs, 1991).

The physical stimulus that elicited discharges in both SA and FA receptors was clearly skin deformation. For instance, during a flexion movement at the MCP joint of the index finger, the skin strain increases in some areas whereas it decreases in others areas on the radial surface of the MCP joint. Accordingly, recordings from units within this area showed responses compatible with these strain changes: some increased their discharge during flexion, other increased during extension.

Skin mechanoreceptors located in the hairy skin of the back of the hand thus seem to be able to encode both static and dynamic joint positions and this is due to their sensitivity to changes in skin strain. The next step was to estimate the capacity of the skin receptors to encode joint positions. It would, of course, have been possible to resolve this issue in a number of different ways. One seemingly attractive way would have been to simply record the discharge of individual mechanoreceptors as a function of individual joint angles. The interpretation of the results from such experiments would, however, be quite complicated. First, the response of an individual mechanoreceptors is a function not only of its inherent receptive properties but also on its location: skin areas close to a joint are subjected to larger strain changes than skin areas at a distance. Second, the strain at a particular spot in the skin is influenced by more than one joint: it is obvious, for instance, that the midsection of the dorsum of the hand is mechanically affected by movements at all MCP joints as well as the wrist joint. Whereas such a study would provide 'quantitative data' on individual receptors, the result would not easily be applied to arbitrary patterns of skin strain. To obtain data that more easily could be generalised to any strain pattern and therefore any joint configuration, I opted for an approach that at first may seem like a detour.

Static strain sensitivity in SA mechanoreceptive units

A device was constructed with which well-controlled skin strains could be applied to a skin patch (Edin, 1992). The device basically consisted of a step motor that acted on one of two beams that rested on the skin. By moving one of the beams, the strain in the skin patch between the two beams could be controlled. The upper limit of skin strains observed during natural MCP joint movements distally on the dorsum was determined to be less than 20 %. Accordingly, microneurographic recordings from SA units were obtained during trapezoidal skin stretches with amplitudes ranging from 1.25 to 20 %.

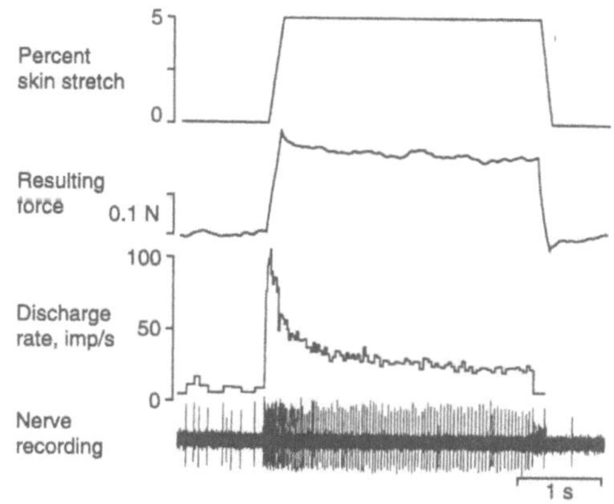

Figure 3. Response of an SAII unit to 5 % skin stretch. This unit was representative not only by its dynamic response to the increased stretch but also by displaying a static strain sensitivity close to the median for SAII units. (Redrawn from Edin, 1992).

Figure 3 illustrates the response of an SA II to a 5 % skin stretch. This recording is typical in several respects. First, in common with all SA units, the unit displayed a conspicuous dynamic response during the phase of increased skin strain. Second, the discharge gradually decreased during the plateau phase to reach a fairly steady level within 1-1.5 seconds. Third, the subsequent analysis showed that this unit displayed a static strain sensitivity that was close to the median sensitivity of SA II units.

For each orientation of the stimulation device, two or more replications at each of 5 predetermined amplitudes were obtained (Fig. 4). The averaged discharge rate during the last 1.5 s of the plateau phase was defined as a unit's *static response*. For all units, there was a more or less pronounced curvilinear relationship between the static response and the stretch amplitude. It is thus a simplification to attribute low-threshold skin mechanoreceptors a specific sensitivity to changes in static strain. Rather, their sensitivity decreases with increased skin stretch. If we only consider the range of stretches up to 10% skin stretch, however, most units showed an approximately linear relationship. Accordingly, the linear equation relating the static response to the stretch amplitude displayed high correlation coefficients. The obvious interpretation of the slope of these equations is *static strain sensitivity*, i.e., the increase in discharge rate for each unit of increased skin strain.

The static strain sensitivity defined in this manner was significant for all recorded SAI and SAII units (lower panels of Fig. 4). Moreover, although the population of SAII units clearly showed a larger strain sensitivity than the group of SAI units, individual SAI units could display a higher strain sensitivity than some SAII units.

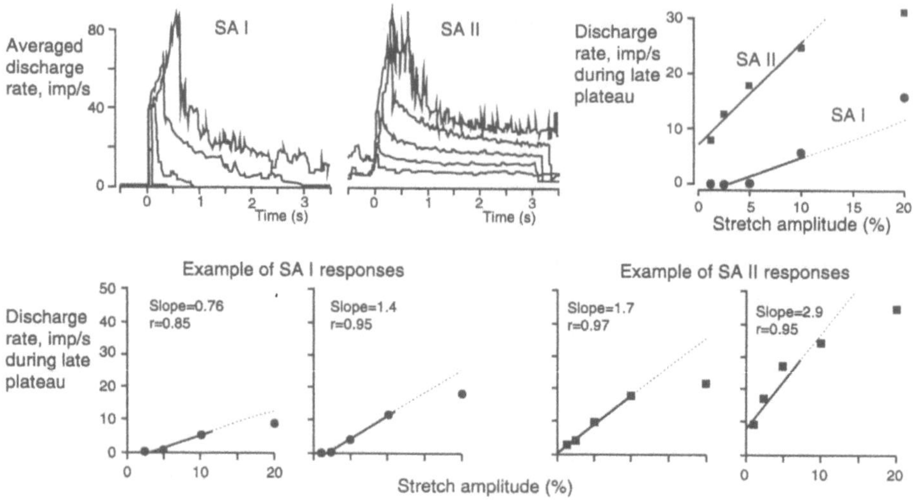

Figure 4. Top panels show averaged discharge rates during ramp-and-hold stretches with amplitudes 1.25, 2.5, 5, 10 and 20%. The prominent dynamic responses of the SA I unit (e.g., 40 Hz at 2.5 % stretch) was not exceptional among SA I units. The SA II unit in the top panels showed a sustained static response to all amplitudes whereas the SAI unit adapted to stretches less than 10 %. The discharge rate during the last 1.5 second of the plateau phase was defined as a unit's 'static response'. Top right panel show averaged static responses for the two units at top left panels plotted against the stretch amplitudes. For most units the static response increased reasonably linearly with the stretch amplitude up to 10% stretches (thick straight lines) and then levelled off at larger stretches. The slopes correspond to the increase in discharge rate per increase in percent stretch; *r* is the correlation coefficient. The median slope for all SAII units was 1.8, for SAI units 1.0. (Redrawn from Edin, 1992).

At the current stage it would be inappropriate to advance a detailed analysis of population responses based on recorded skin strains and known response properties. It is, for instance, clear that both SA and FA units show pronounced dynamic sensitivity to skin strain changes although rigorous data is not yet available. The unfiltered strain record may appear smooth in Figure 3, but the discharge of FA afferents in particular but also of both SAI and SAII afferents were often phase-locked to the 25 μm steps produced every 4 ms in that study (Edin, 1992). This obviously precluded an analysis of all FA units but also the dynamic responses of SA units. The microneurographic studies required for further analysis and modelling is, however, currently undertaken in our laboratory.

It is possible, however, to obtain a rough estimate of the sensitivity to joint position of the MCP joint of the index finger of the 'typical' SAI and SAII afferent located 2-3 cm from the MCP joint (Edin, 1992). These estimates, 0.20-0.30 and 0.36-0.54 imp·s^{-1} per degree joint flexion for type I and type II SA afferents, respectively, compare favourable with the joint position sensitivity reported for the muscle spindle afferents in the finger extensor muscles: 0.25 imp·s^{-1}·deg^{-1} for primary and 0.35 for secondary afferents (Edin and Vallbo, 1990). Notably, these estimates most likely grossly underestimates the positional sensitivity of SA afferents located closer to the MCP joint. The capacity of SA mechanoreceptors to encode finger joint positions thus seem to be at least on par with the capacity of spindle afferents in finger muscles.

Under some circumstances, skin mechanoreceptors in the human hand has precedence over other proprioceptive sources

Skin mechanoreceptors thus are capable of transmitting information about skin strains and thereby joint configurations. Whether the brain make use of this information or not is another matter. Strain sensitive mechanoreceptors certainly play a role in encoding skin strain patterns induced by moving tactile stimuli (Olausson and Norrsell, 1993; Norrsell and Olausson, 1994; Edin *et al.*, 1995; Essick and Edin, 1995); their responsivity to joint movements *per se* may be of no functional significance at all.

If skin mechanoreceptors indeed play a proprioceptive role during perceptual and motor tasks it must be assumed that they encode the strain patterns (cf. Cohen, Prud'Homme and Kalaska, 1994). As such, one would expect that it would be easy to differentiate between a strain pattern associated with joint movements and one that is caused by an external stimuli. For instance, if an experimenter stretches the skin on the back of the hand to induce increased skin strain proximal to the MCP joints, the more proximal parts would be under reduced strain or even compression; such a pattern is certainly never observed during a normal flexion movement at the MCP joints. To be able to manipulate both a single joint and the part of the skin that is affected by movements at this joint, we anaesthetised the distal part of the index finger including the PIP joint leaving the proximal part of the proximal phalanx sentient (dotted skin areas on left index finger of Fig. 5). With successful anaesthesia, the experimenter was able to manipulate the position at the PIP joint itself or the skin in the distal part of the index finger without providing any direct tactile cues from the manipulated tissues (Edin and Johansson, 1995).

Five subjects were asked to respond to the experimental manipulations by either simply mimicking the perceived position at the PIP joint of the anaesthetised index (Fig. 5A, B), or, without moving the anaesthetised finger, touching its tip with the tip of the contralateral normal finger (Fig. 5C, D). All subjects without exception were able to mimic the position at the anaesthetised PIP joint when it was moved into a flexed or extended position (Fig. 5A). Likewise, they found it easy to find the spatial location of the anaesthetised finger tip by moving the contralateral finger into an appropriate position (Fig. 5C). These findings were, of course, not very surprising. The crucial experimental manipulation was, however, to induce skin stretch patterns in the dorsal and palmar skin of the proximal phalanx of the index finger, strain patterns that either were similar to those observed when the PIP joint was flexed or when it was extended. When a strain pattern that corresponded to PIP joint flexion was induced, the subject perceived the PIP joint as flexed irrespective of whether the PIP actually joint was in a flexed or an extended position (Fig. 5B). Moreover, when the induced strain pattern was similar to the one normally associated with PIP joint extension subjects consistently perceived the joint as extended (Fig. 5D). This was the case even if the anaesthetised PIP joint actually was in a flexed position. Notably, with this latter position, the muscle spindle inputs from the finger extensor and flexor muscles can only have been consistent with PIP joint *flexion* and yet the subjects interpreted the incoming

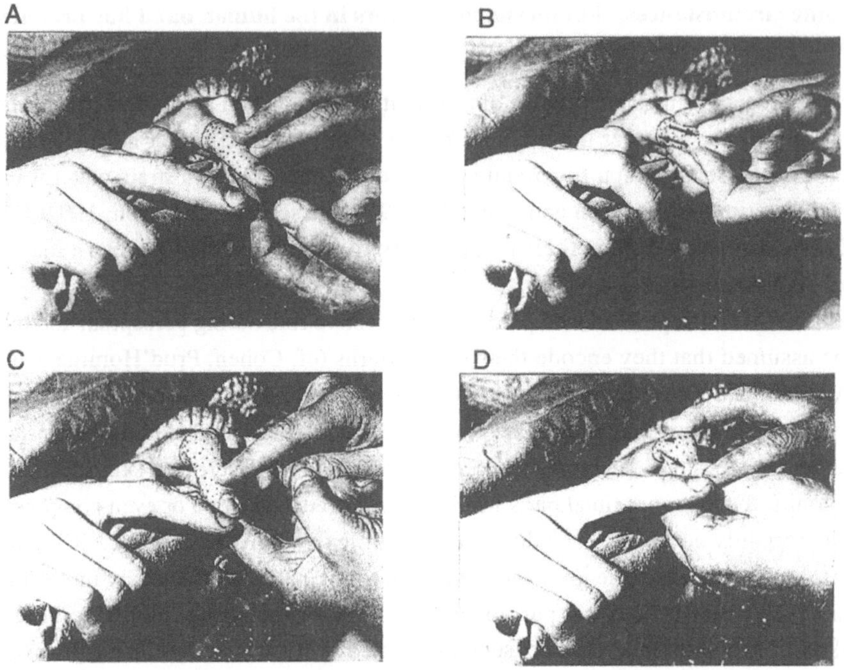

Figure 5. All subjects found it easy to mimic (*A*) the position at the proximal interphalangeal (PIP) joint of the anaesthetised digit, or to touch the tip of the insentient finger (*C*) using the contralateral normal index finger. When the PIP joint of the anaesthetised finger was brought into extension but the strain pattern induced in its proximal part was similar to the one observed during PIP joint flexion, subjects moved the normal finger into flexion (*B*). Conversely, when the PIP joint was kept flexed but the induced skin strain pattern in the proximal, sentient part of the anaesthetised index finger was similar to the one observed during PIP joint extension, the subjects touched the experimenters hand instead of if the tip of his flexed anaesthetised index finger; evidently, the subjects perceived the PIP joint to be extended (*D*). Notably, the sensory messages conveyed by muscle spindles and skin mechanoreceptors must have been in conflict in (*B, D*). Since the subjects evidently relied on the kinaesthetic information originating in the skin, it was concluded that skin afferents has 'precedence' over muscle spindle inputs. (From Edin and Johansson, 1995).

sensory pattern as compatible with a PIP joint *extension*. Interestingly, none of subjects realised that sometimes skin deformations only and no actual PIP joint movement had taken place, indeed, it was even difficult to convince two of them that only skin deformations had taken place in some trials.

There seems to be a number of valid conclusions that can be drawn from this study. First and perhaps most importantly, the brain can interpret skin strain patterns as joint movements. Second, the perception associated with certain changes in strain patterns are felt as *movements* rather than as 'skin strain changes'. Third, at least under some conditions, cutaneous information has precedence over sensory information conveyed by muscle afferents in providing joint position information.

Multiple sensory sources contribute to kinaesthesia

Numerous studies in which muscle vibrations have been employed to artificially excite muscle spindle afferents demonstrate beyond doubt that these afferents can evoke perceptual illusions of movements (Goodwin, McCloskey and Matthews, 1972; Roll and Vedel, 1982). Skin deformation that excites skin afferents seem to have similar effects. But other sensory sources are undoubtedly involved as well. It is, for instance, well established that visual information in the de-afferented man partly compensates for the lack of proprioceptive information (Rothwell *et al.*, 1982). Likewise, it has been argued that joint receptors may play a role for proprioception although the evidence for this is perhaps weaker than for any other sensory source (Schaible and Schmidt, 1983; Proske, Schaible and Schmidt, 1988; Burke, Gandevia and Macefield, 1988; Clark, Grigg and Chapin, 1989; Edin, 1990). In short, a number of different sources of afferent information seem to be simultaneously available to the central nervous systems involved in kinaesthesia and motor control. What can then be said about the functional relationship between the various sources of proprioceptive information?

One prominent feature of the sensorimotor systems is that the afferent sources seem to be organised in a certain hierarchical order. If a subject, for instance, is able to see the experimental limb, muscle vibrations fail to evoke any illusions as if the visual information had precedence over the artificial muscle spindle afferent inputs (Goodwin, McCloskey and Matthews, 1972). Likewise, visual information suppress the illusions of PIP joint movements when skin strains were induced as described above. And under at least some experimental conditions, afferent information originating in the skin has precedence over muscle spindle inputs (Fig. 5B, D).

But is not the illusions induced by muscle vibrations an example of muscle spindle afferents having precedence over cutaneous inputs? Not necessarily. What has rarely if ever been considered in the discussion of vibration induced illusions is the fact that the muscle spindle discharges required for eliciting the perceptual illusions are roughly a magnitude larger than what should be necessary. Roll and co-worker report, for instance, that vibratory stimuli at 80 Hz entrain the majority of muscle spindle afferents resulting in more or less synchronous spindle afferent activation (Roll and Vedel, 1982; Roll, Vedel and Ribot, 1989). Vibratory stimuli at 60-80 Hz elicits illusory movements with an angular velocity of less than 4 $degr \cdot s^{-1}$ (e.g., Gilhodes, Roll and Tardy-Gervet, 1986), 100 Hz stimuli perceived velocities of 3-8 $degr \cdot s^{-1}$ (Goodwin, McCloskey and Matthews, 1972). In microneurographic experiments, however, the typical discharge rates at such low angular velocities is much lower, viz., some 5-15 Hz (Hulliger and Vallbo, 1979; Vallbo, Hulliger and Nordh, 1981; Roll and Vedel, 1982). The relation between cutaneous and muscle spindle afferents suggested above may thus very well hold true under 'normal' conditions but in the presence of unphysiologically high muscle spindle discharge rates, the impact of the muscle spindle inputs may overwhelm the conflicting and perhaps also distorted inputs from skin afferents (Ribot-Ciscar, Vedel and Roll, 1989). A more detailed discussion of several influential studies that may have over-estimated the importance of muscle spindles and under-estimated the importance of skin mechanoreceptors in relation to kinaesthesia and motor control can be found elsewhere (Edin and Johansson, 1995).

Importantly, although muscle spindle afferents may play a comparatively small role in kinaesthesia and in controlling movements *per se*, they may play a crucial role for the control of the skeletal muscles themselves. Muscle length is just one of a large number of factors — fatigue, potentiation, temperature, activation pattern and history to mention a few (Houk, Singer and Goldman, 1970; Bigland-Ritchie *et al.*, 1983; Maton and Gamet, 1989; Thomas *et al.*, 1990a-b; Thomas, Bigland-Ritchie and Johansson, 1991) — that are known to influence the complex biophysical effector properties of skeletal muscles. Important new insights about the roles of muscle spindles in motor control may be obtained if their capacity to inform the central nervous systems about the contractile behavior of a muscle, or perhaps primarily a compartment of the muscle (Botterman, Binder and Stuart, 1978; Cameron *et al.*, 1981; Lucas and Binder, 1984), is considered rather than their capacity to encode one single factor, viz., muscle length or its time derivatives.

The proposed hierarchical structure of various sensory sources does certainly not imply that this hierarchy is static. Rather, it seems reasonable to assume that the input from each source is weighted in a context-dependent manner, e.g., depending on the motor task or contextual factors. Hopefully, the proposed hiearchial hypothesis in its presently loosely defined form may inspire researchers to explore the relationship between and importance of various sensory difference in a range of functional tasks.

Acknowledgements
The microneurographic studies were carried out in the laboratory of Professor J.H. Abbs, Department of Neurology, Madison University and partially analyzed at the Umeå University, Sweden. The studies were supported by the National Institute of Neurological Disorders and Stroke Grant NS-13274, the Swedish Medical Research Council (project 08667), Office of Naval Research, Arlington, VA, USA (grant no. N00014-90-J-1838), and Umeå University, Sweden.

References

Abbs, J.H. and Cole, K.J. Neural mechanisms of motor equivalence and goal achievement. *In:* S.P. Wise (ed.): *Higher brain functions: recent explorations of the brain's emergent properties.* John Wiley & Sons, Inc., pp 15-43.

Abbs, J.H. and Gracco, V.L. (1984) Control of complex motor gestures: orofacial muscle responses to load perturbations of lip during speech. *Journal of Neurophysiology* 51:705-723.

Bigland-Ritchie, B., Johansson, R.S., Lippold, O.C.J., Smith, S. and Woods, J.J. (1983) Changes in motoneurone firing rates during sustained maximal voluntary contractions. *Journal of Physiology* 340:335-346.

Botterman, B.R., Binder, M.D. and Stuart, D.D. (1978) Functional anatomy of the association between motor units and muscle receptors. *American Zoologist* 18:135-152.

Burke, D., Gandevia, S.C. and Macefield, G. (1988) Responses to passive movement of receptors in joint, skin and muscle of the human hand. *Journal of Physiology* 402:347-361.

Cameron, W.E., Binder, M.D., Botterman, B.R., Reinking, R.M. and Stuart, D.G. (1981) "Sensory partitioning" of cat medial gastrocnemius muscle by its muscle spindles and tendon organs. *Journal of Neurophysiology* 46:32-47.

Clark, F.J., Grigg, P. and Chapin, J.W. (1989) The contribution of articular receptors to proprioception with the fingers in humans. *Journal of Neurophysiology* 61:186-193.

Cohen, D.A.D., Prud'Homme, M.J.L. and Kalaska, J.F. (1994) Tactile activity in primate primary somatosensory cortex during active arm movements: correlation with receptive field properties. *Journal of Neurophysiology* 71:161-172.

Cole, K.J., Gracco, V.L. and Abbs, J.H. (1984) Autogenic and nonautogenic sensorimotor actions in the control of multiarticulate hand movements. *Experimental Brain Research* 56:582-585.

Edin, B.B. (1990) Finger joint movement sensitivity of non-cutaneous mechanoreceptor afferents in the human radial nerve. *Experimental Brain Research* 82:417-422.

Edin, B.B. (1992) Quantitative analysis of static strain sensitivity in human mechanoreceptors from hairy skin. *Journal of Neurophysiology* 67:1105-1113.

Edin, B.B. and Abbs, J.H. (1991) Finger movement responses of cutaneous mechanoreceptors in the dorsal skin of the human hand. *Journal of Neurophysiology* 65:657-670.

Edin, B.B., Essick, G.K., Trulsson, M. and Olsson, K.Å. (1995) Receptor encoding of moving tactile stimuli in humans. I. Temporal pattern of discharge of individual low-threshold mechanoreceptors. *Journal of Neuroscience* 15:830-847.

Edin, B.B. and Johansson, N. (1995) Skin strain patterns provide kinaesthetic information to the human central nervous system. *Journal of Physiology* 487:243-251.

Edin, B.B. and Vallbo, Å.B. (1990) Dynamic response of human muscle spindle afferent to stretch. *Journal of Neurophysiology* 63:1297-1306.

Essick, G.K. and Edin, B.B. (1995) Receptor encoding of moving tactile stimuli in humans. II. The mean response of individual low-threshold mechanoreceptors to motion across the receptive field. *Journal of Neuroscience* 15:848-864.

Folkins, J.W. and Larson, C.R. (1978) In search of a tonic vibration reflex in the human lip. *Brain Research* 151:409-412.

Gilhodes, J.C., Roll, J.P. and Tardy-Gervet, M.F. (1986) Perceptual and motor effects of agonist-antagonist muscle vibration in man. *Experimental Brain Research* 61:395-402.

Goodwin, G.M., McCloskey, D.I. and Matthews, P.B.C. (1972) The contribution of muscle afferents to kinesthesia shown by vibration induced illusions of movement and by the effects of paralysing joint afferents. *Brain* 95:705-748.

Halata, Z. and Munger, B.L. (1983) The sensory innervation of primate facial skin. II. Vermillion border and mucosa of lip. *Brain Research Reviews* 5:81-107.

Houk, J.C., Singer, J.J. and Goldman, M.R. (1970) An evaluation of length and force feedback to soleus muscles of decerebrate cats. *Journal of Neurophysiology* 33:784-811.

Hulliger, M., Nordh, E., Thelin, A.-E. and Vallbo, Å.B. (1979) The response of afferent fibres from the glabrous skin of the hand during voluntary finger movements in man. *Journal of Physiology* 291:233-249.

Hulliger, M. and Vallbo, Å.B. (1979) The response of muscle spindle afferents during voluntary tracking movements in man. Load dependent servo assistance? *Brain Research* 166:401-404.

Johansson, R.S., Trulsson, M., Olsson, K.Å. and Westberg, K.-G. (1988) Mechanoreceptor activity from the human face and oral mucosa. *Experimental Brain Research* 72:204-208.

Lucas, S.M. and Binder, M.D. (1984) Topographic factors in distribution of homonymous group Ia-afferent input to cat medial gastrocnemius motoneurones. *Journal of Neurophysiology* 51:50-63.

Maton, B. and Gamet, D. (1989) The fatigability of two agonistic muscles in human isometric voluntary submaximal contraction: and EMG study. II. Motor unit firing rate and recruitment. *European Journal of Applied Physiology and Occupational Physiology* 58:369-374.

Munger, B.L. and Halata, Z. (1983) The sensory innervation of primate facial skin. I. Hairy skin. *Brain Research Reviews* 5:45-80.

Norrsell, U. and Olausson, H. (1994) Spatial cues serving the tactile directional sensibility of the human forearm. *Journal of Physiology* 478:533-540.

Olausson, H. and Norrsell, U. (1993) Observations on human tactile directional sensibility. *Journal of Physiology* 464:545-559.

Proske, U., Schaible, H.-G. and Schmidt, R.F. (1988) Joint receptors and kinaesthesia. *Experimental Brain Research* 72:219-224.

Ribot-Ciscar, E., Vedel, J.P. and Roll, J.P. (1989) Vibration sensitivity of slowly and rapidly adapting cutaneous mechanoreceptors in the human foot and leg. *Neuroscience Letters* 104:130-135.

Roll, J.P. and Vedel, J.P. (1982) Kinaesthetic role of muscle afferents in man studied by tendon vibration and microneurography. *Experimental Brain Research* 47:177-190.

Roll, J.P., Vedel, J.P. and Ribot, E. (1989) Alteration of proprioceptive messages induced by tendon vibration in man: a microneurographic study. *Experimental Brain Research* 76:213-222.

Rothwell, J.C., Traub, M.M., Day, B.L., Obeso, J.A., Thomas, P.K. and Marsden, C.D. (1982) Manual motor performance in a deafferented man. *Brain* 105:515-542.

Schaible, H.-G. and Schmidt, R.F. (1983) Responses of fine medial articular nerve afferents to passive movements of knee joint. *Journal of Neurophysiology* 49:1118-1126.

Thomas, C.K., Bigland-Ritche, B., Westling, G. and Johansson, R.S. (1990a) A comparison of human thenar motor-unit properties studied by intraneural motor-axon stimulation and spike-triggered averaging. *Journal of Neurophysiology* 64:1347-1351.

Thomas, C.K., Johansson, R.S., Westling, G. and Bigland-Ritchie, B. (1990b) Twitch properties of human thenar motor units measured in response to intraneural motor-axon stimulation. *Journal of Neurophysiology* 64:1339-1346.

Thomas, C.K., Bigland-Ritchie, B., Johansson, R.S. (1991) Force-frequence relationships of human thenar motor units. *Journal of Neurophysiology* 65:1509-1516.

Vallbo, Å.B., Hulliger, M. and Nordh, E. (1981) Do spindle afferents monitor joint position in man? A study with active position holding. *Brain Research* 204:209-213.

Somesthesis and the Neurobiology of the Somatosensory Cortex
O. Franzén, R. Johansson and L. Terenius (eds.)

A second tactile system in the human skin with unmyelinated primary afferents

Åke B. Vallbo, Håkan Olausson, Johan Wessberg and Ulf Norrsell

Department of Physiology, Göteborg University, Medicinaregatan 11, S-413 90 Göteborg, Sweden

Summary. In man, unmyelinated afferents from the skin have been identified exclusively with pain and temperature senses on the basis of the stimuli which excite them. In other mammals, on the other hand, many C-afferents respond to light touch. This difference has lead to the suggestion that a phylogenetically old system of unmyelinated mechanoreceptors has largely faded out in the evolutionary process from mammals to man. However, a few touch sensitive C-afferents have recently been described in the human facial skin, suggesting a vestige in a specialized skin region. In the present study we have shown, with microneurography, that touch sensitive C-afferents are present in the human forearm. Their conduction velocities, thresholds to skin indentation, response characteristics to moving stimuli, and receptive field geography have been studied. The findings suggest that man, as other mammals, is equipped with a general system for cutaneous mechanoreception subserved by C-afferents, although it seems to be lacking in the peripheral parts of the extremities.

Introduction

Since Torebjörk and Hallin (1974) started to explore unmyelinated afferents from the human skin it seems to be generally accepted that these slowly conducting fibres are concerned either with nociception or temperature sensibility but not light touch. This interpretation is largely based on the finding that the C-afferents from the human skin respond preferentially to noxious stimuli and temperature changes but not much to light touch. (Torebjörk, 1974; Torebjörk and Hallin, 1974).

The presumed lack of touch receptors with unmyelinated afferents in the human skin was unexpected considering previous studies in other species. Already in the nineteen-fifties Douglas and Ritchie (1957) demonstrated that there are large numbers of unmyelinated afferents that respond to light touch in the cat skin. This system was later explored, at the single unit level, particularly by Iggo (Iggo, 1960; Iggo and Kornhuber, 1977) and Perl (Kumazawa and Perl, 1977) who described the response characteristics of the afferents.

However, the canonical view of a distinct difference in cutaneous innervation between man, on the one hand, and other species, i.e. rat, rabbit, cat, and monkey, on the other, was partially questioned around 1990. First, Johansson, Trulsson, Olsson and Westberg (1988) reported one tactile C-unit from the face while Nordin (1990) soon afterwards presented convincing and beautiful recordings of touch units with low conduction velocities from the face of human subjects. These findings led to the presumption that unmyelinated touch units are present in the face area alone but not in other areas of the human body.

In this chapter, evidence will be presented that the hairy skin of the human forearm is equipped with a system of low-threshold mechanoreceptors which are connected to umyelinated nerve fibres (Vallbo et al., 1993). The findings suggest that the human hairy skin has a second system for the collection of information of light touch in addition to the system of large myelinated units which conduct impulses at high speed from corpuscular endings and hair follicles (Vallbo et al., 1995).

Methods

Twenty-one unmyelinated afferents responding to light touch were recorded from the lateral antebrachial nerve at the elbow, using the microneurography technique. This nerve innervates a strip of skin along the forearm and may occasionally supply parts of hand dorsum. Since the nerve was very thin and electrical stimulation during the search procedure elicited indistinct paraesthesias, it was particularly difficult to localize and impale.

Once an intrafascicular position of the recording electrode had been attained, the skin area was explored with light touch, i.e. the experimenter moved his finger over the skin area. It deserves emphasis that the search procedure was strictly limited to light touch whereas no other stimuli were used to assess whether afferent activity or single unit impulses were seen by the electrode or not.

Mechanical stimuli

The approximate threshold of the individual unit to local indentations was assessed using a series of von Frey hairs made of nylon strings. To study suprathreshold stimuli and response latency to tactile stimuli, a hand-held device was employed, basically consisting of a metal tongue with strain gauge elements. At the end of the metal tongue, a Perspex probe having a rounded tip with a diameter of 2 mm was attached. Finally, responses to moving stimuli as well as the receptive field geography of individual units were analysed in detail with a scanning method (Looft, 1986). Moving stimuli were delivered with a probe attached to an arm extending from the head of a x-y-plotter which in turn was controlled by a small computer. The probe was a metal tube ending with a 2 mm diameter half-sphere. It indented the skin with 4 g. The computer was programmed to move the probe over the skin surface at a speed of 4 mm/s in a series a closely adjacent tracks within a predetermined box of the skin surface. Once it had moved back and forth along one track it was shifted sideways by a quarter of a mm or half a mm.

Results

Conduction velocity

Figure 1 illustrates the two approaches employed to assess the conduction velocity, i.e. electrical stimulation and mechanical stimulation. The upper panel in Fig. 1A shows

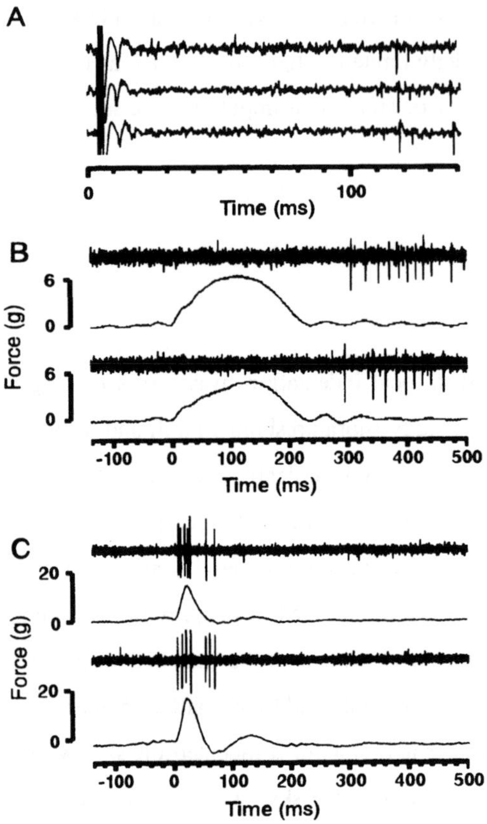

Figure 1. Assessment of fast or slow conduction velocity in 3 different tactile afferent units using electrical and mechanical stimulation. A, Response of a slow conducting afferent to electrical stimulation applied with cushion electrodes on the receptive field of the unit. B, Response of a slow conducting afferent to mechanical stimulations applied to the receptive field of the unit using a strain gauge device. C, Response of a fast conducting afferent to similar mechanical stimulations as in B. Calculations of the conduction velocities on the basis of the delays and conduction distances yielded propagation speeds of 1.0 m/s, 0.9 m/s and 80 m/s in A, B, and C respectively. (From Vallbo et al, 1993, with kind permission).

responses to electrical stimulation. The pulse was applied within the receptive field of the unit with a surface electrode. It may be seen that the latency to the unit's response was about 110 ms giving a conduction velocity of 1 m/s.

The procedure of mechanical stimulation is illustrated in Fig. 1 B. Light touch stimuli were delivered with a rounded probe attached to a strain gauge device to excite the unit. It may be seen that the latency from skin contact, i.e. time zero, was about 300 ms, yielding a conduction velocity of 0.9 m/s. It should be pointed out that the threshold of this unit to mechanical stimulation was not more than 10 mg, implying that the very first part of the indentation excited the unit.

Using these two methods which provided very similar values, it was found that the average conduction velocity in the sample (n=11) was 0.9 m/s.

As a comparison, Fig. 1 C shows the response of a myelinated tactile afferent to mechanical stimulation, presented on the same time scale as in Fig. 1B, to demonstrate the pronounced difference in latency.

The difference in latency to mechanical stimuli between unmyelinated and myelinated tactile afferents was clear not only in off-line recordings. Qualitatively, it was also very obvious during the experiment. Actually, whether a tactile afferent was a fast conducting myelinated fibre or a slowly conducting unmyelinated fibre was immediately evident to the experimenter from the delay between the moment of skin touch and impulse response, as heard through the loudspeakers.

It seems that the proportion of unmyelinated afferents responding to light touch in the lateral antebrachial nerve is substantial because in the present series of experiments as much as 40% of the tactile units collected were unmyelinated (n=21). To what extent this proportion is representative or whether a substantial bias was introduced is difficult to say in retrospect. However, since large fibres usually are easier to record than small fibres it seems unlikely that the unmyelinated afferents are heavily overrepresented.

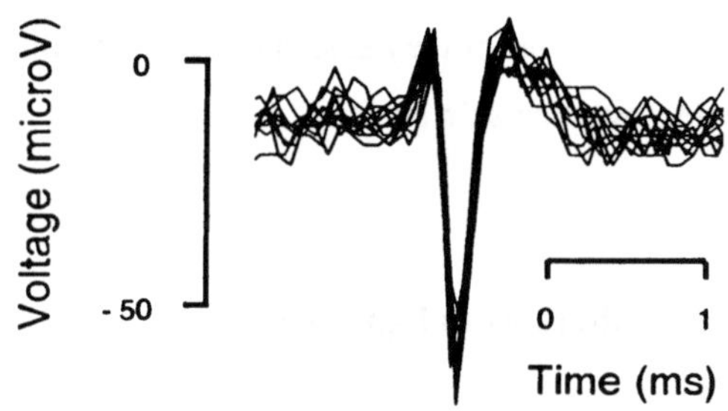

Figure 2. Shape of nerve impulses recorded from an umyelinated mechanoreceptive afferent with the microneurography technique. Recordings of 10 impulses from one afferent are superimposed.

Impulse shape

The impulse shape provided a supportive evidence that the units were unmyelinated. The shape was regularly as shown in Fig. 2, i.e. a triphasic deflection with the most prominent phase in the negative direction. This is the shape expected from an extracellular recording close to the current generator, displaying basically the membrane current at the recording site. It has been shown in many studies that this is regularly the shape of C-fibres whereas this impulse shape is uncommon with myelinated afferents (Torebjörk et al., 1970; Hallin and Torebjörk, 1974; Torebjörk, 1974; Torebjörk and Hallin, 1974, Vallbo, 1976; Nordin, 1990; Macefield et al., 1994). Although the shape of the impulse is not an absolute confirmation of type of afferent nerve fibre, it is obvious that the invariant negative shape within this group of afferents is a strong supportive evidence that they are non-myelinated.

Various characteristics

With regard to general response characteristics these units had similar properties as previously described for unmyelinated mechanoreceptors in other species; i.e. the adaptation was usually intermediate, they were relatively easy to fatigue by repeated stimuli, many of them produced a strong off-response, and a lasting after discharge when a skin deformation was withdrawn. Some of them responded to movements of the hairs, and they were all particularly well excited by stimuli that were moving over the skin surface.

Sensitivity of unmyelinated afferents to light touch

As already described above, the search procedure was limited to light touch. Hence only units which readily responded when the experimenter moved his finger over the skin area were at all noticed.

The thresholds to local indentations were assessed using von Frey hairs. Figure 3 shows the threshold distribution of 18 units. The median value was 0.25 g and a fair proportion of units had a threshold below 0.1 g. Except for two outliers with thresholds

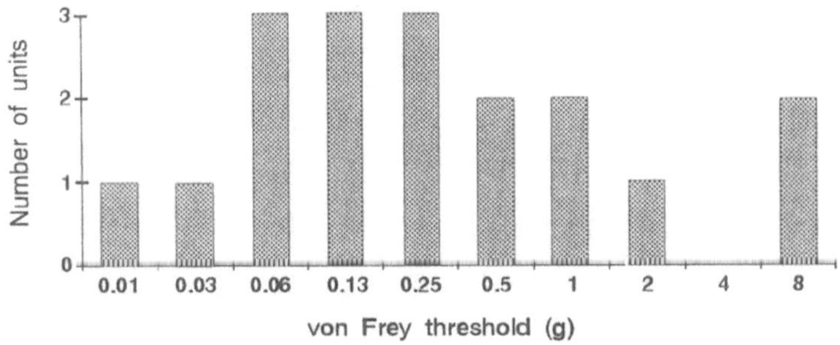

Figure 3. Approximate thresholds of a sample of C-afferents responding to light tactile stimuli, i.e. the experimenter moving his finger over the skin surface. Thresholds were assessed with a set of von Frey hairs yielding the reading indicated on the x-axis when held steadily against a precision balance. (From Vallbo et al, 1993, with kind permission).

of 8 g the distribution seems reasonably Gaussian on a logarithmic x-scale.

Figure 4A demonstrates that high impulse rates were induced by innocuous moving stimuli. The records display nerve impulses and corresponding instantaneous impulse rates when light indentations, i.e. 4 g transmitted with a 2 mm half-sphere, moving over the skin surface.

The vigorous discharge, amounting to impulse rates of 100 impulses s^{-1}, in response to these faint stimuli is particularly relevant. In animal experiments it has actually been found that 100 impulses s^{-1} is very close to the maximal firing rate that C-afferents can produce (Iggo, 1960; Kumazawa and Perl, 1977). It seems therefore obvious that the full working range of this unit with regard to impulse rates was within the innocuous range of mechanical stimuli.

Hence three kinds of experimental evidence were collected for the conclusion that the units are touch receptors rather than mechanical nociceptors. One is that they responded readily to light strokings of the skin with the experimenter's finger tip. Second, their von Frey thresholds are, in most cases low. Third, and probably most convincingly, it was documented that they responded with high impulse rates to innocuous stimuli moving over the skin surface.

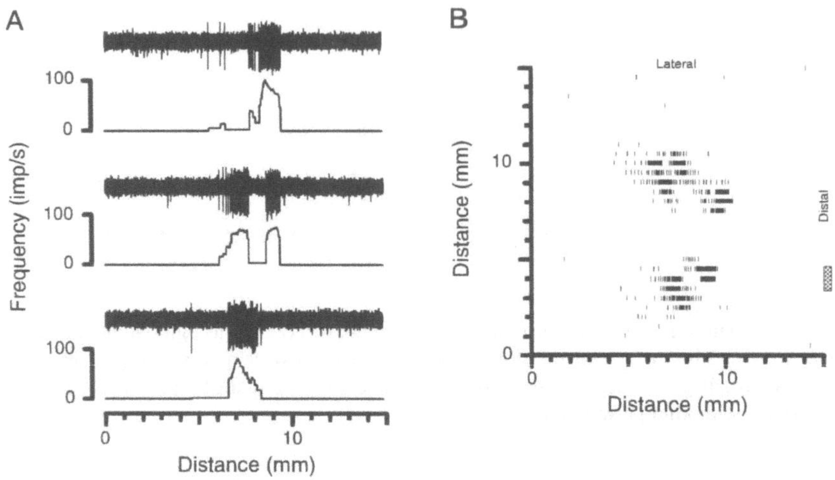

Figure 4. A, Responses of a tactile C-unit to moving stimuli. A light probe (4 g weight) with a rounded tip was moved at a speed of 4 mm/s over the skin surface in three parallel tracks separated by 0.5 mm. Nerve impulses and instantaneous impulse rates are shown as a function of probe position along 15 mm long tracks in proximo-distal direction. B, Geography of the unit's receptive field, based on a scanning of a 15 x 15 mm box covering the receptive field of the unit with moving stimuli as in A. Individual nerve impulses are represented by the short vertical bars. The three tracks of A are indicated by the right hand grey bar in B. (From Vallbo et al, 1993, with kind permission).

Receptive fields

Moving indentations as described above were used to explore the receptive field geography of individual units. A light probe was made to move, first from proximal to distal and then back along the same track. Then it was shifted 0.5 mm medially and moved along a parallel track. Responses to movements in one direction along three adjacent tracks are illustrated in Fig. 4A. By scanning a box of appropriate size, say 15 times 15 mm, with such moving stimuli, data were collected to produce a spatial event plot of the unit's response, i.e. a plot showing the occurrence of impulses as a function of probe position. This spatial event plot, in turn, represents a map of the receptive field of the unit.

Already Fig. 4A demonstrates that responses from closely adjacent tracks may be very different suggesting a delicate microstructure of the receptive field. In the top pair of traces, an intense discharge appeared at 8-10 mm, whereas two separate bursts of impulses occurred in the adjacent track illustrated in the middle. Again, there was only one burst in the next track illustrated at the bottom. The single bursts in the top and bottom traces have similar locations in the proximo-distal direction as the two bursts in the middle trace, indicating that two separate hot spots were defined by the three runs.

Figure 4B shows the spatial event plot of this unit based on a large number of tracks in addition to the three displayed in Fig. 4A, which, in turn, are indicated by the bar to the right in Fig. 4B.

It may be seen that the field is quite complex, having several local maxima which are clearly separated from each other, suggesting several endings. The field is elongated and irregular in shape with a size of about 25 mm^2. This size seems representative because mapping of fields with a handheld instrument indicated that the long axes of the fields were 8.7 ± 4.4 mm (n=8).

Discussion

A second tactile system is present in the human skin as in other species

It seems that the present findings provide convincing evidence that the skin of the human forearm is supplied with highly sensitive mechanoreceptive units with unmyelinated afferents. Their low thresholds and, particularly, their high impulse rates to light tactile stimuli are strong support for the conclusion that the role of the system is to code innocuous skin deformation rather than noxious stimuli. On the basis of the information available so far, there is good reason to regard the system as equivalent to the unmyelinated touch system described in other mammals.

Until a few years ago it appeared that this system was altogether lacking in man. When

Nordin (1990) demonstrated the presence of tactile C-units in the facial skin of human subjects, a reasonable interpretation was that a mere fragment is present in man, localized within a specialized cutaneous region. It was also speculated that the system might have a role in trigeminal neuralgia.

It may be asked why this second tactile system has not been identified before in other skin regions in man. The issue was discussed by Kumazawa and Perl (1977) who pointed out the tactile C-afferents are more common in proximal than distal parts of the extremities in the monkey, concluding that "The present work provides a possible explanation for the absence of C-mechanoreceptors in human studies. The work on man has utilized nerves with distribution to distal limb; in monkey, such body areas were found to have few of these elements. In addition, there is the possibility that the evolutionary progression to man has resulted in a still further elimination of the low-threshold mechanoreceptive class from the unmyelinated population."

For long, it seemed generally accepted that the system of C-mechanoreceptive units has faded during the evolution, i.e. either vanished altogether in man, or left a vestige in the face. Now it seems likely that previous microneurography studies in man have failed to identify C-mechanoreceptors in the arm or leg because the studies were largely limited to the distal parts where such units appear to be lacking. Since the forearm skin is probably more representative of the hairy skin which covers most parts of the body, the identification of unmyelinated afferents in this area supports the presumption that the tactile C-system in man has essentially a general distribution.

A few units in our sample had high thresholds which at first sight might suggest nociceptive rather than tactile function. Whether these units were misclassified or not is difficult to assess in retrospect as scant additional information is available on them. However, it was obvious that they responded to light moving stimuli. An explanation might be that too high von Frey thresholds were recorded if the experimenter failed to hit the most sensitive target point within the receptive field.

Functional role of the system of unmyelinated tactile afferents

If it is assumed that C-mechanoreceptors are present in most skin regions in man, it becomes interesting to speculate about their functional role. Incidentally, it is noteworthy that hypotheses on this issue seem to be lacking altogether in the literature.

A number of groups have attempted differential nerve block in combination with psychophysical tests of tactile sensibility in man (Torebjörk and Hallin, 1973; Mackenzie et al., 1975; Yarnitsky and Ochoa, 1991). The usual interpretation of selective block experiments has been that tactile sensations vanished when propagation failed in myelinated afferents, suggesting that the unmyelinated touch units alone do not provide a sufficient input to produce sensations of touch. It seems still possible that they support tactile sensibility in conjunction with the myelinated afferents. On the other hand, the relevence of the blocking experiments seems questionable because many of them explored the distal parts of the extremities where these units seem to be lacking anyway.

Considering the poor capacity of the system to provide swift and detailed information of important aspects of tactile stimuli, particularly temporal aspects, one might suggest that the system is related to non-discriminative functions of cutaneous sensibility, and therefore have closer connections with older parts of the brain, i.e. limbic and hypothalamic structures, rather than with neocortex. This interpretation seems to fit with the plausible assumption that the system is phylogenetically older than the myelinated tactile system. On the basis of such speculations it seems reasonable to consider functions not only in the area of touch perception, but in the realms of emotions, hormonal control, and tissue maintenance as well.

Although it is impossible to define the functional significance of the unmyelinated touch system at present, it seems necessary to incorporate the system in experimental design and interpretations of data related to somatosensation in man.

Acknowledgements

This work was supported by the Swedish Medical Research council (projects No. 2857 and 3548) and Torsten and Ragnar Söderberg's foundation. We thank Mr. Sven-Öjvind Swahn for technical assistance.

References

Douglas, W.W. and Ritchie, J.M. (1957) Non-medullated fibres in the saphenous nerve which signal touch. *Journal of Physiology* 139: 385-399.

Hallin, R. G. & Torebjörk, H. E. (1974) Single unit sympathetic activity in human skin nerves during rest and various manoevres. *Acta Physiologica Scandinavica* 92: 303-317.

Iggo, A. (1960) Cutaneous mechanoreceptors with afferent C fibres. *Journal of Physiology* 152: 337-353.

Iggo, A. and Kornhuber, H.H. (1977) A quantitative study of C-mechanoreceptors in hairy skin of the cat. *Journal of Physiology* 271: 549-565.

Johansson, R.S., Trulsson, M., Olsson, K.Å. and Westberg, K.-G. (1988) Mechanoreceptor activity from the human face and oral mucosa. *Experimental Brain Research* 72: 204-208.

Kumazawa, T. and Perl, E.R. (1977) Primate cutaneous sensory units with unmyelinated (C) afferent fibers. *Journal of Neurophysiology* 40: 1325-1338.

Looft, F. J. (1986) Response of cat cutaneous mechanoreceptors to punctate and grating stimuli. *Journal of Neurophysiology* 56: 208-220.

Macefield, V. G., Wallin, B. G. & Vallbo, Å. B. (1994) The discharge behaviour of single vasoconstrictor motoneurones in human muscle nerves. *Journal of Physiology* 481: 799-809.

Mackenzie, R. A., Burke, D., Skuse, N. F. & Lethlean, A. K. (1975) Fibre function and perception during cutaneous nerve block. *Journal of Neurology, Neurosurgery, and Psychiatry* 38: 865-873.

Nordin, M. (1990) Low-threshold mechanoreceptive and nociceptive units with unmyelinated (C) fibres in the human supraorbital nerve. *Journal of Physiology* 426: 229-240.

Torebjörk, H.E. (1974) Afferent C units responding to mechanical, thermal and chemical stimuli in human non-glabrous skin. *Acta Physiologica Scandinavica* 92: 374-390.

Torebjörk, H.E. and Hallin, R.G. (1973) Perceptual changes accompanying controlled preferential blocking of A and C fibre responses in intact human skin nerves. *Experimental Brain Research* 16: 321-332.

Torebjörk, H.E. and Hallin, R.G. (1974) Identification of afferent C units in intact human skin nerves. *Brain Research* 67: 387-403.

Torebjörk, H.E., Hallin, R.G., Hongell, A. and Hagbarth, K.-E. (1970) Single unit potentials with complex waveform seen in microelectrode recordings from human median nerve. *Brain Research* 24: 443-450.

Vallbo, Å.B. (1976) Prediction of propagation block on the basis of impulse shape in single unit recordings from human nerves. *Acta Physiologica Scandinavica* 97: 66-74.

Vallbo, Å.B. Olausson, H. Wessberg, J. and Kakuda, N. (1995) Receptive field characteristics of tactile units with myelinated afferents in hairy skin of human subjects. *Journal of Physiology* 483: 783-795.

Vallbo, Å.B., Olausson, H. Wessberg, J. and Norrsell, U. (1993) A system of unmyelinated afferents for innocuous mechanoreception in the human skin. *Brain Research* 628: 301-304.

Yarnitsky, D. & Ochoa, J. L. (1991) Differential effect of compression-ischemia block on warm sensation and heat-induced pain. *Brain* 114: 907-913.

Somesthesis and the Neurobiology of the Somatosensory Cortex
O. Franzén, R. Johansson and L. Terenius (eds.)

Factors influencing the perception of tactile stimuli during movement

C. Elaine Chapman[1,2], Iolanda C. Zompa[1], Stephan R. Williams[1], Jafar Shenasa[1] and Wan Jiang[1]

Centre de recherche en sciences neurologiques[1], École de réadaptation[2] and Département de physiologie[1], Université de Montréal, Montréal, Québec H3C 3J7, Canada

Summary. This chapter provides an overview of the factors that influence the perception of cutaneous stimuli during the course of movement. The inhibitory influences responsible for movement-related reductions in the ability to detect near-threshold stimli and to appreciate the intensity of suprathreshold stimuli are characterized as showing a spatial gradient and sensitivity to the movement kinematics and kinetics. Different models of the signals responsible for the gating effects are considered. The closest approximation to the experimental data was obtained with a model including a nonlinear inhibitory surround. Two other models, masking and a linearly augmenting inhibitory surround, failed to produce results consistent with all experimental observations.

Introduction

The skin is an important sensory organ, playing a major role in subserving discriminative touch, as well as pain, temperature, and probably also proprioception. Like vision and audition, however, its sensory functions are aided by the somato-motor apparatus which serves to orient the sensory organ so as to optimize the acquisition of information about the occurrence of events in the surrounding environment. Paradoxically, however, the act of movement itself reduces the transmission of tactile inputs to primary somatosensory cortex, a phenomenon that has been termed movement-related "gating" of sensory transmission (Ghez and Lenzi, 1971; Chapman, Jiang and Lamarre, 1988; reviewed in Chapman, 1994). A similar phenomenon also occurs in the visual system, since it has been observed that luminance detection thresholds are elevated during saccadic eye movements (Latour, 1962; Zuber and Stark, 1966). This article will review the influence of movement on the perception of tactile stimuli, with special emphasis upon the factors that modify the degree of perceptual gating.

Movement impairs detection of near-threshold stimuli

Coquery, Malcuit and Coulmance (1971) demonstrated that the ability to detect the occurrence

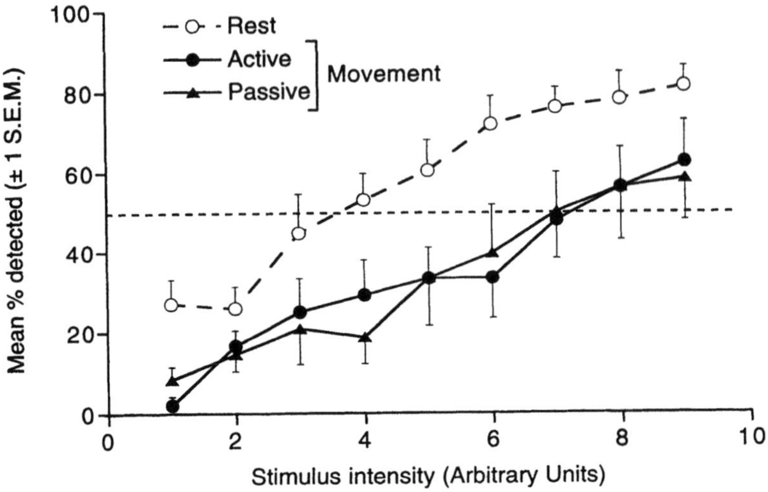

Fig. 1. Effects of active and passive movement on the ability of subjects to detect near-threshold electrical stimuli (single, constant current, square wave pulses of 2 ms duration) applied to the ventral surface of the forearm. Averaged stimulus-response curves (n=8 subjects) plot the mean percent of stimuli detected as a function of the intensity of the stimulus during movement and at rest. For the active movements, subjects performed rythmical flexions and extensions about the elbow (60° total displacement) at a frequency of 0.5 Hz. Passive movements mimicked the speed and amplitude of the active movements. The dotted line indicates detection threshold (50% detected). Data replotted from Chapman et al. (1987).

of weak, near-threshold cutaneous stimuli (trains of electrical shocks) is diminished during voluntary movements of the stimulated part. These observations were later confirmed by Garland and Angel (1974). Subsequently, Chapman et al. (1987) demonstrated that detection thresholds are, correspondingly, elevated during movement. This effect is illustrated in Fig. 1. Weak electrical stimuli were applied to the ventral surface of the forearm (see legend for details) under two conditions, namely with the subjects either at rest or during movement. After first estimating the detection threshold (50% detected), a series of nine current intensities were chosen that bracketed the estimated threshold value. Six alternating blocks of 9 trials at rest and 9 trials during movement were employed. It can be seen that performance (% detected) was diminished at all intensities of stimulation during movement, an effect that was significant for both active and passive movements (repeated measures analyses of variance (ANOVAs), $P<0.02$ in each case, *df* 1,7). Detection threshold (point where the curves cross the interrupted line) was, correspondingly, increased.

One criticism that could be directed at the latter studies is related to the choice of the stimulus, namely electrical stimulation. This has the advantage of ease of application, i.e. the electrodes can be firmly affixed to the limb and so their relation to the skin will not change during movement. The disadvantage with this type of stimulation is, however, the fact that the input is unnatural, being synchronous and activating a variety of different afferents irrespective of their role in tactile discrimination. The sensation elicited by such stimulation generally bears little resemblance to a natural sensation. However, as Dyhre-Poulson (1978) originally demonstrated, the *detection of natural tactile stimuli* is also impaired during movement. He applied a weak vibration (200 Hz, 8 repetitions) to the tip of the index finger during a tracking movement, and observed that performance declined from 75% detected at rest, to 65% during movement. Subsequently, Post, Zompa and Chapman (1994) demonstrated that detection threshold for vibratory stimuli (arrays of vibrating pins in an Optacon that varied in the number of active pins, 4 repetitions) is also elevated during a different motor task, rythmical isometric contractions about the elbow (Fig. 2A).

These effects are not, however, generalisable to all types of cutaneous afferent input. Thus, Feine et al. (1990) found that the threshold for detecting pain, elicited using thermal heat stimulation, was unchanged during the same task of dynamic, isometric contractions about the elbow. Taken together, these observations suggest that movement-related gating influences may be directed more towards diminishing the transmission of information from large-fibre, as compared to small-fibre, afferents. It remains to be determined, however, whether all sensory abilities within the range of large-fibre afferents are equally affected by movement. In this regard, Schmidt, Schady and Torebjörk (1990) reported that movement effectively diminishes referred sensations of both flutter (low frequency vibration) and pressure elicited by intraneural microstimulation (intensity, near threshold).

Influence of movement upon the perception of suprathreshold, cutaneous stimuli

While most studies have concentrated upon examining the effects of movement upon the

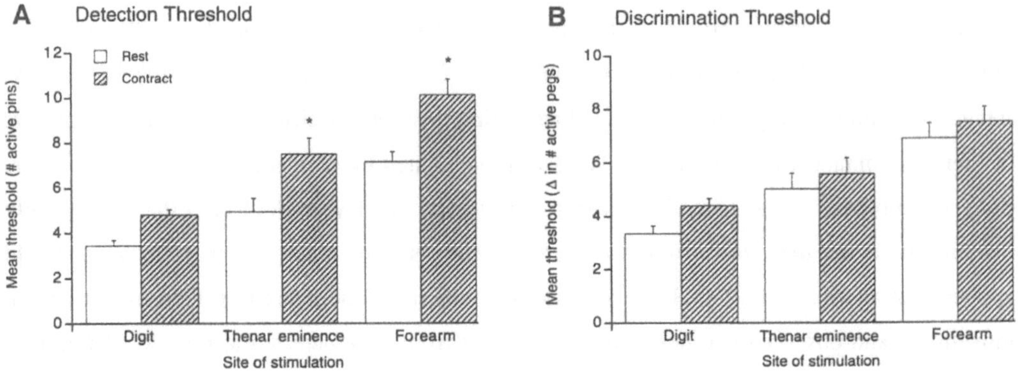

Fig. 2. Effects of rythmical isometric elbow flexions and extensions (0.5 Hz, 20 N in either direction) on vibrotactile detection (**A**) and discrimination threshold (**B**) measured at three sites, the distal phalanx of the index finger, the thenar eminence and the ventral forearm in 16 subjects. Threshold was measured at rest and during the motor task (±1 S.E.M.). Detection threshold is the number of activated in pegs in the Optacon array that could be detected on 50% of the trials. Discrimination threshold is the difference in the number of activated pegs in two sequentially presented arrays that could be perceived on 50% of the trials (Standard stimulus: 6 pegs activated, 11 repetitions). * $P<0.001$ (results of pair-wise comparison. Scheffé procedure, rest vs. contract, *df* 15). All other differences were not significant. Data replotted from Post, Zompa and Chapman (1994).

detection of near-threshold stimuli, a number of more recent studies have examined the influence of movement on the ability to perceive suprathreshold stimuli. Two sensory abilities have been evaluated, discrimination threshold (the "just noticeable difference") and subjective intensity (scaling or magnitude estimation).

It has uniformly been reported that discrimination thresholds are unaffected by the presence of movement, i.e. the relative differences between inputs of the same modality are preserved. Thus, the discrimination threshold for innocuous electrical stimuli is unchanged during movement (Chapman et al., 1987). As shown in Fig. 2B, the ability to discriminate small differences in the spatial extent of vibrotactile arrays is likewise not modified in the presence of motor activity (Post, Zompa and Chapman, 1994). Finally, tactual discrimination of texture (Lamb, 1983), embossed letters (Vega-Bermudez et al., 1991) or softness (Srinivasan and

LaMotte, 1995) is unaffected by active movement, although in each case movement between the skin and the discriminandum was an important component and so the stimulus had to be passively displaced over the skin during the "no movement" trials.

The subjective intensity of suprathreshold electrical stimuli has variously been reported to be diminished (Coquery, Malcuit and Coulmance, 1971; Milne et al., 1988; Post, Zompa and Chapman, 1994) or unaffected (Chapman et al., 1987) during movement of the stimulated part. Scaling of texture is unchanged during movement (Lederman, 1981). More recently, Post, Zompa and Chapman (1994) showed that the subjective intensity of vibrotactile stimuli is significantly reduced during isometric contractions about the elbow (Fig. 3). Using the same task, Feine et al. (1990) found that the subjective intensity of noxious and innocuous heat stimuli shows no modulation during motor activity. Thus, the modality of the sensory input is an important factor in determining the perceptual consequences of motor activity. In addition, differences in experimental protocols likely contributed to the contradictory results reported above. When we attempted to replicate our early observation of no change in magnitude

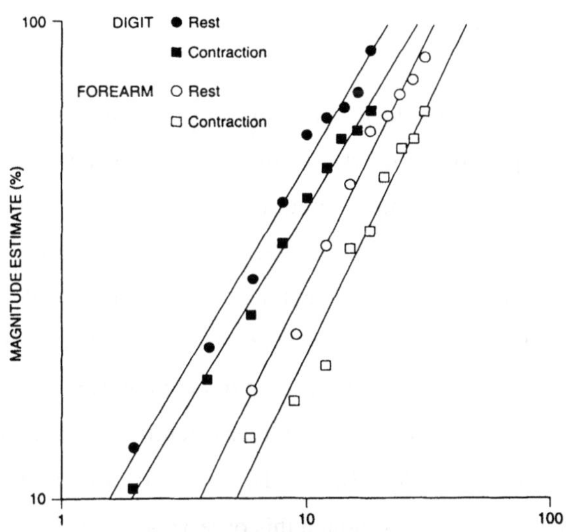

Fig. 3. Influence of motor activity (rythmical isometric contractions about the elbow at 0.5 Hz and a target force of 20 N in each direction) on the ability of subjects to scale the intensity of vibrotactile stimuli applied to the ventral forearm or the digit. Data are taken from Post, Zompa and Chapman (1994) and plotted on logarithmic scales (pooled data from 12 subjects). Note that the least-squares best-fit curve was in each case linear, and that the curves were shifted to the right during the motor task with their slopes unchanged.

estimates of innocuous electrical stimuli during movement (Chapman et al., 1987), we found that the subjective intensity (electrical and vibrotactile stimuli) was *reduced* during both isotonic and isometric contractions (Post, Zompa and Chapman, 1994). We suggested that the results of our two studies could be reconciled by the fact that the experimental protocol had been changed. Originally, we had minimized cutaneous feedback from the moving limb by supporting the upper arm in a sling (Chapman et al., 1987). But, this approach was not practical in the study using vibration because it was not possible to maintain contact between the Optacon and the skin. Consequently, the elbow rested on a rigid support, and there was increased cutaneous feedback from this region during the motor tasks. The suggestion that changes in peripheral feedback explained the difference in the results obtained is supported by the observation that tactile stimulation alone can mimic the movement-related gating effects on cortical somatosensory evoked potentials (Jones, Halonen and Shawkat, 1989) and can elevate the detection threshold (Pertovaara, Kemppainen and Leppänen, 1992).

Factors influencing gating

The *kinematics of the movement* can influence the degree of movement-related perceptual gating. Several laboratories have reported that the degree of movement-related gating of the perception of near-threshold stimuli augments with a modest increase in the average speed of finger movement (Angel and Malenka, 1982; Schmidt, Schady and Torebjörk, 1990). We have also investigated this phenomenon using two different motor tasks, rythmical flexions and extensions about the elbow performed either isotonically (Fig. 4) or isometrically (not shown) at two different speeds, 0.5 Hz and 1.0 Hz. When the stimuli were applied to the forearm (Fig. 4A), the proportion of stimuli detected during movement was significantly diminished at both speeds (ANOVAs, $P<0.001$ in each case, *df* 1,7), but there was no difference as a function of the speed of movement ($P=0.85$; *df* 1,7). Although this observation suggests that speed was not an important factor in the determining the degree of gating of sensory perception, it should be pointed out that the sensory suppression was so profound in both of these experiments that task performance during movement was below threshold level (dotted line) even at the highest intensity employed. When the stimuli were applied to a more remote site, the hand (Fig. 4B), there was now no significant change in performance at the lowest velocity (P=0.07; *df* 1,7). Only at the higher speed was detection significantly reduced (P<0.0001; *df* 1,7). Consequently performance was significantly reduced when speed was increased (P=0.0068; *df* 1,7).

Movement kinetics also modulate the degree of movement-related perceptual gating. Pertovaara, Kemppainen and Leppänen (1992) demonstrated that detection thresholds on the hand increase, in a load-dependent manner, during isometric finger flexions. Using a different motor task, Post, Zompa and Chapman (1994) made a similar observation for isometric contractions about the elbow. Moreover, the results indicated that the spatial extent of the gating effects showed a wider distribution when the force level was augmented. As "movement time" remained constant, the increase in the target force level was accompanied by a concommitant change in dF/dt. In companion experiments to those shown in Fig. 4, however, modest changes in dF/dt (cycle speed increasing from 0.5 to 1.0 Hz) while keeping the target force constant had no effect on the degree of movement-related gating. In summary, the degree and the extent of movement-related gating of sensory perception changes as a function of the kinematics and kinetics of the

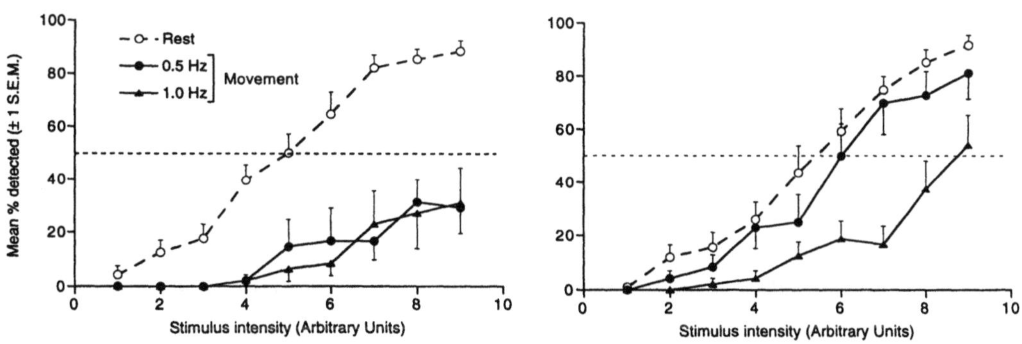

Fig. 4. Effects of two speeds (0.5 Hz, 1.0 Hz) of rythmical isotonic elbow flexions and extensions on the ability of 8 subjects to detect near-threshold electrical stimuli (see legend Fig. 1 for details) applied to the forearm (**A**) or the dorsum of the hand (**B**). Some of these data appeared in Feine et al. (1990).

movements.

There is also general agreement that the *proximity* between the site of stimulation and the movement influences the degree of movement-related gating of perception of innocuous, electrical stimuli. Thus, movements of nonstimulated limbs have no effect on detection threshold, discrimination threshold or magnitude estimates (Chapman et al., 1987; Milne et al., 1988; Schmidt, Torebjörk and Schady, 1990), i.e. the gating effects are limited to the moving limb. The exact spatial distribution of these effects is, however, less clearly established. Coquery, Malcuit and Coulmance (1971) reported that the perception of near-threshold electrical shocks was strongly diminished when the stimulus was applied to the moving hand or digits. Weaker gating effects were observed when the elbow or shoulder was moved. When the situation was reversed, i.e. when the stimuli were applied to the proximal limb segments, the results were less clear. For example, the perception of stimuli applied to the forearm was equally affected by movements of the digits, wrist, elbow and shoulder. More recent observations indicate that the spatial extent of the modulation may be more finely tuned. Thus, Schmidt, Torebjörk and Schady (1990) reported that there is a spatial gradient in the degree of gating across the hand, with the effects being maximal when the stimulated digit was moved, and gradually decreasing as successively more distant fingers were moved. Similar observations have been made for movements involving more proximal joints. Figure 2A compares the effects of motor activity on vibrotactile detection thresholds at the forearm, hand and digit during dynamic, rythmic isometric contractions about the elbow. Detection thresholds showed, as expected, a proximo-distal gradient. During the motor task, threshold was significantly elevated at the two proximal sites, but no significant change was seen at the more remote site (digit). Similar observations were also made for isotonic elbow movements (see Fig. 4A and B). In a similar vein, although magnitude estimates at both the digit and the forearm stimulation sites were significantly reduced during motor activity, the effects were again larger at the more proximal site (Fig. 3: overall 15% reduction at the forearm, and 10% at the digit). Thus, in contrast to the results of Coquery, Malcuit and Coulmance (1971), gating effects are spatially distributed even for movements about the elbow.

Although motor activity thus appears to be the key factor in explaining the changes in perception, one other factor needs to be considered, namely *attention*. Effectively, most of the experiments described above required the subjects to perform two tasks, a motor task and a perceptual task, so that the subjects' attention was essentially split during the movement condition. In this regard, Post and Chapman (1991) showed that the detection of vibrotactile stimuli is diminished when attention is split, or diverted away, from the stimulus. More recently, however, Zompa and Chapman (1995) have demonstrated that tactile discrimination thresholds (for texture) are also modified by attention. Since detection but not discrimination is affected by movement-related gating, these observations suggest that any role of attention in movement-related gating can only be relatively minor.

The gating signal

As reviewed above, the inhibitory influences that are responsible for the movement-related reductions in detection and subjective intensity have been characterized as showing a spatial gradient and sensitivity to the movement kinematics and kinetics. The latter observations are, in particular, consistent with the notion that the sources of the gating signals include both peripheral afferent feedback from the moving limb (e.g. Fig. 1, effects of passive movement) and the central motor command (Chapman, Jiang and Lamarre, 1988; Jiang, Chapman and Lamarre 1990), since both of these signals will increase as a function of the movement parameters.

As also pointed out above, the gating influences on perception are such that the relative differences between suprathreshold inputs are preserved, even though their magitude is decreased. We had originally proposed (Chapman et al., 1987) that gating signals act as a form of masking stimulus, interfering with the transmission of weak, but not strong, stimuli. In light of more recent observations, however, we have suggested that this notion needs to be revised (Post, Zompa and Chapman, 1994). A successful model needs to explain two additional observations. First, movement-related gating of the subjective intensity of electrical stimuli depends upon the presence of a strong source of peripheral feedback, and is relatively less sensitive to gating influences than is the perception of natural, spatially distributed stimuli

(Optacon arrays). Second, movement diminishes subjective intensity in a nonlinear fashion (see Fig. 3, noting the logarithmic scales).

Figures 5 and 6 show three potential forms for the gating signal that we have considered (shaded area on insets) and their effects on the amplitude of theoretical central responses to suprathreshold stimuli, assuming that perception is linked to the eventual amplitude of the central response (insets: two series of arbitrary responses to graded stimuli differing in their spatial extent by a factor of two). For all three models, the detection of near-threshold stimuli would be reduced in the presence of the gating signal.

In the first model (Fig. 5A), we chose to evaluate our original suggestion that masking was responsible for the changes in perception, i.e. the signal-to-noise ratio was reduced by a constant during movement. The stimulus-response functions at rest and during movement were, in this representation, parallel. Log-log plots (not shown) gave nonparallel, converging linear stimulus-response curves. This model would, thus, result in no change in the discrimination threshold (preservation of relative differences even in the presence of gating) and a reduction in subjective intensity at low but not high intensities of stimulation. The latter is not consistent with the experimental observations which show greater reductions at higher, as compared to lower, suprathreshold intensities (Fig. 3).

In the second model (Fig. 5B), it was reasoned that the inhibitory influence might show a monotonic and symmetrical increase from the central point of stimulation so that more spatially distributed inputs would be subjected to stronger inhibition. Consistent with the results obtained with magnitude estimation, this model leads to a nonlinear reduction in stimulus intensity, with the greatest reduction associated with the highest intensities. Spatially expanded stimuli are proportionally affected more than are spatially restricted stimuli. In addition, however, the two curves are not parallel, indicating that discrimination threshold would be elevated. The latter observation is not consistent with experimental observations of no change in discrimination threshold.

The closest approximation to the experimental data was obtained with the third model (Fig. 6) in which a nonlinear inhibitory surround was proposed (based on an arbitrarily chosen hyperbolic function: $y = 1/(0.9242 - 0.0302x)$. It can be seen that there is minimal gating of

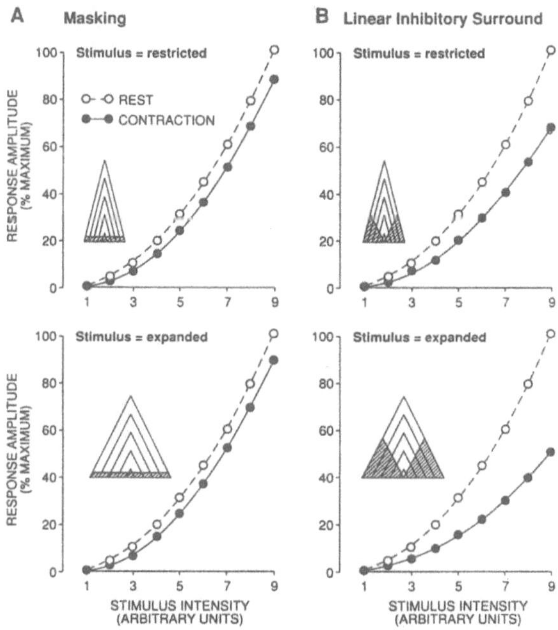

Fig. 5. Comparison of the effects of hypothetical gating signals (shaded areas on insets), **A**, masking and **B**, linear inhibitory surround, on the amplitude of central cortical responses to graded peripheral stimuli that were either spatially restricted (top) or expanded (bottom). The form of the spatial distribution of the theoretical central responses was based on observations that excitatory receptive fields for cutaneous neurones in feline somatosensory cortex are unimodal and generally symmetrically distributed about the centre of the field (Gardner and Spencer, 1972; Laskin and Spencer, 1979). One simplifying assumption was made: the gating signal was considered as unitary and acting at a single site (see Chapman, 1994). The curves show the theoretical central response amplitude (area under the representative curves shown in the insets) plotted as a function of the stimulus intensity at rest and during motor activity (latter after subtracting the gating signal).

the response for spatially restricted stimuli (A), consistent with observations made with electrical stimuli under a condition of minimal peripheral feedback (above). In contrast, gating is much more prominent for stimuli with a greater spatial extent especially at higher intensities (B). At the lower end of the curve, discrimination threshold would be unchanged. At the higher end, an increase in discrimination threshold should become evident. This has not been reported as yet, possibly because the stimulus intensities used in most experiments (including our own) were near the low end of the stimulus-response function. Overall, we suggest that this model gives results consistent with our previous findings (Chapman et al., 1987; Post, Zompa and Chapman, 1994), and provides the testable hypothesis that discrimination threshold measured with higher

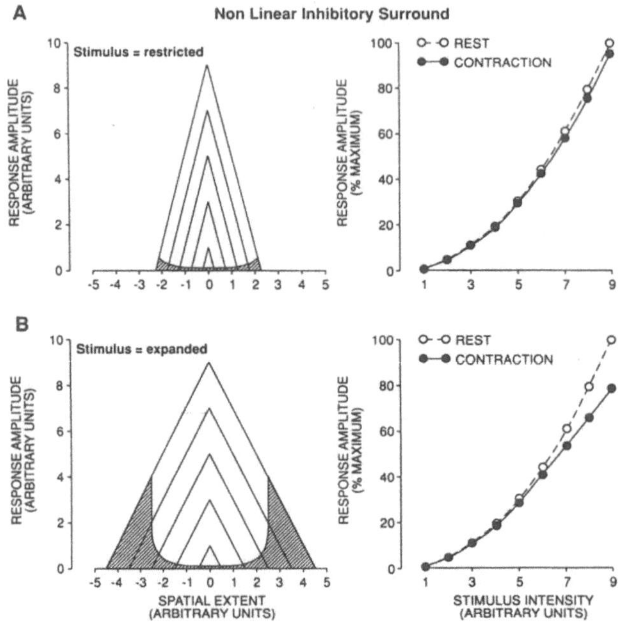

Fig. 6. Nonlinear inhibitory surround gating signal. Data plotted as in Fig. 5. In this model, spatially restricted stimuli are subject to no, or minimal, gating (**A**). Higher intensity, spatially distributed stimuli, on the other hand, are subject to gating controls (**B**).

intensities of stimuli should be elevated during motor activity.

Concluding remarks

Motor activity results in gating of the perception of near-threshold and suprathreshold tactile stimuli. Although we have not discussed the neural mechanisms in this chapter, other work from this laboratory indicates that the underlying mechanisms are complex (Chapman and Ageranioti-Bélanger, 1991; Ageranioti-Bélanger and Chapman, 1992; Chapman and Ageranioti, 1991). Thus, some neurones in primary somatosensory cortex receiving tactile inputs are spared from gating influences. Other neurones have their responsiveness strongly reduced during movement, and this in situations in which the monkeys were attentive to the stimuli because they were engaged in performing a texture discrimination task. The challenge in the future will be to determine the respective roles of these different types of neurones in relation to perceptual processes.

Acknowledgements
We would like to thank the following for the excellent technical assistance provided: the late Richard Bouchoux, Marc Bourdeau, Daniel Cyr, Giovanni Filosi, Claude Gauthier and Christian Valiquette. We thank Dr. Trevor Drew for helpful comments on the manuscript. Funding from the Medical Research Council of Canada, the Fonds de la recherche en santé du Québec (scholarship to C.E.C. and bursary to S.R.W), Groupe de la recherche sur le système nerveux central (FCAR) and Unimedia is gratefully acknowledged.

References

Ageranioti-Bélanger, S.A. and Chapman, C.E. (1992) Discharge properties of neurones in the hand area of primary somatosensory cortex in monkeys in relation to the performance of an active tactile discrimination task. II. Area 2 as compared to areas 3b and 1. *Experimental Brain Research* 91: 207-228.

Angel, R.W. and Malenka, R.C. (1982) Velocity-dependent suppression of cutaneous sensitivity during movement. *Experimental Neurology* 77: 266-274.

Chapman, C.E. (1994) Active versus passive touch: factors influencing the transmission of somatosensory signals to primary somatosensory cortex. *Canadian Journal of Physiology and Pharmacology* 72: 558-570.

Chapman, C.E. and Ageranioti, S.A. (1991) Comparison of the discharge of primary somatosensory cortical (SI) neurones during active and passive tactile discrimination. *Proceedings Third IBRO World Congress of Neuroscience*, p. 317.

Chapman, C.E. and Ageranioti-Bélanger, S.A. (1991) Discharge properties of neurones in the hand area of primary somatosensory cortex in monkeys in relation to the performance of an active tactile discrimination task. I. Areas 3b and 1. *Experimental Brain Research* 87: 319-339.

Chapman, C.E., Bushnell, M.C., Miron, D., Duncan, G.H. and Lund, J.P. (1987) Sensory perception during movement in man. *Experimental Brain Research* 68: 516-524.

Chapman, C.E., Jiang, W. and Lamarre, Y. (1988) Modulation of lemniscal input during conditioned arm movements in the monkey. *Experimental Brain Research* 72: 316-334.

Coquery, J.M., Malcuit, G. and Coulmance, M. (1971) Altérations de la perception d'un stimulus somesthésique durant un mouvement volontaire. *Comptes Rendus Société de Biologie (Paris)* 165: 1946-1951.

Dyhre-Poulsen, P. (1978) Perception of tactile stimuli before ballistic and during tracking movements. *In:* G. Gordon (ed.): *Active Touch*, Pergamon Press, Oxford. pp. 171-176.

Feine, J.S., Chapman, C.E., Lund, J.P., Duncan, G.H. and Bushnell, M.C. (1990) The perception of painful and nonpainful stimuli during voluntary motor activity in man. *Somatosensory and Motor Research* 7: 113-124.

Gardner,E.P. and Spencer, W.A. (1972) Sensory funneling. II. Cortical neuronal representation of patterned cutaneous stimuli. *Journal of Neurophysiology* 35: 954-977.

Garland, H.T. and Angel, R.W. (1974) Modulation of tactile sensitivity during movement *Neurology (NY)* 24: 361.

Ghez,C. and Lenzi, G.L. (1971) Modulation of sensory transmission in cat lemniscal system during voluntary movement. *Pflügers Archiv* 323: 273-278.

Jiang, W., Chapman, C.E. and Lamarre, Y. (1990) Modulation of somatosensory evoked responses in the primary somatosensory cortex produced by intracortical microstimulation of the motor cortex in the monkey. *Experimental Brain Research* 80: 333-344.

Jones, S.J., Halonen, J.-P. and Shawkat, F. (1989) Centrifugal and centripetal mechanisms involved in the 'gating' of cortical SEPs during movement. *Electroencephalography and Clinical Neurophysiology* 74: 36-45.
Lamb, G.D. (1983) Tactile discrimination of textured surfaces: psychophysical performance measurements in humans. *Journal of Physiology (London)* 338: 551-565.
Laskin, S.E. and Spencer, A.W. (1979) Cutaneous masking. II. Geometry of excitatory and inhibitory receptive fields of single units in somatosensory cortex of the cat. *Journal of Neurophysiology* 42: 1061-1082.
Latour, P.L. (1962) Visual threshold during eye movements. *Vision Research* 2: 261-262.
Lederman, S.J. (1981) The perception of surface roughness by active and passive touch. *Bulletin of the Psychonomics Society* 18: 253-255.
Milne, R.J., Aniss, A.M., Kay, N.E. and Gandevia, S.C. (1988) Reduction in perceived intensity of cutaneous stimuli during movement: a quantitative study. *Experimental Brain Research* 70: 369-576.
Pertovaara, A., Kemppainen, P. and Leppänen, H. (1992) Lowered cutaneous sensitivity to nonpainful electrical stimulation during isometric exercise in humans. *Experimental Brain Research* 89: 447-452.
Post, L.J. and Chapman, C.E. (1991) The effects of cross-modal manipulations of attention on the detection of vibrotactile stimuli in humans. *Somatosensory and Motor Research* 8: 149-157.
Post, L.J., Zompa, I.C. and Chapman, C.E. (1994) Perception of vibrotactile stimuli during motor activity in human subjects. *Experimental Brain Research* 100: 107-120.
Schmidt, R.F., Schady, W.J.L. and Torebjörk, H.E. (1990) Gating of tactile input from the hand. I. Effects of finger movement. *Experimental Brain Research* 79: 97-102.
Schmidt, R.F., Torebjörk, H.E. and Schady, W.J.L. (1990) Gating of tactile input from the hand. II. Effects of remote movements and anaesthesia. *Experimental Brain Research* 79: 103-108.
Srinivasan, M.A. and LaMotte, R.H. (1995) Tactual discrimination of softness. *Journal of Neurophysiology* 73 88-101.
Vega-Bermudez, F., Johnson, K.O. and Hsiao, S.S. (1991) Human pattern tactile recognition: active versus passive touch, velocity effects and patterns of confusion. *Journal of Neurophysiology* 65: 531-546.
Zompa, I.C. and Chapman, C.E. (1995) Effect of cross-modal manipulations of attention on the ability of human subjects to discriminate changes in texture. *Somatosensensory and Motor Research* 12: 87-102.
Zuber, B.L. and Stark L. (1966) Saccadic suppression: elevation of visual threshold associated with saccadic eye movements. *Experimental Neurology* 16: 65-79.

Somesthesis and the Neurobiology of the Somatosensory Cortex
O. Franzén, R. Johansson and L. Terenius (eds.)

Changing the intended direction of movement

Giuseppe Pellizzer

Department of Physiology, University of Minnesota Medical School, Minneapolis, MN 55455, USA, and Brain Sciences Center, Veterans Affairs Medical Center, Minneapolis, MN 55417, USA

Summary. The behavioral and neural correlates of processing of motor directional information are described for two different visuomotor tasks: mental rotation and context-recall. Psychological studies in human subjects suggested that these two tasks may involve different mechanisms processing directional information. Analyses of the activity of single cells and neuronal populations in the motor cortex of behaving monkeys performing in the same tasks provided direct insight into the neural mechanisms involved, and confirmed their different nature.

Introduction

The objective of cognitive neuroscience is to define the neural mechanisms underlying cognitive processing. The strategy usually followed in experimental studies is to record neural events (e.g., activity of neurons, electroencephalogram) during the performance of a cognitive task. Two levels of analysis of the data can be performed. A first level of understanding of the cognitive processes consists of describing the association of neural events to relevant variables of the task. This provides a first degree of insight into the processes, but has usually little explanatory power. A second level of analysis is to model the cognitive processes and its development in time from the neural data. Only this second level of analysis provides a potential explanation of the neural mechanisms of the cognitive processes. Two examples of cognitive processing in visuomotor tasks are described in the following sections: mental rotation and context-recall. For each one, behavioral experiments and neural mechanisms as observed in the motor cortex are presented. However, to understand the results of these experiments, it is necessary first to describe how neural activity in the motor cortex can be related to directional motor information.

Neural coding of direction of movement

The relation between direction of movement and single cell activity in the primary motor cortex was studied in awake monkeys (e.g., Georgopoulos et al., 1982; Schwartz et al., 1988). In these studies, neuronal activity was recorded in the arm area of the primary motor cortex contralateral to the performing arm, while the monkey moved the hand from a starting position in the direction of a visual target. Significant change of activity in relation to the direction of movement was observed before the initiation of movement and during its execution. How much a cell is active depends on the cell and on the direction of movement. Cell's activity is broadly tuned with respect to the direction of movement; that is, the activity is highest for movements in a particular direction, and decreases progressively for movements made farther away from this direction. The direction of movement corresponding to the highest activity of the cell is called the preferred direction of the cell. The preferred directions vary from cell to cell, and are distributed in all directions. These results indicate (1) that a given cell contributes in movements in various directions, (2) that movements in a particular direction involves the contribution of a large population of cells, and (3) that the direction of movement is uniquely defined only at the level of a population of cells.

One hypothesis about the population coding of direction of movement is to regard the directionally tuned cells as an ensemble of vectors (Georgopoulos et al., 1983, 1986). Each vector represents the contribution of a directionally tuned cell. A particular vector points in the direction of the cell's preferred direction, and its length is a function of the cell activity for a given direction of movement. The vector sum of this ensemble of vectors is called the neuronal population vector, and points at or near the direction of movement. These results indicate that information concerning the direction of movement can be extracted unambiguously from the neuronal ensemble. What is most important for the next sections of this chapter is that the neuronal population vector provides a tool to monitor the processing of directional information in time before the initiation of movement, that is, when the movement is intended. For example, it was observed that the neuronal population vector computed in time points in the direction of the upcoming movement during the response time, that is, after the visual target is presented and before the movement is initiated (Figure 1; Georgopoulos et al., 1984, 1988); during an instructed delay, that is, when a delay is imposed before the initiation of the movement (Georgopoulos et al., 1989a); and during a memorized delay, that is, when the information about the direction of the movement must be kept in memory (Smyrnis et al., 1992).

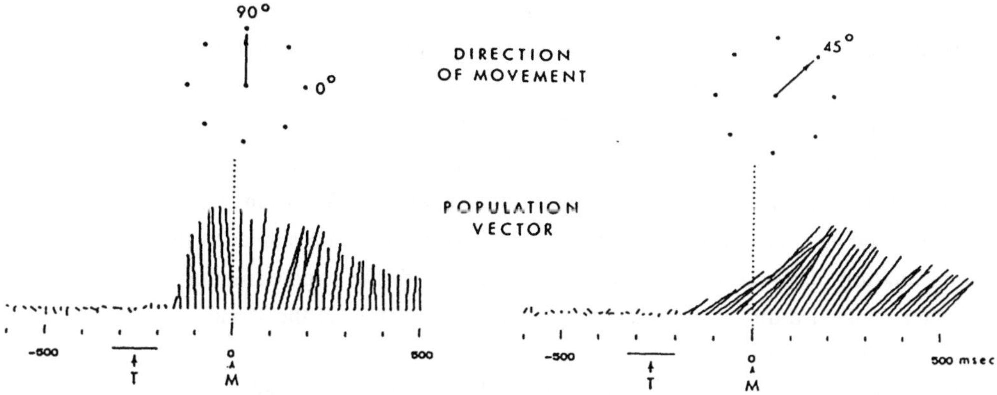

Figure 1. Time course of population vector for movements at 90° and 45°. The population vector was calculated every 20 ms. After the target onset (T) and before the begining of the movement (M) the population vector increases in length and its direction points toward the direction of the upcoming movement. (Georgopoulos et al., 1984; reproduced with permission).

Mental rotation

Shepard and colleagues (Shepard and Metzler, 1971; Shepard and Cooper, 1982) reported a series of experiments that supports the hypothesis of a mental rotation of visual images. Typically, subjects have to judge whether two figures are the same or are mirror images of each other. The finding was that the response time increased as a linear function of the angular difference between the orientation of the two figures. This suggested that subjects mentally rotate the image of one figure into congruence with the other to make their judgment.

The process of mental rotation is not limited to perceptual type of tasks, but can be also inferred in visuomotor tasks (Georgopoulos and Massey, 1987; Pellizzer and Georgopoulos, 1993). In these tasks, human subjects were asked to make a planar arm movement in a direction that was at a specified angle away from the direction of a visual stimulus. The results were that the response time increased as a linear function of the angle. These results are similar to those described for mental rotation of visual images and are compatible with the hypothesis that subjects mentally rotate their intended direction of movement from the direction of the visual stimulus toward the direction of movement.

The hypothesis of mental rotation assumes that there is a neurophysiological variable that represents the orientation of an imagined object, or the direction of the intended movement, and that this variable undergoes a continuous transformation during mental rotation passing through values that represent intermediate directions. Direct evidence for such a process was obtained in an experiment with monkeys that were trained to move a handle on a planar surface either toward a visual stimulus (*direct* task) or 90 degrees counterclockwise from it (*transformation* task) depending on whether the stimulus was dim or bright, respectively (Georgopoulos et al., 1989b; Lurito et al., 1991).

The activity of cells in the arm area of the motor cortex contralateral to the performing arm was recorded extracellularly. The changes of cell activity after the visual stimulus appeared in the *direct* task were as described previously (Georgopoulos et al., 1982, 1984, 1988). The activity of these cells also changed in the *transformation* task. There were no cells the activity of which changed only in one task, which suggests that the representation of the direction of movement involves the same population of motor cortical cells in both tasks. The analysis of the population vector as a function of time after the target onset revealed the process of rotation of the intended direction of movement. During the response time the population vector rotated from the direction of the target counterclockwise through 90 degrees to the direction of the movement (Figure 2). The rates of rotation of the population vector (about 400 deg/s) were within the range of those obtained with human subjects in the same task (Georgopoulos and Massey, 1987; Pellizzer and Georgopoulos, 1993). In addition, it was found that cells with preferred directions intermediate between the direction of the stimulus and the direction of the movement were transiently recruited in the middle of the response time, which supports the hypothesis of a true rotation of the population signal (Lurito et al., 1991).

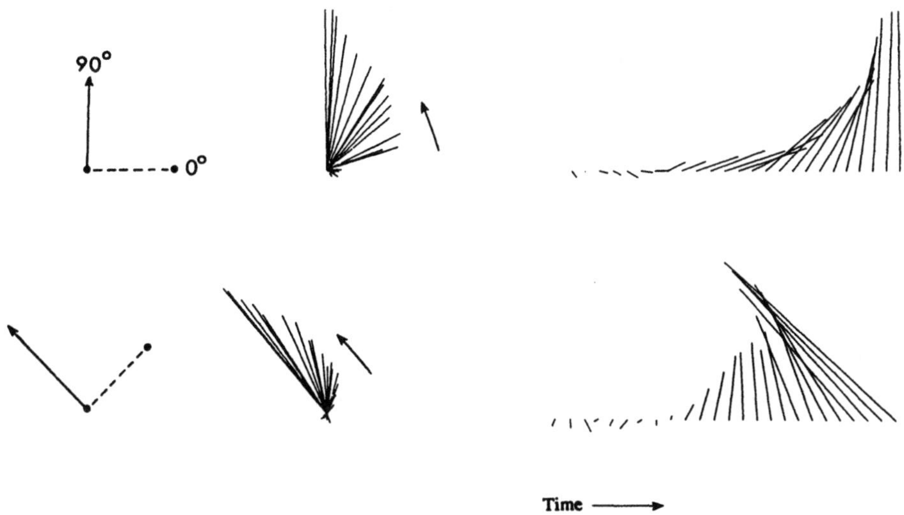

Figure 2. Neuronal population vector in two conditions of the transformation task. Left panel: Stimulus (dashed line) and movement (arrow) directions. Right panel: Neuronal population vector plotted every 10 ms from the onset of the stimulus until after the onset of the movement. Middle panel: Same data as in right panel but plotted in polar fashion. When the population vector lengthens it points initially in the direction of the stimulus and then rotates counter-clockwise and points in the direction of the movement. (Lurito et al., 1991; reproduced with permission).

These results have several remarkable points: (1) There was no a priori reason to expect that the population vector would point to any direction other than the direction of movement, on the simple hypothesis that the motor cortex is only involved in the production of movement. The interpretation that the population vector indicates the intended direction of movement suggests that, in the transformation task, the intention of movement spans a range of directions during the response time. (2) There was no a priori reason to expect the direction of the population vector to shift gradually counterclockwise, for no explicit instruction was given to the animal about how to accomplish the transformation task. This strategy may minimize time and computational load to solve the transformation required.

Mental rotation and memory scanning

Are visual mental rotation and visuomotor mental rotation associated in some way? To answer this question we took advantage of the large diversity of rates of rotation between subjects (Pellizzer and Georgopoulos, 1993). We found a significant correlation of the rates of rotation in human subjects performing in the two tasks. This suggested that the two types of mental rotation either share a common stage of processing, or are the object of similar constraints although performed by separate neural mechanisms. It is also possible that the correlation between processing rates results from general processing rates differences between subjects, that is, some subjects may perform generally faster (or slower) than others. To control for this, we needed a different type of task where a processing rate can also be estimated. We chose a visuomotor memory scanning task (Georgopoulos and Lurito, 1991). In this task, a list of two to five visual stimuli directions were presented sequentially. Then, one of them, except the last one, was displayed again as test stimulus. The subjects had to make an arm movement in the direction of the stimulus that followed the test stimulus in the sequence. The processing rates in this task did not correlate with those in either mental rotation tasks. These results suggested that (1) visual mental rotation and visuomotor mental rotation share common processing constraints that cannot be ascribed to general mental processing capacities; (2) that mental rotation and memory scanning involve different types of processing mechanisms.

Memory scanning

An important aspect of cognitive function relates to retrieval of memorized information. Sternberg (1966, 1967, 1969) devised a series of tasks that requires to find an information among items in a memorized list. Two of these tasks (*item-recognition* and *context recall*) are briefly described here. Typically, subjects were presented with a list of items to memorize, followed with a test stimulus. In the *item-recognition* task, the subjects had to decide whether the test stimulus was contained in the list presented, or was not in the list. The results were that the response time for positive and negative responses increased as a linear function of the number of items contained in the list. This suggested that the task involves a scanning process, that is, a serial comparison of the test stimulus with the memorized items. If the scanning process were self-terminating, the number of comparisons in trials requiring a positive response would be on average half those in trials requiring a negative response, therefore the slope of the response time versus the number of items for the positive responses should be half the one for the negative responses. Instead, if the scanning process is exhaustive, that is, all items are scanned even when a match had occurred, the two slopes should be equal. The results fit the latter pattern, which supports the hypothesis that the *scanning to recognize* is exhaustive.

In the *context-recall* task, the subjects had to locate the test stimulus in the list. Then they had to name the item that followed the test stimulus in the list. Like in the previous task, the response time increased with the number of items in the list, but at a slower rate. Moreover, the response time increased also with the serial position of the test stimulus in the list. Therefore, these results suggest that the retrieval of contextual information, viz. the *scanning to locate*, is a self-terminating scanning process.

These paradigms provide a way to investigate cognitive operations on memorized items. The *context-recall* task was adapted to test memory scanning in the motor system (Georgopoulos and Lurito, 1991; Pellizzer and Georgopoulos, 1993). In these experiments, a sequence of visual stimuli directions was presented to the subjects. They had either to move the hand toward them when they were presented (Georgopoulos and Lurito, 1991), or only to look at them (Pellizzer and Georgopoulos, 1993). Then one of the stimuli, except the last one, was presented again as test stimulus. The subjects had to make a movement in the direction of the stimulus that followed the test stimulus in the sequence. The results showed that the response time increased with the number of stimuli in the list, but not with the serial position of the test stimulus. This suggests that the scanning process in this case is either exhaustive, or self-terminating with a random start within the sequence.

The activity of cells in the motor cortex was recorded recently during a *context-recall* task (Pellizzer et al., 1995). A monkey was trained in two tasks (*control* and *context-recall*) to exert a force pulse on a semi-isometric manipulandum in eight directions. The force exerted was displayed as a feedback cursor. In the *control* task, a yellow stimulus was presented in one direction and stayed on the screen, then after a delay it turned blue. This provided the go signal for the monkey to exert a force pulse such that the feedback cursor crossed the position of the stimulus. In the *context-recall* task, three different stimuli were presented successively and stayed on the screen. Then, the first or second stimulus turned blue. This identified the test stimulus and gave the go signal. The monkey had to move the cursor in the direction of the stimulus that followed the test stimulus in the sequence.

The activity of single cells tuned to the direction of the motor response in the *control* task served as markers to interpret the activity recorded during the *context-recall* task. It was observed that cell activity during the first half of the response time matched the activity associated with a response toward the direction of the second stimulus. Then, what happened to the cell activity during the second half of the response time depended on the serial position of the test stimulus. When the test stimulus was the first stimulus in the sequence, cell activity continued to match the activity associated with a response toward the second stimulus, which was the correct direction of response. However, when the test stimulus was the second in the sequence, cell activity changed abruptly to correspond to the activity associated with the direction of the third stimulus, which was in this condition the correct direction of response (Figure 3). The switching process was visualized at the ensemble level using the neuronal population vector computed in time. When the anticipated response did not have to change, the population vector pointed in the appropriate direction throughout the response time. In contrast, when the response had to be changed, the population vector changed abruptly from the direction of the test stimulus to the direction of the motor response.

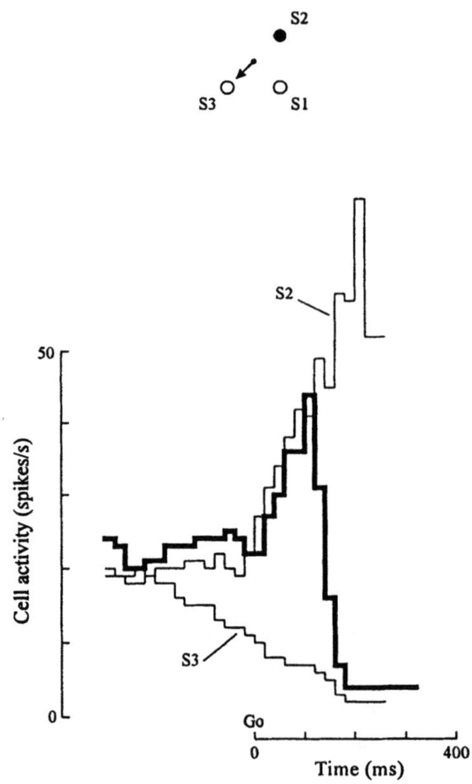

Figure 3. Peristimulus histograms of activity of a motor cortical cell for two directions in the control task, and one condition in the context-recall task. The two thin lines represent the activity of the cell in the control task for motor responses in the upward-right (S2) and downward-left (S3) directions. The thick line represents the activity of the cell in the condition of the context-recall task illustrated at the top. After the go signal, cell activity initially increased in the same way as in the control case for the direction towards the test stimulus (S2) and then changed abruptly to the level corresponding to the control activity.

The abrupt change in the direction of the population vector observed in this study is quite different from the slow rotation observed in the mental rotation experiment. Additional evidence for the different nature of the two processes was that (1) the duration of change of direction of the neuronal population vector did not vary with the angle between the test stimulus and the direction of the response, unlike what would be expected if the transformation were a rotation at a constant speed; (2) cells with preferred directions intermediate between the direction of the test stimulus and the direction of the motor response were not recruited in the middle of the response time, unlike what was observed in the mental rotation experiment.

Conclusions

The experiments presented in this chapter provide an example of interaction between psychological and neurophysiological studies of cognitive function. Psychological studies suggested that the neural mechanisms underlying the processing of directional motor information in visuomotor mental rotation and context-recall tasks are of different nature. The relation between neural activity in the motor cortex and the intended direction of movement gives a tool to investigate the neural mechanisms underlying the processing of directional motor information in both tasks. Obviously other brain areas than motor cortex are likely to be involved, but additional studies are needed to identify them, and to elucidate their relative contributions during these tasks. Neurophysiological studies provided direct insight into the neural mechanisms of changing the direction of the intended direction of movement during mental rotation and context-recall tasks. The analyses of neural activity in the motor cortex showed that during the mental rotation task, the direction of the intended direction of movement is slowly and gradually transformed from the direction of the stimulus toward the direction of the response. Instead, during the context-recall task when a change of the intended direction of movement is required, this direction switches abruptly. Therefore it seems that the time spent to derive the direction of the motor response in the mental rotation task reflects an analogue transformation process, instead in the context-recall task it reflects a selection process.

Acknowledgements

This work was supported by United States Public Health Service grant NS17413.

References

Georgopoulos, A.P., Caminiti, R., Kalaska, J.F., and Massey, J.T. (1983) Spatial coding of movement: A hypothesis concerning the coding of movement direction by motor cortical populations. *Experimental Brain Research Supplement* 7: 327-336.

Georgopoulos, A.P., Crutcher, M.D. and Schwartz, A.B. (1989a) Cognitive spatial-motor processes: 3. Motor cortical prediction of movement direction during an instructed delay period. *Experimental Brain Research* 75: 183-194.

Georgopoulos, A.P., Kalaska, J.F., Caminiti, R. and Massey, J.T. (1982) On the relations between the direction of two-dimensional arm movement and cell discharge in primate motor cortex. *Journal of Neuroscience* 2: 1527-1537.

Georgopoulos, A.P., Kalaska, J.F., Crutcher, R., Caminiti, R. and Massey, J.T. (1984) The representation of movement direction in the motor cortex: Single cell and population studies. *In:* G.M. Edelman, W.E. Gall and W.M. Cowan (eds.): *Dynamic Aspects of Neocortical Functions*, Wiley, New York, pp. 501-524.

Georgopoulos, A.P., Kettner, R.E. and Schwartz, A.B., (1988) Primate motor cortex and free arm movements to visual targets in three-dimensional space: II. Coding of the direction of movement by a neuronal population. *Journal of Neuroscience* 8: 2928-2937.

Georgopoulos, A.P. and Lurito, J.T. (1991) Cognitive spatial-motor processes. 6. Visuomotor memory scanning. *Experimental Brain Research* 83: 453-458.

Georgopoulos, A.P., Lurito, J.T., Petrides, M., Schwartz, A.B. and Massey, J.T. (1989b) Mental rotation of the neuronal population vector. *Science* 243: 234-236.

Georgopoulos, A.P. and Massey, J.T. (1987) Cognitive spatial-motor processes: 1. The making of movements at various angles from a stimulus direction. *Experimental Brain Research* 65: 361-370.

Georgopoulos, A.P., Schwartz, A.B. and Kettner, R.E. (1986) Neuronal population coding of movement direction. *Science* 233:, 1416-1419.

Lurito, J.T., Georgakopoulos, T. and Georgopoulos, A.P. (1991) Cognitive spatial-motor processes. 7. The making of movements at an angle from a stimulus direction: Studies of motor cortical activity at the single cell and population levels. *Experimental Brain Research* 87: 562-580.

Pellizzer, G. and Georgopoulos, A.P. (1993) Common processing constraints for visuomotor and visual mental rotation. *Experimental Brain Research* 93: 165-172.

Pellizzer, G., Sargent, P. and Georgopoulos, A.P. (1995) Motor cortical activity in a context-recall task. *Science* 269: 702-705.

Schwartz, A.B., Kettner, R.E. and Georgopoulos, A.P. (1988) Primate motor cortex and free arm movements to visual targets in three-dimensional space: I. Relations between single cell discharge and direction of movement. *Journal of Neuroscience* 8: 2913-2927.

Shepard, R.N. and Cooper, L.A. (1982) *Mental Images and their Transformations.* MIT Press, Cambridge, MA.

Shepard, R.N. and Metzler, J. (1971) Mental rotation of three-dimensional objects. *Science* 171: 701-703.

Smyrnis, N., Taira, M., Ashe, J. and Georgopoulos, A.P. (1992) Motor cortical activity in a memorized delay task. *Experimental Brain Research* 92: 139-151.

Sternberg, S. (1966) High-speed scanning in human memory. *Science* 153: 652-654.

Sternberg, S. (1967) Retrieval of contextual information from memory. *Psychonomic Science* 8: 55-56.

Sternberg, S. (1969) Memory-scanning: Mental processes revealed by reaction-time experiments. *American Scientist* 57: 421-457.

Somesthesis and the Neurobiology of the Somatosensory Cortex
O. Franzén, R. Johansson and L. Terenius (eds.)

Disturbances of motor behaviour after parietal lobe lesions in the human

Hans-Joachim Freund

Department of Neurology, Heinrich-Heine-University Düsseldorf, D-40225 Düsseldorf, Germany

Introduction

There are many reports on parietal lobe damage in the human and in non-human primates which describe a bewildering multiplicity of phenomena. It is therefore difficult to approach a systematic examination of the question of what the typical deficits are and whether damage to different parts of the parietal lobe produces consistent and specific regional patterns of disturbance.

To begin with, one would expect sensory and cognitive deficits after parietal damage whereas motor functions would be expected to be compromised only by the effects of the sensory disturbances. Indeed, the effects of postcentral excisions (Foerster, 1936) or lesions (Déjérine, 1907; Head and Holmes, 1911; Critchley, 1953) are classically described as closely resembling the effects of deafferentation with prominent ataxia, dysmetria and a poverty of movement. There is, however, ample evidence to show that lesions of the posterior parietal cortex cause major disturbances of motor behaviour, in particular, that of purposive higher order motor functions.

Before discussing this issue a note of caution must be expressed with respect to the interpretation of lesion data in humans. Lesions vary widely in relation to size, site and subcortical extent. The latter may cause disconnection syndromes by interrupting long-distance cortico-cortical connections. Such effects cannot be attributed to parietal lobe dysfunctions.

The impact of a lesion on function depends on a number of determinants. One factor is what Jackson called: "the lesion momentum". An acute lesion causes more severe functional deficiencies than a slowly developing lesion. It is well-known that even large tumours which infiltrate extensive brain areas remain asymptomatic unless they produce focal seizures. In patients with tumours of the precentral gyrus it can be shown that regional cerebral blood flow activation during finger movements only occurs outside the tumour, leading to functional shift effects between 9-43 mm along either the mediolateral body representation of the motor cortex or into the premotor or parietal cortex (Seitz et al., 1995). Furhermore, it is known that developmental lesions are much better compensated for than late acquired lesions.

In addition to these differences in the properties of lesions, the liability of brain areas for functional impairment is also different. When preferentially unilateral systems are affected, the deficiency will be severe, persistent and will affect only the contralateral side. Unilateral damage of bilateral systems usually causes mild and transient deficiencies which are well compensated for by the other side. Moreover, damage of lateralised brain functions such as language or higher order motor control only appears after damage to the major hemisphere.

In most lesion studies these variables are not well defined. This lack of definition makes the interpretation of the results and their comparison extremely difficult. For this reason the present review will focus on studies implying well defined small lesions restricted to different parts of the parietal lobe. This survey will concentrate on three parietal areas of interest: 1. The anterior parietal cortex (postcentral gyrus), 2. The superior posterior parietal cortex (SPL) and 3. The inferior parietal cortex (IPL).

Lesions of the anterior parietal lobe

The sequelae of this lesion type are still the best defined. Foerster (1936) reported that surgical excisions of the postcentral gyrus lead initially to complete anaesthesia, areflexia and hypotonia of the contralateral side and to a disturbance of motility with slightly diminished force and slowing of movement. The execution of accurate movements or the selective muscle activation required for fractionated finger movements becomes impossible. Consequently, isolated movements of the toes cannot be executed without moving the leg at the knee and hip as well. During regression of the mild initial weakness, ataxia becomes prominent, closely resembling that seen after dorsal root section. Excess activation of the agonist and antagonist muscles leads to discontinuous movement patterns. The smoothness and accuracy of the movements can, however, be improved by visual guidance. Dystonic posturing and the control of the fine finger movements during prehension and object manipulation, or exploration, is impaired (tactile paresis). Numerous clinical studies have confirmed this postcentral syndrome.

Lesions of the superior posterior parietal lobule (SPL)

The disturbances of elementary sensation, with their secondary effects on movement-precision which mimic the situation in deafferented patients, contrast with those caused by damage to the SPL. Damage to the SPL leads to the impairment of complex somatosensory functions such as object recognition, and identification of textures and spatial information without deficiencies of elementary sensation. On the motor side, hand function is severely compromised so that active touch, object exploration and manipulation become impossible (Pause et al., 1989). Quantitative analysis of hand and finger movements in these patients showed a derangement of the dynamics in the digital palpation of objects, along with a breakdown of the finely-tuned scanning process of the fingers preventing the sequential sampling of mechanoreceptive information. Fig. 1 shows the normal scan path of the fingers during exploration of an object, contrasted with the loss of the spatial and temporal characteristics of the explorative profile of the finger movements.

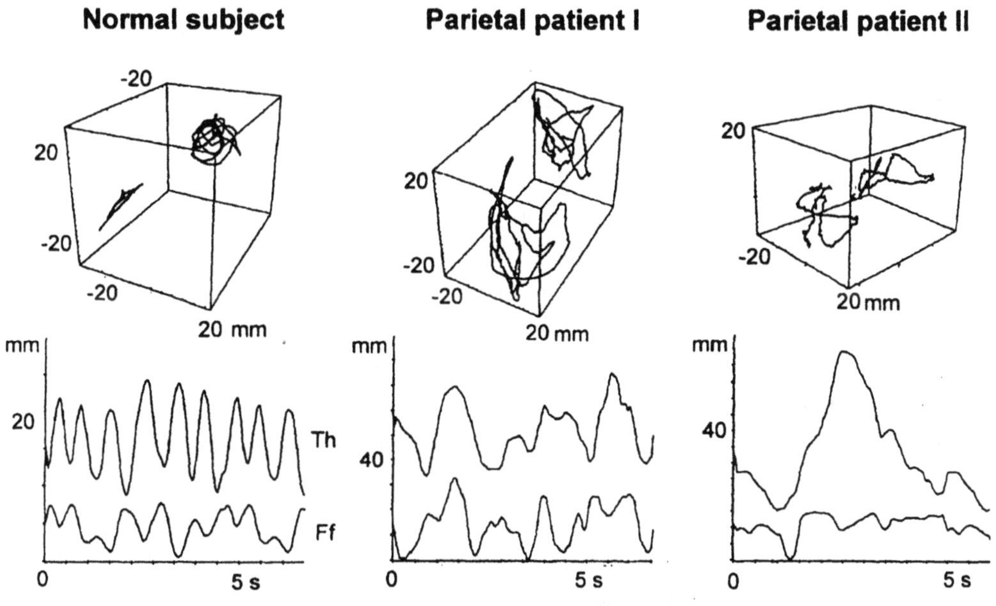

Fig. 1 Spatial and temporal characteristics of finger movements during tactile exploration of a ball, as revealed by an opto-electronic system. The movement trajectories of the forefinger and thumb are shown as viewed from the front, so that the vertical and horizontal movement components are displayed. The left side shows the performance of a normal subject, and the middle and right side the performance of the affected hand of 2 apraxic patients. The workspace of the scanning movements is larger and more irregular for the affected hand. The temporal profiles in B and C show the breakdown of the finely-tuned, regular motion.

The essence of this disturbance lies in the impairment of the conception and generation of the spatial-temporal movement patterns required to bring those receptor sheaths into action that would normally provide the information required for the identification and manipulation of objects. This selective deficiency is strictly unimodal and can partly be compensated for by visual guidance. It is called "tactile apraxia" (Pause et al., 1989; Freund 1987, 1992, 1995) and it illustrates the interdependence of the sensory motor processes. The extraction of sensory features relies on the adequacy of the purposive motor acts. Consequently, the somatosensory association cortex is not only involved in the processing of somatosensory information but also in the elaboration of the movement concepts.

Another unimodal higher order motor disturbance that follows lesions of a sensory association area is visuomotor apraxia. This condition is the inability to direct the arm or the head/eyes adequately towards a target. It is usually seen after parieto-occipital lesions, frequently occurring bilaterally (Balint, 1909; Gowers, 1897; Rondot et al., 1977). The disturbance was originally described as optic ataxia and later as visuomotor ataxia. The movement disturbance affects only the visual domain, representing another unimodal movement disturbance. Again there are perceptive, cognitive and motor aspects of the dysfunction: the imperfect perception of distances and spatial relationships, paralleled by inadequate target-directed adjustment of the arm and eyes and by a grossly deranged movement pattern. This makes the condition different from purely atactic disturbances where the process of aiming is faulty but the movement is otherwise executed with a correct kinematic pattern. Therefore, the apractic motor behaviour is strikingly different from that observed in acutely blind patients who can easily remember target location and perform smooth and accurate movements to remembered targets. In contrast, these patients show grossly deranged trajectories of reaching, aiming, grasping and hand pre-shaping. Their ability to learn to reach for targets with known locations cued by proprioceptive or acoustic input is poor. For these reasons we prefer the term "visuomotor apraxia" in order to characterise not only the deficient visual guidance but also the apractic kinematic features (Jeannerod, 1988; Freund 1993, 1995; Pause et al., 1989, Classen et al., in print).

Tactile apraxia only affects the contralateral side of the body. Visuomotor apraxia may solely affect the opposite hemifield but it also has an ipsilateral component, probably owing to bilateral parieto-pontine-cerebellar-thalamo-frontal cortical projections (Glickstein et al. 1980, 1985; Classen et al., in press).

Lesions of the inferior posterior parietal lobule (IPL)

Damage to the IPL affects motor behaviour differently, depending on the side of the lesion. Damage to the area on the side of the language-dominant hemisphere leads to the classical forms of ideational or ideomotor apraxia. According to Liepmann's (1900, 1920) definition, ideational apraxia is characterised by a deficiency in the conception of the movement so that the patient does not know what to do or how to organise movement sequences. In ideo-kinetic (ideomotor) apraxia, the patient does not know how to perform a particular motor act such as a gesture. The disturbance affects both sides of the body and is often associated with aphasia. In this case, the movement repertoire must be tested by imitation. On the basis of the apraxias, Liepmann postulated that the left hemisphere is not only dominant for language but also for praxis. The apraxias seen after IPL damage are supramodal in nature. This is compatible with the presumed supramodal organisation of the human IPL.

The frequent association of apraxia and aphasia raises these possibility that these two disorders are the expression of a common underlying deficiency of communication which affect linguistic and gestural behaviour in a general sense. There has been a long-standing discussion regarding the nature of the speech deficiency seen in the Wernicke type of sensory aphasia where comprehension is preferentially affected but where speech production is grossly deranged. Aphasia could be conceived as another example illustrating the two aspects of sensory motor dysfunction that have already been described for the tactile and visuomotor apraxias. The plasticity of this general principle of sensory motor processing is illustrated by observations on deaf-mute patients using American Sign Language (ASL) for communication (Bellugi et al., 1989). The fact that a visual-gestural communication system can substitute for the auditory-vocal one shows the capacity of the human nervous system to deal with symbolic and iconic information in a context-specific way, irrespective of the sensory modality or the motor subsystem involved. In three patients with left hemisphere damage, Bellugi and co-workers (1989) observed a breakdown in the capacity to communicate by ASL, affecting comprehension and production. The patients were unable to produce the spatial temporal trajectories required for ASL with their arms and their hands in order to convey the required information. The quantitative analysis of the trajectories combined with linguistic analysis was designed to explore how the brain controls movement at different levels: linguistic, symbolic, spatial and motor. The errors in the spatially organised syntax of ASL stood in contrast to the flawless performance of a range of tasks including drawing, spatial construction, spatial attention, judgement of line orientation and spatial discrimination. Conversely, right hemisphere-damaged deaf-mute ASL patients showed gross impairment in the latter test battery but exhibited a performance of ASL almost free of errors. In terms of motor performance, ASL motor aphasia is also an apraxia. As in the case of the unimodal apraxias such apractic deficiencies are highly specific. This was shown by a series of apraxia tests in the signers with left hemisphere lesions and ASL aphasia. The results showed that the language deficiencies were due to the specific linguistic components of sign language, rather than an underlying motor disorder of the capacity to express and comprehend symbols or of the production and imitation of representational and non-representational movements. This demonstrates that, in aphasic and apractic disturbances, only specific functions are selectively disturbed. This is also the case in agnosias. Furthermore, it raises another interesting point, namely that sign language aphasia is a special type of apraxia in ASL patients, affecting comprehension and the generation of limb movements which convey linguistic information. This is consistent with Liepmann's interpretation of the disorders of verbal expression as apraxia of the speech muscles.

Lesions of the right IPL are frequently associated with the constructional apraxia first described by Kleist (1911). The deficiency is characterised by the difficulty in putting one-dimensional units together to form two-dimensional figures or patterns. It appears in formative activities such as arranging, building or drawing, where the specialised part of the task is faulty although the movement elements are normal. It is also seen after lesions of the dominant hemisphere but is more frequently observed after lesions of the non-dominant side.

Apraxia and the left hemisphere

Unlike language organisation where the minor hemisphere is virtually unable to maintain residual language functions after the age of 10-12, the right hemisphere is fully able to concert the whole movement repertoire for the opposite side of the body. Damage of the corpus callosum can produce an apraxia of the left hand but only because language information is not available to that hemisphere. The reason why a left hemisphere lesion interferes with the otherwise normal control of left sided motor behaviour by the right hemisphere is unclear. Freeman (1984) proposed that the principal function of the left hemisphere is the instrumentation of a priority of action control.

Conclusion

Although the parietal lobe represents a purely sensory projection-and-association cortex, a damage to this part of the brain may interfere grossly with normal motor behaviour. Whereas lesions of the postcentral somatosensory-projection cortex lead to motor disturbances mimicking those seen after deafferentation, thus representing a secondary effect, lesions of the SPL (superior posterior parietal lobule) cause agnostic and apractic deficiencies confined either to the somatosensory (tactile apraxia) or visual (visuomotor apraxia) modalities. All these deficiencies affect contralateral functions.

Lesions of the IPL (interferior posterior parietal lobule), a presumed-polymodal association area, when located in the language-dominant hemisphere, may cause ideational or ideomotor apraxia. These conceptual apraxias are supramodal and affect both sides of the body. The wide range of apractic deficiencies which vary from very selective, unimodal, focal-apractic disturbances to the gross impairment of all daily motor behaviour in the ideational apraxia that is seen after damage of the sensory association cortex, emphasises an important aspect of brain function which was not apparent on the basis of animal studies: the significance of the sensory association cortex for the conception and elaboration of motor behaviour. This capacity is partly lateralised.

References

Balint, R. (1909) Seelenlähmung des Schauens, optische Ataxie, räumliche Störung der Aufmerksamkeit. *Monatsschrift für Psychiatrie und Neurologie* 25: 51-81.

Bellugi, U., Poizner, H., and Klima, E.S. (1989) Language, modality and the brain. *Trends in Neurosciences* 12: 380-388.

Classen, J., Kunesch, E., and Binkofski, F. Subcortical origin of visuomotor apraxia. *Brain.* In press

Critchley, M. (1953) The parietal lobes. Hafner, New York.

Déjérine, J. (1907) A propos de l'agnosie tactile. *Revue Neurologique* 15: 781-784.

Foerster, O. (1936) Motorische Felder und Bahnen: sensible corticale Felder. In: H. Bumke und O. Foerster (eds): *Handbuch der Neurologie*, Springer-Verlag, Berlin, 6, pp 358-448.

Freeman, R.B., Jr. (1984) The apraxias, purposeful motor behavior, and left-hemisphere function. In: W. Prinz and A.F. Sanders (eds): *Cognition and Motor Processes.* Springer-Verlag, Berlin, pp 29-50.

Freund, H.-J. (1987) Abnormalities of motor behaviour after cortical lesions in humans. In: F. Plum (ed): *Handbook of Physiology.* Section 1. The Nervous System. Vol. V, Higher Functions of the Brain. Part 2. American Physiology Society, Bethesda, pp 763-810.

Freund, H.-J. (1992) The Apraxias. In: A.K. Asbury, G.M. McKhann, W.I. McDonald (eds): Diseases of the Nervous System. Clinical Neurobiology, Vol. II, Wiley, Chichester, pp 751-767.

Freund, H.-J. (1995) The apraxias. In: Ch. Kennard (ed): *Recent advances in clinical neurology.* Churchill Livingstone, New York, 1995, pp. 29-49

Glickstein, M., Cohen, J.L., and Dixon, B., Gibson, A., Hollins, M., LaBossière, E., and Robinson, F. (1980) Corticopontine visual projections in macaque monkeys. *Journal of Comparative Neurology.* 190: 209-229.

Glickstein, M., May, J., Mercer, B. (1985) Cortico-pontine projections in the macaque: the distribution of labelled cortical cells after large injections of horseradish peroxidase in the pontine nuclei. *Journal of Comparative Neurology.* 235: 343-359.

Gowers, W. (1887) Lectures on the diagnosis of diseases of the brain. Churchill, London.

Head, H., and Holmes, G. (1911) Sensory disturbances from cerebral lesions. *Brain* 34: 102-254.

Jeannerod, M. (1988) *The Neural and Behavioural Organization of Goal-Directed Movements.* Claredon Press, Oxford.

Kleist, K. (1911) Der Gang und der gegenwärtige Stand der Apraxieforschung. *Ergebnisse der Neurologie und Psychiatrie* 1: 342-452.

Liepmann, H. (1900) Das Krankheitsbild der Apraxie ("motorische Asymbolie"), auf Grund eines Falles von einseitiger Apraxie. *Monatsschrift für Psychiatrie und Neurologie* 8, 182-197.

Liepmann, H. (1920) Apraxie. In: *Brugsch's Ergebnisse der Gesamten Medizin.* Urban & Schwarzenberg, Berlin, pp 518-543.

Pause, M., Kunesch, E., Binkofski, F. and Freund, H.-J. (1989) Sensorimotor disturbances in patients with lesions of the parietal cortex. Brain 112: 1599-1625.

Rondot, P., de Recondo, J., and Ribadeau Dumas, L. (1977) Visuomotor apraxia. *Brain* 100: 355-376.

Seitz, R. J., Huang, Y., Knorr, U., Tellmann, L., Herzog, H., and Freund H.-J. (1995) Large-scale plasticity of the human motor cortex. *NeuroReport* 6: 742-744.

Somesthesis and the Neurobiology of the Somatosensory Cortex
O. Franzén, R. Johansson and L. Terenius (eds.)

PET and fMRI scans of the cerebral cortex in humans and single neuron responses from SI in monkeys to rubbing embossed dot and grating patterns across a fingerpad

H. Burton[1], R.J. Sinclair[1], W. Lin[3], and A.K. MacLeod[3]

Washington University, [1]Department of Anatomy and Neurobiology, 660 S. Euclid Ave. and [3]Department of Radiological Sciences 4525 Scott Ave., St. Louis, MO 63110

Summary. PET images averaged across nine humans and fMRI scans from single subjects showed activation of comparable regions in parietal cortex following stimulation of fingerpads with embossed dot and grating surfaces. Human magnitude estimates of roughness to an extended range of similar texture patterns matched linear, inverted U and plateau neural response functions to rubbing the same surfaces in monkey SI cortex.

Introduction

Neuroimaging has shown blood flow changes in parietal and frontal cortical regions during somatosensory stimulation. Understanding this network should help develop neural models of tactile object recognition. Reaching this goal requires integrating data obtained from single neuron recordings in monkeys with the slower temporal resolution of neuroimaging techniques. Several problems currently limit success with this endeavor. First, the stimulus factors previously manipulated in these two types of studies differed too much. In published PET studies it is impossible to associate different classes of peripheral receptors with activated cortical regions because stimulation often involved large amplitude mechanical vibrations that indiscriminately activated cutaneous and deep receptors over an indeterminate extent of selected body regions (Burton et al., 1993; Fox et al., 1987). Second, several neuroimaging studies confounded regions for motor behaviors with those for somatosensory processes because subjects performed tactile discrimination tasks that required elaborate exploratory manipulations (Roland and Seitz, 1991). The location and types of activated peripheral receptors again could not be identified in these studies. We examined PET and functional MRI scans to find only those regions activated by stimuli identical to embossed dot or grating patterns used in neurophysiology studies (Connor et al., 1990; Connor and Johnson, 1992; Sinclair and Burton, 1991; Sinclair and Burton, 1993). We assumed that the seconds of activation needed for neuroimaging techniques engaged neurons in sustained examples of similar response characteristics.

Another issue that ultimately affects the design of neuroimaging studies of the cortical areas responsible for discriminating surface textures is a model for perceiving roughness that matches psychophysical functions with neurophysiological observations in the cortex. We studied human magnitude estimates of roughness and recorded from SI cortical neurons to stimulation with an extended range of dot and grating surfaces having spacings between 1 and 8mm. The objective was to assess predictions about the psychophysics and underlying neural activity patterns related to roughness perceptions (Connor et al., 1990; Connor and Johnson, 1992; Johnson and Hsiao, 1992).

Materials and Methods

Each experiment involved rubbing one or more fingerpads with a photo-etched, polyimide plastic (Nyloprint) embossed surface. Table 1 summarizes the various surface patterns used in the different studies. Surfaces were applied passively, using mechanical devices at the speeds specified in Table 1, and manually by the experimenters for the fMRI scans. In these experiments motor behavior was minimized because individually fashioned orthopedic casts held fingers in position for stimulation. Subjects actively touched the surfaces in the magnitude estimation studies. The direction of rubbing the rows of dots and gratings was perpendicular to the long axis of the finger for PET studies and parallel to the long axis for all others. Surface patterns and stimulation timing varied to accomodate the temporal resolution of the different techniques (Table 1). Additional details about methods are in the figure legends and in previous papers (Burton et al., 1993; Ogawa et al., 1993; Sinclair and Burton, 1993).

Table 1. Surface parameters and stimulus timing for different studies

Study	Patterns	Surface Dimensions (mm) spacing	rise	ht.	Stimulus Timing Speed mm/s	Contact Time	Inter-Stimuli	Trials
PET	gratings s=2+	0.5 2.75	.25 .25	.5 .5	100	40s	20-40 min	4
fMRI	dots s=1	2X4	.5	.5	.5-1#	70s*	70s*	2
Single Cell	dots s=8	1-8X4	.5	.5	50	~1.8s	2-4s	10
	gratings s=8	1-8	.5	.5	50	~1.8s	2-4s	10
Magnitude Estimates	dots s=8	1-8X4	.5	.5	self-paced	1-2s	self-paced	5
	gratings s=8	1-8	.5	.5	self-paced	1-2s	self-paced	5

+Number of different surfaces (s)
#Repetition frequency
*Five repeated scans with each 14s in duration

Results and Discussion

Cortical Areas Activated by Textured Surfaces

The averaged regional cerebral blood flow (rCBF) difference images in Figure 1 show foci

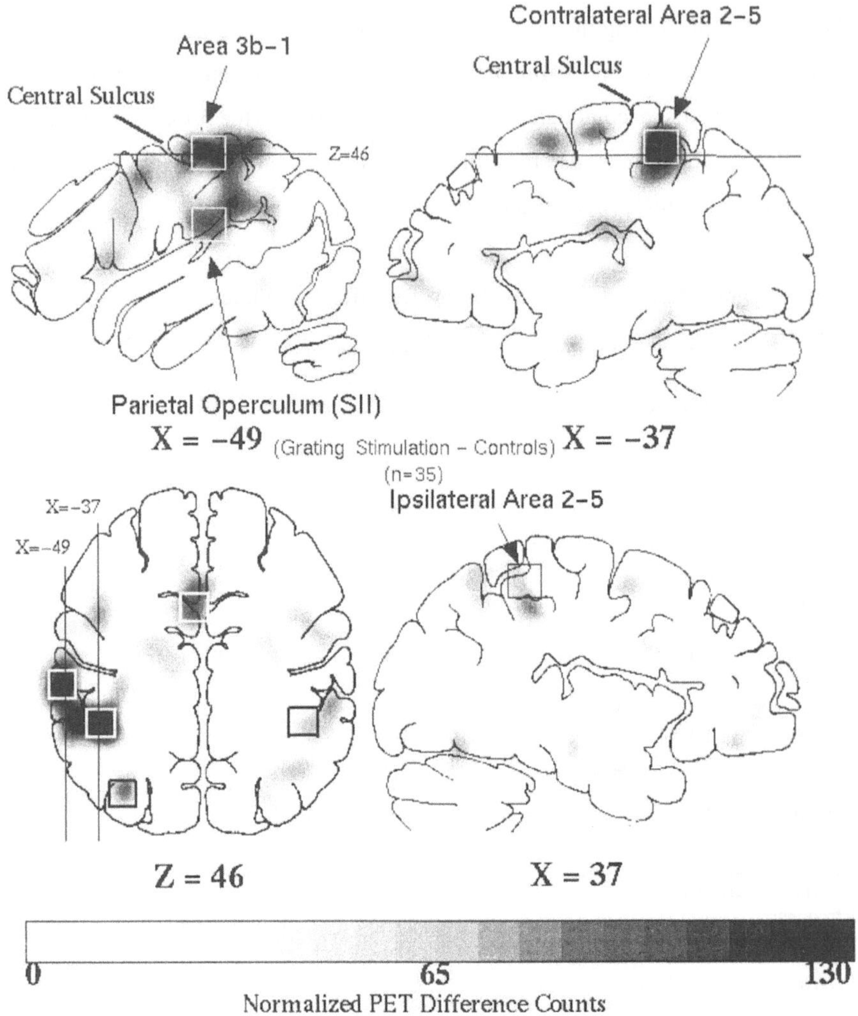

Figure 1. Average difference images in sagittal and axial planes for tactile activation minus no-stimulation controls. Horizontal and vertical lines mark location of axial (Z=46) and contralateral sagittal (X=-49 & -37) slices on opposing planes. Boxes show locations of statistically significant increases in CBF. Nine subjects received intravenous bolus injections of O^{15} labeled water before scans. PET counts obtained with a Seimens 953B scanner using 3-D acquisition, back projection, 31 axial slices (spaced 3.375mm), and in-plane resolution of 2mm. Scans normalized to 1000 for global blood flow. Slices linearly interpolated to Tailarach stereotactic space (Fox et al., 1985). Images filtered to yield full-width at half-maximum resolution of 10.5mm. During activation scans, a wheel, rotating at ~100mm/sec, applied a counterbalanced force of ~100gms against restrained fingertips 2 and/or 3. Six pairs of smooth-rough stripes of embossed gratings encirled the wheel (Table 1). With eyes closed, subjects attended to transitions in and magnitudes of roughness throughout activation scans.

of increased CBF from 35 scans taken from 9 subjects. Random division of the data into two scan sets allowed statistical cross-validation of several regions in parietal cortex (Burton et al., 1993; Drevets et al., 1995). By inference from studies in monkeys, the anatomical locations of these regions suggest similar functional subdivisions of human somatosensory cortex (Table 2). The smaller yet similarly located rCBF over the ipsilateral intraparietal sulcus suggests possible bilateral activation of areas 2-5 as recently shown in monkeys (Iwamura et al., 1994). A statistically significant peak over SII was anterior but CBF increases also lay posterior. This suggests more extensive activation in this location. Recent anatomical findings show multiple somatosensory areas along the parietal operculum in primates (Burton et al., 1995).

Additional foci appeared in frontal cortex in the PET images. Only the site over the medial frontal cortex for SMA was significant. Smaller CBF changes occurred over the lateral frontal gyrus near area 6 ventral. The precentral gyrus in PET and fMRI images showed little activation. Thus only higher order motor areas displayed activations.

Table 2. Location of statistically validated centers of mass in PET images

Regions		Coordinates			Counts
Anatomical	Functional	X	Y	Z	
Contralateral to Stimulated Finger					
s. intraparietalis	area 2-5	-37	-37	48	168
g. frontalis medialis	SMA	- 3	5	50	144
g. postcentralis	area 3b	-51	-23	46	135
parietalis opercularis	SII	-45	-23	20	117
lobulus parietalis	area 7	-29	-63	50	102
Ipsilateral to Stimulated Finger					
s. intraparietalis	area 2-5	47	-37	48	95

*Terminology and coordinates from Talairach and Tournoux, 1988.

Magnetic resonance imaging revealed similar activation foci in parietal cortex from rubbing the fingertips with an embossed dot pattern (Figure 2). Multiple foci of increased signal intensity included at least three regions along the curvature of the postcentral gyrus. The most anterior lay close to the central sulcus, possibly in area 3b. More posterior foci occupied the intraparietal sulcus and adjoining parietal cortex bilaterally. These possibly show activation of areas 1-2 and

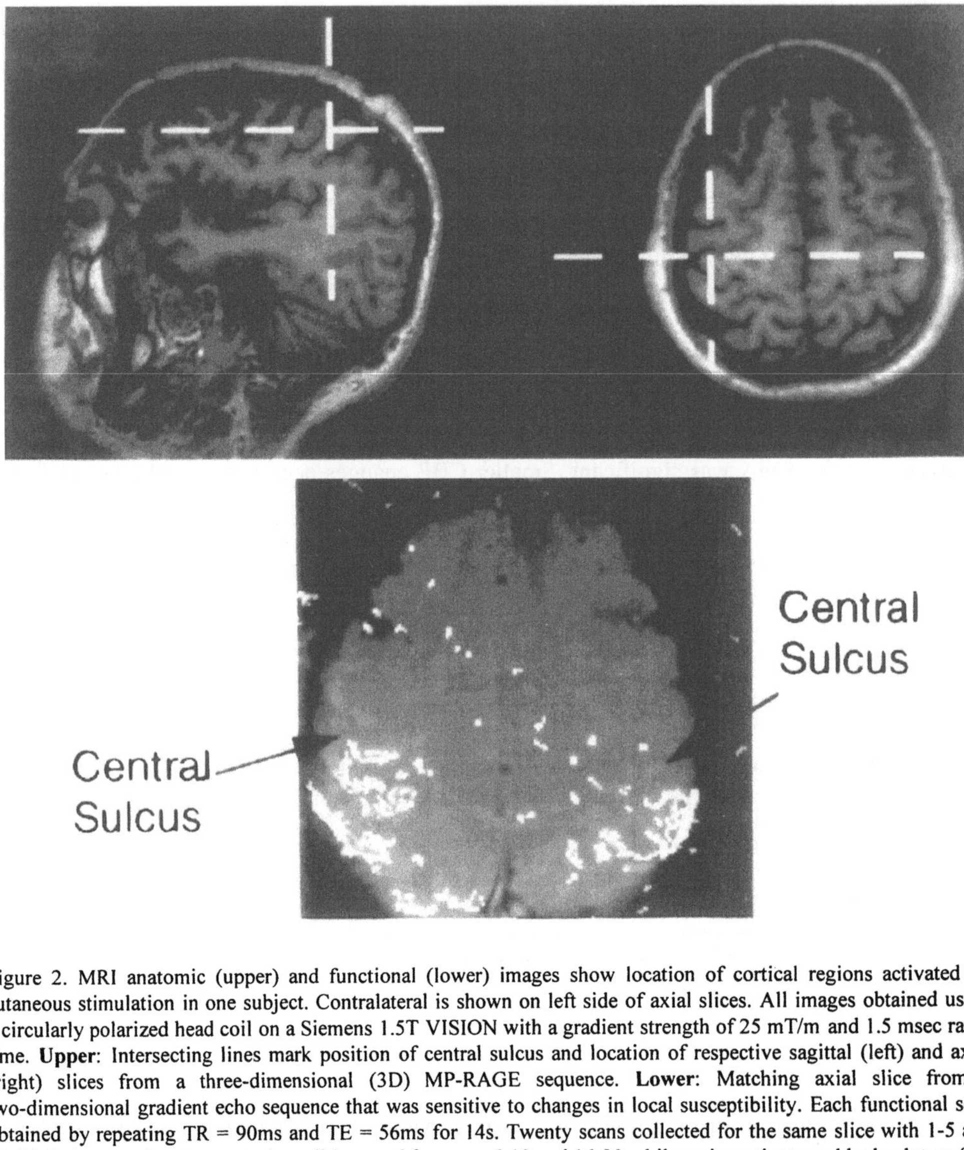

Figure 2. MRI anatomic (upper) and functional (lower) images show location of cortical regions activated by cutaneous stimulation in one subject. Contralateral is shown on left side of axial slices. All images obtained using a circularly polarized head coil on a Siemens 1.5T VISION with a gradient strength of 25 mT/m and 1.5 msec ramp time. **Upper**: Intersecting lines mark position of central sulcus and location of respective sagittal (left) and axial (right) slices from a three-dimensional (3D) MP-RAGE sequence. **Lower**: Matching axial slice from a two-dimensional gradient echo sequence that was sensitive to changes in local susceptibility. Each functional scan obtained by repeating TR = 90ms and TE = 56ms for 14s. Twenty scans collected for the same slice with 1-5 and 11-15 during no-stimulus control conditions and for scans 6-10 and 16-20 while an investigator rubbed a dot surface across fingertips 2, 3, and 4. The image shows a by-pixel distribution of significant cross correlation values in time for response magnitudes over the 20 scans. Response intensity in each pixel correlated with a hypothetical rectangular shaped reference distribution that assumed maximum response magnitudes only during activation scans 6-10 and 16-20. Perfect cross correlation suggests that activity recorded at that pixel increased and decreased across the temporal order of the 20 scans in phase with the reference model. Significant cross-correlation values of highlighted pixels exceeded 0.5.

5. Another region appeared further posterior in the superior parietal lobule in the likely territory of area 7.

The observed foci show that cutaneous stimulation confined to the fingerpads activates an extensive network of cortical areas. Some of these regions probably correspond to separate subdivisions of SI and SII previously identified in monkeys. In addition, several higher order areas in parietal and frontal cortex are also activated when just the fingerpads are stroked with textured surfaces.

Psychophysical and Neural Response Functions to Textured Surfaces

Roughness perceptions may depend on the spatial distribution of discharges in a population of peripheral receptors. Specifically, Johnson and colleagues (Connor et al., 1990; Connor and Johnson, 1992; Johnson and Hsiao, 1992) showed that the spatial variation in neural activity across a population of receptors in monkeys created a neural spatial pattern that isomorphically reproduced the geometry of dot surfaces with spacings between 1 and 6mm. Peripheral, especially slowly adapting, receptor response functions that were based on a spatial variation metric, closely matched inverted "U" functions obtained from human magnitude estimates of roughness for the same dot spacings. Maximum roughness and the peak of the neural metric occurred at spacings of ~3mm. These investigators predicted that through central convergence of excitatory and inhibitory components of a cell's receptive field, the average firing rates of cortical neurons would represent spatial variations in the periphery with similarly shaped functions (Connor and Johnson, 1992). We tested this hypothesis by recording average firing rates of SI neurons when their receptive fields were stimulated with dot and grating surfaces (Table 1). As the range of spacings studied exceeded those used previously, we re-examined human magnitude estimates of roughness to the same extended range of surfaces.

Figure 3 shows that average firing rates of some SI cells followed the predicted inverted U-shaped function to stimulation with dot surfaces whose spacing extended to 8mm. Peak firing rates and judgements of roughness in some subjects also occurred at 3mm for dot surfaces (Figure 3C,D). Stimulation with gratings did not produce comparable inverted U functions. Figure 4 shows examples of additional exceptions. The average firing rate for some cells reached a plateau for dot spacings between 3-4mm. Some humans judged both kinds of surfaces as having similar roughness for all spacings greater than 4mm (Figure 4A,B). Other cells displayed linearly increased firing rates for greater spacings on both types of surfaces. Some humans similarly reported progressive increases in roughness sensations over the same range of spacings (Figure 4 C,D).

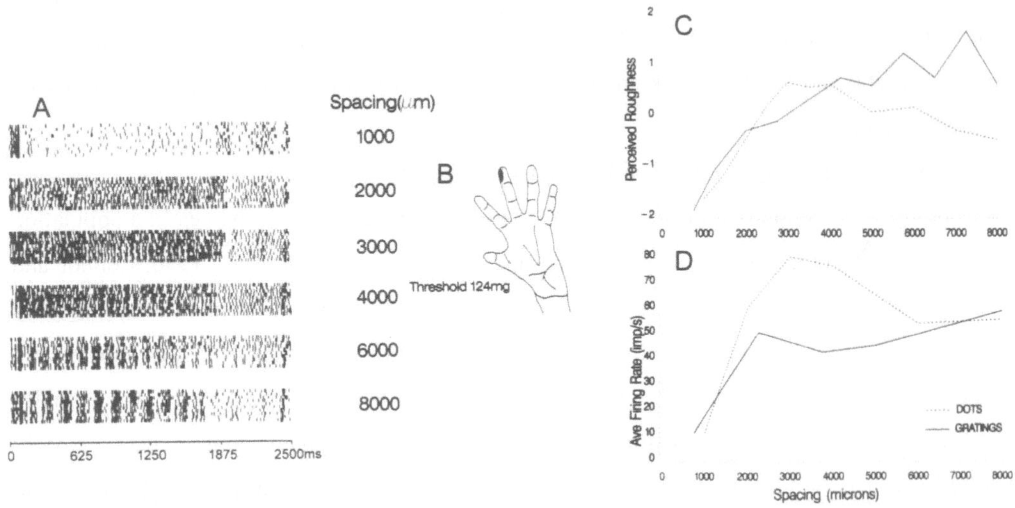

Figure 3. Comparison between single neuron responses from the distal digit region of monkey area 3b-1 and human roughness estimates of the same embossed surfaces. Each dot surface had a different spacing between the rows of dots (1 to 8mm) but a constant 4mm spacing within rows. Dot widths were ~.5mm. Each grating surface had different groove widths that also spanned 1-8mm but constant ridge widths of .25mm. **A.** Raster lines show single trial responses to rubbing this cell's receptive field with dot surfaces. A robotics device moved the surfaces against the fingerpad at constant velocity (50mm/s) and force (50gm). The stroke was horizontal and parallel to the distal-proximal axis of the finger. Each group of raster lines illustrates replicate responses to the dot surfaces with the spacings listed on the right. **B.** Receptive field for this cell was along the radial side of the distal pad for digit three. The cell had a rapidly adapting response to manually applied stimulation and, based on force-calibrated nylon monofilaments, had a force threshold of 124mg. **C.** The graph shows standardized magnitude estimation functions from one subject for dot and grating surfaces. Subjects rubbed one glabrous fingerpad from distal to proximal across the rows of gratings or dots. Subjects were instructed to maintain constant stroke velocity and force for all surfaces. They reported numerical estimates of roughness for each trial. No baseline modulus was provided. **D.** The graph shows single cell average firing rates (AFR) for the interval when different dot and grating surfaces were rubbed over the receptive field. The data for the curve for dot surfaces summarize the responses shown in A and peaks at about the same spacing as roughness magnitude estimates in C. Data for the graph showing distribution of activity to grating surfaces were obtained from the same cell.

Finding that response functions of different SI cortical neurons resemble judgements of roughness supports the notion that mean levels of activity in SI cortex code roughness sensations. Further, as these functions reflect spacings for both dot and grating patterns, they support prior suggestions that roughness sensations principally arise from an intensive continuum of spacing between elements on a surface. The latter possibly relate to the volume of skin distorted by falling into the spaces (Taylor and Lederman, 1975). Use of a greater range of surfaces and both dot and grating patterns revealed a greater variety of psychophysical and neural functions than previous findings of only inverted U functions (Connor et al., 1990; Connor and Johnson, 1992). These data are therefore not completely consistent with a spatial variation metric across the whole extent of human estimates of surface roughness nor with the prediction that central neurons

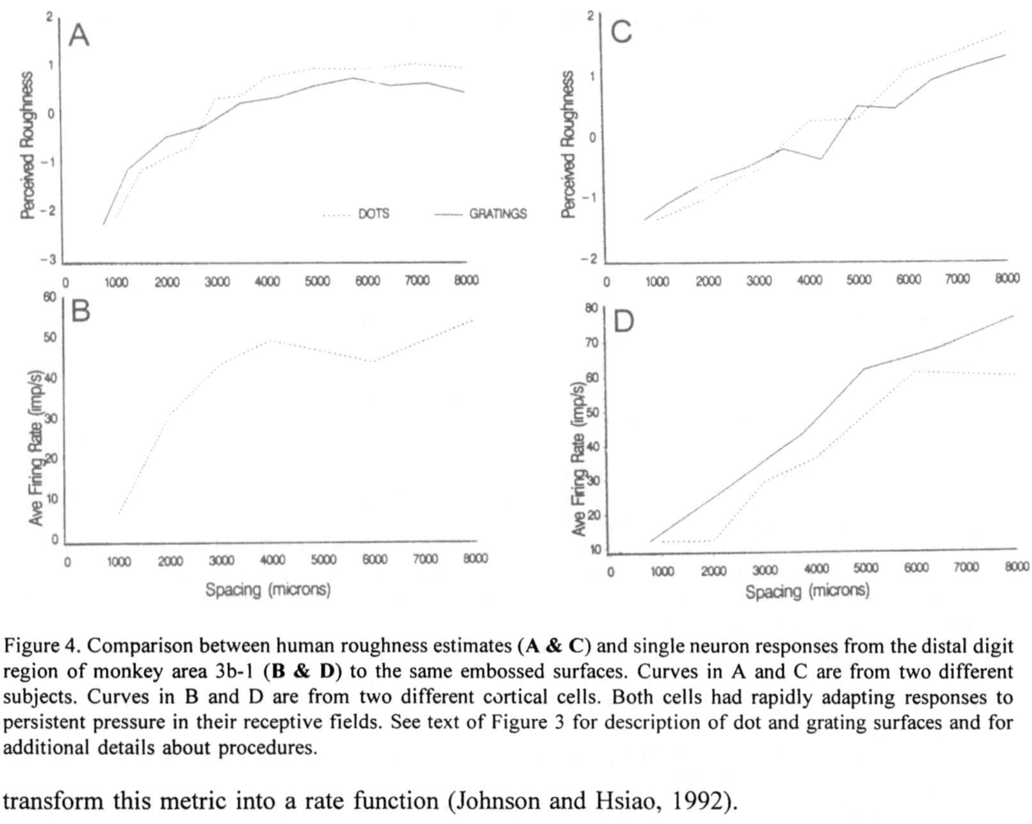

Figure 4. Comparison between human roughness estimates (**A & C**) and single neuron responses from the distal digit region of monkey area 3b-1 (**B & D**) to the same embossed surfaces. Curves in A and C are from two different subjects. Curves in B and D are from two different cortical cells. Both cells had rapidly adapting responses to persistent pressure in their receptive fields. See text of Figure 3 for description of dot and grating surfaces and for additional details about procedures.

transform this metric into a rate function (Johnson and Hsiao, 1992).

Knowledge of the micromorphic structure of surfaces is important for tactile object recognition (Loomis and Lederman, 1986). Our studies show that this sensitivity to surface texture activates a vast and subdivided cortical network. Understanding the processes responsible for object recognition requires knowing the transformations through this circuit. The first stage occurs in SI. However, as our results show, the underlying mechanisms remain ambiguously tied to neural metrics such as spatial variation, which is a response measure obtained from recordings in the peripheral nervous system.

Acknowledgements: Supported by research funds from NIH NS 31005, HL 13851 and Siemens Medical Systems.

References:

Burton, H., Fabri, M. and Alloway, K. (1995) Cortical areas within the lateral sulcus connected to cutaneous representations in areas 3b and 1: A revised interpretation of the second somatosensory area in macaque monkeys. *Journal of Comparative Neurology* 355: 539-562.

Burton, H., Videen, T. O. and Raichle, M. E. (1993) Tactile vibration activated foci in insular and parietal opercular cortex studied with positron emission tomography: mapping the second somatosensory area in humans. *Somatosensory and Motor Research* 10: 297-308.

Connor, C. E., Hsiao, S. S., Phillips, J. R. and Johnson, K. O. (1990) Tactile roughness: Neural codes that account for psychophysical magnitude estimates. *Journal of Neuroscience* 10: 3823-3836.

Connor, C. E. and Johnson, K. O. (1992) Neural coding of tactile texture: comparison of spatial and temporal mechanisms for roughness perception. *Journal of Neuroscience* 12: 3414-3426.

Drevets, W. C., Burton, H., Videen, T. O., Snyder, A. Z., Simpson, J., J.R. and Raichle, M. E. (1995) Blood flow changes in human somatosensory cortex during anticipated stimulation. *Nature* 373: 249-252.

Fox, P. T., Burton, H. and Raichle, M. E. (1987) Mapping human somatosensory cortex with positron emission tomography. *Journal Neurosurgery* 67: 34-43.

Fox, P. T., Perlmutter, J. S. and Raichle, M. E. (1985) A stereotactic method of anatomical localization for positron emission tomography. *Journal Computer Assisted Tomography* 9: 141-153.

Iwamura, Y., Iriki, A. and Tanaka, M. (1994) Bilateral hand representation in the postcentral somatosensory cortex. *Nature* 369: 554-6.

Johnson, K. O. and Hsiao, S. S. (1992) Neural mechanisms of tactual form and texture perception. *Annual Review Neuroscience* 15: 277-350.

Loomis, J. M. and Lederman, S. J. (1986) Tactual Perception. *In:* K. R. Boff, L. Kaufman, & J. P. Thomas (ed.): *Handbook of Perception and Human Performance,* vol. 2, John Wiley, New York, pp. 1-41.

Ogawa, S., Menon, R. S., Tank, D. W., Kim, S. G., Merkle, H., Ellermann, J. M. and Ugurbil, K. (1993) Functional brain mapping by blood oxygenation level-dependent contrast magnetic resonance imaging. *Biophysical Journal* 64: 803-812.

Roland, P. E. and Seitz, R. J. (1991) The physiological anatomy of a single brain function: Somatosensory discrimination of shape. *In:* O. Frazén, & J. Westmen (ed.): *Information Processing in the Somatosensory*

System, vol. 57, Stockton Press, New York, pp. 275-288.
Sinclair, R. J. and Burton, H. (1991) Neuronal activity in the primary somatosensory cortex in monkeys (*Macaca mulatta*) during active touch of textured surface gratings: responses to groove width, applied force and velocity of motion. *Journal of Neurophysiology* 66: 153-169.
Sinclair, R. J. and Burton, H. (1993) Neuronal activity in the second somatosensory cortex of monkeys (*Macaca mulatta*) during active touch of gratings. *Journal of Neurophysiology* 70: 331-350.
Talairach, J. and Tournoux, P. (1988) *Co-Planar Stereotaxic Atlas of the Human Brain.* Thieme, Stuttgart.
Taylor, M. M. and Lederman, S. J. (1975) Tactile roughness of grooved surfaces: A model and the effect of friction. *Perception and Psychophysics* 17: 23-36.

Somesthesis and the Neurobiology of the Somatosensory Cortex
O. Franzén, R. Johansson and L. Terenius (eds.)

Magnetic resonance functional mapping of cortical activation associated with differing sensorimotor hand paradigms

H. Boecker[1] and J. Frahm

Biomedizinische NMR Forschungs GmbH, Am Faßberg 11, D-37077 Göttingen, Germany
[1]Neurologische Klinik der Technischen Universität München, Moehlstr. 28, D-81675 München, Germany. Current address: M.R.C. Cyclotron Unit, Hammersmith Hospital, DuCane Road, London W12 0HS, UK

Introduction

Gradient-echo magnetic resonance imaging (MRI) sequences that emphasize contrasts due to localized changes in cerebral blood oxygenation (CBO) have been introduced as a new technique for noninvasive functional mapping of the human brain. The underlying physiology is a transient hyperoxygenation of the venous blood pool subsequent to a regional increase in cerebral blood flow surpassing oxygen consumption. This understanding has been confirmed by simultaneous recordings of MRI and near-infrared optical spectroscopy (Obrig et al., 1994). Previous work has demonstrated the unique potential of CBO-sensitized MRI to localize sites of functional brain activation at high spatial resolution similar to that of anatomic MR images (Frahm et al., 1993, Frahm et al., 1994). Moreover, using correlational analysis of MRI signal time courses (Rogowska et al., 1992, Bandettini et al., 1993), recent applications were extended to evaluations of (dys-) functional cooperativity of distant brain sites within primary pathways and subcortical networks (e.g., see Kleinschmidt et al., 1994, Boecker et al., 1994a).

Initial MRI studies of human motor activation during sequential finger opposition reported signal intensity increases not only in contralateral but also ipsilateral primary motor cortex (Kim et al., 1993a,b). In another study activation in the supplementary motor area was observed during complex but not simple finger movements (Rao et al., 1993). The purpose of this work was to reexamine these issues with the aforementioned improvements in methodology and to define specific patterns of task-related activation during motor hand paradigms of varying complexity. In extension to primarily mapping motor-associated activation, tactile exploration of macrogeometric objects was included as a challenging motor paradigm with a major component of somatosensory information processing.

Methods and Subjects

Investigations were performed in healthy individuals (mean age: 30 years, range: 23-42 years, explicit right hand dominance in all volunteers but one) and included self-paced unilateral tapping of the middle finger ("FT", 5 subjects, frequency about 2 Hz), self-paced unilateral sequential opposition of fingers to the thumb ("SO", 4 subjects, frequency about 2 Hz), and unilateral tactile exploration ("TE", 6 subjects) of surface, shape, and geometric features of different macrogeometric wooden objects (spheres, cubes, prisms, rectangular parallelepipeds, and more complex irregular shapes with up to eight angles; edge length: 2.5-6.5 cm; weight: 4-16 g) with use of all fingers and the palm. Objects for TE were exchanged every 3 s.

In all experimental conditions both sides were investigated sequentially. The protocol during dynamic MRI comprised four to six periods of 54 s duration each with 18 s of activation and 36 s of rest. Onset and end of performance were triggered by a brief light flash (FT), softly touching the ipsilateral foot (SO), or object presentation (TE). Head fixation was accomplished by a vacuum head holder. Subjects had ear plugs and were advised to keep their eyes closed. Task execution was controlled by on-line video monitoring and excluded associated movements of the other hand.

All studies were performed at 2.0 T (Siemens Magnetom SP4000) using the standard headcoil and motion-compensated radiofrequency-spoiled FLASH (fast low angle shot) MRI sequences. Delineation of anatomy and macrovasculature by three-dimensional and single-slice flow-sensitized T1-weighted MRI, respectively, was followed by CBO-sensitive functional imaging (6 s temporal resolution, 0.8 x 1.6 x 4.0 mm^3 spatial resolution) using dynamic acquisitions of spin density-weighted images (TR/TE = 62.5/30 ms, flip angle 10°) with T2* sensitivity (Frahm et al., 1993, Frahm et al., 1994). Four to six contiguous sections parallel to the bicommissural line covered the primary sensorimotor hand area (M1, S1), the supplementary motor area (SMA), parts of premotor areas (PMA) and superior parietal cortex (PC) bilaterally.

Functional cooperativity of multiple brain sites was determined by correlational analysis of task-related activation. Briefly, signal intensity time courses of a dynamic data set were correlated with a box-car waveform on a pixel-by-pixel basis (Bandettini et al., 1993). To account for hemodynamic latencies and rise times the reference time course was shifted by one image (6 s) relative to the periodic stimulation protocol. In its simplest form the evaluation strategy results in a map of correlation coefficients ranging from -1 to +1. Identification and area delineation of task-related activation involved thresholding of positive correlation coefficients exceeding +0.5, incorporation of pixels with intermediate correlation coefficients but in direct neigborhood to peak values, and color-coded overlay of such activation maps onto flow-sensitized anatomic images. The procedure ensures both high specificity and sensitivity in mapping clusters of specifically activated pixels (for further details concerning data analysis see Kleinschmidt et al., 1996).

Results

The present investigations identified specific, i.e. task-dependent patterns of cooperative cortical activation related to hand paradigms that differ in complexity and somatosensory involvement. Observations were made in single subjects, but the patterns were highly reproducible both intra- and interindividually. The findings may be summarized as follows:

(i) Activation of contralateral M1, contralateral S1, and bilateral SMA was seen in each task.

(ii) Ipsilateral activation of M1 was nearly absent in unilateral FT and seen only in 1/5 volunteers when performing the right hand task.

(iii) Weak ipsilateral activation of M1 occurred in about 50 % of the examinations performed with SO and TE,

(iv) Strong bilateral activation of posterior S1 was consistently seen with TE.

(v) TE was further characterized by accentuating weak activation of the superior PC observed in SO.

(vi) Activation of PMA was inconsistent with respect to uni- or bilateral involvement throughout all tasks.

Typical examples are shown in Figs. 1 and 2 emphasizing activation of SMA and contralateral M1 in FT and bilateral S1 in TE, respectively. In a few cases additional studies were performed to investigate coactivation of thalamic relay nuclei during TE. Pertinent coronal sections either covered the entire thalamus or included both the posterior part of the ventral group of thalamic nuclei and the primary sensorimotor hand area. A preliminary example is given in Fig. 3 depicting foci of contralateral posterior thalamic activation within the ventral group of nuclei.

Discussion

CBO-sensitive MRI revealed robust patterns of task-specific cortical activation for unilateral self-paced FT and SO as well as for TE (Boecker et al., 1994b, 1996). The results complement recent positron emission tomography (PET) and MRI studies on human sensorimotor processing in

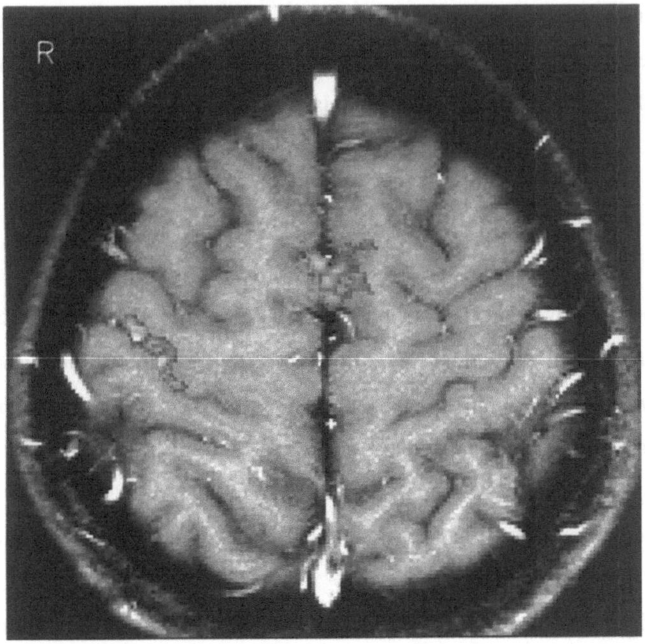

a)

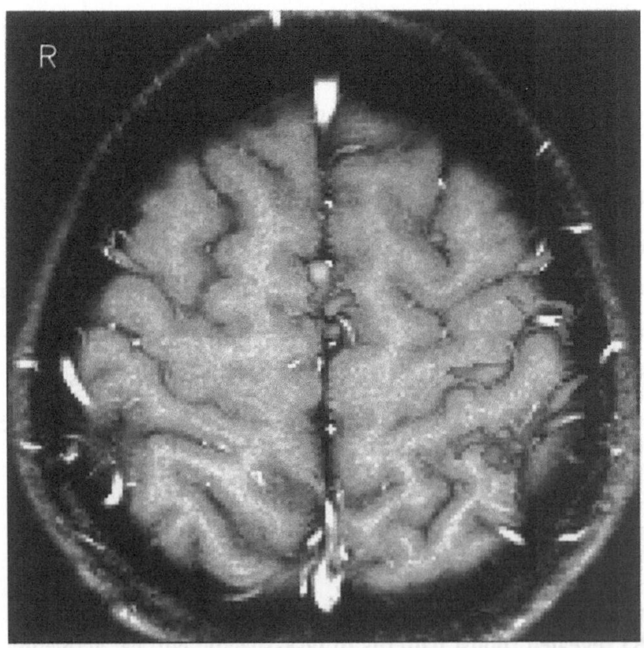

b)

Fig. 1. Correlation maps demonstrating task-related activation in the hand area of contralateral primary motor cortex and of the supplementary motor area during self-paced unilateral tapping of the middle finger of (a) the left and (b) the right hand (R: right side). For further details see (Boecker et al., 1994b).

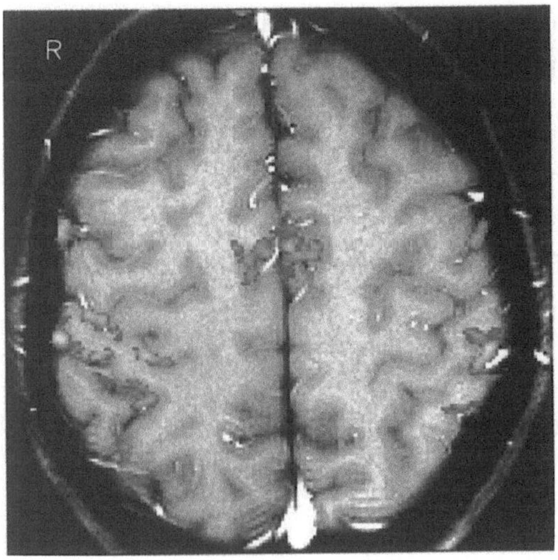

a)

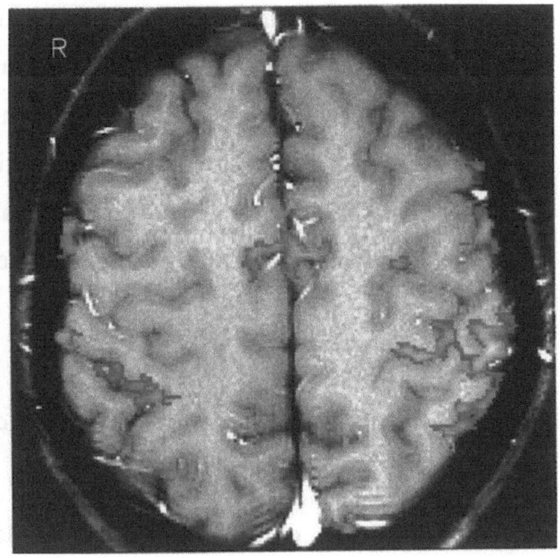

b)

Fig. 2. Correlation maps demonstrating task-related activation of contralateral primary motor cortex and supplementary motor area as well as contralateral and ipsilateral primary sensory motor cortex during tactile exploration with (a) the left and (b) the right hand. For further details see (Boecker et al., 1996).

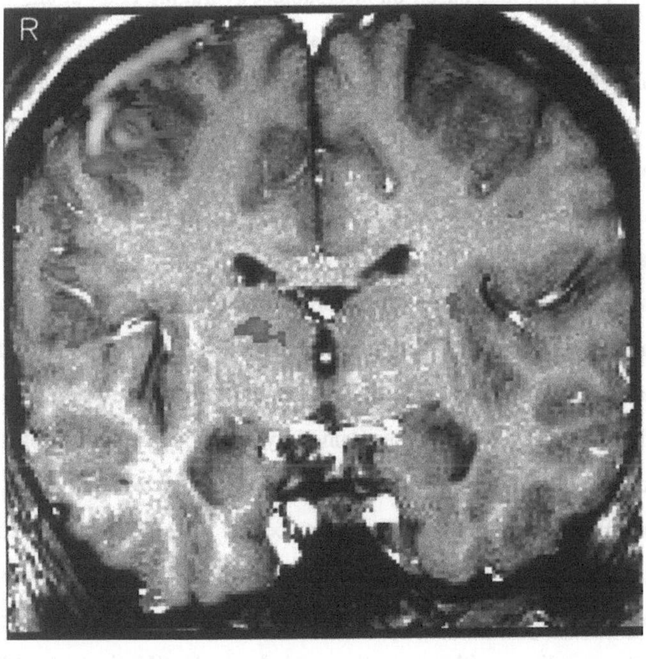

Fig. 3. Correlation map demonstrating task-related activation in the contralateral posterior thalamus during tactile exploration with the left hand. Functional coupling of subcortical nuclei to primary sensorimotor cortex was established by temporal cross-correlations of signal time courses with a reference time course derived from M1, e.g. compare (Kleinschmidt et al., 1994).

various respects: first, the use of weak excitation pulses, i.e. the elimination of T1 saturation, emphasizes changes in CBO rather than artifactual signal increases due to flow or motion; second, the high spatial resolution of the FLASH imaging approach provides so far unsurpassed accuracy in localizing sites of human cortical and subcortical brain activation; and third, the specificity and sensitivity achievable by optimized correlational analyses of the dynamics of task-related MRI signal alterations allow a reliable determination of temporally coherent activation in multiple brain sites.

Unilateral finger tapping and sequential finger opposition

Beyond involvement of the contralateral primary sensorimotor hand area, activation in other structures reflected the nature of the functional challenge. SMA activation was present throughout all tasks in single individuals contrasting the notion that the SMA is exclusively reserved for complex motor tasks as proposed in earlier PET studies (Orgogozo and Larson, 1979; Roland et al., 1980) and an MRI study using echo planar imaging (Rao et al., 1993). The present result is in line with more recent PET findings based on intersubject averaging (Fox et al., 1985; Colebatch et al., 1991) and with less reproducibility by individual analysis (Shibasaki et al., 1993). In fact, high-resolution functional MRI seems promising to elucidate further aspects of human motor processing such as the relationship between task complexity and activation of SMA or PMA (Kim et al., 1994). Here, no definite conclusions regarding lateral premotor activation in FT and SO tasks could be drawn as activation of respective sites was inconsistent in terms of unilateral or bilateral signal enhancement.

In contrast to other MRI studies of hand function (Kim et al., 1993a,b), the absence of ipsilateral M1 activation in about 50 % of the volunteers examined by FT and SO (and TE) is striking. Nevertheless, there is a clear-cut tendency towards enhanced ipsilateral activation from FT to SO (and TE). This finding is most likely to be explained by enhanced proximal coinnervation with the more complex tasks. Support comes from PET findings (Colebatch et al., 1991) where significant bilateral M1 activation was found for unilateral proximal (shoulder) movements but not distal (finger) movements. This interpretation reflects the view that ipsilateral control of upper limb innervation is strongest for axial and proximal limbs and weakest for individual fingers. Alternatively, recent recordings with magnetoencephalography (Hari et al., 1993), echo planar MRI (Rao et al., 1993), and PET (Shibasaki et al., 1993) have assigned enhanced ipsilateral M1 activation during skilled complex finger movements as opposed to simple self-paced index finger flections to the increased "complexity" of the task. Obviously, further studies using carefully controlled paradigms are required to reliably resolve the physiologic contributions of motor complexity and proximal coinnervation to ipsilateral activation of the primary motor cortex.

Tactile exploration

Tactile exploration of macrogeometric objects induced a pattern of distributed cortical activation clearly distinct from those of the motor tasks discussed above. In addition to PET findings of contralateral S1 activation during sensory stimulation (Fox et al., 1987; Seitz et al., 1991; Burton et al., 1993; O'Sullivan et al., 1994), this study showed strong activation of ipsilateral posterior S1 (postcentral sulcus extending into the anterior bank), presumably corresponding to area 2, in all volunteers performing TE. This new finding of bilateral posterior S1 activation is likely to reflect transcallosal activation as suggested by anatomic and electrophysiologic data. The anatomic basis for the interpretation is given by Killackey and collaborators who were able to demonstrate callosal connections from the hand representations of area 2 and hypothesized "that there is a strong tendency for the callosal transfer of information to be delayed in the processing chain across the three fields, so that most of the transfer occurs after reaching area 2, or even after further processing in other representations" (Killackey et al., 1983). This is supported by ipsilateral median nerve stimulation in humans, where intraoperatively somatosensory evoked potentials (SEPs) with long latency (40-50 ms) were recorded in areas 4, 1, 2, and 7 as opposed to strictly contralateral SEPs of short latency (20-40 ms) from area 3b (Allison et al., 1989a,b). The delay suggests that the ipsilateral potentials are generated by transcallosal input. More recent physiologic evidence stems from single cell recordings in monkey somatosensory cortex which identified neurons with bilateral hand receptive fields in area 2 and adjoining parietal cortex (Iwamura et al., 1994). The absence of statistically significant activation of ipsilateral S1 in a variety of sensory paradigms studied by PET (Seitz et al., 1991; Burton et al., 1993; O'Sullivan et al., 1994) may be due to limited spatial resolution in conjunction with a lack of "spillover" from non-activated ipsilateral M1.

Interpretation of PC activation seen with TE is hampered since measurements were restricted to the superior parts of PC. However, activation foci in the superior parietal lobule are compatible with the encoding of kinematics of upper limb movements (Kalaska et al., 1990). Lesion studies in patients with localized damage in this region have revealed severe deficits in tactile exploration and discrimination of macrogeometric objects (Roland et al., 1987; Pause et al., 1989). Sensorimotor-related mapping of small subnuclei in contralateral posterior thalamus underlines the potential of high-resolution MRI to generally improve delineation of cortico-subcortical interactions in the human brain.

Concluding Remarks

The detection of robust and task-specific activation patterns for unilateral finger tapping, sequential finger opposition, and tactile exploration in single subjects revealed considerable interindividual variability. While partly attributable to corresponding anatomic differences, the spatial extent of activated sites may have varied due to a rigid evaluation protocol using correlation coefficients as absolute thresholds for the definition of "activation". Future improvements therefore clearly demand consideration of individual noise-corrected maps that account for intertrial differences due to (systemic) alterations of hemodynamic responsiveness or global changes between task and control states (Kleinschmidt et al., 1996). In general, the ability of MRI techniques to reproducibly detect task-related cortical and subcortical activation in individual subjects is expected to further our understanding of central motor control and somatosensory information processing.

Acknowledgements
Figures 1 to 3 were generated by using the facilities of the Gesellschaft für wissenschaftliche Datenverarbeitung Göttingen. We thank Andreas Kleinschmidt, M.D., for many stimulating discussions and his help with the preparation of the manuscript.

References

Allison, T., McCarthy, G., Wood, C.C., Darcey, T.M., Spencer, D.D. and Williamson, P.D. (1989a) Human cortical potentials evoked by stimulation of the median nerve. I. Cytoarchitectonic areas generating short-latency activity. *Journal of Neurophysiology* 62:694-710 .

Allison, T., McCarthy, G., Wood, C.C., Williamson, P.D. and Spencer, D.D. (1989b) Human cortical potentials evoked by stimulation of the median nerve. II. Cytoarchitectonic areas generating long-latency activity. *Journal of Neurophysiology* 62:711-722 .

Bandettini, P.A., Jesmanowicz, A., Wong, E.C. and Hyde, J.S. (1993) Processing strategies for time-course data sets in functional MRI of the human brain. *Magnetic Resonance in Medicine* 30:161-173 .

Boecker, H., Kleinschmidt, A., Weindl, A., Conrad, B., Hänicke, W. and Frahm J. (1994a) Dysfunctional activation of subcortical nuclei in palatal myoclonus detected by high-resolution MRI. *NMR in Biomedicine* 7:327-329 .

Boecker, H., Kleinschmidt, A., Requardt, M., Hänicke, W., Merboldt, K.D. and Frahm, J. (1994b) Functional cooperativity of human cortical motor areas during simple finger movements. A high-resolution magnetic resonance imaging study. *Brain* 117:1231-1239 .

Boecker, H., Khorram-Sefat, D., Kleinschmidt, A., Merboldt, K.D., Hänicke, W., Requardt, M. and Frahm, J. (1996) High-resolution functional magnetic resonance imaging of cortical activation during tactile exploration. *Human Brain Mapping* 3:000-000 .

Burton, H., Videen, T.O. and Raichle, M.E. (1993) Tactile-vibration-activated foci in insular and parietal-opercular cortex studied with positron emission tomography: mapping the second somatosensory area in humans. *Somatosensory Motor Research* 10:297-308 .

Colebatch, J.G., Deiber, M.P., Passingham, R.E., Friston, K.J. and Frackowiak, R.S.J. (1991) Regional cerebral blood flow during voluntary arm and hand movements in human subjects. *Journal of Neurophysiology* 65:1392-1401 .

Fox, P.T., Fox, J.M., Raichle, M.E. and Burde, R.M. (1985) The role of cerebral cortex in the generation of voluntary saccades: a positron emission tomographic study. *Journal of Neurophysiology* 54:348-69 .

Fox, P.T., Burton, H. and Raichle, M.E. (1987) Mapping human somatosensory cortex with positron emission tomography. *Journal of Neurosurgery* 67:34-43.

Frahm, J., Merboldt, K.D. and Hänicke, W. (1993) Functional MRI of human brain activation at high spatial resolution. *Magnetic Resonance in Medicine* 29:139-44 .

Frahm, J., Merboldt, K.D., Hänicke, W., Kleinschmidt, A. and Boecker, H. (1994) Brain or vein - oxygenation or flow? On signal physiology in functional MRI of human brain activation. *NMR in Biomedicine* 7:45-53 .

Hari, R., Salmelin, R., Forss, N. and Knuutila J. (1993) Bilateral activation of the human somatomotor cortex by distal hand movements. *Society of Neuroscience Abstracts* 19:1497 .

Iwamura, Y., Iriki, A., Tanaka, M. (1994) Bilateral hand representation in the postcentral somatosensory cortex. *Nature* 369:554-556 .

Kalaska, J.F., Cohen, D.A.D., Prud'homme, M. and Hyde, M.L. (1990) Parietal area 5 neuronal activity encodes movement kinematics, not movement dynamics. *Experimental Brain Research* 80:351-364 .

Killackey, H.P., Gould, H.J., Cusick, C.G., Pons, T.P. and Kaas, J.H. (1983) The relation of corpus callosum connections to architectonic fields and body surface maps in sensorimotor cortex of new and old world monkeys. *Journal of Comparative Neurology* 219:384-419 .

Kim, S.G., Ashe, J., Georgopoulos, A.P., Merkle, H., Ellermann, J.M., Menon, R.S., Ogawa, S., Ugurbil, K. (1993a) Functional imaging of human motor cortex at high magnetic field. *Journal of Neurophysiology* 69:297-302.

Kim, S.G., Ashe, J., Hendrich, K., Ellermann, J.M., Merkle, H., Ugurbil, K., Georgopoulos, A.P. (1993b) Functional magnetic resonance imaging of motor cortex: hemispheric asymmetry and handedness. *Science* 261:615-7 .

Kim, S.G., Tagaris, G., Hu, X., Menon, R.S., Strupp, J., Golub, A., Ugurbil, K., Ashe, J. (1994) Functional MRI of the lateral premotor area during simple and complex finger movements. *Society of Neuroscience Abstracts* 20:1396 .

Kleinschmidt, A., Merboldt, K.D., Hänicke, W., Steinmetz, H. and Frahm, J. (1994) Correlational imaging of thalamocortical coupling in the primary visual pathway of the human brain. *Journal of Cerebral Blood Flow and Metabolism* 14:952-957 .

Kleinschmidt, A., Requardt, M., Merboldt, K.D., and Frahm, J. (1996) On the use of temporal correlation coefficients for magnetic resonance mapping of functional brain activation. Individualized thresholds and spatial response delineation. *International Journal of Imaging Systems and Technology*, 000-000 .

Obrig, H., Kleinschmidt, A., Merboldt, K.D., Dirnagl, U., Frahm, J., Villringer, A. (1994) Monitoring of cerebral blood oxygenation during human brain activation by simultaneous high-resolution MRI and near-infrared spectroscopy. 2nd Annual Meeting, Society of Magnetic Resonance, Book of Abstracts p. 67 .

Orgogozo, J.M. and Larsen, B. (1979) Activation of the supplementary motor area during voluntary movement in man suggests it works as a supramotor area. *Science* 206:847-50 .

O'Sullivan, B.T., Roland, P.E. and Kawashima, R. (1994) A PET study of somatosensory discrimination in man. Microgeometry versus macrogeometry. *European Journal of Neuroscience* 6:137-148 .

Pause, M., Kunesch, E., Binkofski, E., Freund, H.J. (1989) Sensorimotor disturbances in patients with lesions of the parietal cortex. *Brain* 112:1599-1625 .

Rao, S.M., Binder, J.R., Bandettini, P.A., Hammeke, T.A., Yetkin, F.Z., Jesmanowicz, A., Lisk, L.M., Morris, G.L., Mueller, W.M., Estkowski, L.D., Wong, E.C., Haughton, V.M., Hyde, J.S. (1993) Functional magnetic resonance imaging of complex human movements. *Neurology* 43:2311-2318 .

Rogowska, J., Hamberg, L.M., Hunter, G.J., Aronen, H.J. and Wolf, G.L. (1992) Biological application of correlation maps to the analysis of dynamic susceptibility-contrast MRI series data. Eleventh Annual Meeting, Society of Magnetic Resonance in Medicine, Book of Abstracts p. 477 .

Roland, P.E., Larsen, B., Lassen, N.A. and Skinhøj, E. (1980) Supplementary motor area and other cortical areas in organization of voluntary movements in man. *Journal of Neurophysiology* 43:118-36 .

Roland, P.E. (1987) Somatosensory detection of microgeometry, macrogeometry and kinesthesia after localized lesions of the cerebral hemispheres in man. *Brain Research Reviews* 12:43-94 .

Seitz, R.J., Roland, P.E., Bohm, C., Greitz, T. and Stone-Elander, S. (1991) Somatosensory discrimination of shape: tactile exploration and cerebral activation. *European Journal of Neuroscience* 3:481-492 .

Shibasaki, H., Sadato, N., Lyshkow, H., Yonekura, Y., Honda, M., Nagamine, T., Suwazono, S., Magata, Y., Ikeda, A., Miyazaki, M. Fukuyama, H., Asato, R., Konishi, J. (1993) Both primary motor cortex and supplementary motor area play an important role in complex finger movement. *Brain* 116:1387-98.

Somesthesis and the Neurobiology of the Somatosensory Cortex
O. Franzén, R. Johansson and L. Terenius (eds.)

Whole-head neuromagnetic recordings of human somatosensory cortical functions

Nina Forss

Low Temperature Laboratory, Helsinki University of Technology

Summary Whole-head neuromagnetic recordings reveal functions of at least 4 cortical areas activated by somatosensory stimulation: the contralateral SI area, contra- and ipsilateral SII areas, and the posterior parietal cortex. The short-latency signals are generated solely in the contralateral SI cortex, whereas long-latency responses arise from all these 4 areas. Short- and long-latency responses behaved differently: the early responses were significantly smaller and appeared later to airpuff than to electric stimulation, implying that SI encodes faithfully the details of stimulus, such as location and type. The long-latency responses from SII and PPC were less sensitive to the stimulus type, suggesting that these areas are involved in more integrative functions. SEFs also have clinical applications; they may serve as functional landmarks in localisation of primary somatomotor cortex.

Spontaneous somatomotor μ-rhythm can be recorded over the Rolandic fissures. It is dampened during the movement of contralateral limbs, and to a lesser extent of ipsilateral limbs; μ-rhythm is increased substantially after the movement. μ-rhythm consists of 10 and 20 Hz components; 10-Hz oscillations occur over primary somatosensory cortex, whereas the sources of the 20-Hz signals may receive contribution from the motor cortex as well.

Introduction

Brain activation generates in cortical pyramidal cells electric currents which are the source of cerebral magnetic fields. The magnetic fields can be measured outside the head with a neuromagnetometer with a method called magnetoencephalography (MEG; for a recent review, see Hämäläinen et al. 1993).

With MEG it is possible to monitor totally noninvasively local activity of fissural cortex in awake humans. The spatial resolution of MEG for cortical sources under favourable conditions is a few millimeters, and the temporal resolution is better than a millisecond. The modern measuring devices cover the whole scalp and therefore allow detection of simultaneous activity of several sources all over the cortex, which makes this method very suitable in studies of somatosensory cortical functions. Here I report on recent whole-head measurements of somatosensory evoked fields (SEFs), and of spontaneous brain activity in healthy control subjects.

Somatosensory evoked fields

The somatotopic organization of the primary somatosensory cortex SI is well reflected in the distribution of somatosensory evoked fields (Brenner et al. 1978; Hari et al. 1984; Okada et al. 1984). MEG recordings have also indicated signs of somatotopic arrangement in the second somatosensory cortex (SII) in the upper lip of the Sylvian fissure. In contrast to SI, the SII areas are activated bilaterally after peripheral somatosensory stimuli (Hari et al. 1983, 1984, 1994; Teszner et al. 1983; Kaukoranta et al. 1986). More recently, a novel source area in the posterior parietal cortex was found in whole-head neuromagnetic recordings (Forss et al. 1994a).

SEPs and SEFs are typically elicited by electric stimuli applied on peripheral nerves. Such stimuli produce strongly synchronized afferent volley and large cortical responses, but are unnatural, since they activate simultaneously a large number of fibers with different conduction velocities, both from deep and superficial receptors. Airpuff stimulation activates selectively rapidly adapting cutaneous mechanoreceptors (Gardner et al. 1984) and is therefore suitable for studies of cortical areas involved in processing of tactile stimuli.

Our helmet-shaped magnetometer array has 122 planar SQUID (superconducting quantum interference device) sensors (Ahonen et al. 1993). Each sensor unit contains a pair of gradiometers that measure two orthogonal tangential derivatives of the magnetic field component normal to the helmet surface. The planar gradiometers detect the largest signal just above the local source area, where the field gradient has its maximum. Therefore, the probable source area can be approximated on the basis of the overall response distribution. During the MEG recordings, the subject sits in a magnetically shielded room with the head supported against the helmet-shaped bottom of the magnetometer. The exact location of the head with respect to the sensors is measured with the aid of three coils placed at known sites on the scalp. The locations of the coils with respect to anatomical landmarks on the head are determined with a 3-D digitizer to allow alignment of the MEG and magnetic resonance (MR) image coordinate systems.

The subjects used in the following studies were experienced, healthy volunteers (age range 22-38 yrs). Figure 1 shows SEFs of one subject to airpuff stimuli presented to dorsum of the left middle finger once every 3 s. The rise time was 7 ms and peak pressure 10 kPa. The response distribution suggests that several brain areas are activated even by this simple sensory stimulus; the earliest deflections (A) peak over the right anterior parietal cortex, whereas longer-latency responses (B, C) occur over the temporal areas. An additional deflection (D) is observed medial and posterior to the contralateral SI area.

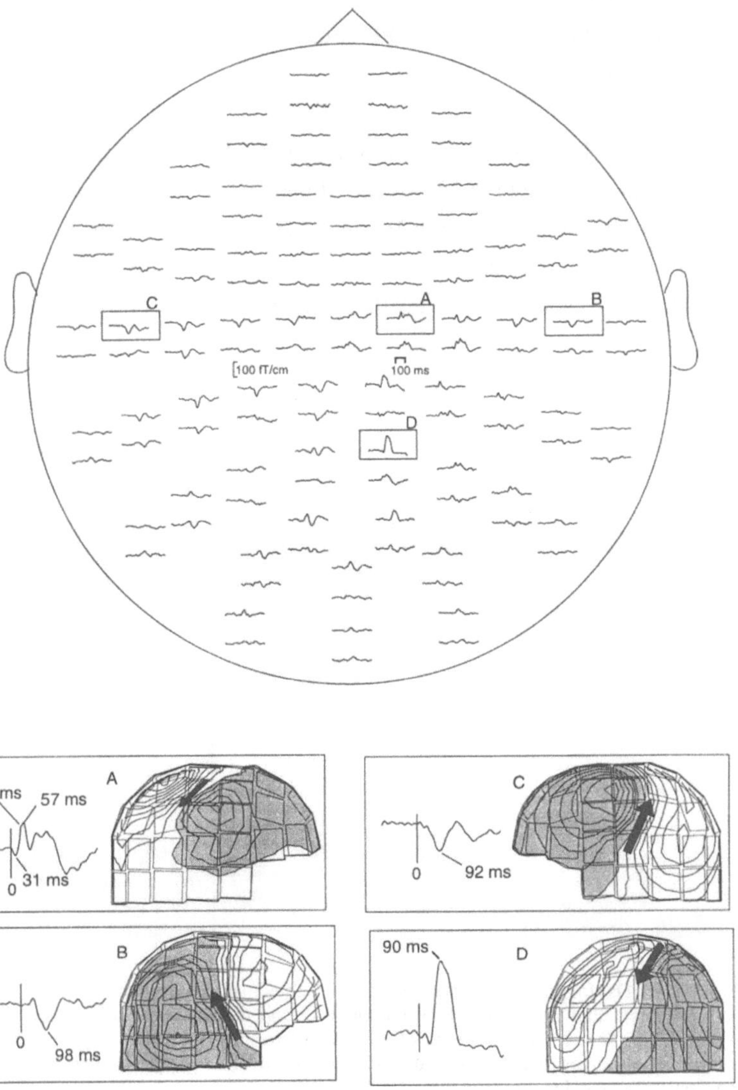

Figure 1. SEFs of one subject to left-sided airpuffs. The head is viewed from the top, and in each response pair, the upper trace illustrates the field derivative along the latitude and the lower trace along the longitude. The field patterns during the main deflections (shaded areas) are displayed over sensor arrays viewed from the sides and back. The squares show the locations of the sensor units, and the arrows indicate the ECDs; the isocontours are separated by 40 fT, and the shaded areas indicate magnetic field emerging from the head (adapted from Forss et al., 1994).

The magnetic field patterns during these deflections and the corresponding equivalent current dipoles (ECDs; arrows) are shown in the lower part of Fig.1. Each ECD, best describing the source area at a selected time period, was found by a least-squares search using a subset of channels over the response area. The earliest responses, labelled as N20m and P35m, arise from the primary somatosensory cortex. The later responses originate from second somatosensory cortices (SII) and from the posterior parietal cortex (PPC). The source areas in the human SII and PPC areas were first noninvasively observed in neuromagnetic recordings.

Figure 2 shows the ECD sites superimposed on the subject's MR images. The sources of the earliest cortical response (1 at 31 ms) and of the next deflection (2 at 45 ms) both agree with activation of the Broadmann area 3b; the latter source is more medial by about 1 cm. The PPC activation occurs at 90 ms in the postcentral fissure, probably reflecting activity in areas 5 or 7. The SII sources at 92 and 98 ms lie in the cortex of Sylvian fissures.

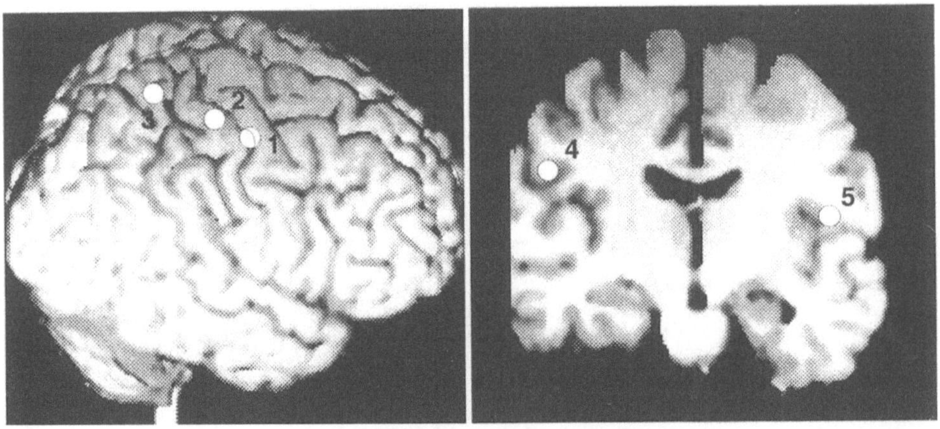

Figure 2. *Left*: ECDs for airpuff responses of one subject superimposed on her 3-D MRI surface reconstruction. Sites 1 and 2 refer to SI responses at 42 and 69 ms, respectively, and site 3 shows the posterior source at 104 ms. The locations of the ECDs are projected to overlying surface in viewing direction. *Right*: The sources (4 and 5) of the signals recorded over the temporal areas lie in the upper lips of the Sylvian fissures (adapted from Forss et al., 1994).

In another session, SEFs were recorded to electric median nerve stimulation (Forss et al. 1994a). The left and right median nerves were stimulated at the wrists using 0.3-ms constant-current pulses with interstimulus intervals (ISIs) of 1, 3, and 5 s; intensity varied from 4 to 6 mA and exceeded the motor threshold. Figure 3 shows that SEFs recorded over the SI cortex to airpuff and electric stimuli. are similar in waveforms. The responses are quite consistent in their morphology for all subjects. In agreement with previous studies, the first deflections to electric stimuli peaked on average at 21–22 ms (N20m), at 36 ms (P35m), and at 58 ms (P60m). The corresponding mean latencies to airpuffs were clearly longer: 28 ms, 43–46 ms, and 66–71 ms, respectively. Even when the extra 2.5-ms delay caused by the more distal stimulation site for the airpuffs was taken into account, the latencies of the earliest responses were significantly longer ($p < 0.008$) to airpuffs than to electric stimuli. The source areas were the same; the earliest responses to electric stimuli were generated in the SI cortex, and the long-latency responses in parietal operculum and posterior parietal cortex. The mean dipole moments for the SI responses were significantly weaker for airpuffs than for electric stimuli, the airpuff responses being on average 44% ($p < 0.005$) and 52% ($p < 0.05$) of the corresponding P35m responses to the right- and left-sided electric stimuli, respectively (Forss et al.,1994b).

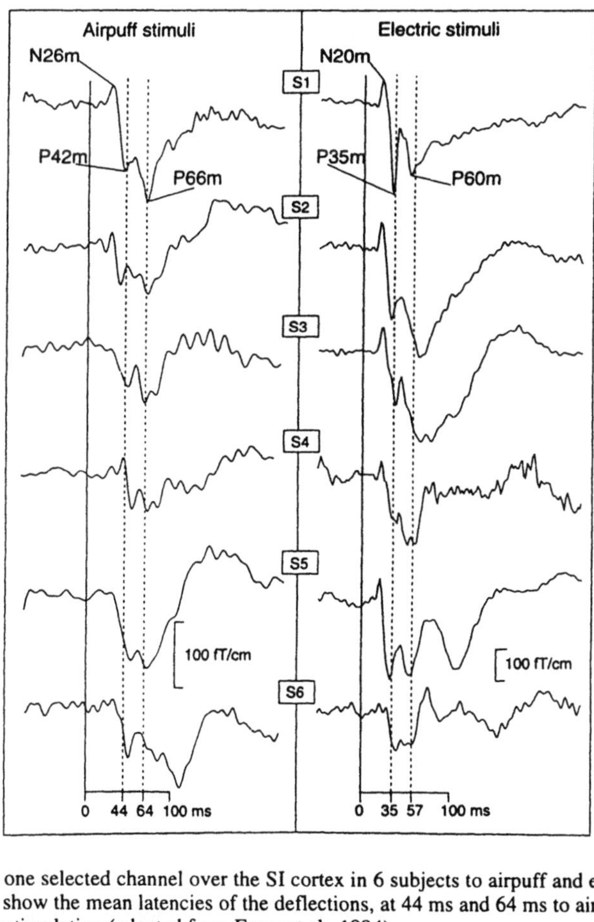

Figure 3. SEFs from one selected channel over the SI cortex in 6 subjects to airpuff and electric stimuli. The dashed vertical lines show the mean latencies of the deflections, at 44 ms and 64 ms to airpuffs, and at 35 ms and 57 ms to electric stimulation (adapted from Forss et al., 1994).

Airpuff responses from SII and PPC were observed in all subjects, but they were more variable both in latency and amplitude. In general, the latencies or the dipole moments of the long-latency responses arising from areas outside SI did not differ significantly between airpuffs and electric stimuli.

The different behavior of the early and late airpuff responses agrees with the findings of Hashimoto (1987) about differences in early and late SI components. The earliest responses arising from SI seem to reflect differences in stimulation, suggesting that SI encodes details of tactile stimuli, such as the size of the stimulated area and the type of receptors that were activated. The longer latency responses originating from other somatosensory areas were less sensitive to stimulus type, implying that these areas have more integrative functions. These results agree with animal data showing that neurons in SI, particularly in area 3b, reproduce almost exactly the stimulus features encoded by the receptor in the skin (Mountcastle 1957; Phillips et al. 1988). Animal studies also imply that precise encoding of stimulus properties, such as location, is not a likely role for SII nor for PPC (Hyvärinen 1982; Burton 1986).

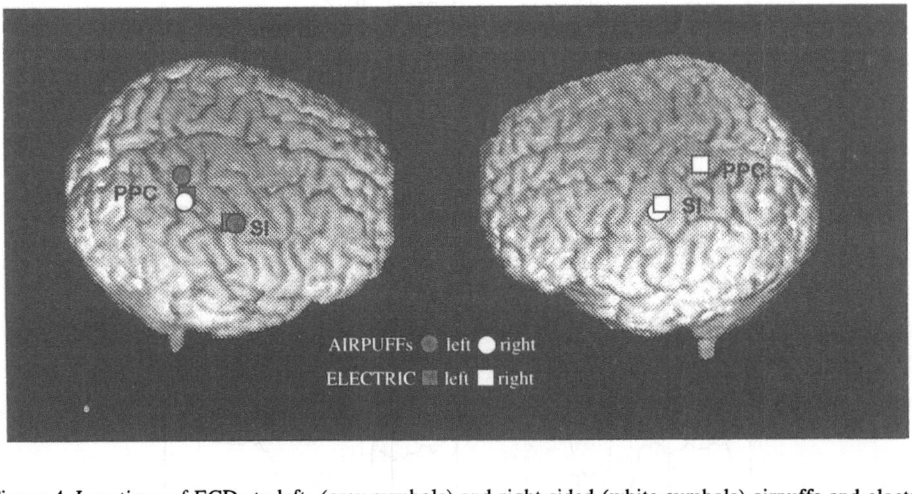

Figure 4. Locations of ECDs to left- (gray symbols) and right-sided (white symbols) airpuffs and electric stimuli in one subject. Note that airpuffs activate PPC only in the right hemisphere (adapted from Forss et al., 1994).

Figure 4 illustrates the locations of the ECDs to airpuffs and electric stimuli in one subject. The SI responses are generated symmetrically in the right and left hemispheres, contralateral to stimulation, but airpuffs activate the PPC source constantly in the right hemisphere, regardless of the side of stimulation. The amplitudes of PPC responses were significantly smaller for right-sided (ipsilateral) than left-sided (contralateral) stimuli. This indicates dominance of the right PPC in the processing of tactile stimuli. Such a hemispheric difference is of particular interest since lesions of the right PPC lead to the well-known neglect syndrome, characterized by ignorance of stimuli presented to the left side of space.

Somatosensory evoked fields have also clinical uses as functional landmarks (*cf.* Paetau et al. 1992). For example, localization of central fissure and primary sensory and motor areas with respect to epileptic foci is important in presurgical evaluation of epileptic patients. In patients with brain tumours in vicinity of central fissure it is possible to locate the central fissure with the aid of SEFs, and to evaluate whether the tumor mass has infiltrated into the primary somatomotor area.

Spontaneous somatomotor rhythms

MEG recordings also reveal spontaneous rhythms which resemble the well known EEG rhythms. The 8-13 Hz posterior α-activity, which is dampened by opening of the eyes, is seen in MEG as well (Cohen 1968; Hari 1993; Salmelin and Hari 1994). The somatomotor μ-rhythm, with its characteristic comb-like shape consists of 10 and 20 Hz components (Tiihonen et al. 1989). It can be recorded over the Rolandic fissures and is dampened during the movement of contralateral limbs, and to a lesser extent of ipsilateral limbs. This rhythm is increased substantially after the end of both voluntary and passive limb movements (Salmelin and Hari 1994a).

Interestingly, source modelling suggests that the two frequency ranges in μ-rhythm have slightly different source areas; the 10 Hz oscillations occur predominantly over the primary somatosensory cortex, but the sources of the 20 Hz signals are slightly more anterior (Salmelin and Hari 1994a). It can be suggested that the 10-Hz rhythm is purely somatosensory in origin, whereas the 20 Hz signals also receive contribution from the motor cortex (Salmelin and Hari 1994b).

Acknowledgements:
This study was financially supported by the Academy of Finland. I thank professor Riitta Hari for constructive criticism on the manuscript.

References

Ahonen, A., Hämäläinen, M., Kajola, M., Knuutila, J., Laine, P., Lounasmaa, O., Simola, J., and Tesche, C. (1993) A 122-channel SQUID instrument for investigating the magnetic signals from the human brain. *Physica Scripta* T49: 198-205.

Brenner, D., Lipton, J., Kaufman, L., and Williamson, S. (1978) Somatically evoked magnetic fields of the human brain. *Science* 199: 81-83.

Burton H. (1986) Second somatosensory cortex and related areas. *In*: E.G. Jones and A Peters (eds): *Cerebral Cortex*. Plenum, pp 31-98.

Cohen, D. (1968) Magnetoencephalography: evidence of magnetic field produced by alpha-rhythm currents. *Science* 161: 784-786.

Forss, N., Hari, R., Salmelin, R., Ahonen, A., Hämäläinen, M., Kajola, M., Knuutila, J., and Simola, J. (1994a) Activation of the human posterior parietal cortex by median nerve stimulation. *Experimental Brain Res.*earch 99: 309-315.

Forss, N., Salmelin, R., and Hari, R. (1994b) Comparison of somatosensory evoked fields to airpuff and electric stimulation. *Electroencephalography and Clinical Neurophysiol.ology* 92: 510-517.

Gardner EP, Hämäläinen HA, Warren S, Davis J, and Young W. (1984) Somatosensory evoked potentials and cortical single unit responses elicited by mechanical tactile stimuli in awake monkeys. *Electroencephalography and Clinical Neurophysiol.ology* 58: 537-552.

Hämäläinen, M., Hari, R., Ilmoniemi, R., Knuutila, J., and Lounasmaa, O. (1993)Magnetoencephalography - theory, instrumentation, and applications to noninvasive studies of the working human brain. *Rev.iews of Modern Physics* 65:413-498.

Hari, R., Hämäläinen, M., Kaukoranta, E., Reinikainen, K., and Teszner, D. (1983) Neuromagnetic responses from the second somatosensory cortex in man. *Acta Neurol.ogica Scandinavica* 68: 207-212.

Hari, R., Reinikainen, K., Kaukoranta, E., Hämäläinen, M., Ilmoniemi, R., Penttinen, A., Salminen, J., and Teszner, D. (1984) Somatosensory evoked cerebral magnetic fields from SI and SII in man. *Electroencephalography and Clinical Neurophysiol.ology* 57: 254-263.

Hari R. (1993) Magnetoencephalography as a tool of clinical neurophysiology. In: E Niedermeyer and F Lopes da Silva (eds): *Electroencephalography. Basic Principles, Clinical Applications and Related Fields.* 3rd edition. Williams & Wilkins, pp 1035-1061.

Hari, R., Karhu, J., Hämäläinen, M., Knuutila, J., Salonen, O., Sams, M., and Vilkman, V. (1993) Functional organization of the human first and second somatosensory cortices: A neuromagnetic study. *European J.ournal of Neurosci.ence* 5: 724-734..

Hashimoto I. (1987) Somatosensory evoked potentials elicited by air-puff stimuli generated by a new high-speed air control system. *Electroencephalography and Clinical Neurophysiol.ology* 67: 231-237.

Hyvärinen J, (1982) *The Parietal Cortex of Monkey and Man.* Springer, Heidelberg Berlin New York.

Kaukoranta, E., Hämäläinen, M., Sarvas, J., and Hari, R. (1986) Mixed and sensory nerve stimulations activate different cytoarchitectonic areas in the human primary somatosensory cortex SI. *Exp.erimental Brain Res.earch* 63: 60-66.

Mountcastle VB. (1957) Modality and topographic properties of single neurons of cat's somatic sensory cortex. *J.ournal of Neurophysiology* 20: 408-434.

Okada, Y., Tanenbaum, R., Williamson, S., and Kaufman, L. (1984) Somatotopic organization of the human somatosensory cortex as revealed by magnetic measurements. *Exp.erimental Brain Research* 56: 197-205.

Paetau, R., Kajola, M., Karhu, J., Nousiainen, U., Partanen, J., Tiihonen, J., Vapalahti, M., and Hari, R. (1992) Magnetoencephalographic localization of epileptic cortex—impact on surgical treatment. *Annals of Neurol.ogy* 32: 106-109.

Phillips J, Johnson K, and Hsiao S. (1988) Spatial pattern representation and transformation in monkey somatosensory cortex. *Proceedings of National Academy of Sci ecesUSA* 85: 1317-1321.

Salmelin, R., and Hari, R. (1994a) Spatiotemporal characteristics of sensorimotor MEG rhythms related to thumb movement. *Neurosci.ence* 60: 537-550.

Salmelin, R., and Hari, R. (1994b) Characterization of spontaneous MEG rhythms in healthy adults. *Electroencephalography and Clinical Neurophysiol.ology* 237-248.

Teszner, D., Hari, R., Nicolas, P., and Varpula, T. (1983) Somatosensory evoked magnetic fields: mapping and the influence of stimulus repetition rate. *Nuovo Cimento* 2D: 429-437.

Tiihonen, J., Kajola, M., and Hari, R. (1989) Magnetic mu rhythm in man. *Neurosci.ence* 32: 793-800.

O. Franzén, R. Johansson and L. Terenius (eds.)

Optical imaging of intrinsic signals in somatosensory cortex

Mark Tommerdahl, Ph.D., and Barry Whitsel, Ph.D.

Departments of Biomedical Engineering and Physiology, University of North Carolina at Chapel Hill, USA

Summary. The method of optical imaging of intrinsic signals emitted in the near-infrared (IR; 820-846 nm) was used to study the response of somatosensory cortex (SI) of anesthetized cats and squirrel monkeys to precisely controlled vibrotactile stimuli. The cortical territory exhibiting decreased average IR reflectance in response to a low frequency, modest amplitude (25 Hz; 200 - 400 microns peak-to-peak, sinusoidal) vibrotactile stimulus shifted location in a manner consistent with a published map of SI topographical organization generated from receptive field mapping observations. The size of the SI territory responding with a decrease in reflectance was greatly reduced by intravenous administration of even small doses of a selective NMDA receptor blocker (ketamine). Extracellular single and multiunit recordings obtained during microelectrode penetrations of the same SI region studied with the IR imaging method revealed that (i) the spike discharge activity of neurons sampled within territories exhibiting decreased average IR reflectance values in response to a vibrotactile stimulus was consistently elevated by the stimulus, and (ii) the discharge activity of neurons sampled in regions exhibiting stimulus-evoked increases in average IR reflectance was consistently depressed. Neurons located in regions which did not exhibit a reflectance change rarely were influenced by the stimulus. Analysis of the time course of the IR reflectance changes evoked by both short (500-1000 msec) and long-duration (6-10 sec) stimuli revealed considerable dynamics in the two types of SI territories (excitatory vs. inhibitory) that modify their average IR reflectance in response to vibrotactile stimulation. In a final series of experiments we evaluated the IR intrinsic signals evoked in SI of anesthetized subjects in whom the spinal dorsal column pathway had been surgically transected. The findings show that in the absence of the spinal dorsal column pathway, the dynamics of the IR intrinsic signals evoked in SI by a low frequency vibrotactile stimulus are strikingly different than observed in normal subjects. In a number of respects, the abnormal dynamics of the IR intrinsic signals observed in SI after dorsal column transection appear to parallel the disturbances in vibrotactile detection and discriminative capacity reported for monkeys with a selective dorsal column lesion.

Our laboratory's interest in applying the method of optical imaging of cerebral cortical intrinsic signals to somatosensory cortex was motivated by the possibility that this approach might allow us to acquire information difficult, if not impossible to obtain using alternative (e.g., neurophysiological or neurometabolic) methods for studying the spatially distributed cortical response to a natural skin stimulus. With the IR imaging method we hoped to observe the spatially distributed global SI response to a natural skin stimulus at high spatial resolution. In addition, we sought to experimentally evaluate our proposal that (i) the spatially patterned cortical response to a skin stimulus changes systematically and substantially during a period of repetitive skin stimulation, and (ii) the changes in SI response are accompanied by changes in tactile detection and discriminative capacity (Whitsel et al., 1991; Lee and Whitsel, 1992; Lee et al., 1992; Tommerdahl et al., 1993).

Compared to the method we have used previously to map the global response of somatosensory cortex to natural skin stimuli - the 2DG metabolic mapping method - the IR imaging method has two primary advantages. First, one can obtain multiple images of the response of the same cortical region (to the same or to different stimuli) in the same subject, each with a spatial resolution comparable to or even exceeding that of a 2DG image. Second, it allows analysis of the cortical response to stimulation at relatively high temporal resolution. Other strengths of the IR imaging method are: (a) the cortical intrinsic signal detected in the IR range is relatively independent of blood flow changes (Haglund et al., 1993); (b) it reflects changes in the volume of the extracellular fluid compartment attributable to stimulus-evoked changes in extracellular $[K^+]$ and/or neurotransmitter release (Lieke et al., 1989; Cohen, 1973); (c) it is attenuated in a dose-dependent manner by cortical NMDA receptor block (Haglund et al., 1992; 1993; Tommerdahl, personal observation); (d) the principal source of the signal is the dendrites of cortical neurons (Grinvald et al., 1994); and (e) the signal is not a direct reflection of either neuronal spike discharge activity or, more generally, the local cortical voltage changes evoked by a sensory stimulus (the intrinsic signal has a much slower onset and decay than the neuroelectrical responses of single neurons or neuron populations, and it can be eliminated by drugs - e.g., by NMDA receptor blockers - that do not prevent stimulus-evoked cortical neuron spike discharge activity). Since all of these attributes of the intrinsic signal are also attributes of the hypothetical stimulus-directed dynamic process we believe is important for normal somatosensory cortical information processing and representation (Whitsel et al., 1991), the method of intrinsic signal imaging seemed uniquely appropriate for use in experimental studies addressing the following questions: (i) Does a skin stimulus trigger a dynamic cortical process?, and if such a dynamic process is evoked by skin stimulation, (ii) What are its temporal characteristics?, (iii) What cortical neural mechanisms are involved?, (iv) Does the process dynamically modify the capacity of SI to process and represent subsequent skin stimuli?, and (v) Are stimulus-evoked modifications in the capacity of SI to process and represent skin stimuli paralleled by changes in somatosensory perceptual capacity?

This paper presents findings obtained in preliminary studies which utilized the IR optical imaging method. A more comprehensive paper describing the detailed findings obtained with this method under very different conditions of natural skin stimulation (i.e., vibrotactile, brushing, and thermal) is in preparation.

Figure 1 shows the optical imaging system developed in our laboratory. A hydraulically-sealed chamber capped with an optical window is cemented to the skull over an opening that exposes the region of somatosensory cortex to be imaged. Use of a closed chamber greatly reduces the noise contributed to images by brain fluctuations associated with the cardiovascular and respiratory cycles. A modified (OMPI-1) Zeiss operation microscope attaches directly (via a cylindrical light guide) to the chamber. This allows images of the exposed cortex to be acquired (using a slow scan CCD camera; Photometrics, Inc.) at variable magnification (between 1 and 100X). A filter with FWHM range of 820-846 nm is inserted between the light source and light guide. Images are collected with a 486 PC. A second 486 PC controls the stimulator and triggers the collection of images by the first 486 PC. In all the experiments described in this paper a sinusoidal vertical displacement skin stimulus was delivered with a Chubbuck-type stimulator (Chubbuck, 1966). Images were obtained before (prestimulus), during, and after vibrotactile stimulation of the skin. Difference images were formed by subtracting each prestimulus image from an image acquired at a fixed time after stimulus onset (the temporal resolution was determined by the spatial resolution of the image acquired). Interpretation of the dark and light areas in a difference image formed in this way is straight-forward. Because an increase in neuroelectrical activity is accompanied by increased IR absorption, less light is reflected back to the camera from a cortical region activated by stimulation - the result is that cortical regions which are activated by a stimulus appear darker in such an image than those regions which are not. Similarly, cortical regions whose neuroelectrical activity is decreased by a stimulus appear lighter than regions whose activity level is increased or unaffected.

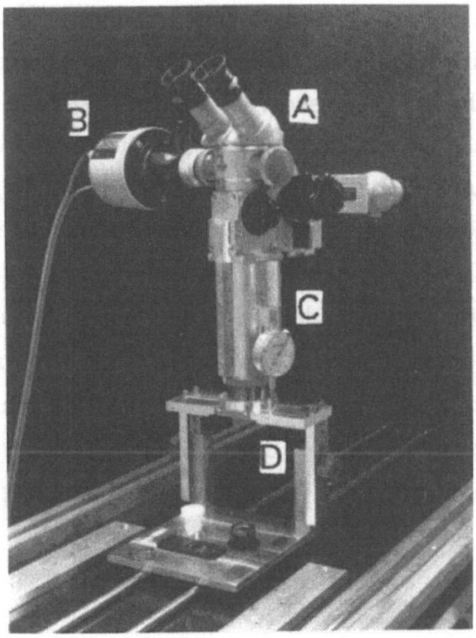

Figure 1. A: Zeiss OMPI-1 operation microscope. B: Photometrics slow-scan CCD camera. C: Light guide and focusing tube. D: Bridge that clamps to cortical recording chamber and supports the microscope and camera.

One of our first goals was to assess the sensitivity of the IR imaging method to changes in the parameters of skin stimulation. As an example, we sought to determine if a change in the locus of the SI optical response to a vibrotactile stimulus occurred when the position of the stimulus was shifted on the skin and, if there was such a change, was the change in locus consistent with what is known about the effects of a change in skin position on the locus of the responding SI neurons? In an experiment whose results are illustrated in Figure 2, a vibrotactile stimulus (25 Hz, 0.5 mm peak-to-peak amplitude, sinusoidal, 2 mm diameter probe) was applied with a Chubbuck-type stimulator to different loci on the volar surface of the contralateral hand of a halothane/nitrous oxide anesthetized squirrel monkey. The duration of a single stimulus trial was 4 seconds; interstimulus interval (ISI) was 35 seconds. An IR image of the exposed cortical region was acquired prior to each stimulus application, and at successive 2 second intervals after stimulus onset. For each stimulus a "difference" IR image was computed by subtracting the prestimulus image from each of the images obtained after stimulus onset. Figure 2 is intended to encourage direct comparison of the cortical locus of the IR response detected at 4 seconds after the onset of vibrotactile stimulation at each of the four skin sites with the locus of activity predicted by the map of squirrel monkey SI topographical organization published by Sur et al. (1982; Figure 2H). Figure 2E shows the vascular pattern of the SI region imaged (obtained without IR filter); 2A - 2D are images of the stimulus-evoked IR reflectance changes - the darker an area in such difference images, the greater the intrinsic signal in that area. Figure 2F shows the major cortical source of the intrinsic signal evoked by each of the four different stimuli - the bounded regions in Figure 2F were obtained by smoothing and thresholding the distribution of reflectance values (boundaries define regions occupied by the most significant 2% of the reflectance distribution). IR imaging results like those shown in Figure 2 are fully consistent with the idea that the cortical areas of decreased IR reflectance correspond to regions of stimulus-evoked neuronal excitation - a suggestion that will be strongly supported by findings obtained in combined single neuron recording and IR imaging experiments (described in a later section of this paper).

The substantial discrepancy between the size of the regions of stimulus-evoked decreased IR reflectance and the regions indicated by the map of Sur et al. (1982) to contain neurons with RF and modal properties consistent with those of the vibrotactile stimulus that was applied deserves further comment. Specifically, we suspected that the actual size of the responding cortical region is underestimated by the map of Sur et al. (1982) because (i) the RF mapping experiments of Sur et al. (1982) were carried out under supra-anesthetic levels (35 - 55 mg/kg) of ketamine, an agent known to be a relatively selective and potent NMDA receptor blocker, and (ii) NMDA receptors have been shown, both by our laboratory (Duncan et al., 1982; McKenna et al., 1982) and by others (see Daw et al., 1993 for review) to mediate a major fraction of the SI neuron response to peripheral sensory drive, especially that fraction originating from non-central regions of the RF.

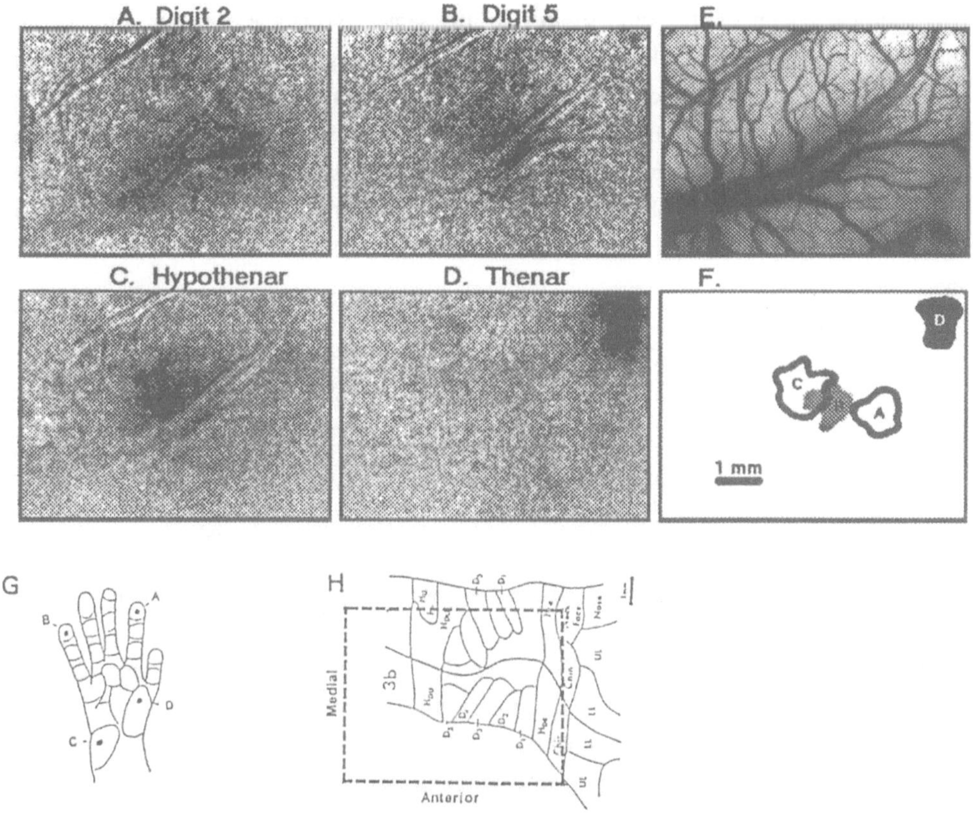

Figure 2. A - D: averaged difference (stimulus - minus prestimulus) images. Note that vibrotactile stimulation leads to an increased absorption of IR light (less reflectance; indicated by dark areas): a change which, based on the findings of others (e.g., Grinvald, 1985; Grinvald et al., 1991; 1995; Narayan et al., 1994), accompanies neuronal activation. E: vascular pattern. The observed sequence of shifts in the locus of the region of decreased IR reflectance corresponds generally with the shifts in the sites of neuronal activation predicted by the map of SI topographical organization published by Sur et al. (1982). However, for each stimulus site even the prinicipal locus of the SI region activated by the vibrotactile stimulus is substantially larger (5 - 20X) than the size of the area predicted by the map of Sur et al. (1982).

Figure 3 shows results (obtained from the same region in SI of the same squirrel monkey that provided the observations in Figure 2) which confirm this suspicion. The image on the left in Figure 3 shows the IR reflectance decrease that occurred in SI in response to vibrotactile stimulation of a site on the contralateral hypothenar eminence (same pattern as is shown in Figure 2C); whereas the image on the right in Figure 3 shows the IR reflectance pattern evoked in the same cortical region of the same subject by the same vibrotactile stimulus after intravenous administration of a relatively low dose of ketamine (7.5 mg/kg). The shrinkage of the cortical area responding to the vibrotactile stimulus is obvious. This and similar results lead us to conclude that the discrepancy between the size of the responding SI areas predicted by the RF map of Sur et al. (1982) and the results obtained by IR optical imaging is, to a substantial extent, attributable to the very different functional states of the NMDA receptor system when the cortical response to tactile drive is evaluated under the conditions of ketamine and halothane/nitrous oxide anesthesia.

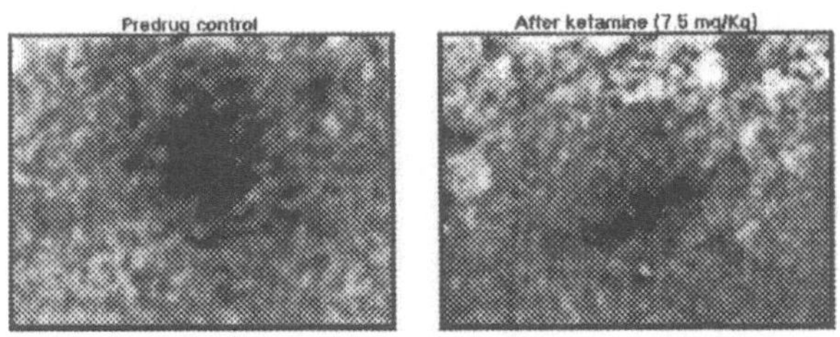

Figure 3. Ketamine's actions on the intrinsic signal evoked in SI by vibrotactile stimulation.

The IR images shown in Figure 4 show that the IR reflectance pattern also changes systematically with time after onset of a continuous vibrotactile stimulus, and that the time course and the nature of the time-dependent changes in the pattern can be detected and quantified. The IR images in Figure 4 were obtained from a halothane/nitrous oxide anesthetized cat that received a series of 10-second vibrotactile stimulus presentations to the ulnar eminence on the contralateral foreleg (25 Hz, peak-to-peak amplitude 400 microns). ISI was 35 seconds. On each stimulus trial a prestimulus image was obtained 2 seconds prior to stimulus onset and poststimulus images at 2, 4, 7, and 10 seconds after stimulus onset. Figure 4A shows the results, averaged over all 25 trials, at 2 seconds after stimulus onset; and Figure 4B shows the pattern at 7 seconds after stimulus onset. Comparison of the difference images shown in Figures 4A and 4B shows that the SI region exhibiting IR reflectance decrease (activation) is large and relatively spatially homogeneous at 2 seconds, but consists of multiple, smaller, nonoverlapping regions of decreased IR reflectance at 7 seconds after stimulus onset.

To enable objective visualization of the relationship between the cortical vascular pattern (Figure 4C) and the two prominent regions exhibiting peak activation in Figure 4B, the contours of the vascular pattern are overlaid with the thresholded IR pattern (Figure 4D). It is clear that the regions of peak reflectance decrease do not correspond to the cortical vascular pattern (i.e., they do not reflect stimulus-evoked changes in the cortical macro-vasculature). The reflectance patterns observed at 2 and 7 seconds after stimulus onset are compared quantitatively (using an image segmentation procedure) in Figure 4F (Figure 4E indicates the orientation of image segmentation). The major differences between the reflectance versus distance plots obtained at 2 and 7 seconds after stimulus onset (T2 vs. T7; Figure 4F) provide strong support for the claim that the IR pattern changes dramatically within a few seconds of the onset of a temporally extended vibrotactile stimulus.

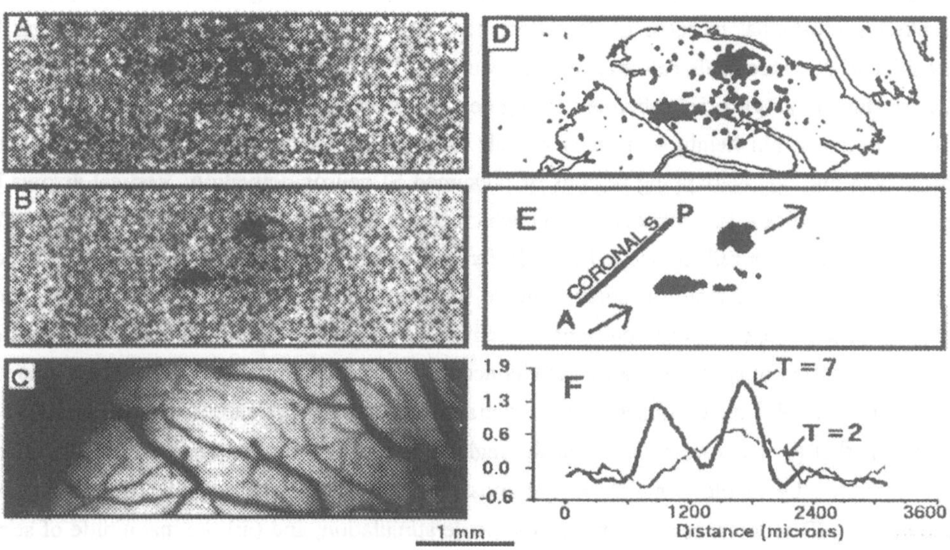

Figure 4. The stimulus-evoked IR activity pattern in somatosensory cortex changes with time after stimulus onset. A-F: Described in text. Y axis in Panel F is proportional decrease in reflectance $*10^{-3}$.

Although we found the IR imaging method to yield substantial information about the SI response to skin stimulation, we believe the full potential of the method can only be achieved if it is used in combination (in the same experiment) with neurophysiological recording methods. In such combined experiments (using anesthetized squirrel monkeys), we initially acquired IR images of the cortical IR response to a vibrotactile stimulus, and then used those images to guide the placement of microelectrode penetrations for acquisition of single neuron recording data. This approach enabled us (i) to place microelectrode penetrations in known and precise relationship to the stimulus-evoked IR activity pattern, and (ii) to record the activity evoked by the same stimulus from single neurons and small neuron aggregates located at sites in SI distinguishable from each other on the basis of the intrinsic signal measured at those sites.

Figure 5 shows representative average IR difference images (stimulus - prestimulus; 40 trials; B&C), and examples of spike train recordings (D) obtained in the course of microelectrode penetrations performed at three different locations (sites 1, 2, and 3) in SI of a squirrel monkey exposed to a vibrotactile stimulus on the contralateral hand. The cortical sites sampled by the microelectrodes were selected based on their average IR reflectance values. For example, although at site 3 the stimulus caused a large decrease in average IR reflectance, at site 2 it evoked a substantial increase in IR reflectance; and at site 1 only a modest, but nevertheless significant, decrease in IR reflectance was recorded. The spatial histogram of average reflectance values shown in Figure 5H (obtained by segmenting IR image 5C from left to right at a bin width of 30 microns; bin height = 1 mm) reveals the average mediolateral dimensions of the two prominent and nonoverlapping regions of presumed SI neuron activation. Vertical arrows below plots in 5H indicate the cortical sites at which single neuron activity was recorded; the Y-axis indicates changes in reflectance x 10^{-3} ([reference - stimulus] / reference). E, F, and G are PST histograms computed from the spike trains of single neurons recorded at the three sites, and they clearly show that elevated spike discharge activity accompanied the stimulus at sites 3 and 1, and that a depression of spike discharge activity accompanied the stimulus at site 2. The single neuron recording and IR imaging observations illustrated in Figure 5 are representative, and lead us to conclude that (i) most neurons within the regions of stimulus-evoked decreases in IR reflectance are activated by tactile stimulation; (ii) most neurons within the regions of stimulus-evoked increases in IR reflectance are inhibited by tactile stimulation; and (iii) the magnitude of stimulus-evoked neuronal activation (inhibition) covaries with the magnitude of the reflectance decrease (increase).

Figure 5. A: Vascular Pattern. B-H: IR imaging and neurophysiological recording data obtained from SI cortex of the same squirrel monkey. Vibrotactile stimulus parameters: 25 Hz, 400 microns peak-to-peak, 7 seconds duration, 45 seconds interstimulus interval, stimulator probe contacted 2 mm site on contralateral radial interdigital pad. B: Difference image at 2 seconds after stimulus onset. C: 7 seconds after stimulus onset. E - H: Described in text.

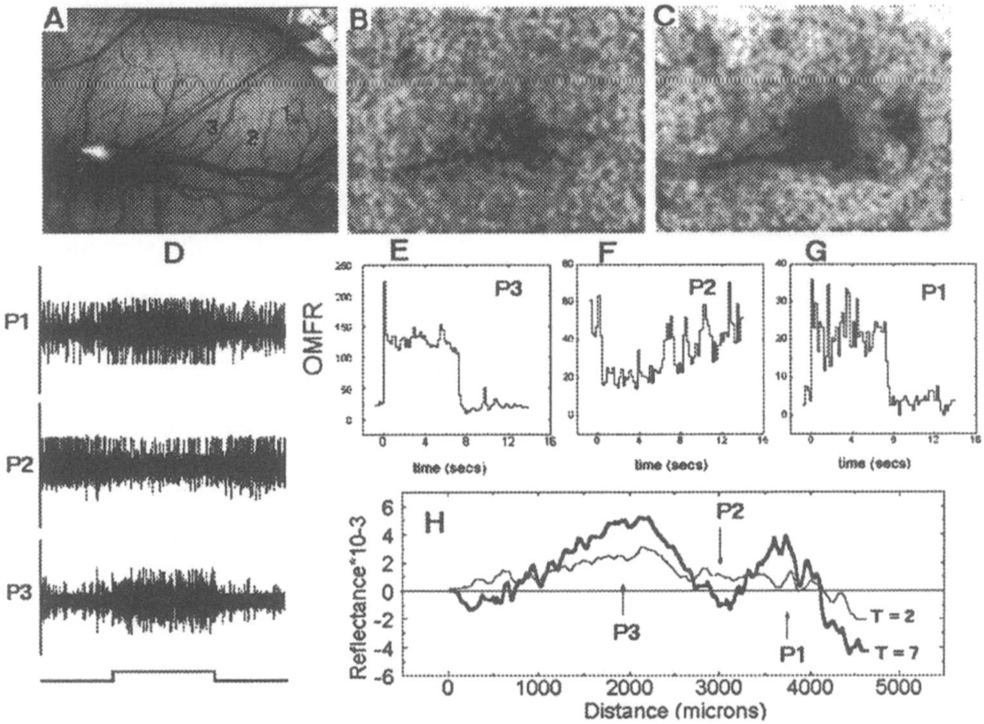

To obtain more direct information about the effects of prior exposure to vibrotactile stimulation on the somatosensory cortical response to a vibrotactile stimulus, experiments were carried out in which stimuli 1 second in duration were interleaved (on a trial by trial basis) with stimuli of longer duration (4 - 10 seconds; both stimuli were delivered to the same skin site and differed only in duration). Figures 6 and 7 present representative observations obtained from an anesthetized cat studied in this way. In both figures averaged difference images computed from trials using a 1-second stimulus are shown in the left column, whereas the averaged difference images computed from trials using a 10-second stimulus are shown on the right. The results shown in Figure 6 reveal that at 2 seconds after stimulus onset (T=2), the activated region is extensive and very similar for both the 1-second and 10-second stimuli. With the 10-second stimulus (panels on

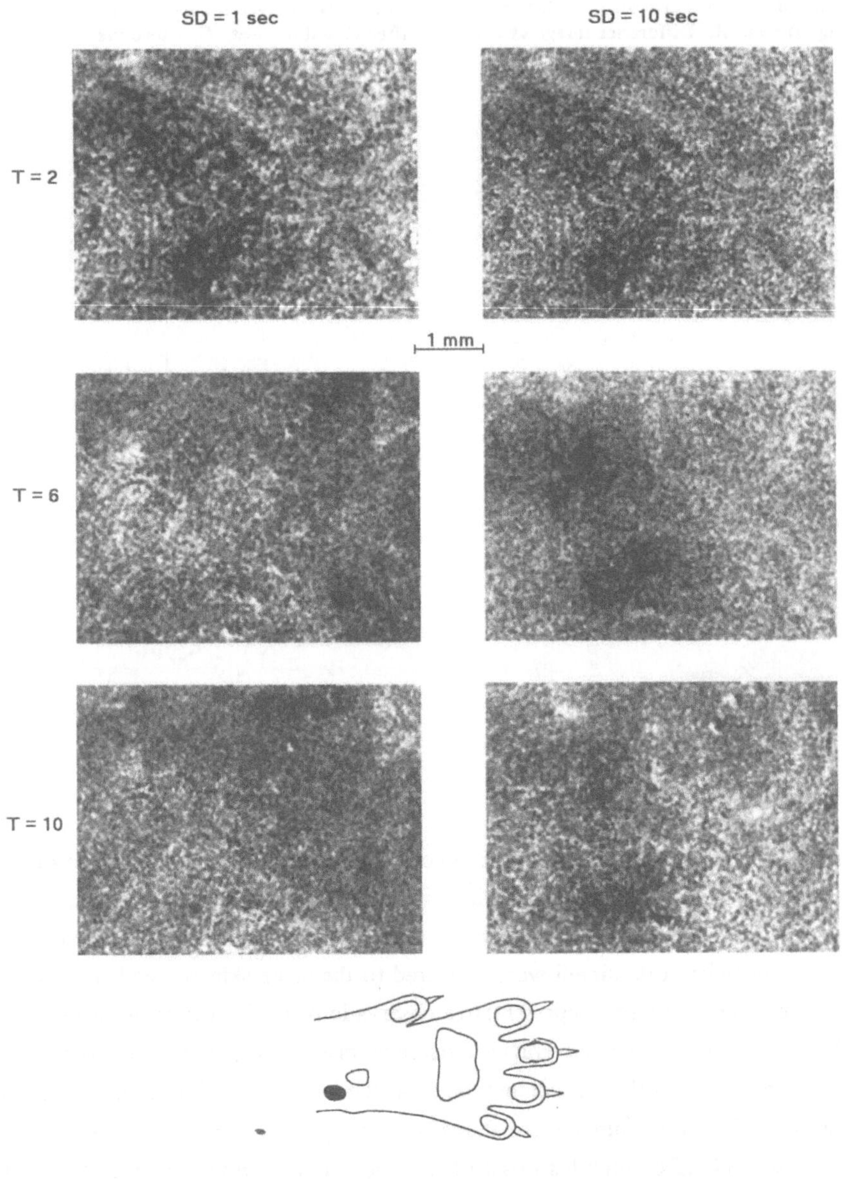

Figure 6. Temporal dynamics of IR reflectance pattern evoked in cat SI by vibrotactile stimulus. SD = Stimulus Duration. Stimulus site indicated by filled region on figurine.

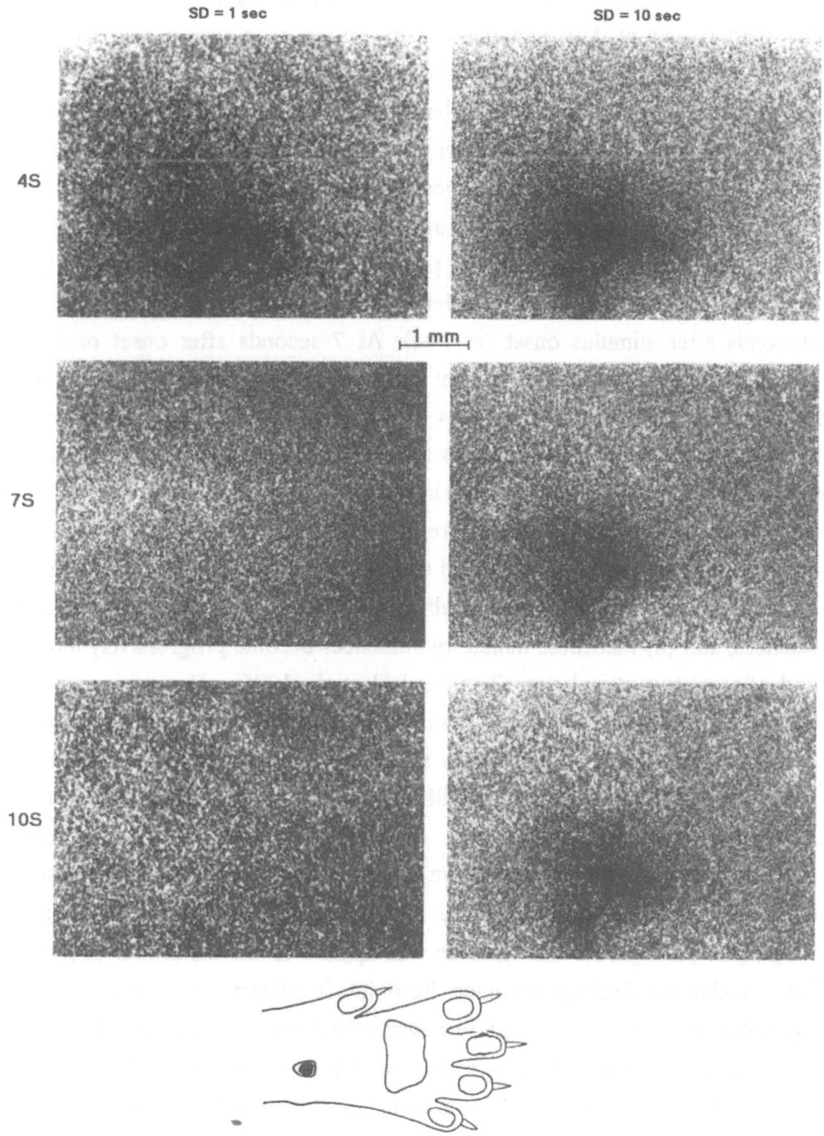

Figure 7. Temporal dynamics of IR reflectance pattern evoked in cat SI by vibrotactile stimulus. Stimulus site on contralateral pisiform pad.

right), however, at 6 seconds after onset (T=6) two distinct regions of activation are evident. Although these same two regions remain discernable at 10 seconds after stimulus onset (T=10), each is smaller than it was at T=6 and the activity within each region has become spatially inhomogeneous. In the case of the 1-second stimulus (panels on left), a most striking observation is that at both T=6 and T=10 there is no above-background activity observed in the vicinity of the region that is activated at T=2. Instead, the activity throughout this region at T=6 and T=10 is actually ***below*** the levels observed in the same locations prior to stimulation.

Figure 7 illustrates results obtained from the same subject, but this time using a stimulus to a different skin site (figurines indicating stimulus placement shown in Figures 6 and 7). Note that the responses to the 1-second stimulus (on the left) and the 10-second stimulus (on the right) are similar at 4 seconds after stimulus onset (at T=4). At 7 seconds after onset of the 1-second stimulus (at T=7), however, there is a prominent suppression of activity in the region in the very same region that exhibits elevated activity at T=4. It should be noted that for this same subject the differences between the early and late responses to the 10-second stimulus (shown on the right), while present, are more subtle than those shown in Figure 6.

We recently advanced the idea (i) that the response of areas 3b and 1 of monkey anterior parietal cortex to tactile afferent drive conveyed centrally over pathways which ascend the spinal cord outside the dorsal columns is subject to abnormal ("maladaptive") dynamic corticocortical inhibitory influences, and (ii) that these inhibitory influences become progressively more powerful during a period of repetitive stimulation (Tommerdahl et al., 1995). Because our proposal that corticocortical interactions are abnormal in the absence of the dorsal columns was based primarily on 2DG metabolic observations, we initiated a series of studies in which we are using the IR imaging method to study the dynamics of the SI response to vibrotactile stimulation in subjects (anesthetized squirrel monkeys) with a dorsal column lesion. The findings obtained to date (exemplary data from a single subject are shown in Figure 8) indicate that in the absence of the dorsal column pathway the IR intrinsic signal evoked in SI by vibrotactile stimulation of a contralateral digit pad (in Figure 8 the stimulus was applied to the distal pad of D3) is grossly abnormal. First, unlike the findings obtained from intact subjects, a robust stimulus-evoked decrease in IR reflectance centered on the representational zones of the stimulated skin site in areas 3b and 1 is discernable ***only*** when very brief stimuli are used (Figure 8C). Second, with a longer-duration vibrotactile stimulus to the same skin site (Figure 8D) an abnormally intense and long-lasting increase in IR reflectance occupies essentially all the regions of areas 3b and 1 devoted to the representation of the volar surface of the foot. Third, contrary to the imaging results obtained in intact subjects, the average IR reflectance in the regions located anterior (area

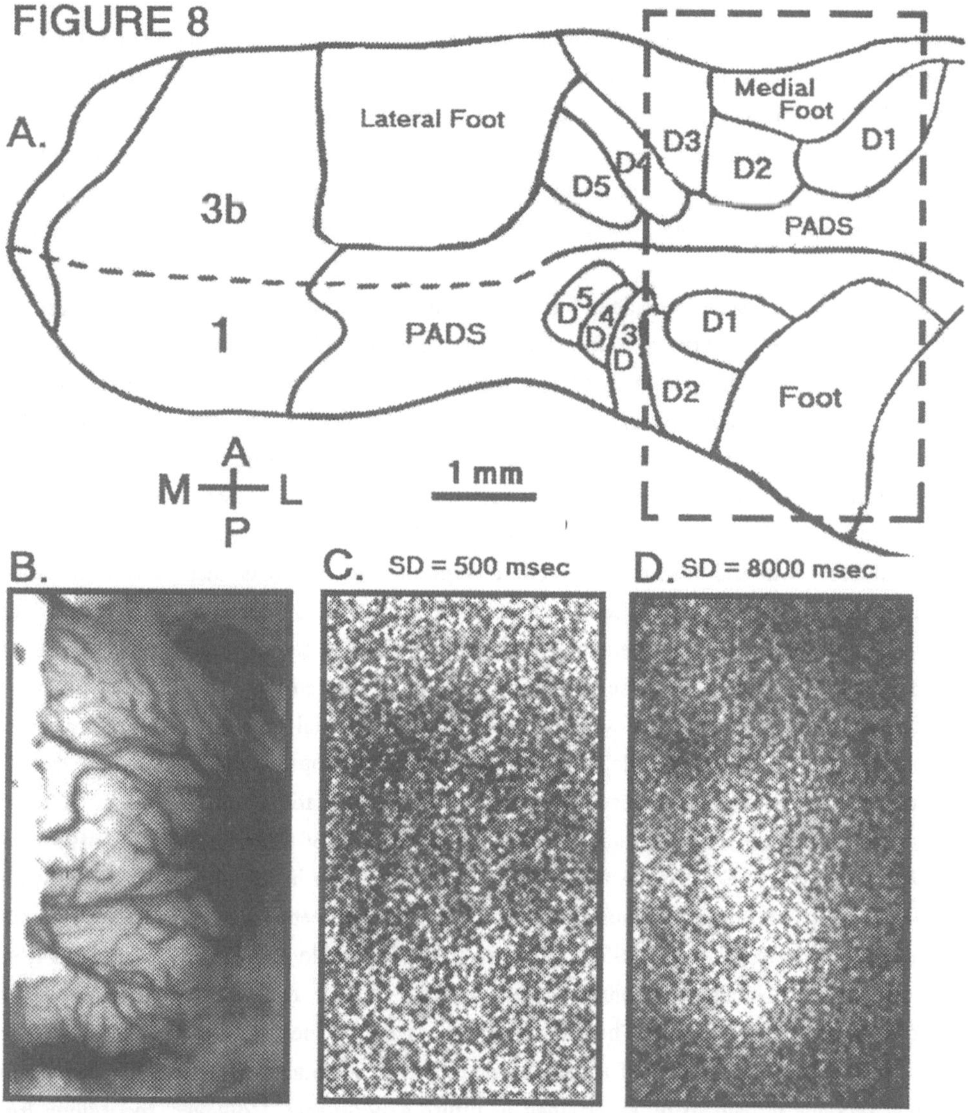

IR responses to vibrotactile stimulation (25 Hz, 400 micron amplitude sinusoid, 4 mm diameter probe) of the distal pad of contralateral digit 3 squirrel monkey. Dorsal column pathway was lesioned at T6 four months prior to IR experiment. A: Map (modified from Sur et al., 1982) of squirrel monkey representation in foot cytoarchitectural areas 3B and 1. Rectangular region enclosed by dotted line indicates field studied. B: Vascular pattern. C & D: Averaged difference images of responses evoked in the same cortical region by a brief (C: duration = 500 msec) and prolonged (D: duration = 8 sec) stimulus. There is a clearcut excitatory response in areas 3b and 1 (dark regions in C) to the brief stimulus, the 3b-1 response to the long duration stimulus is almost exclusively inhibitory.

3a) and posterior (area 2) to the cutaneous foot representation in areas 3b and 1 decreases in response to vibrotactile stimulation. These findings parallel in many respects the SI distribution of 2DG labeling in subjects with dorsal column lesions. They also appear to be in fundamental accord with the published reports that although monkeys with a dorsal column transection perform as well as intact subjects on cutaneous detection and discrimination tasks when brief skin stimuli are used, performance deteriorates sharply when the same subjects are required to discriminate temporally extended skin stimuli (see Tommerdahl et al., 1995 for recent review of relevant literature).

To summarize, we have developed an optical imaging system capable of detecting and quantifying the spatially distributed IR reflectance patterns evoked by natural skin stimuli in somatosensory cortex. The data indicate that the SI reflectance patterns shift with changes in stimulus position in a manner generally consistent with the maps of SI topographical organization obtained in prior receptive field mapping studies. Our observations (unpublished) also indicate that the patterns are highly sensitive to vibrotactile stimulus parameters other than position (e.g., frequency and amplitude). The single neuron recording data obtained at selected sites in the same regions evaluated with the IR imaging method suggest that SI areas that exhibit decreased reflectance in the IR range in response to a vibrotactile stimulus are occupied by neurons excited by that stimulus, and that areas exhibiting stimulus-evoked increases in IR reflectance are occupied by neurons inhibited by the stimulus. Analysis of the temporal progression of the IR reflectance patterns are consistent with the idea that the spatial distribution of SI neuronal excitation evoked by a vibrotactile stimulus is initially widespread and relatively spatially homogeneous, but within a few seconds of continued stimulation the SI region activated fractionates into multiple, smaller regions separated by regions of lesser activity (Whitsel et al, 1991). We consider it likely that the temporal changes that occur in the pattern of SI activation during a continuous vibrotactile stimulus are accompanied by perceptual changes of the type that Hollins and co-workers (Hollins, 1995; Hollins, et al., 1990; Goble and Hollins, 1993, 1994; see Hollins, Delemos & Goble, this volume) have reported in their recent human psychophysical studies of vibrotactile adaptation. The nature and locus of the mechanisms responsible for the observed time dependency of the SI activation pattern to a vibrotactile stimulus that lasts more than a few seconds remain to be ascertained in future experiments. However, recordings from monkey cutaneous primary mechanoreceptive afferents studied under stimulus conditions identical to those used in the imaging studies of this paper (Delemos et al., 1995; in preparation) indicate strongly that the responsible mechanisms are located within the CNS.

The experiments described in this paper were supported, in part, by grants from the Whitehall Foundation, NS32358 and MH48654. We gratefully acknowledge the contributions of O. Favorov, K. Delemos, and C. Vierck to both the experiments and the ideas expressed in this paper.

References

Chubbuck, J.G. (1966). Small-motion biological stimulator. APL Technical Digest, May-June: 18-23.

Cohen, L.B. (1973). Changes in neuron structure during action potential propagation and synaptic transmission. Physiological Reviews 53: 373-418.

Daw, N.W., Stein, P.S.G., and Fox, K. (1993). The role of NMDA receptors in information processing. Ann Rev Neurosci 16:207-222.

Delemos, K.A., Whitsel, B.L., Favorov, O. and Tommerdahl, M. (1995). Effects of vibrotactile adaptation on responses of cutaneous mechanoreceptive afferents and somatosensory cortical neurons. In preparation.

Duncan, G. H., Dreyer, D.A., McKenna, T.M. and Whitsel, B.L. (1982). Dose- and time- dependent effects of ketamine on SI neurons with cutaneous receptive fields. J. Neurophysiol. 47:677-699.

Grinvald, A., Frostig, R. D., Siegal, R. M., and Barfeld, E. (1991). High resolution optical imaging of functional architecture in the awake monkey. Proc. Natl. Acad. Sci. U.S.A. 88: 11559-11563.

Grinvald, A., Lieke, E. E., Frostig, R.D. and Hildesheim, R. (1994). Cortical point-spread function and long-range lateral interactions revealed by real-time optical imaging of macaque monkey primary visual cortex. J. Neuroscience 14: 2545-2568.

Grinvald, A. (1985). Real-time optical mapping of neuronal activity: From single growth cones to the intact mammalian brain. Ann. Rev. Neuroscience 8: 263-305.

Goble, A.K. and Hollins, M. (1993). Vibrotactile adaptation enhances amplitude discrimination. J. Acoustical Soc. America 93: 418-424.

Goble, A. and Hollins, M. (1994) Vibrotactile adaptation enhances frequency discrimination. J. Acoust. Soc. Am. 96: 771-780.

Haglund, M., Ojemann, G. and Hochman, D. (1992). Optical imaging of epileptiform and functional activity in human cerebral cortex. Nature 358: 668-671.

Haglund, M., Ojemann, G and Blasdel G (1993). Optical imaging of bipolar cortical stimulation. J Neurosurgery 78: 785-793.

Hollins, M., Goble, A.K., Whitsel, B.L. and Tommerdahl, M. (1990). Time course and action spectrum of vibrotactile adaptation. Somatosensory and Motor Research 7: 205-221.

Hollins, M., Delemos, K.A. and Goble, A.K. (1995). Vibrotactile adaptation of the RA system: A psychophysical analysis. This volume.

Lee C-J, Whitsel BL (1992) Mechanisms underlying somatosensory cortical dynamics: I. *In vivo* studies. Cerebral Cortex 2:81-106.

Lee, C.-J., Whitsel, B. L., and Tommerdahl, M. (1992) Mechanisms underlying somatosensory cortical dynamics: II. In vitro studies. Cerebral Cortex 2:107-133.

Lieke, E., Frostig, R., Arieli, A., Ts'o, D., Hildesheim, R., and Grinvald, A. (1989). Optical imaging of cortical activity: Real-time imaging using extrinsic dye-signals and high resolution imaging based on slow intrinsic - signals. Annual Rev. Physiol. 51: 543-59.

McKenna TM, Whitsel BL, Dreyer DA (1982) Anterior parietal cortical topographic organization in macaque monkey: a re-evaluation. J. Neurophysiol 48:289-317.

Narayan, S. M., Santori,, E. M., and Toga, A. W. (1994) Mapping functional activity in rodent cortex using optical intrinsic signals. Cerebral Cortex 4: 195 - 204.

Sur, M., Nelson, R.J., and Kaas, J. H. (1982) Representations of the body surface in cortical areas 3b and 1 of squirrel monkeys: Comparisons with other primates. J. Comp. Neurol. 211: 177-192.

Tommerdahl, M., Whitsel, B.L., Nakhle, B. and Favorov, O. (1993). Minicolumnar activation patterns in cat and monkey SI cortex. Cerebral Cortex 3: 399-411.

Tommerdahl, M., Vierck, C.,Whitsel, B., Cooper, B., Juliano, S. and Nahkle, B. (1995). Effects of dorsal column lesions on metabolic activity patterns in somatosensory cortex. Cerebral Cortex. In Press.

Whitsel, B.L., Favorov, O., Kelly, D.G., and Tommerdahl, M. (1991). Mechanisms of dynamic peri- and intra-columnar interactions in somatosensory cortex: Stimulus-specific contrast enhancement by NMDA receptor activation. In: Information Processing in the Somatosensory System Franzen, O. and Westman, J. (eds.), New York, Stockton Press, pp. 353-369.

O. Franzén, R. Johansson and L. Terenius (eds.)

Somatosensory and frontal cortical processing during pain experience

Stuart WG Derbyshire[1,2], Andrew CN Chen[1] and Anthony KP Jones[1]

[1]*Human Physiology and Pain Research Laboratory, Rheumatic Diseases Centre, Hope Hospital, Manchester University School of Medicine, Eccles Old Road, Salford M6 8HD, UK*

[2]*MRC Cyclotron Unit, Hammersmith Hospital, Du Cane Road, London W12 0HS, UK*

Summary. This paper describes the use of Positron Emission Tomography (PET) neuroimaging and EEG/ERP brain mapping to understand somatosensory and frontal cortical processing during human pain experience. Using PET in response to pain it has so far been observed that the main cortical projection site is the anterior cingulate cortex (area 24) (Talbot et al., 1991; Jones et al., 1991). Responses in the somatosensory cortex and prefrontal cortex, by comparison, are more variably reported (Roland, 1992; Derbyshire et al., 1994). We believe these differences are a consequence of variability in 'secondary' pain responses such as attention and localization. Using brain-potential mapping and electrical source analysis to map the processing of pain has revealed three major laser evoked brain potential (LEP) components at about 100, 150 and 230 msec. Multi spatio-temporal dipole modelling revealed an activation from the contralateral somatosensory cortex area peaking at 106 msec. A frontal activation started at 90 msec and peaked at 130 msec. In addition the scalp map showed a bilateral SII activation. The pain related EP was also evident at the vertex with a peak intensity at 150 msec and reversed polarity at 230 msec. This vertex activation was probably associated with cingulate activity. Using EEG coherence mapping with cold tonic pain a contralateral C3 area exhibited higher coherence activation, in addition to the central vertex coherence enhancement in middle EEG frequency and the bi-frontal activation of high frequency beta-activity.

Introduction

The specific involvement of the somatosensory cortex (S1) in pain processing has been long assumed (Guyton, 1991), but has recently been discussed and reassessed in light of the conflicting findings using PET (Roland, 1992ab; Stea & Apkarian, 1992; Evans et al., 1992). Talbot et al. (1991), Casey et al. (1994) and Coghill et al. (1994) have demonstrated increases in S1 activity in response to phasic pain using PET, Jones et al. (1991) and Derbyshire et al. (1994), however, have reported no changes, and Apkarian et al. (1992) have reported decreases in S1 activity in response to tonic pain using SPECT. There is no clear explanation for this pattern of results.

Stea and Apkarian have suggested that this variability may be a consequence of the varying function of S1 as a stimulus moves from non-noxious to noxious. It is suggested that the quality of low intensity, non-noxious, input is analysed in S1 but that as the intensity increases towards pain the quality analysis is compromised and inhibited to facilitate localisation. If, for painful stimuli, the main function of S1 is to register location via inhibition of quality analysis, as they suggest, then net blood flow to S1 is liable to decrease as pain increases and localisation becomes more precise (Apkarian & Stea, 1992). This framework provides a potential explanation for the variability in results from the somatosensory area when using a phasic stimulus as with our past work and Talbot's work. When the temperature of the non-painful hot (NPH) stimulation is much lower than the temperature of the painful hot (PH) stimulation the resulting subtraction (PH-NPH) includes a non-nociceptive component and rCBF increases in S1 can be expected. This non-nociceptive component is reduced as the temperature of the non-painful stimulation approaches pain threshold and rCBF decreases in S1 can be expected.

While somatosensory activation in response to pain was expected by most researchers in the field, prefrontal cortex activity was not widely expected. There is great paucity of literature and understanding as to the involvement of frontal cortex in pain processing. However prefrontal activation is a reminder of earlier reports of relief of suffering in advanced cancer by prefrontal lobotomy and is complementary to ours and others suggestion that, although pain localization and discrimination might be a function of spinothalamic projections to somatosensory cortex, suffering is more likely to depend on a spinoreticulomedial thalamic projection. The projections of the so-called 'medial pain system' have long been associated with the suffering of long term chronic pain, here we demonstrate that these medial projections, including prefrontal cortex, are also involved in the processing of experimental acute pain. Perhaps surprisingly, activation of prefrontal cortex is disrupted by on-going chronic pain.

To supplement the limitation of temporal resolution in PET imaging, we have sought to understand the spatial and temporal sequence of ERP mapping during human pain. ERP mapping provides a msec range of cerebral activity changes in response to painful stimulation. A recent report (Chen and Bromm, 1995) illustrated precise frontal, somatosensory and vertex activations during a set of noxious laser stimulation to the facial area. Laser stimulation activates primarily the small skin fibres, such as the Aδ or c-fibres which are often associated with nociception. We also reported (Chen and Rappelsbereger, 1994) how different brain areas may synchronize during human pain processing using the EEG coherence mapping. These electrophysiological studies are aimed to facilitate our understanding of different cortical participation in pain activation.

Material and Methods

Twelve normal subjects including six right handed male volunteers, age 28-50 years (mean age 35 years) and six right handed female volunteers, age 47-69 years (mean age 54.2 years), and six atypical facial pain (AFP) patients, age range 42-65 years (mean age 53), took part in the studies described. Each subject was compared in their response to painful (mean temp. 46.4 ^{0}C) and non-painful heat (meant temp. 41.5 ^{0}C) stimuli following inhalation of the freely diffusible positron emmiting ^{15}O-labelled tracer, carbon dioxide ($C^{15}O_2$). Monitoring the emissions gives an image of regional cerebral blood flow (rCBF) which can be used as an indication of central neuronal activity and allows the assessment of a painful stimulus versus a non-painful stimulus.

Permission to carry out these studies was obtained from the ARSAC-UK (Administration of Radioactive Substances Advisory Committee, UK) and the Research Ethics Committee of Hammersmith Hospital, fully informed signed consent was obtained from each subject prior to scanning. Each subject underwent six two minute scans, three painful and three non-painful, over the course of a single two hour session.

Intermittent ramps of increasing heat, hot or painful hot, were delivered every 15 seconds using a Marstock thermal threshold stimulator. Scans were obtained from PET scanner; CTI model 931-08/12 Knoxville, U.S.A. described elsewhere.

To make the comparison of painful thermal stimulation with the effect of increasing heat intensity without pain, the following procedures were carried out: correction for head movement between scans; reorientation into a standardised stereotactic anatomical space; and a correction made for global changes in blood flow between scans. These procedures allow flow values for each stimulus condition to be pooled across subjects. Finally a statistical comparison of blood flow distributions between stimulus conditions was performed to identify sites of significantly changed regional flow.

The topographic brain mapping in response to phasic painful laser stimulation was conducted in 10 male subjects (Bromm and Chen, 1995). Brief 2 msec Tulium infrared laser (λ 1.8 υm, beam size 20 mm^2) was repeatedly applied (40 stimuli, 20 sec ISI) to the right temple area using a painful intensity (400 mj). The cerebral responses to the laser stimuli were then averaged off-line and the spatio-temporal convolution of evoked potentials were sequentially interpolated into brain maps at different time points (Chen and Bromm, 1995). To examine the cerebral responses to tonic noxious stimulation, on going EEG coherence mapping (Chen and Rappelsbereger, 1994) was used with 19 subjects (11 male, 8 female). An ice cube was placed in the palm of either the subject's right or left hand, under eyes-closed or eyes-open conditions. Two minutes of EEG were gathered during this noxious stimulation and epoched into segments of 2 sec for frequency transformation into power spectrum averagings. Local intrahemispheric and distal interhemispheric coherences were calculated in each of the 6 major broad EEG bands. Interpolation of these coherence values was conducted to establish the coherence maps at resting vs noxious stimulation. Differences in coherence values between these two conditions were mapped and the associated statistical significance probability mappings were generated.

Results

Figure 1 overleaf shows increased rCBF in response to pain as compared to heat for the twelve normal volunteers. Increased rCBF in the contralateral thalamus and basal ganglia can be clearly observed. Ipsilatateral increases can be observed in the prefrontal cortex (area 10) and inferior parietal cortex, increased rCBF to the anterior cingulate is bilateral. There was no change in rCBF in the region of the somatosensory cortex on either side. Figure 2 shows the same comparison for the group of six chronic pain patients. Again there are clear contralateral rCBF increases in the thalamus and basal ganglia and a bilateral increase in the anterior cingulate cortex. However, there is no significant change in the prefrontal cortex on either side.

Further investigation of the somatosensory changes.

One explanation of the variable somatosensory response across studies may be fluctuations in the intensity of the stimulus received. One possibility is that large differences between a subject's painful heat stimulus and non-painful heat stimulus result in somatosensory rCBF increases because of the increased non-nociceptive information processing. Whereas smaller differences produce rCBF decreases. This hypothesis was investigated further by splitting the twelve male and female subjects into three groups of four based on their heat and painful heat differentials. The results are illustrated in figures 3 and 4 overleaf.

As the search for the somatosensory cortex was constrained by a knowledge of its location, Z scores significant at the $P<0.05$ level could be accepted as confirming the hypothesis of change in this region. As can be seen from figure 3 rCBF shows a significant increase in the somatosensory cortex only with the smallest heat difference, opposite to that expected. Figure 4 shows the reverse relationship. Decreases in rCBF are apparent with both the larger heat differences.

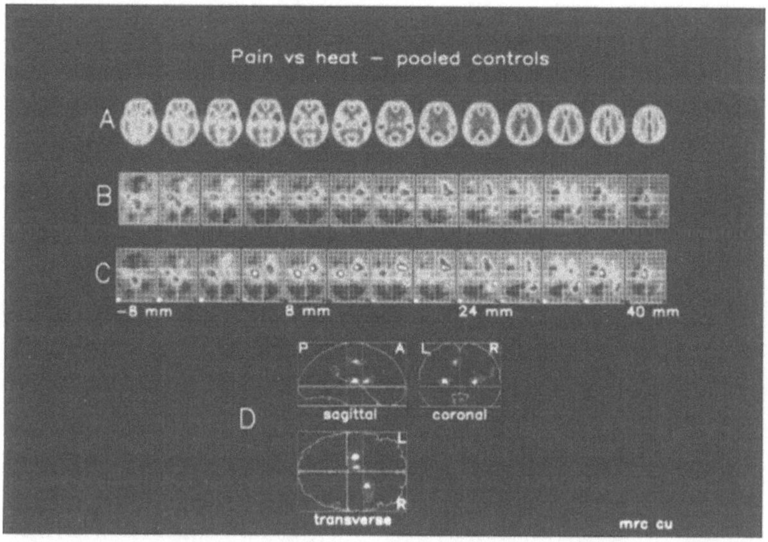

Figure 1 Data averaged from the group of normal volunteers. At the top are transverse images of the brain after stereotaxic normalization, with the distances from the AC-PC plane indicated. A, Anatomical features obtained by averaging all blood flow scans from the twelve subjects. B, the arithmetical difference between adjusted mean blood flows for painful hot and non-painful hot stimuli. C, The SPM{t} values derived from the formal pixel by pixel comparison of the adjusted mean blood flows and variances for each of the two conditions. The colour scale is arbitrary, threshold significance is indicated by the lower left pixel for each plane. D, The orthogonal projections of the statistical comparison at a $P<0.001$ (Z threshold 3.09). The areas showing significant increases in blood flow are an area within the region of the periaqueductal gray, thalamus/basal ganglia, thalamus/insula, prefrontal area 10, inferior parietal area 40 and anterior cingulate cortex (area 24).

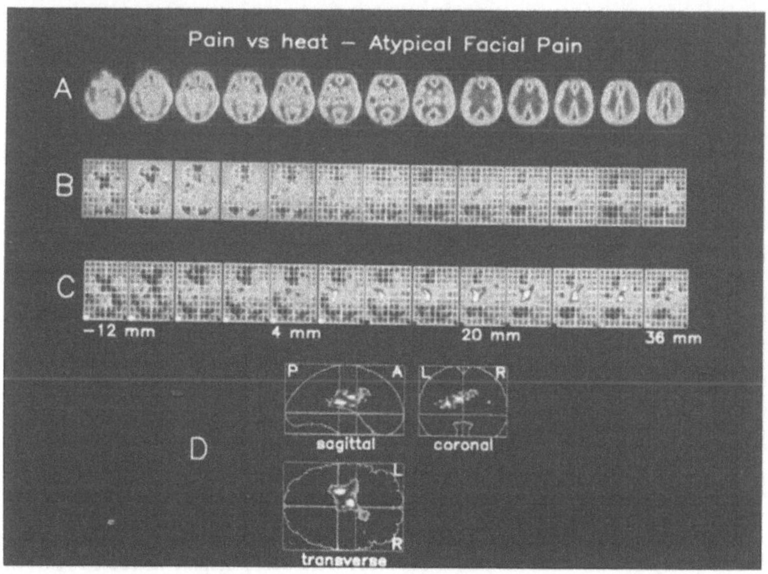

Figure 2 Data averaged from the group of six AFP patients. A, B, C and D are as above, presented here using the data from the AFP patients The areas showing significant increases in blood flow are lentiform nucleus, insula, thalamus and anterior cingulate cortex.

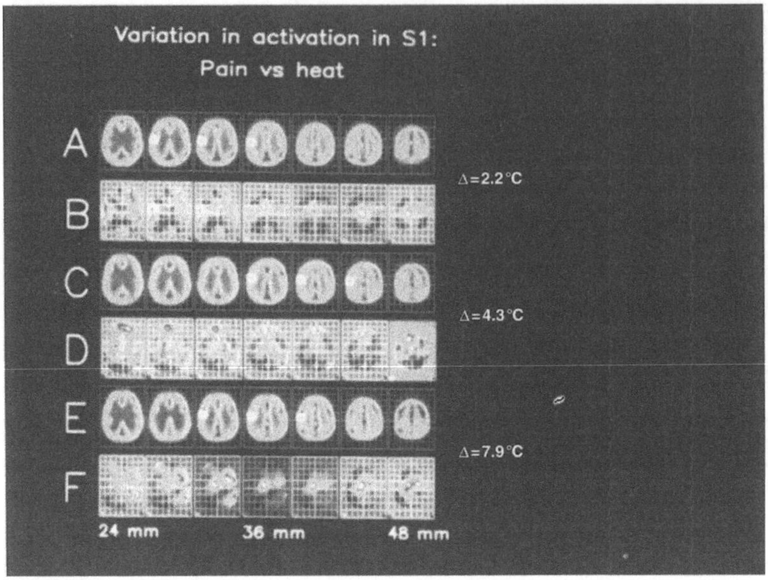

Figure 3 A, C and E indicate the position of somatosensory cortex (circled) on the averaged blood flow images for all runs in all the four subjects. B, D and F show the significant increases in blood flow in response to pain which are circled in the region of the somatosensory cortex. Only the smallest PH-NPH difference, indicated in red, shows a significant response.

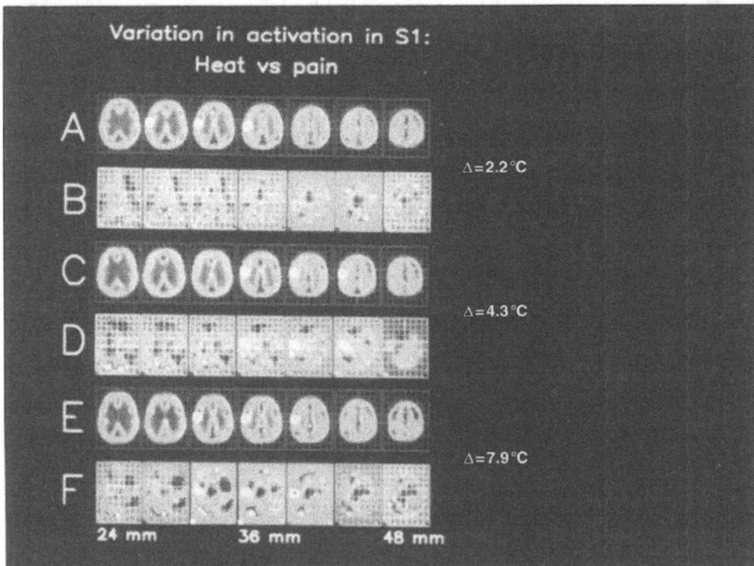

Figure 4 A, C and E as above. B, D and F show the decreases in blood flow in response to pain for the region of somatosensory cortex. Both the larger PH-NPH differences display increased rCBF in the region of interest

ERP and EEG coherence mapping

At 90 msec, a prefrontal ERP positivity emerged which is followed by a clear contralateral activation (focal negativity) in the S-II area at 105 ms and an ipsilateral activation in a similar site at 110 ms. At 130 ms, the frontal positivity was co-immerged with the vertex negativity. However, the largest potential was a negative response followed by a positive response at the vertex site, peaking at 150 ms and 220 ms respectively. In contrast to the PET results, these results suggest a frontal and somatosensory response during the processing of a phasic painful stimulus in man.

EEG coherence mapping using tonic pain revealed pronounced decreased alpha amplitude and some increased beta activities in the central area of the brain. At the prefrontal sites, bilateral beta coherence activations were also observed during painful stimulation. In addition, this study also demonstrated a contralateral coherence activation at the temporal-parietal site. This activation was consistent in each of the stimulation conditions. Thus, coherence mapping results strongly indicated the participation of the somatosensory area, prefrontal area and vertex cortical area.

Discussion

Apkarian's work with tonic stimulation has led to the intriguing possibility that nociceptive input to the somatosensory cortex inhibits neurons which primarily process innocuous inputs (Apkarian et al., 1992). However, the results of the subgroup analysis reveal the opposite possibility. The pain stimuli used here and by others consists of a sharp pain experience preceded and followed by no input or innocuous (heat) input. By contrast, Apkarian's pain stimulus was a tonic heat stimulus. Apkarian's result implies, therefore, that the somatosensory activation seen in other reports may reflect net activation of non-nociceptive neurons. This conclusion is not supported by the separation of the group of twelve into three subgroups of four. Subgroup 1 had a high heat threshold but barely increased the stimulus for the experience of pain. Thus this small heat difference may reflect a larger nociceptive input, the non-painful temperature bordering close to pain with the painful temperature clearly breaching pain threshold. The opposite is true of subgroups 2 and 3 with larger heat differences due to a lower heat threshold. Here the subjects may have experienced a non-pain input as part of the cognitive subtraction. Thus when the difference between non-painful hot and painful hot contained less innocuous input the net activation gave a somatosensory increase, suggesting somatosensory responses to be pain specific. However, it has recently been suggested that both increases and decreases may be occuring within the somatosensory cortex depending on the region to be stimulated and the level of anticipatory anxiety associated with the experiment (Drevets et al., 1995). Thus, while it is clear is that the presence of pain is likely to result in a change in rCBF to the somatosensory cortex, the mere presence of pain is not sufficient to predict the direction of the rCBF changes.

Both the ERP and EEG coherence results indicate the involvement of S1 cortex in the processing of pain, which is in apparent contradiction of our findings with PET. However, these activations may reflect the sensory-discriminatory processing of pain activation. Although the PET results are of a high spatial resolution, there is no temporal resolution - results are averaged across a time period of between 30 sec and 150sec. Thus any sequence of noxious processing in the brain, which may indicate early rCBF increases and then decreases for example, will be lost using PET. Our results suggest that responses in the S1 region are early and of short duration, this is in contrast to the forebrain structures which respond later and last longer. We speculate that the interaction of the prefrontal and cingulate area may reflect the cognitive-evaluative component and affective-motivational component, respectively, of human pain processing. This places the emphasis for pain processing onto the frontal region as suggested here and elsewhere.

The prefrontal cortex has important projections to the anterior cingulate cortex and basal ganglia (Groenewegen et al., 1990) as well as weaker connections with insula cortex. Prefrontal cortical rCBF was significantly increased in all subjects studied. Shallice (1982) has proposed that the process by which complex behavioural units or schemas are brought to conscious attention is the function of the 'supervisory attentional system' - part of the 'programming, regulation and verification of human activity' (Luria, 1973) by the frontal lobes. The frontal activation observed in the control subjects may therefore reflect the normal 'supervision' of attention and emotional schemas in response to pain, which breaks down in the pain patients. This is consistent with the perception of physical symptoms proposed by Pennebaker (1982). Pain sensations are increased when subjects anticipate pain in an attended area but decreased when they anticipate pain in a non-attended area. This suggests a possible interaction between S1 and prefrontal areas such that the S1 cortex initially localizes a stimulus, down-regulating other areas in the process, and the prefrontal cortex regulates the attentional response. The conscious appreciation of pain occurs as other parts of the pain 'neuromatrix', including the anterior cingulate cortex, are incorporated. The failure of the prefrontal cortex to regulate the attentional response to pain means that pain is more likely to come 'on-line' during the experience of chronic pain.

Conclusion

It is still not clear why there is such discrepancy in the reports of S1 activity in response to pain. However, it is now established that both increases and decreases in rCBF are occuring in the S1 region during the normal pain response. This finding, plus the finding from ERP that the somatosensory response is earlier than the other cortical responses and of shorter duration, leads us to the suggestion that somatosensory cortex is involved in the localization of pain stimuli and not the processing of pain per se. In order to understand the suffering sensation of pain it is important to understand the responses of the prefrontal and anterior cingulate cortex. Both these areas are clearly active in a group of normal volunteers, however the prefrontal response is not apparent in a group of patients suffering chronic atypical facial pain. This could represent the breakdown of the 'supervisory' role of the prefrontal region allowing continual priming of pain networks and the establishment of constant pain experience mediated by the anterior cingulate cortex.

References

Apkarian, V.A., Stea, R.A., Manglos, S.H., Szevernyi, N.M., King, R.B. and Thomas, F.D. (1992) Persistent pain inhibits contralateral somatosensory cortical activity in humans. *Neuroscience Letters* 140; 141-147.

Bromm, B. and Chen, A.C.N. (1995) Brain electrical source analysis of laser evoked potentials in response to painful trigeminal nerve stimulation. Electroencephalography and Clinical Neurophysiology, 95, 14-26.

Casey, K.L., Minoshima, S., Berger, K. L., Koeppe, R.A., Morrow, T.J. and Frey, K. A. (1994) Positron Emission Tomographic Analysis of Cerebral Structures Activated Specifically by Repetitive Noxious Heat Stimuli. *Journal of Neurophysiology* 71; 802-807.

Chen, A.C.N. and Bromm, B. (1995) Pain related generators of laser-evoked brain potentials: Brain mapping and dipole modelling. *Advances in Pain Research and Therapy* 22; 245-266.

Chen, A.C.N. and Rappelsbereger, P. (1994) Brain and human pain: Topographic EEG amplitude and coherence mapping. *Brain Topography* 7; 129-140.

Coghill, R.C., Talbot, J.D., Evans, A.C., Meyer, E., Gjedde, A., Bushnell, M.C. and Duncan, G.H. (1994) Distributed processing of pain and vibration by the human brain. *The Journal of Neuroscience* 14; 4095-4108.

Derbyshire, S.W.G., Jones, A.K.P., Devani, P., Friston, K.J., Feinmann, C., Harris, M., Pearce, S., Watson J.D.G. and Frackowiak R.S.J. (1994) Cerebral responses to pain in patients with atypical facial pain measured by Positron Emission Tomography. *Journal of Neurology, Neurosurgery and Psychiatry* 57; 1166-1173.

Drevets, W.C., Burton, H., Videen, T.O., Snyder, A.Z., Simpson, J.R. and Raichle, M.E. (1995) Blood flow changes in human somatosensory cortex during anticipated stimulation. *Nature* 373; 249-252.

Evans, A.C., Meyer, E. and Marret, S. (1992) Pain and activation in the thalamus. *Trends in Neuroscience* 15; 252.

Groenewegen, H.J., Berendse, H.W., Wolters, J.G. and Lohman, A.H.M. (1990) The anatomical relationship of the prefrontal cortex with the striatopallidal system, the thalamus and the amygdala: evidence for a parallel organization.' In H.B.M. Uylings, C.G. Van Eden, J.P.C. De

Bruin, M.A. Corner and M.G.P. Feenstra (eds):*The Prefrontal Cortex,* Elsevier Science Publishers.

Guyton, A.C. (1991) *Basic Neuroscience: Anatomy and Physiology.* W.B. Saunders Company.

Jones, A.P.K., Brown, W.D., Friston, K.J., Qi, L.Y. and Frackowiak, R.S.J. (1991) Cortical and subcortical localization of response to pain in man using Positron Emission Tomography. *Proceedings of the Royal Society of London* 244; 39-44.

Luria, A.R. (1973) *The working brain. An introduction to neuropsychology.* Translated by B. Haigh, New York, Basic Books, pp. 187.

Pennebaker, J. (1982) *The psychology of physical symptoms.* Springer-Verlag New York Inc.

Roland, P.E. (1992a) Cortical representation of pain. *Trends in Neuroscience* 15; 3-5.

Roland, P.E. (1992b) Reply. *Trends in Neuroscience* 15; 252-253.

Shallice, T. (1982) Specific impairements of planning. *Philosophical Transactions of the Royal Society of London B* 298; 199-209.

Stea, R.A. and Apkarian, A.V. (1992) Pain and somatosensory activation. *Trends in Neuroscience* 15; 250-251.

Talbot, J.D., Marret, S., Evans, A.C., Meyer, E., Bushnell, M.C. and Duncan, G.H. (1991) Multiple representations of pain in human cerebral cortex. *Science* 251; 1355-1358.

Somesthesis and the Neurobiology of the Somatosensory Cortex
O. Franzén, R. Johansson and L. Terenius (eds.)

Local receptive field diversity within cortical neuronal populations

Oleg V. Favorov and Douglas G. Kelly

Departments of Biomedical Engineering and Statistics, University of North Carolina at Chapel Hill, USA

Summary. The aim of this review paper is to draw attention to a little-appreciated but ubiquitous feature of cerebral cortical organization that we propose is fundamental to an understanding of cortical functioning. Specifically, we (i) describe evidence supporting the idea that neighboring cells in the cortex have richly diverse receptive fields, and (ii) suggest some of the possible mechanisms that generate this receptive field diversity. The possible roles of local receptive field diversity in cortical information processing are discussed.

Introduction

The receptive fields (RFs) of neighboring neurons in sensory cortex typically overlap without being identical. Traditionally the similarity has been emphasized, while differences between such RFs have been ignored on the assumption that they are insignificant in both their magnitude and their functional importance. How similar are the RFs of neighboring cortical neurons? Figure 1 shows RFs collected in a typical radial microelectrode penetration of Area 3b in cat. In this penetration 21 single units were isolated and their RFs were mapped. Differences in the sizes, shapes, and skin positions among the 21 RFs are most prominent (Fig. 1B). This is true across the entire depth of this radial penetration, even among near neighbors. There is only one very small skin locus that is common to all 21 RFs (Fig. 1C). Thus, neurons in this radial penetration all share a common input from one skin locus, but also each neuron receives afferent input from some additional–and frequently very extensive–surrounding territories, which are different for different neurons. This organization of RFs is very characteristic of somatosensory cortical columns (Favorov and Whitsel, 1988a,b; Favorov and Diamond, 1990).

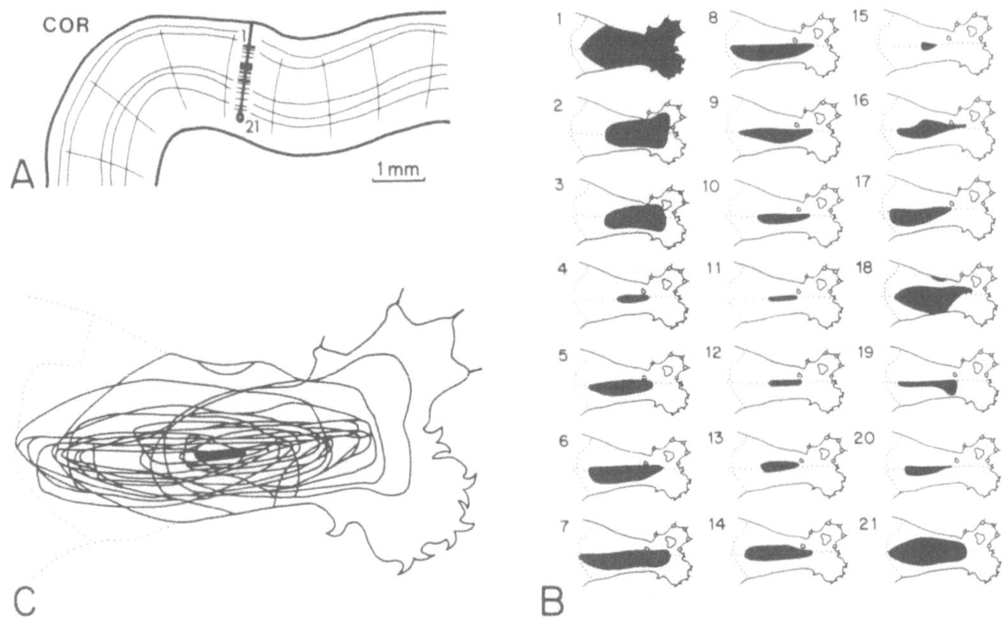

Figure 1. Single unit RFs mapped in a radial penetration of primary somatosensory cortex of a cat anesthetized with nitrous oxide/oxygen mixture. A: Histological reconstruction of the electrode track. Sites at which single units were isolated are indicated by dashes. COR–coronal sulcus. B: RFs of 21 single units. C: Superimposed outlines of all 21 RFs. Blackened area is the skin site included in each of the RFs. Reproduced, with permission, from Favorov and Diamond (1990).

Prominent differences in RFs among neighboring cortical neurons, such as those shown in Figure 1, have been documented extensively in the experimental literature. For example, reproduced in Figure 2 are some RF data taken from papers of Powell and Mountcastle (1959) on somatosensory cortex and Hubel and Wiesel (1974b) on visual cortex. This figure shows a degree of local RF diversity which is very comparable to that in Figure 1. RF diversity of such magnitude appears to be a general characteristic of the cerebral cortex, observed in a variety of cortical areas. In particular, it has been described in somatosensory cortex (*e.g.* McKenna *et al.*, 1982; Iwamura *et al.*, 1985; Favorov and Whitsel, 1988a,b; Favorov and Diamond, 1990), in visual cortex (*e.g.* Hubel and Wiesel, 1962, 1974b; Creutzfeldt *et al.*, 1974; Albus, 1975a), and in auditory cortex (*e.g.* Abeles and Goldstein, 1970; Vaadia and Aertsen, 1992; South and Weinberger, 1994). It has also been found in motor cortex (*e.g.* Brooks *et al.*, 1961; Welt *et al.*, 1967; Lemon and Porter, 1976). Interestingly, neighboring cells in motor cortex can differ markedly not only in the sources of their afferent inputs, but also on the efferent side in their involvement in motor acts (*e.g.* Evarts, 1967; Lemon *et al.*, 1976). Besides primary sensory and motor cortical areas, prominent local diversity in response properties has been described in association cortex, *e.g.* in area MT, which processes visual information about stimulus motion (Zohary *et al.*, 1994); in area IT, which processes information about shape attributes of visual stimuli (Gochin *et al.*, 1991; Fujita *et al.*, 1992; Gawne and Richmond, 1993), and in the frontal cortex (Vaadia and Aertsen, 1992).

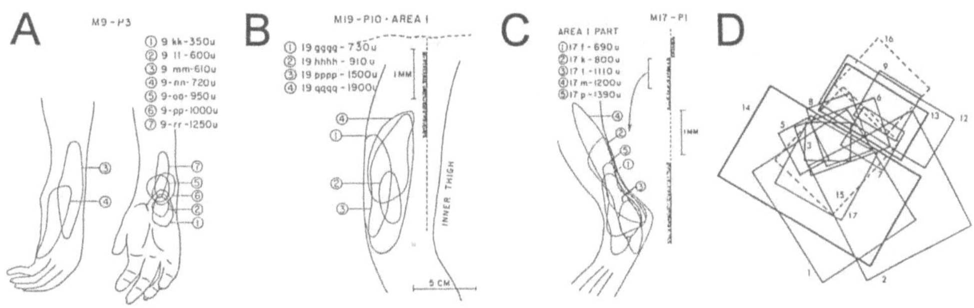

Figure 2. Single unit RFs mapped in three radial penetrations of monkey primary somatosensory cortex (A, B, C; reproduced, with modifications, from Powell and Mountcastle, 1959), and in a radial penetration of monkey primary visual cortex (D; reproduced, with permission, from Hubel and Wiesel, 1974b). Each panel shows RF data collected in one penetration, with RF outlines shown superimposed on a single drawing, revealing the presence of prominent RF diversity in these radial penetrations.

The traditional emphasis on RF uniformity within local groups of neurons has logically led to the view that neighboring neurons in the cortex must behave essentially the same way, or "speak in a single voice" in response to peripheral stimulation. However, Figures 1 and 2, and more generally the literature cited above, make it clear that while neighboring neurons can show a remarkable uniformity in some of their RF properties (such as the afferent input submodality in somatosensory cortex, or preferred stimulus orientation in visual cortex), in other RF properties (RF location, for example) they can differ prominently. When RFs are considered *in toto*, in all their dimensions, then it turns out that neighboring neurons have only little in common: a stimulus which is effective in driving one cell will be much less effective in driving its neighbor, even though it is quite similar to that neighbor's preferred stimulus. This is borne out by experimental findings that neighboring cells correlate very poorly (average correlation coefficient less than 0.2) in their responses to peripheral stimuli (Gochin *et al.*, 1991; Gawne and Richmond, 1993; Gawne *et al.*, 1993; Ghose *et al.*, 1993; Zohary *et al.*, 1994). Neighboring cells do deal with the same stimuli, but each cell in its own way.

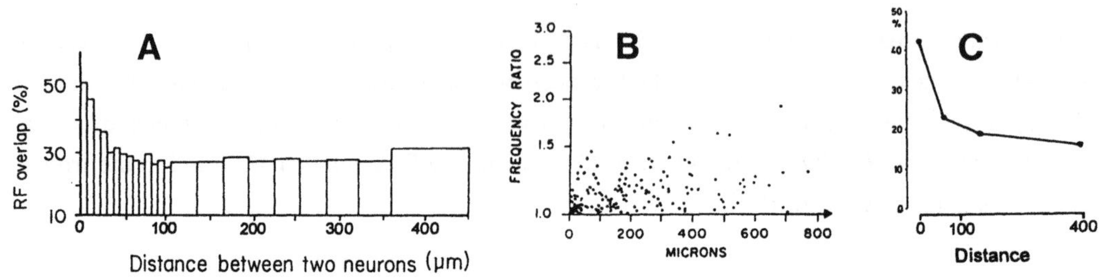

Figure 3. RF similarity as a function of distance separating 2 neurons in the plane of the cortical surface. In A, the ordinate is the average degree of RF overlap of 2 cells in somatosensory cortex (reproduced, with modifications, from Favorov and Whitsel, 1988a). In B, the ordinate is the ratio of optimal stimulus frequencies of 2 cells in auditory cortex (higher frequency/lower frequency; reproduced, with permission, from Abeles and Goldstein, 1970). In C, the ordinate is the frequency of encountering 2 cells in visual cortex with optimal stimulus orientations differing by less than 7.5° (reproduced, with modifications, from Albus, 1975b). The three plots show that in tangential cortical penetrations, cells that are near neighbors have very similar RF properties, and that the similarity declines with the distance separating the pair. What is most interesting is that this decline has two distinct phases: in the first phase, at cell separations less than 50 μm, the average RF similarity declines very quickly with distance, while in the second phase, at separations greater than 50 μm, the rate of decline is much slower. In contrast to tangential penetrations, in near-radial penetrations cells with very similar RFs are found at much greater separations, thus suggesting the presence in the cortex of radially oriented strands of cells, < 50 μm in diameter, with greatly reduced RF diversity (see Favorov and Whitsel, 1988a; Favorov and Diamond, 1990 for in-depth discussion).

Minicolumnar contributions to RF diversity

The prominence of local RF diversity suggests that it is not simply the result of imprecision in neuronal connections. We want to further this suggestion by considering two sets of questions: (i) What mechanisms might be responsible for local diversity, and (ii) What might be its functional implications or benefits? We have approached these questions by considering the relationship between local RF diversity and cortical columnar organization. The columnar principle suggests that local RF diversity should be less prominent in the radial cortical dimension. Indeed, experimental evidence obtained in auditory (Abeles and Goldstein, 1970), visual (Hubel and Wiesel, 1974a; Albus, 1975b), and somatosensory (Merzenich *et al.*, 1981; Favorov and Whitsel, 1988a; Favorov and Diamond, 1990) cortices suggests (see Fig. 3) that *within* minicolumns, which are radially oriented cell cords *ca.* 50 μm in diameter (Mountcastle, 1978), cells have very similar RF properties, and the prominent differences in RF properties observed among neighboring neurons are mostly confined to neurons that belong to different, albeit neighboring, minicolumns. In other words, local RF diversity is mostly attributable to diversity *among*, and not within, minicolumns. Minicolumns most likely correspond to the 50 μm-diameter radially-oriented cords of neuronal cell bodies that are very noticeable in Nissl-stained cortical sections. They are also likely to be related to the *ontogenetic columns* of Rakic (1988), and also to the *pyramidal cell modules* defined by bundling of apical dendrites of pyramidal cells (Peters and Yilmaz, 1993). The radial-*vs.*-tangential anisotropy in distribution of local RF diversity within cortical tissue has also manifested itself in 2-deoxyglucose (2DG) metabolic labeling studies (McCasland and Woolsey, 1988; Tommerdahl *et al.*, 1993): the macrocolumn-sized patches of 2DG label evoked in somatosensory cortex by peripheral stimuli, when viewed at higher spatial resolution, turned out to consist of a patchwork of interdigitated radial strands of enhanced and reduced 2DG label, suggesting that neighboring minicolumns, having prominently diverse RFs, were driven to markedly different degrees by the stimuli used.

Because of the predominance of radially oriented intrinsic connectivity within cerebral cortex, it is to be expected that cells lying in a particular minicolumn will have very similar RFs. But why are RFs so much more different among adjacent minicolumns? Our hypothesis is that these differences emerge during perinatal development. The ontogenetic development of the cortex is a dynamic, self-organizational process during which minicolumns actively drive their neighbors to establish afferent connections with different sets of thalamic neurons (Favorov and Kelly, 1994a,b).

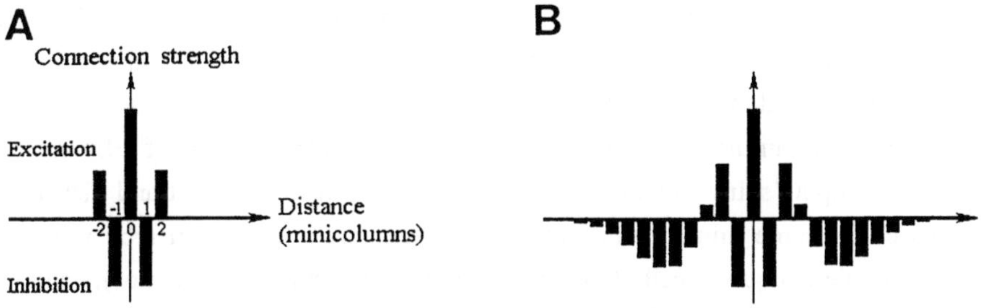

Figure 4. A. Pattern of lateral cortical connections that can generate local RF diversity. B. Combination of the pattern shown in A with a more extensive Mexican-hat pattern, proposed to reflect more accurately the organization of lateral connectivity in the cortex, because it would generate both local RF diversity and a larger-scale topographic orderliness (for details, see Favorov and Kelly, 1994b).

How can they do it? A possible answer is: Via lateral interactions among adjacent minicolumns. A traditional view of the organization of lateral connections in the cortex, based on the belief in local RF uniformity, is that cells that are near neighbors have predominantly excitatory interconnections, while more distant neighbors have inhibitory interconnections (a so-called "Mexican-hat" pattern). However, such a pattern cannot generate local RF diversity. What is needed for that outcome, and for preservation of topographic orderliness on a larger scale, is for adjacent minicolumns to have inhibitory lateral interactions, and for more distant minicolumns to have excitatory interactions (see Fig. 4). A likely candidate to mediate the inhibitory interactions among adjacent minicolumns is the class of interneurons called double-bouquet cells. These are GABAergic cells whose bodies and local dendritic fields are confined to layer 2 and the upper part of layer 3. Their most distinctive characteristic is that their axon collaterals descend in a tight bundle down through layers 3 and 4 and into layer 5, making synapses all along the way on dendrites of pyramidal and spiny-stellate cells (Jones, 1975). A recent study of DeFelipe *et al.* (1989) suggests that double-bouquet cells are likely to inhibit cells in adjacent minicolumns but not in their own.

In order to test the idea that neighboring minicolumns actively drive each other to develop different afferent connections, we set up a model of a somatosensory cortical region consisting of 61 minicolumns arranged in a hexagonal pattern (Favorov and Kelly, 1994a). These 61 minicolumns were intended to represent a larger-scale somatosensory cortical column, a segregate (Favorov *et al.*, 1987; Favorov and Whitsel, 1988a; Favorov and Diamond, 1990). Segregates are discrete cortical macrocolumns 0.3-0.6 mm in diameter that are found in somatosensory cortex of cats and monkeys. These columns have sharp boundaries defined by their RF positions on the skin, and they compartmentalize somatosensory cortex into a honeycomb-like mosaic. Within a segregate, as one moves from one minicolumn to the next, the RFs of a linear sequence of neurons shift back and forth on the skin, displaying local RF diversity, without yielding an overall RF shift across the entire segregate. Only at a border separating adjacent segregates do RFs shift *en masse* in a single step to a new skin position (for details see Favorov and Whitsel, 1988a; Favorov and Diamond, 1990).

Each minicolumn in the model consisted of three cells representative of different cell types in the cortex: an input "spiny-stellate" cell (which received thalamic input and distributed it to all other cells of the same minicolumn and, to a lesser degree, to other nearby minicolumns), an intrinsic "double-bouquet" cell (which inhibited adjacent minicolumns), and an output "pyramidal" cell. Connections from thalamus to minicolumns were plastic; they were allowed to self-organize in accordance with a Hebbian rule during a "developmental" period in which the network was driven by punctate "skin" stimuli. A key feature of this model was that the thalamic connections to minicolumns were not prescribed explicitly according to some preconceived idea about the mapping relationship between skin and cortex, but were selected by the minicolumns themselves, driven by the prior history of skin stimulation. As we anticipated, during the period of the network's self-organization each minicolumn was pressured by inhibitory interactions with

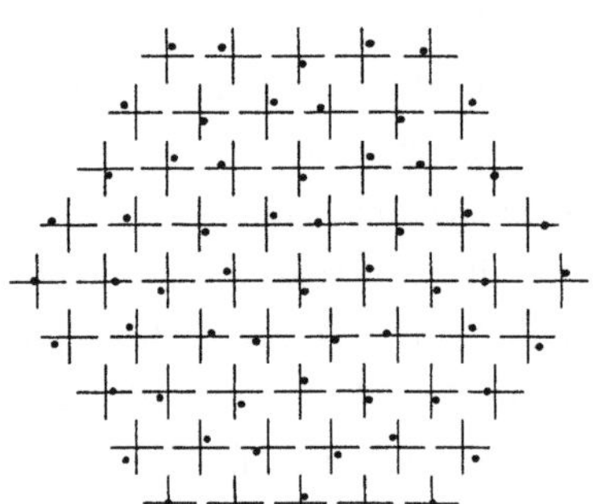

Figure 5. RF centers of the spiny-stellate cells of the 61 minicolumns of the fully developed model segregate. Each RF center is shown on a separate two-dimensional graph representing the skin field, with coordinate origin at the center of the skin field. The hexagonal arrangement of the plots in the figure corresponds to the hexagonal arrangement of their minicolumns in the segregate. This allows one to appreciate the diversity in RF position encountered as one moves from one minicolumn to the next across the segregate. Note the shuffled arrangement of RFs across the segregate and the absence of a topographic gradient.

adjacent minicolumns to acquire afferent connections different from those of its immediate neighbors. On the other hand, it was also pressured by the excitatory interactions with a larger group of its neighbors to acquire afferent connections similar to theirs. To satisfy these contradictory pressures, minicolumns acquired a shuffled, highly permuted arrangement of thalamic-minicolumnar connections, and consequently the RFs from one minicolumn to the next shifted back and forth on the skin (see Fig. 5). However, across the entire segregate they stayed close together.

Having a cortical model with locally diverse RFs, we could next ask whether this diversity confers any functional benefits on the network. Indeed, new functional properties emerged in the network coincident with the establishment of shuffled RF organization. In particular, cells acquired the ability to discriminate the orientation of bar stimuli, as well as the direction of moving stimuli (see Fig. 6). Different minicolumns in the network came to prefer different stimulus directions and orientations, and the maps of these stimulus preferences had many features in common with maps in visual cortex (see Favorov and Kelly, 1994b). These stimulus feature-extracting properties arose from the inhibitory interactions among adjacent minicolumns with markedly shifted RFs: when minicolumns in the model were given very similar RFs or somatotopically arranged RFs, neurons lost their directional selectivity and orientation tuning, even though inhibition was still present.

To summarize, our modeling results suggest that local RF diversity arises, in part, among neighboring minicolumns due to the effects of local inhibitory interactions among adjacent minicolumns on their thalamocortical connections. These local RF differences, in turn, result in emergence of new functional properties which make neurons sensitive to the shape and temporal features of peripheral stimuli.

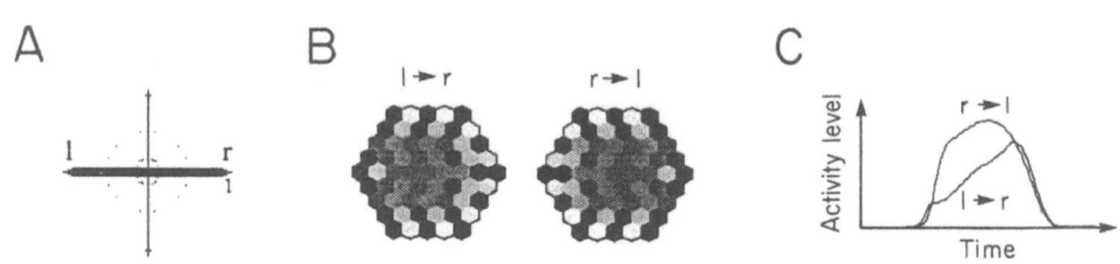

Figure 6. Network's responses to point stimuli moving in opposite directions. A: The path traversed on the skin by the stimuli; the end-points of the path are marked *l* and *r*. The distance traversed on the skin corresponds to about 15 mm on the tip of the monkey index finger. B: Activity patterns evoked in the segregate (shown as a large hexagon) by two stimuli, one moving from point *l* to point *r* and the other from *r* to *l*. Shading of each small hexagon indicates average activity level of the spiny-stellate cell in the corresponding minicolumn during the stimulus period. Note that, although the two stimuli traveled through exactly the same set of skin points and differed only in the temporal sequence of activation of these points, they evoked clearly different patterns of activity in the segregate. C: Superimposed time courses of the response of a representative spiny-stellate cell, taken from the leftmost minicolumn in the segregate, to the two stimuli. The stimulus that traveled from *r* to *l* evoked a significantly stronger response in this cell than did the same stimulus traveling in the opposite direction, from *l* to *r*.

Complex nonlinear dynamics as a source of local diversity

Recently we expanded our model—essentially a model of cortical layer 4—by adding upper cortical layers. Since a very prominent feature of the organization of the upper cortical layers is their extensive system of lateral connections integrating activity within and among macrocolumns, we expanded our model to contain three segregates, each made up of 61 minicolumns (see Fig. 7). Layer 4 of this new model is an exact replica of the earlier network (as regards both internal and afferent connections). In the upper layer, each minicolumn has excitatory pyramidal-like cells and inhibitory basket-like cells. Afferent input to the upper layer comes from layer 4; each pyramidal cell receives its afferent connections from the layer 4 spiny-stellate cell of its own minicolumn. Within the upper layer, the output connections of each pyramidal or basket cell extend to a randomly-chosen set of pyramidal and basket cells in the minicolumns of all three segregates. The strength of each of these connections is determined, during the period of the network's self-organization, by a Hebbian rule based on the correlation in stimulus response properties of the two minicolumns. Another feature of the upper layer is that the membrane potential of its cells is controlled by a family of membrane conductances that operate on markedly different time scales. They include fast AMPA and $GABA_A$ conductances, a slower NMDA conductance, and even slower K^+ conductance of slow afterhyperpolarization.

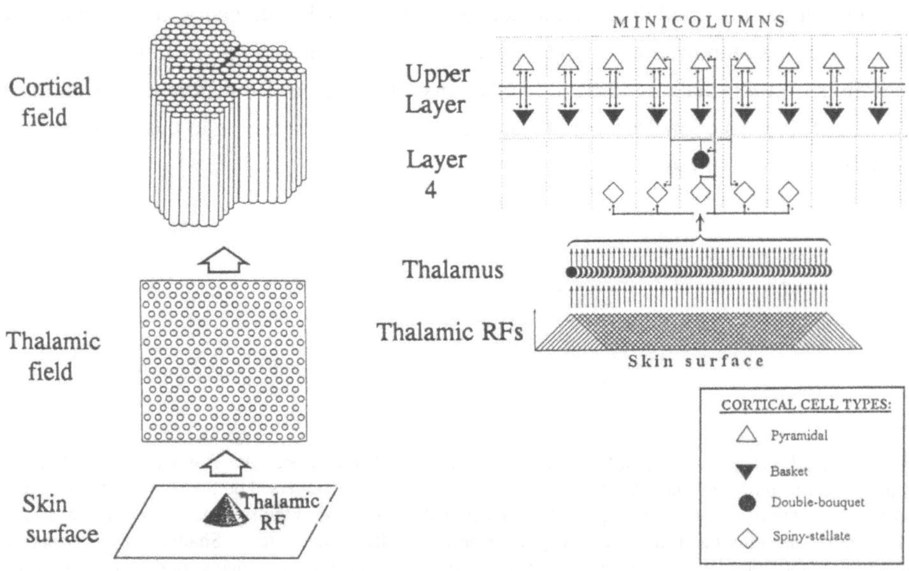

Figure 7. A: The general plan of the model. Stimulus patterns are applied to the skin field. The skin field provides stimulus-evoked input to the thalamic field, an array of cells shown as small circles. Thalamic cells have somatotopically arranged RFs on the skin. Cells of the thalamic field, in turn, provide input to the cortical field that is organized as a set of 3 segregates, each containing 61 minicolumns (shown as small-diameter cylinders). B: The connectional pattern of the model. For clarity, connections are shown only for one minicolumn.

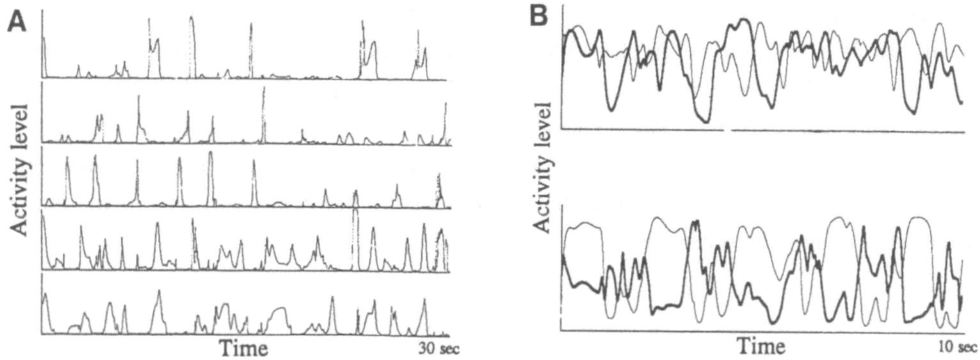

Figure 8. Complex, nonstationary dynamics in the model's upper layer. A: Time courses of activities of 5 pyramidal cells taken at random, but representative of the entire layer. During 30,000 time updates of the network's state (1 update corresponds approximately to 1 msec of real biological time), the network received a constant input drive, but the activity of cells fluctuated wildly. These fluctuations are deterministic, since no noise was added at any point to the system (and connection strengths were kept constant). B: Time courses of 2 pairs of pyramidal cells, each pair taken from the *same* minicolumn. In the top plot the 2 cells have similar average activity levels, but their fluctuations are not related. In the bottom plot they have different average activity levels, and they also are anticorrelated in their fluctuations.

The combination of multi-time-scale membrane processes, extensive lateral connectivity, and local minicolumnar RF diversity, makes this network dynamically very complex. This is manifested in the emergence of chaotic deterministic fluctuations in the state of the network's cells. Figure 8A shows that the model's upper-layer cells have very irregular activity patterns, each cell displaying its own unique pattern, even when the input drive remains constant. As seen in Figure 8B, upper-layer cells even within the same minicolumn can exhibit different behaviors. It is noteworthy that such cells receive exactly the same afferent input from layer 4, and their lateral connections are very similar in that they are drawn at random from functionally the same pool of neurons chosen by the Hebbian rule.

To conclude, the behavior of this cortical model suggests that complex nonlinear dynamics, which are characteristic of living cortex, might provide another mechanism for local diversity in the behaviors of neighboring cortical cells by amplifying small differences in inputs into greater differences in stimulus-evoked behaviors.

What might be the functional benefits of such complex, chaotic dynamics? Two examples will illustrate the potential of such dynamics to impart novel and useful information-processing capabilities to the cortical network. As a first example, we find that some upper-layer cells in the model can exhibit temporal correlations in their activity fluctuations, but that this coherence is very stimulus-specific (see Fig. 9), similar to what is found in real cortex (*e.g.* Aertsen *et al.*, 1989). As Figure 9 suggests, in the presence of any given stimulus, a particular group of cells, which are specific to that stimulus and behave coherently, will emerge in the network. The more or less synchronous activity of this group of cells will set it apart from other cells in the network that will also be active, but not in phase with this group, and will make the group's influence on the target networks much more effective (von der Malsburg, 1981; Abeles, 1982). Thus, complex nonlinear dynamics can expand the information-coding capabilities of this, and presumably cortical networks, by reflecting information about stimulus characteristics not only in the average response activity of the cells, but also in the transient coherence in the activity fluctuations among individual cells.

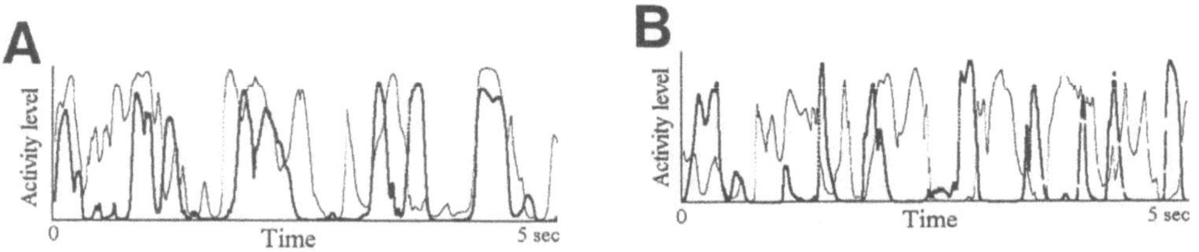

Figure 9. Stimulus-dependent coherence in activity fluctuations of individual cells in the network. A: Time courses of activities of two pyramidal cells taken from 2 different minicolumns during the continuous presence of a particular stimulus (a horizontal, centrally-positioned bar), revealing that these two cells had a clear tendency to be active simultaneously (the correlation coefficient of their activities during a much longer period of time than is plotted here was 0.73). B: The same two cells were monitored in the presence of a different stimulus, also centrally positioned, but orthogonal to the one in A. This time the two cells showed much less tendency to be active simultaneously (correlation coefficient was 0.17). It is interesting to note that there was very little change, from one stimulus to the other, in the *average* activity levels of the two cells. Thus these two cells would be useless in distinguishing between the two stimuli, if only their average firing rates were considered. However, in a real cortex they would be effective in signaling the difference between the two stimuli, because their synchronous firing in the presence of the horizontal bar would drive the cells on which their axons converge much more effectively than the out-of-phase firing in the presence of the vertical bar.

As a second example of the potential functional benefits of chaotic dynamics, we find that the upper layer of the model shows sensitivity to temporal order in which a set of stimuli are presented. Different sequences of stimuli generate, at the time of their completion, different *spatial* patterns of activity in the network's upper layer, even when the last stimuli in these sequences were identical. This makes it possible, for example, to recognize, by spatial pattern alone, a temporal sequence of letters as a particular word, as opposed to its anagrams (see Fig. 10). This temporal sensitivity is brought to the upper layer by the sensitive dependence of chaotic dynamics on initial conditions. Layer 4, with its much simpler dynamics, on the other hand, does not have such sensitivity to temporal order. Thus, at the level of layer 4, information about a sequence of stimuli is represented in a *temporal* sequence of spatial patterns of activity, each evoked by a particular stimulus in that sequence. The upper layer, however, can represent this spatially-and-temporally encoded information about the stimulus sequence by a purely spatial pattern that is specific for the entire time sequence, not just for the current stimulus. This makes temporal information more accessible for use in perception and decision making.

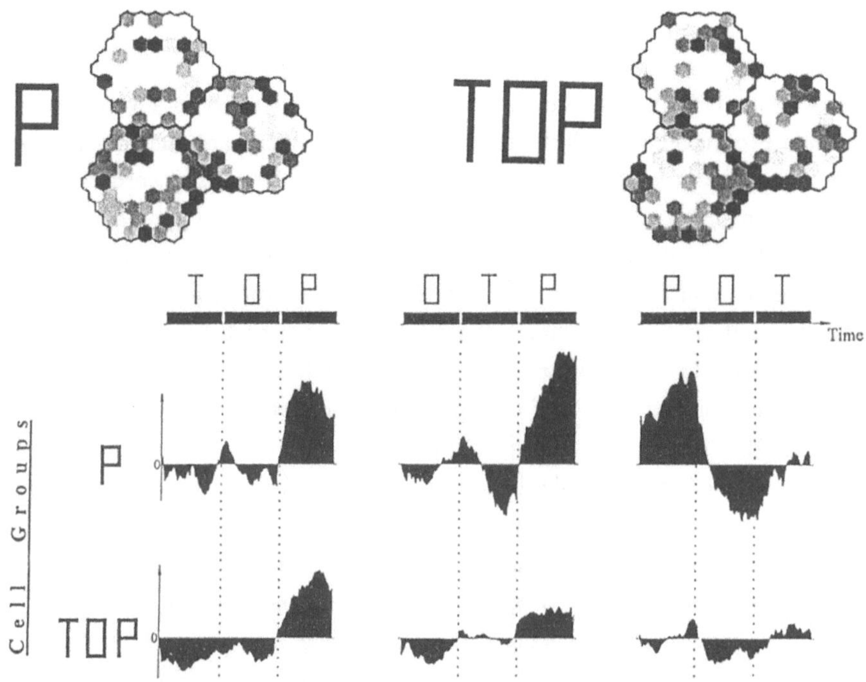

Figure 10. Word recognition by the model network. Four stimuli, in the form of letters O, P, T, and X, were presented to the network in all 24 possible sequences of three different letters ('three-letter words'), read one letter at a time. Each letter was presented at the center of the skin field for 500 msec, followed immediately by the next letter. For each pyramidal cell in the network, the correlation coefficient was computed between its activity and the presence (0) or absence (1) of individual letters, as well as entire words. The top panel shows correlations of the pyramidal cells in the 183 minicolumns (shown as small hexagons) with the presence/absence of the letter P and the word TOP. The degree of shading of a hexagon shows the magnitude of the correlation coefficient (white--none; black--strongest). As this panel shows, the network contains distinct groups of cells that are selectively sensitive, at least to some degree, to the presence or absence of individual letters and even entire words. Next, the activities of these letter- and word-specific groups were monitored during presentation of various words to the network. Since each group was identified by the correlations of its constituent cells with a specific letter or word, the activity of this group could be simply represented as the weighted sum of the activities of all the pyramidal cells in the network, the weights being the correlation coefficients with the letter or word represented. The bottom panel shows the behaviors of two groups–one specific to the letter P, the other to the word TOP–during presentation of three words: TOP, OTP, and POT. Shown are the activities of the two groups averaged over 20 trials of each word. The P group responded strongly every time the letter P was present on the skin, irrespective of its place in the word. The behavior of the TOP group requires some explanation. When the word TOP was being presented, the group did not respond to the first two letters, T and O. This is a correct behavior, because the first two letters are consistent with both words TOP and TOX, and the decision that the word is TOP could not be made until the third letter turned out to be P, at which point the group responded strongly. To show that the TOP group responds to the entire word, and not simply to the letter P, the two letters T and O were interchanged to spell OTP. Now the TOP group responded much more weakly to the terminal letter P than when the same first two letters were in the desired order. Next, because the TOP group had a small but noticeable response to the letter P in OTP, it might still appear that this group is primarily detecting the presence of P, with some additional modulation by the preceding letters. However, when the word POT was presented, the letter P failed to elicit any response from the TOP group, showing that it is truly sensitive to the entire sequence, not simply to any single letter. In conclusion, this figure shows that the model's upper layer carries, in its spatial pattern of *instantaneous* activity, information both about what is happening in the sensory periphery at that particular moment (*e.g.* the P group) and also what has happened previously (*e.g.* the TOP group).

Conclusions

The primary idea put forth in this review is that local RF diversity is a very prominent and common feature of cerebral cortical organization, with significant functional implications. Indeed, we believe that it is of fundamental importance for understanding cortical function. We also outlined two particular putative mechanisms by which local RF diversity might be generated in the cortex. One is a developmental mechanism, centered on inhibitory interactions among adjacent minicolumns, and responsible both for RF differences *among* neighboring minicolumns, and for relative uniformity *within* minicolumns. The second mechanism is based on complex nonlinear dynamics. This mechanism might be responsible for some of the differences both among and within minicolumns. Finally, we provided examples of functional benefits that might arise from local RF diversity; specifically, RF diversity might be a necessary condition for the generation of new functional properties in cortical neurons, and for novel ways to encode sensory information using complex nonlinear dynamics (McKenna *et al.*, 1994).

Recent years have shown increasing use of simultaneous recording from multiple neighboring neurons, which has in turn led to growing realization of the prominence of local RF diversity and hence to growing interest in its functional importance (*e.g.* Gochin *et al.*, 1991; Gawne and Richmond, 1993; Gawne *et al.*, 1993; Ghose *et al.*, 1993; Maldonado and Gray, 1994; South and Weinberger, 1994; Zohary *et al.*, 1994). We can only expect still greater interest in this most promising area of cortical studies.

Acknowledgements
We gratefully acknowledge the helpful suggestions of Dr. Barry L. Whitsel during the preparation of this paper. This work was supported, in part, by NINDS Grant NS30686, NIMH Grant MH48654, and ONR grant N00014-95-1-0113.

References

Abeles, M. (1982) Local cortical circuits. An electrophysiological study. Springer-Verlag, Berlin.
Abeles, M. and Goldstein, M. H., Jr. (1970) Functional architecture in cat primary auditory cortex: columnar organization and organization according to depth. *Journal of Neurophysiology* 33: 172-187.
Aertsen, A., Gerstein, G.L., Habib, M.K. and Palm, G. (1989) Dynamics of neuronal fringe correlation: Modulation of "effective connectivity" *Journal of Neurophysiology* 61: 900-917.
Albus, K. (1975a) A quantitative study of the projection area of the central and the paracentral visual field in area 17 of the cat: I. The precision of the topography. *Experimental Brain Research* 24:159-179.
Albus, K. (1975b) A quantitative study of the projection area of the central and the paracentral visual field in area 17 of the cat: I. The spatial organization of the orientation domain. *Experimental Brain Research* 24:181-202.
Brooks, V.B., Rudomin, P. and Slayman, C.L. (1961) Peripheral receptive fields of neurons in the cat's cerebral cortex. *Journal of Neurophysiology* 24: 302-325.
Creutzfeldt, O., Innocenti, G.M. and Brooks, D. (1974) Vertical organization in the visual cortex (area 17) in the cat. *Experimental Brain Research* 21: 315-336.

DeFelipe, J., Hendry, M.C. and Jones, E.G. (1989) Synapses of double bouquet cells in monkey cerebral cortex visualized by calbindin immunoreactivity. *Brain Research* 503:49-54.

Evarts, E.V. (1967) Representation of movements and muscles by pyramidal tract neurons of the precentral motor cortex. *In*: M.D. Yahr and D.P. Purpura (eds): *Neurophysiological Bases of Normal and Abnormal Motor Activities,* Raven Press, Hewlett, NY.

Favorov, O.V., Diamond, M.E. and Whitsel, B.L. (1987) Evidence for a mosaic representation of the body surface in area 3b of the somatic cortex of cat. *Proceedings of the National Academy of Sciences of the United States of America* 84: 6606-6610.

Favorov, O.V. and Whitsel, B.L. (1988a) Spatial organization of the peripheral input to area 1 cell columns: I. The detection of "segregates." *Brain Research Reviews* 13: 25-42.

Favorov, O.V. and Whitsel, B.L. (1988b) Spatial organization of the peripheral input to area 1 cell columns: II. The forelimb representation achieved by a mosaic of segregates. *Brain Research Reviews* 13: 43-56.

Favorov, O.V. and Diamond, M.E. (1990) Demonstration of discrete place-defined columns -- segregates -- in the cat SI. *Journal of Comparative Neurology* 298: 97-112.

Favorov, O.V. and Kelly, D.G. (1994a) Minicolumnar organization within somatosensory cortical segregates: I. Development of afferent connections. *Cerebral Cortex* 4: 408-427.

Favorov, O.V. and Kelly, D.G. (1994b) Minicolumnar organization within somatosensory cortical segregates: II. Emergent functional properties. *Cerebral Cortex* 4: 428-442.

Fujita, I., Tanaka, K., Ito, M. and Cheng, K. (1992) Columns for visual features of objects in monkey inferotemporal cortex. *Nature* 360: 343-346.

Gawne, T.J., Kjaer, T.W., Hertz, J.A. and Richmond, B.J. (1993) The responses of adjacent cortical neurons are highly independent throughout the visual system. *Society for Neuroscience Abstracts* 19: 424.

Gawne, T.J. and Richmond, B.J. (1993) How independent are the messages carried by adjacent inferior temporal cortical neurons? *Journal of Neuroscience* 13: 2758-2771.

Ghose, I., Ohzawa, I. and Freeman, R.D. (1994) Receptive-field maps of correlated discharge between pairs of neurons in the cat's visual cortex. *Journal of Neurophysiology* 71: 330-346.

Gochin, P.M., Miller, E.K., Gross, C.G. and Gerstein, G.L. (1991) Functional interactions among neurons in inferior temporal cortex of the awake macaque. *Experimental Brain Research* 84: 505-516.

Hubel, D.H. and Wiesel, T.N. (1962) Receptive fields, binocular interaction, and functional architecture in the cat's visual cortex. *Journal of Physiology* 160: 106-154.

Hubel, D.H. and Wiesel, T.N. (1974a) Sequence regularity and geometry of orientation columns in the monkey striate cortex. *Journal of Comparative Neurology* 158: 267-294.

Hubel, D.H. and Wiesel, T.N. (1974b) Uniformity of monkey striate cortex: A parallel relationship between field size, scatter, and magnification factor. *Journal of Comparative Neurology* 158: 295-305.

Iwamura, J., Tanaka, M., Sakamoto, M. and Hikosaka, O. (1985) Diversity in receptive field properties of vertical neuronal arrays in the crown of the postcentral gyrus of the conscious monkey. *Experimental Brain Research* 58: 400-411.

Jones, E.G. (1975) Varieties and distribution of non-pyramidal cells in the somatic sensory cortex of the squirrel monkey. *Journal of Comparative Neurology* 160: 205-267.

Lemon, R.N. and Porter, R. (1976) Afferent input to movement-related precentral neurons in conscious monkeys. *Proceedings of the Royal Society of London, series B* 194: 313-339.

Lemon, R.N., Hanby, J.A. and Porter, R. (1976) Relationship between the activity of precentral neurons during active and passive movements in conscious monkey. *Proceedings of the Royal Society of London, series B* 194: 341-373.

Maldonado, P.E. and Gray, C.M. (1994) Heterogeneity in local distributions of receptive field properties in the cat primary visual cortex. *Society for Neuroscience Abstracts* 20: 625.

McCasland, J.S. and Woolsey, T.A. (1988) High-resolution 2-deoxyglucose mapping of functional cortical columns in mouse barrel cortex. *Journal of Comparative Neurology* 278: 555-569.

McKenna, T.M. and Whitsel, B.L., Dreyer, D.A. (1982) Anterior parietal cortical topographic organization in macaque monkey: a re-evaluation. *Journal of Neurophysiology* 48: 289-317.

McKenna, T.M., McMullen, T.A. and M.F. Schlesinger (1994) The brain as a dynamic physical system. *Neuroscience* 60: 587-605.

Merzenich, M. M., Sur, M., Nelson, R. J. and Kaas, J.H. (1981) Organization of the SI cortex: multiple cutaneous representations in areas 3b and 1 of the owl monkey. *In*: C. N Woolsey (ed.), *Cortical Sensory Organization,* vol. 1, Humana Press, Clifton, N.J., pp. 47-66.

Mountcastle, V.B. (1978) An organizing principle for cerebral function: the unit module of the distributed system. *In*: Mountcastle VB and Edelman GM (eds.): *The Mindful Brain*, MIT Press, Cambridge, MA, pp. 7-50.

Peters, A. and Yilmaz, E. (1993) Neuronal organization in area 17 of cat visual cortex. *Cerebral Cortex* 3: 49-68.

Powell, T.P.S. and Mountcastle, V.B. (1959) Some aspects of the functional organization of the cortex of the postcentral gyrus of the monkey: A correlation of findings obtained in a single unit analysis with cytoarchitecture. *Bulletin of the Johns Hopkins Hospital* 105: 133-162.

Rakic, P. (1988) Specification of cerebral cortical areas. *Science* 241: 170-176.

South, D.A. and Weinberger, N.M. (1994) Tunning characteristic heterogeneity of constituent single units within auditory cortical cluster recordings. *Society for Neuroscience Abstracts* 20: 324.

Tommerdahl, M., Favorov, O.V., Whitsel, B.L., Nakhle, B. and Gonchar, Y.A. (1993) Minicolumnar activation patterns in cat and monkey SI cortex. *Cerebral Cortex* 3: 399-411.

Vaadia, E. and Aertsen, A. (1992) Coding and computation in the cortex: single-neuron activity and cooperative phenomena. *In*: A. Aertsen and V. Braitenberg (eds), *Information processing in the cortex. Experiments and theory*, Berlin Springer-Verlag, pp. 81-122.

von der Marlsburg, C. (1981) The correlation theory of brain function. Internal report 81-2. Max-Planck-Institute for Biophysical Chemistry, Gottingen, (FRG).

Welt, C., Aschoff, J., Kameda, K. and Brooks, V.B. (1967) Intracortical organization of cat's sensory motor neurons. *In*: M.D. Yahr and D.P. Purpura (eds): *Neurophysiological Bases of Normal and Abnormal Motor Activities*, Hewlett, NY, Raven Press.

Zohary E., Shadlen M.N. and Newsome, W.T. (1994) Correlated neuronal discharge rate and its implications for psychophysical performance. *Nature* 370: 140-143.

Somesthesis and the Neurobiology of the Somatosensory Cortex
O. Franzén, R. Johansson and L. Terenius (eds.)

Functional segregation and integration in the nervous system: Theory and models

Giulio Tononi

The Neurosciences Institute, 10640 John J. Hopkins Drive, San Diego, California, 92121, USA

Summary. This chapter summarizes some recent work that addresses the problem of integrating the activity and plasticity of functionally segregated areas in the brain. It is proposed that such integration is achieved in large part through the process of reentry, the recurrent parallel exchange of signals among neuronal groups. It is then argued that value systems, i.e. diffuse projection systems that release neuromodulatory substances in response to salient events, allow global saliency signals to interact with local temporal correlations in determining plastic changes. Finally, by introducing the notion of neural complexity, a theoretical perspective is presented that offers an "*in principle*" resolution of the apparent counterposition of functional segregation and integration.

Keywords: integration, segregation, complexity, reentry, value system, computer model

Introduction

One of the most longstanding problem in neurobiology is the relationship between segregation and integration of function in the brain. This problem emerges within different sensory systems, such as the somatosensory, auditory, or visual system, within the motor system, as well as in connection with the interactions among sensory and motor systems. The generality of the problem suggests that there may be a general solution, or at least some unifying concepts as well as some common mechanisms. Rather than focussing on aspects that are specific to a particular modality or submodality, this chapter will concentrate on issues that are common to all systems studied. In what follows, after introducing the problem of functional integration in the brain, we discuss the notion of reentry, a process that has been suggested as central to its solution. We then address another aspect of integration in the brain, namely the interaction of global saliency signals with local synaptic changes, by resorting to the notion of value-dependent learning. A set of computer simulations designed to illustrate these notions is briefly reviewed. Finally, we provide a precise definition of integration as well as a measure, called neural complexity, of how well integration coexists with segregation. For a detailed account of this work, the reader should consult the original publications (Edelman, 1987; Sporns et al., 1991; Tononi et al., 1992b; 1994; Friston et al., 1994).

The problem of integration

The problem of integration emerged in parallel with the discovery of localization and segregation of function in the brain. It became clear that different modalities were represented in different portions of the cerebral cortex and of subcortical structures. Both "horizontal" and "vertical" subdivisions were proposed: different submodalities were localized to specific cortical areas, and different levels of processing were attributed to primary, secondary, and tertiary areas. More recently, neurophysiology has demonstrated finer and finer parcellations within each cortical area; for example, neighboring neuronal groups tend to be specialized for different stimulus attributes or different values of the same attribute. In terms of development and activity-dependent processes, it appears that localization of function results from the tendency of axons carrying correlated signals to group together and segregate from other axons (according to the *dictum*, "what fires together, wires together"). This process of grouping and segregation seems to be a key organizational principle that allows temporal correlations in the input to be converted into spatial proximity of synaptic terminals leading to the formation of axonal patches (cf. Pearson et al., 1987, for a theoretical analysis applied to the somatosensory system).

While the presence of specialized neuronal groups is advantageous from a functional point of view, it is essential that the activity of these specialized groups and areas be integrated to yield a coherent perceptual scene. This is evident phenomenologically: despite the diversity of elements and relations, each perceptual scene has an undeniable unity and coherence. Functionally, a unified perceptual scene is essential to drive adaptive behavior, which is necessarily integrated. In addition, a glance at the anatomical organization of the brain, especially of the thalamocortical system, reveals that reciprocal and parallel connectivity is the rule rather than the exception, indicating that ongoing mutual interactions among neuronal groups and areas are not just possible but unavoidable. As would be expected form such an anatomical arrangement, many studies have demonstrated the pervasive presence of short-term temporal correlations among cortical and thalamic structures (for reviews, see Aertsen, 1993; Singer, 1993; Bressler, 1995; König and Engel, 1995). Based on these considerations, a solution of the problem of integration has been proposed that takes into account the cooperative interactions within and among brain areas as mediated by a dynamic process which has been called *reentry* (Edelman, 1987, 1989; Tononi et al., 1992b).

In addition to the integration of neural activity, there is another level at which the issue of segregation and integration of function should be considered. According to our present understanding, synaptic changes take place in circumscribed regions of the brain, based largely on temporal correlations between pre- and postsynaptic activity that are registered locally. As we will argue below, it is important to realize that such local changes need to be modulated by the occurrence of events that are salient and that have adaptive significance to the entire organism. Without such a global signal, plastic changes occurring in local, functionally specialized regions of the brain would not be coordinated among themselves and would not be tied to the overall needs and values of the organism. How this further level of integration may take place will be discussed in a subsequent section dealing with a process that has been called *value-dependent learning* (Edelman, 1987, Reeke et al., 1990; Tononi et al, 1992a,b; Friston et al., 1994).

Reentry: cooperative interactions among groups of neurons

Many early attempts at explaining neural integration have been based on the notion that the brain is organized hierarchically with a progressive increase in the specificity of its neurons from the sensory periphery to more central areas. According to this model, perceptual integration is achieved by the confluence of diverse processing streams at a very high hierarchical level or "master area". Although feedforward convergence is an important anatomical feature of the cortex, as demonstrated by multimodal areas and by neurons with complex response properties (e.g. "face" cells), a master area, the activity of which represents entire perceptual or mental states, has not been found. Such a possibility is also conceptually unlikely, given the combinatorial explosion required by convergence in a single place. A neurally-based theory of spatiotemporal integration that does not require a master area was proposed by Edelman (1978) within the framework of the theory of neuronal group selection and has since been extended and refined (Edelman 1987; 1989; Sporns et al. 1989; 1991; Tononi et al. 1992a,b). Two of the main tenets of this theory are that neurons act together in local collectives called neuronal groups and that they communicate with each other and correlate their activity by the process of reentry. Reentry is defined as ongoing parallel signalling occurring between separate neuronal groups along large numbers of ordered anatomical connections in a bidirectional and recursive fashion. Reentrant signalling can take place via reciprocal connections between and within maps (as seen in cortico-cortical, corticothalamic, and thalamocortical radiations) and via more complex arrangements seen in the connections among cortex, basal ganglia, and cerebellum. Reentrant signalling has several general properties that are of fundamental importance to the function of the brain. These properties can be schematically subdivided in constructive, correlative, and associative. As we will show, the overall result is that reentry allows the integration, at many different levels, of functionally segregated maps.

The *constructive properties* of reentry refer to the ability of reciprocal and recursive interactions across multiple areas to influence the response properties of neurons within each area. Reentry can give rise to new properties of the system, such as simultaneous satisfaction of multiple constraints; cross-modal construction, as when one area uses the outputs of another area, which may be specialized for a different function, for its own operation; conflict resolution, for example between the responses of different areas to the same stimulus; and recursive synthesis, as when a stable response pattern requires several cycles of reentry between specialized areas (Finkel and Edelman, 1989).

In many cases, rather than radically altering the response properties of neuronal units, reentry may primarily act to *correlate* their activity, both within and between areas. This is of fundamental importance in the brain, because there are no specific tags or labels that could possibly indicate whether disjunct signals in different brain areas refer to the same object. Nor is there evidence for a master area where the many attributes of a scene are synthesized. By allowing the establishment of specific patterns of correlations across multiple areas, reentry is the key process that guarantees the unity of a perceptual scene.

The *associative properties* of reentry refer to the consequences of reentrant signaling on synaptic changes that are constantly occurring in the brain. Finding appropriate parameters to successfully learn how to grasp, manipulate, and recognize objects may look like an impossible dynamical problem given the enormous numbers of degrees of freedom that the nervous sytem must control. Yet every infant does it quite naturally (see Thelen and Smith, 1994) and every adult is easily able to adjust to constantly changing conditions (see e.g. Merzenich and Sameshima, 1993). We have suggested that the exchange of signals through reentry is advantageous because distant groups of neuron become sensitive to correlations over different modalities and submodalities, both motor and perceptual. Plastic changes that are sensitive to correlations sampled across different sensory modalities and submodalities, across perceptual and motor activities, and between past and present events help reduce the number of degrees of freedom. Thus, during interactions with the environment, functionally segregated perceptual and motor maps undergo coordinated synaptic changes reflecting intermodal correlations that are due to invariances in the environment. In addition, reentrant loops allow the effects of previous activity to influence synaptic changes, thus assuring the integration of behavioral and perceptual sequences. The overall outcome of this process is that, whenever there are correlations between the activity of neuronal groups, and these correlations match those sampled by interacting with the environment, connections subserving those correlations become selectively enhanced over others. Coherent patterns of activity that fit well with correlations in the input would be strongly favored. Ultimately, these statistical correlations will reflect the spatiotemporal properties of signals arising in the real world and can thus serve adaptive behavior.

Value dependent learning: global signals that modulate local signals

Although reentry guarantees that differential amplification at the level of populations of synapses reflects multimodal correlations existing among functionally segregated areas, the resulting selectional events if unconstrained would purely reflect the changes in the patterns of correlated neural activity that occur during the interaction with the environment. They would lack reference to the saliency of events or their significance to the organism. A further set of mechanisms constrain *value-dependent learning*. By these means, evolutionarily selected phenotypic structures can exert a global influence on local synaptic events depending on the adaptive value of a situation (Reeke et al., 1990; Tononi et al., 1992a,b; Friston et al., 1994).

The notion of value was developed in an attempt to understand brain functioning from a selectionist perspective (Edelman, 1987). From such a perspective, there are no programs, sets of instructions, or teachers explicitly controlling synaptic changes in neuronal systems. There are, however, *values, i.e.* structures or events involving the nervous system that have been selected during evolution because they contribute to adaptive behavior and to phenotypic fitness (Reeke et al., 1990). Such values in turn can constrain or guide selective events occurring during experience. A value, for example, may be the presence of bilateral symmetry, the particular kind of receptors possessed by an organism and their position on skin, joints, and muscles, the speed at which sensory events are sampled, the particular way a limb can or cannot move, or some gross anatomical aspects of the structure in which synaptic plasticity takes place. All these phenotypic aspects were selected during evolution and they impose essential constraints on selective events occurring in the nervous system in somatic time. Selective events therefore occur not in an impossibly large statistical search space, but in a space closely predisposed, in its main outlines, by evolution. It has been convincingly argued that only by virtue of these constraints or values somatic selection can converge towards adaptive solutions in sufficient time (Reeke et al., 1990).

As a more specific solution to constraining neural selective events in somatic time, evolution has also endowed organisms with specific systems that can sense and signal to the entire brain the occurrence of behaviors or events having adaptive value and thereby to modulate changes in neural activity and plasticity so that the organism may appropriately adapt to them. We have called these systems *value systems*. In synthesis, the key characteristic of value systems is that of allowing salient events on a global (organismic) scale to affect local changes in neuronal activity or in plasticity. More specifically, value systems are defined as local groups of neurons that: i) have diffuse projections; ii) display tonic firing that may correlate with long-lasting behavioral states; iii) display phasic firing which is triggered by salient intrinsic or extrinsic events, at first innately, then via acquired connections; iv) diffusely relase neuromodulatory substances, by which they v) globally affect neural activity and excitability, both tonically and phasically; vi) globally affect plasticity, both tonically and phasically; vii) different combinations of value systems may be activated under different conditions and may have differential effects on the activity and plasticity of target cells.

This definition applies well to diffuse projection systems capable of releasing neuromodulatory substances, in particular to the cholinergic, noradrenergic, serotoninergic, dopaminergic, and histaminergic systems. Much evidence indicates that these neuromodulatory systems possess at least certain properties of value systems, such as the ability to give a transient but strong response to the occurrence of events having adaptive value, to signal such an occurrence to wide areas of the brain through diffuse projections, and to release substances that modulate neural excitability as well as changes in synaptic strength (for refs see e.g. Hasselmo, 1995). The defining properties of value systems and the degree to which the neurochemical systems mentioned above satisfy such requirements cannot be discussed in detail here. It should be noted, however, that salient events triggering the activity of value systems include not only primary values, such as pain, strong sensory stimulation, or the obtainment of a reward, but also subtler events. Among these are secondary values, i.e. events that reliably predict the occurrence of primary or innate values (Friston et al., 1994), and global aspects of neural activity, such as the degree of coherence in cortical networks, that may signal the presence of novelty.

The modulation of local synaptic changes by global signals that are associated directly or indirectly with evolutionarily selected values constitutes a major means of ensuring value-dependent learning. The latter occurs in the context of cooperative interactions among groups of neurons brought about by reentry which yields a unified perceptual and behavioral performance. To show how these processes might operate together, we choose a recently analyzed example - perceptual and behavioral integration in the operation of the cerebral cortex.

How it all may work: a computer model

The processes of reentry and of value-dependent learning are supported by an extensive body of experimental evidence (for refs, see Edelman, 1993). Such evidence can be significantly complemented by large scale simulations that serve to illustrate and test the role of these processes in the integration of brain function (Sporns et al., 1989, 1991). Here we briefly summarize some results obtained from a model that exemplifies many of the issues discussed above. The model is a large-scale simulation of the visual system which illustrates a solution to the problem of integrating or "binding" the activity of functionally segregated brain areas dealing with different attributes of an object, such as form, color, and location (Tononi et al., 1992b). Due to the complexity of the model, the reader is referred to the original publication for a description of how anatomical and physiological properties of the visual system were simulated. Despite the choice of the visual system, many of the principles embodied in the model are equally applicable to the somatosensory system or to other modalities, since they reflect general properties of cortical organization. The model makes specific suggestions as to the neural basis of certain classes of psychological phenomena, paradigmatic examples of which are form-from-motion and motion capture.

In these simulations, reentry occurs among nine functionally segregated areas, divided into three anatomical streams for form, color, and motion. Consistent with functional segregation in the visual cortex, units within each separate area of the model responded to different properties (motion, form, and color) of the stimuli. Reentrant interactions within and between the multiple areas gave rise to short-term temporal correlations between the units responding to different attributes of the same stimulus, while units responding to attributes of different stimuli were less correlated. In this way, several features of a single object were "bound" together across different streams and hierarchical levels and at the same time that object was differentiated from other objects present in the same visual scene. The model also suggests that such *correlative properties* of reentry are at the basis of an entire class of perceptual illusions, such as motion capture and other capture phenomena (Ramachandran, 1990).

These simulations have also been used to make specific suggestions about the *constructive properties* of reentry. Although neuronal response properties can be determined by a hierarchical arrangement of forward connections, various backward, lateral, intra-areal and recursive interactions may be equally important. For instance, reciprocal and recursive interactions between multiple visual areas can influence the response properties of neurons within each area, and give rise to new properties of the system. The constructive aspect of reentry is particularly evident when reentrant signals from a hierarchically "higher" area (*V5*) interact with signals coming from the periphery to construct responses consistent with form from motion in area *V1/V2*; these responses disappear when reentrant inputs are cut. Some experimental evidence consistent with this prediction has been recently provided by PET studies in humans (Zeki, 1993).

The model also illustrates the *associative properties* of reentry. Due to reentry, each functionally segregated group or map can sample correlations across different modalities and submodalities. Selective events can thus occur at several different levels in terms of differential amplification of synapses. For example, a correct response requires the registration of intermodal correlations between color (red) and shape (cross). There are no units that are directly selective for such an arbitrary conjunction of properties. However, specialized units selective for red can respond to such a correlation if reentrantly connected to specialized units selective for cross-like patterns. Fast local synaptic changes occurring at one functionally specialized area will thus be influenced by events occurring in some other specialized part of the system. In this way, both a high specificity of response and a wide range of associations are accommodated. The advantages that such an arrangement provides are obvious. For example, the task presented to the model required the discrimination of a red cross, no matter at which location. It would take a very long time for the system to learn from scratch to respond to a particular configuration of pixels, irrespective of their location in the visual field, but only if they are red. In addition, the required connectivity would be practically useless (actually harmful) if the system were retrained to respond to green crosses, or to red squares. The strategy followed by the brain, and modelled in these simulations, is based instead on the development and selection of functionally specialized neuronal groups and areas that respond to statistical regularities in the environment, e.g. different shapes (crosses and squares) and different colors (red and green). It thus becomes easy to identify a particular, arbitrary conjunction of these regularities. The actual synaptic changes, occurring among specialized units communicating by reentry, are relatively minor and can easily be reversed.

Finally, these simulations strongly illustrate the role of a *value system* in providing a global saliency signal that modulates local synaptic events in a way that depends upon the adaptive value of the response. In the simulations, the specific patterns of activity and correlations determined by reentrant interactions across functionally segregated neuronal groups and areas produced a focus of synchronous activity in an area controlling eye movements. This activity often gave rise to a foveation response towards one of the objects presented. Such a response was used as a basis for conditioning the model to foveate an object that was characterized by a certain conjunction of shape and color, e.g. a red cross, in any position of the visual field. Operant conditioning was achieved, whenever the model foveated the red cross, through the activation of diffusely projecting units of the value system that signaled saliency (e.g. the delivery of a reward in a conditioning paradigm) and thereby influenced the modification of cortico-cortical connections. After the training period, the model foveated the red cross almost exclusively, although that simulation was presented simultaneously with other objects sharing the same attributes, e.g. a red square and a green cross. The correct response, which depended on the integration of color, form, and location, was achieved without the need for a hierarchically superordinate area, for units that responded to that particular conjunction, or for a preexisting program. By these means, selective events over a relatively short time scale can be constrained in the direction of evolutionarily selected values. By projecting to the entire cortical mantle, and by releasing neuromodulatory signals, neural value systems can affect whatever complex set of interactions leads to an adaptive response.

The neural model discussed here also illustrates that, whereas synaptic changes occurr in many specialized neuronal groups, there is no place where a specific "memory trace" is located. Changes in synaptic weights occur in many different locations, but generally it is difficult if not impossible to attribute a specific memorial function to any particular synapse or cell. The model's memory, expressed as increased ability to re-categorize the surrounding world of stimulus objects, is not replicative or localized in a precise fashion, but it is itself of an integrated nature. Behavioral performance (i.e. the measurable effect of memory) can be degraded in various ways and to various extents by "lesions" in different parts of the model (some of them far removed from sites of synaptic change). Note that, although neural network models have often been used to show the distributed nature of memory, in the present models memory is not just the delocalized property of a single network, but it is an integrated function distributed among several functionally distinct brain areas.

A synthesis: functional segregation and integration reconciled

We have seen that a general principle of brain organization is the coexistence of functional segregation and specialization at a small scale with integration and cooperative interactions at a larger scale. That is, individual areas or neuronal groups are functionally segregated, yet the cortex as a whole gives rise to an integrated and cooperative activity. While the chosen modality to illustrate these points in the computer simulations was visual, there is every reason to expect that similar principles operate in other modalities throughout the brain.

In order to be able to make some experimental predictions, we have recently attempted to provide a more general theoretical perspective as well as a measure of functional segregation and integration in the brain (Tononi et al., 1994). By making certain simplifying assumptions, we have shown that these two organizational aspects can be formulated within a unified framework. In such framework, functional segregation and integration are characterized in terms of deviations from statistical independence among the components of a neural system, measured using the concepts of statistical entropy and mutual information (Papoulis, 1991). Here, we will outline the key aspects of this approach.

Consider a bipartition of a neural system X into a jth subset X^k_j composed of k components (which can be taken to represent neuronal groups) and its complement $X-X^k_j$. The deviation from statistical independence between X^k_j and $X-X^k_j$ is measured by their mutual information MI:

$$MI(X^k_j;X-X^k_j) = H(X^k_j) + H(X-X^k_j) - H(X)$$

where $H(X^k_j)$ and $H(X-X^k_j)$ are the entropies of X^k_j and $X-X^k_j$ considered independently, and H(X) is the entropy of the system as a whole. $MI = 0$ if X^k_j and $X-X^k_j$ are statistically independent and $MI > 0$ otherwise.

The *integration* I(X) of a neural system is then defined as the total deviation from independence among all its n components and can be measured through a single measure that is a generalization of the notion of mutual information. I(X) is defined as the difference between the sum of the entropies of all individual components $\{x_i\}$ considered independently and the entropy of X considered as a whole:

$$I(X) = \sum_{i=1}^{n} H(x_i) - H(X)$$

If the n components of the system are completely independent, as when neuronal groups are completely desynchronized, integration is zero. If, on the other hand, such groups show strong deviations form independence, as when they are fully synchronized, integration is high.

Measuring integration in terms of deviation from statistical independence allows us simultaneously to characterize functional segregation and integration within a neural system, by considering the average integration for subsets of increasing size. Functional segregation is expressed by the relative statistical independence of the activity of individual neuronal groups if these groups are considered in small subsets, i.e. a few at a time (low average integration for small subsets). Conversely, functional integration is expressed by a high degree of statistical dependence when neuronal groups are considered in large subsets, i.e. many at a time (high average integration for large subsets).

This leads to the formulation of a measure, called *neural complexity* (C_N), which reflects the interplay between functional segregation and integration within a neural system, and which is defined as follows:

$$C_N(X) = \sum_{k=1}^{n} \left((k/n) I(X) - \langle I(X^k_j) \rangle \right)$$

where we consider all subsets X^k composed of k-out-of-n components of the system ($1 \leq k \leq n$) and the average integration for subsets of size k is denoted as $\langle I(X^k_j) \rangle$ (the index j indicates that the average is taken over all n!/(k!(n-k)!) combinations of k components). In essence, C_N measures how much the increase of average integration with increasing subset size deviates from linearity, i.e., roughly speaking, how much the whole is more integrated than its parts. C_N can be shown to be equivalent to the mutual information between each part of a neural system and the rest, summed over all possible bipartitions:

$$C_N(X) = \sum_{k=1}^{n/2} \langle MI(X^k_j; X-X^k_j) \rangle.$$

Thus, for complexity to be high, the average mutual information between each component and the rest of the brain must be high, indicating that the system is very integrated. At the same time, the average mutual information must be higher the larger the subset one is considering. This indicates that each component has a rather specialized function, in that considering more components provides additional mutual information. Consistent with intuitive notions and with current attempts in physics and biology to conceptualize complex systems, C_N is high for systems such as the vertebrate brain that conjoin local specialization with global integration. On the other hand, C_N is low for systems that are composed either of completely independent parts or of parts that show completely homogeneous behavior.

As an illustration, we calculated the complexity associated with a primary cortical area simulated in detail in a previous model (Sporns et al., 1991). The connectivity embodied in this model highlights principles of organization that are fairly general in the cortex and that apply to many areas and modalities. We considered the intrinsic (spontaneous) activity of the network (triggered by uncorrelated Gaussian noise) rather than any specific stimulus, in order to avoid extrinsic sources of mutual information. By systematically varying some of the parameters of the model, we observed that certain structural characteristics of cortical connectivity were associated with high values of C_N. These characteristics include a high density of connections, strong local connectivity helping to organize cells into neuronal groups, patchiness in the connectivity among neuronal groups, and a large number of short reentrant circuits (Tononi et al, 1994). Although these conclusions were drawn on the basis of computer simulations, preliminary observations indicate that it is possible to validate this measure of complexity by applying it to data obtained from functional neuroimaging.

Conclusion

In this chapter, we have mentioned several times the counterposition between functional segregation and integration, between local and global signals, between spatially localized synaptic changes and the integrated nature of memory as recategorization, between the enormous variability observed at the microscopic scale and the relative constancy which is pervasive at macroscopic levels. The coexistence of local variation and global coherency, of short-term unpredictability and long-term regularity is thus one of the hallmarks of complex systems. Not surprisingly, such complexity is both the substrate and the result of selectional processes. Here, we have seen that the process of reentry, by favoring cooperative interactions among groups of neurons, leads to the emergence of an integrated behavior from the activity of functionally specialized areas. We have then shown that the activity of value systems allows the interaction of global saliency signals with local temporal correlations in determining plastic changes. Finally, by introducing the notion of neural complexity, we have presented a theoretical perspective that offers an "*in principle*" resolution or synthesis of the dialectic of local and global, or of localizationism and holism.

Acknowledgements. The work by the author reviewed in this paper was carried out as part of the Institute Fellows in Theoretical Neurobiology program at The Neurosciences Institute, which is supported through the Neurosciences Research Foundation.

References

Aertsen, A. and Arndt, M. (1993) Response synchronization in the visual cortex. *Current Biology*, 3:586:594.
Bressler, S.L. (1995) Large-scale cortical networks and cognition. *Brain Research Reviews* 20:288-304.
Edelman, G. M. (1978). Group selection and phasic re-entrant signalling: a theory of higher brain function. *In*: G. M. Edelman and V. B. Mountcastle (eds.), *The Mindful Brain*, Cambridge, Massachusetts: MIT Press.
Edelman, G. M. (1987). *Neural Darwinism: The Theory of Neuronal Group Selection*. New York: Basic Books.
Edelman, G. M. (1989). *The Remembered Present: A Biological Theory of Consciousness*. New York: Basic Books.
Edelman, G. M. (1993). Neural Darwinism: Selection and reentrant signaling in higher brain function. *Neuron*, 10:1-20.
Finkel, L. H., and Edelman, G. M. (1989). The integration of distributed cortical systems by reentry: a computer simulation of interactive functionally segregated visual areas. *Journal of Neuroscience*, 9, 3188-3208.
Friston, K. J., Tononi, G., Reeke, G. N., Jr., Sporns, O., and Edelman, G. M. (1994). Value-dependent selection in the brain: Simulation in a synthetic neural model. *Neuroscience*, 59, 229-243.
Hasselmo, M.E. (1995). Neuromodulation and cortical function: modeling the physiological basis of behavior. *Behavioral Brain Research*, 67, 1-27.
König, P and Engel, A. (1995) Correlated firing in sensory-motor systems. *Current Opinions in Neurobiology, in press.*
Merzenich, M.M. and Sameshima, K. (1993) Cortical plasticity and memory. *Current Biology*, 3:187-196.
Papoulis, A. (1991). *Probability, Random Variables, and Stochastic Processes* (Third ed.). New York: McGraw-Hill.
Pearson, J.C., Finkel, L. H., and Edelman, G. M. (1987). Plasticity in the organization of adult cerebral cortical maps: A computer simulation based on neuronal group selection. *Journal of Neuroscience*, 7, 4209-4223.

Ramachandran, V. S. (1990). Visual perception in people and machines. *In*: A. Blake and T. Troscianko (eds.), *AI and the Eye* (pp. 21-77). New York: Wiley.

Reeke, G. N., Jr., Finkel, L. H., Sporns, O., and Edelman, G. M. (1990). Synthetic neural modeling: A multilevel approach to the analysis of brain complexity. *In:* G. M. Edelman, W. E. Gall, and W. M. Cowan (eds.), *Signal and Sense: Local and Global Order in Perceptual Maps,* New York: Wiley.

Singer, W. (1993) Synchronization of cortical activity and its putative role in information processing and learning. *Annual Reviews of Physiology*, 55:349-374..

Sporns, O., Gally, J. A., Reeke, G. N., Jr., and Edelman, G. M. (1989). Reentrant signaling among simulated neuronal group leads to coherency in their oscillatory activity. *Proceedings of the National Academy of Science USA*, 86, 7265-7269.

Sporns, O., Tononi, G., and Edelman, G. M. (1991). Modeling perceptual grouping and figure-ground segregation by means of active reentrant conditions. *Proceedings of the National Academy of Science USA*, **88, 129-133.**

Thelen, E., and Smith, L. B. (1994). *A Dynamic Systems Approach to the Development of Cognition and Action.* Cambridge: The MIT Press.

Tononi, G., Sporns, O., and Edelman, G. M. (1992a). The problem of neural integration: induced rhythms and short-term correlations. *In:* E. Basar & T. Bullock (eds.), *Induced Rhythms in the Brain* (pp. 365-393). Boston: Birkhauser.

Tononi, G., Sporns, O., and Edelman, G. M. (1992b). Reentry and the problem of integrating multiple cortical areas: Simulation of dynamic integration in the visual system. *Cerebral Cortex*, 2, 310-335.

Tononi, G., Sporns, O., and Edelman, G. M. (1994). A measure for brain complexity: Relating functional segregation and integration in the nervous system. *Proceedings of the National Academy of Science USA*, 91, 5033-5037.

Zeki, S. (1993). *A Vision of the Brain*. Oxford: Blackwell Scientific Publications.

Subject index

(The page number refers to the first page of the chapter in which the keyword occurs)

MIX
Papier aus verantwortungsvollen Quellen
Paper from responsible sources
FSC® C105338

If you have any concerns about our products,
you can contact us on
ProductSafety@springernature.com

In case Publisher is established outside the EU,
the EU authorized representative is:
Springer Nature Customer Service Center GmbH
Europaplatz 3, 69115 Heidelberg, Germany

Printed by Libri Plureos GmbH
in Hamburg, Germany